Friends of the earth, we too care.
This book has been printed on 100% recycled paper.
It is estimated that in the total printing of 55,000 copies, 327 trees have been saved.

AMOS TURK, Ph.D.

Professor, Department of Chemistry,
The City College of the City University of New York

JONATHAN TURK, Ph.D.

Naturalist

JANET T. WITTES, Ph.D.

Division of Epidemiology,
Columbia School of Public Health

ROBERT WITTES, M.D.

Department of Medicine,
Memorial Sloan-Kettering Medical Center, New York

Environmental Science

W. B. SAUNDERS COMPANY PHILADELPHIA • LONDON • TORONTO

W. B. Saunders Company: West Washington Square
Philadelphia, Pa. 19105

12 Dyott Street
London, WC1A 1DB

833 Oxford Street
Toronto, Ontario M8Z 5T9, Canada

Cover illustration courtesy of the PHOTOGRAPHY OF H. ARMSTRONG ROBERTS

Environmental Science ISBN 0-7216-8928-0

Print No.: 9 8 7 6 5 4 3 2

PREFACE

The separate scientific disciplines that are embodied in modern academic departments evolved gradually, along with technological developments, the establishment of scientific societies, and the publication of textbooks. Traditionally, textbooks have been the culmination of such educational evolutions. They were the edited lecture notes of the master teachers, later revised and kept up to date by successive generations of their students. Popular sequences of instruction evolved; they were often recapitulations of the chronology of discovery, with blind alleys erased and a logically appealing progression of knowledge laid before the student. Chemistry texts, for example, started with Boyle's and the other gas laws, from which were derived molecular and atomic weights, hence valences and the unfolding of chemical periodicity, which, in turn, provided the framework for the study of chemical properties and reactions. Such instructional patterns have been stable for long periods of time.

Environmental science is in part a new integration of old disciplines, and in part a new discipline of its own. Consequently, traditional, student-tested patterns of instruction have not yet been developed. It is therefore particularly important at this juncture that a textbook of environmental science should work well as a teaching instrument. To this end, we have aimed to develop concepts and subject matter in a continuous, logical progression, so that the student's knowledge builds upon itself.

Some of the pedagogic tools that we have used are described in the following paragraphs.

The Order of Topics

The order of chapters and topics has been selected with study sequences in mind, not with the idea that one form of environmental stress is any more or less critical than another. Thus, the central concepts of self-regulating systems and the flow of energy in the biosphere are introduced very early and are re-emphasized throughout the book. A background of the relevant basic principles is provided for each individual chapter.

Economics and Social Policy

Should social and economic issues in environmental studies be given separate attention, or be integrated with the technological aspects? Both, we believe. To discuss, say, radioactive wastes or emissions from automobiles as purely technical subjects is to miss much of the heart of the matter, and we have therefore chosen the integrated approach. The last chapter, however, devotes itself exclusively to economics and social policy.

Curriculum

This text provides material that would be appropriate for any of a number of one-semester courses. The accompanying instructor's manual suggests various specific selections for curricula that emphasize biological, chemical, physical, geological, or social science topics. It would also be possible to combine some of these curricula and extend the material for a full-year course.

Study Problems

There is hardly a better aid to learning than the execution of study problems. Straightforward, tricky, informative, computational, graphical, short-answer, true-false, essay—all are valuable. We have tried to offer a generous variety.

As a matter of convenience for making assignments, the problems are titled, and the computational ones are separated from the others. Numerical answers are provided for all the computational problems.

Language

New disciplines engender new vocabularies. We have therefore taken pains to deal in some depth with newly fashionable terms like "organic foods," "noise pollution," "natural balance," and even "ecology."

Bibliography

An annotated bibliography appears at the end of each chapter.

Appendix

The Appendix provides information on units of measurement and on chemical formulas, which will be useful to students who have not had a course in chemistry. A table of atomic weights and a three-place log table are also provided.

ACKNOWLEDGMENTS

Professor Samuel S. Butcher of the Department of Chemistry of Bowdoin College and Professor Peter W. Frank of the Department of Biology of the University of Oregon reviewed the entire manuscript most thoroughly and incisively, and we thank them for their many helpful comments.

In addition to the overall reviews by Professors Butcher and Frank, various other colleagues examined particular chapters that dealt with their fields of special experience and expertise. These exacting reviews were extremely valuable, and highly appreciated. We list them below:

CHAPTER	***REVIEWERS***
1. Natural Ecology	Eugene P. Odum, Alumni Professor of Zoology, University of Georgia Professor Armando de la Cruz, Mississippi State University Professor Stanley Wecker, Department of Biology, The City College of The City University of New York
2. Human Adaptation	Dr. Ellen Solomon, University of California Medical Center, San Francisco
3. Extinction of Species	Dr. Beryl Simpson, Associate Curator, Department of Botany, Smithsonian Institution
4. Human Population	Dr. Judith Goldberg, Research Statistician of the Health Insurance Plan of Greater New York Dr. Elizabeth Whelan, Executive Director, Demographic Materials, Inc.
5. Energy	Professor John Fowler, Department of Physics, University of Maryland
6. Agriculture and 7. Pests and Weeds	Mr. Robert Josephy, Proprietor of Blue Jay Orchards, Bethel, Connecticut; Vice-Chairman of the Connecticut State Board of Agriculture

8. Radioactive Wastes	Professor William Miller, Department of Physics, The City College of The City University of New York
9. Air Pollution	Dr. Stephen Stoldt, Apollo Chemical Corporation
10. Water Pollution	Dr. Hussein Naimie, Director of Research, Environmental Research and Applications, Inc. Professor John Fillos, Department of Civil Engineering, The City College of The City University of New York
12. Noise	Mr. Milton Rosenstein, New York Institute of Technology
13. Social, Legal and Economic Aspects	Mr. R. David Flesh, Director of Environmental Economics, Copley International Corporation. Mr. Paul Snyder, Jr., attorney, Boulder, Colorado. Ms. Kate Bernstein, political consultant, Great Neck, New York

Many others helped in many other ways. A partial list includes Miriam Bergman, who gathered and classified reference sources; Evelyn Manacek, who typed most of the manuscript; and the staff of the W. B. Saunders Company, who put it all together. We are indebted to them all.

Finally, we want to acknowledge how much we learned from Mr. Grant Lashbrook, who brings to his art and design a humane knowledge of environmental subjects, a conceptual clarity, and an insight into the moods that illustrative matter can evoke.

AMOS TURK

JONATHAN TURK

JANET T. WITTES

ROBERT E. WITTES

CONTENTS

1. THE ECOLOGY OF NATURAL SYSTEMS 1
1.1 What Is Ecology? 1
1.2 Energy 2
1.3 Energy Relationships within an Ecosystem 4
1.4 Nutrient Cycles 10
1.5 Ecosystems and Natural Balance 17
1.6 Major Ecosystems of the Earth 20
1.7 The Ecological Niche 28
1.8 Overlap and Diversity of Niches 29
1.9 Interactions among Species 33
1.10 Sporadic and Cyclic Disruptions 39
1.11 Global Interactions among Ecosystems 40
1.12 Natural Succession 43
1.13 The Role of People 47

2. HUMAN ADAPTATION TO ENVIRONMENTAL CHANGE 55
2.1 Introduction 55
2.2 The Biological Basis of Genetic Adaptation ... 56
2.3 The Influence of the Environment on the Gene Pool—Natural Selection 67
2.4 Adaptation by Learning 71
2.5 The Interaction between Cultural and Genetic Mechanisms 75

3. THE EXTINCTION OF SPECIES 81
3.1 Evolution and the Extinction of Species in Prehistoric Times 81
3.2 Factors Leading to the Destruction of Species in Modern Times 86
3.3 How Species Become Extinct—The Critical Level 93
3.4 Two Case Histories of Endangered Species 96
3.5 Current Trends in Species Extinction 98
3.6 The Displacement of Species—Effects on Ecosystems 100
3.7 The Nature of the Loss 101

4. THE GROWTH OF HUMAN POPULATIONS 106

4.1 Introduction 106
4.2 Extrapolation of Population Growth Curves 109
4.3 An Introduction to Demography 112
4.4 Measurements of Mortality 117
4.5 Measurements of Birth Rates 125
4.6 Measurements of Growth 127
4.7 Accuracy of the Data 135
4.8 Demographic Trends in the United States 138
4.9 Consequences of Population Density 141
4.10 The Demographic Transition 144
4.11 Worldwide Demographic Trends 145
4.12 Consequences of Rapid Population Growth in Underdeveloped Countries 147
4.13 Arguments against Stabilization of Population 149
4.14 Methods of Population Control 150
4.15 The Ultimate Human Population 153

5. ENERGY: RESOURCES, CONSUMPTION, AND POLLUTION 161

5.1 Energy Consumption 161
5.2 Energy and Power 164
5.3 Work by Man, Beast, and Machine 165
5.4 The First Law of Thermodynamics (or "You Can't Win") 166
5.5 Heat Engines and the Second Law of Thermodynamics (or "You Can't Break Even") 167
5.6 Our Fossil Fuel Supply 170
5.7 Environmental Problems Caused by Coal Mining and Oil Drilling 175
5.8 Transportation of Fuels: A Case Study of the Alaska Pipeline 178
5.9 Nuclear Fuels 181
5.10 Renewable Energy Sources 182
5.11 The Energy Crisis 187
5.12 Energy Consumption for Transportation 188
5.13 Energy Consumption for Industry 190
5.14 Energy Consumption for Household and Commercial Use 191
5.15 Electricity 195
5.16 Transmission of Electricity 197
5.17 Thermal Pollution 198
5.18 Waste Heat and Climate 203
5.19 Solutions to the Problem of Thermal Pollution 204
5.20 Energy—An Overview 209

6. AGRICULTURAL SYSTEMS 217

6.1 Introduction 217
6.2 Agricultural Disruptions 219

6.3 Energy Flow in Industrial Agriculture........ 227
6.4 Chemical Fertilization 230
6.5 Mechanization in Agriculture..................... 236
6.6 Irrigation 237
6.7 Agriculture in the Less-Developed Nations and The Green Revolution 239
6.8 Food Supplies for a Growing Population..... 244

7. CONTROL OF PESTS AND WEEDS 257
7.1 Competition for Food between Small Herbivores and Man................................ 257
7.2 Insecticides – Introduction and General Survey 259
7.3 The Action of Chemical Pesticides – Broad-Spectrum Poisoning 260
7.4 The Action of Chemical Pesticides – Persistence in the Environment................ 264
7.5 Physical Dispersion of Organochloride Insecticides 265
7.6 Biological Concentration of Organochloride Insecticides – Effects on Non-Target Species 266
7.7 Effects of Insecticides on Human Health 271
7.8 Other Methods of Pest Control 274
7.9 Herbicides... 283
7.10 Economic Factors in Pest Control 285

8. RADIOACTIVE POLLUTANTS 290
8.1 Introduction.. 290
8.2 Atomic Structure 292
8.3 Natural Radioactivity 294
8.4 How Man Has Produced More Radioactive Matter 297
8.5 Nuclear Fission Reactors.......................... 300
8.6 Environmental Hazards from Fission Reactors 308
8.7 Nuclear Fusion 316
8.8 Biological Effects of Radiation 320
8.9 Problems and Issues Related to Radioactive Pollutants.............................. 330

9. AIR POLLUTION 335
9.1 Introduction.. 335
9.2 The Physical Nature of Air Pollutants......... 335
9.3 Pure Air and Polluted Air......................... 339
9.4 Gaseous Air Pollutants 340
9.5 Particulate Air Pollutants......................... 346
9.6 The Meteorology of Air Pollution.............. 349
9.7 The Effects of Air Pollution – Introduction.. 354
9.8 The Effects of Air Pollution on the Atmosphere and on Climate 356
9.9 Effects of Air Pollution on Human Health.. 361

9.10 Damage to Vegetation 370
9.11 Injury to Animals 371
9.12 Deterioration of Materials 372
9.13 Aesthetic Effects 373
9.14 Air Pollution Indices 373
9.15 The Control of Air Pollution—Introduction 376
9.16 Control of Air Pollution Emissions 377
9.17 Control of Air Pollution in Enclosed Spaces 381
9.18 The Automobile 383

10. WATER POLLUTION 399

10.1 The Nature of Water Pollution 399
10.2 Water 402
10.3 The Hydrological Cycle (Water Cycle) 404
10.4 Types of Impurities in Water 405
10.5 The Components of Natural Waters 407
10.6 Nutrients, Microorganisms, and Oxygen 411
10.7 The Pollution of Inland Waters by Nutrients 417
10.8 Industrial Wastes in Water 425
10.9 The Pollution of the Oceans 429
10.10 The Effects of Water Pollution on Human Health 435
10.11 Water Purification 438
10.12 Economics, Social Choices, and Strategy in Water Pollution Control 446

11. SOLID WASTES 455

11.1 Solid Waste Cycles 455
11.2 Sources and Quantities 456
11.3 The Nature of the Solid Waste Problem 459
11.4 Land and Ocean Disposal 461
11.5 Incineration 464
11.6 Recycling—An Introduction 465
11.7 Glass Containers 467
11.8 Junk Automobiles 469
11.9 Old Paper 472
11.10 Recycling—Techniques and Technology 472
11.11 The Future of Recycling 481

12. NOISE 487

12.1 Sound 487
12.2 Pure and Impure Tones 489
12.3 Hearing 491
12.4 Noise 492
12.5 The Decibel Scale 493
12.6 Loudness 497
12.7 The Effects of Noise 499
12.8 Noise Control 503
12.9 A Particular Case: The Supersonic Transport (SST) 507
12.10 Epilogue 511

13. SOCIAL, LEGAL, AND ECONOMIC ASPECTS OF ENVIRONMENTAL DEGRADATION 516

13.1 Introduction 516
13.2 Cost of Pollution Control 521
13.3 Pollution Control: Altruism, Legislation, and Enforcement 529
13.4 Economic Incentives for Pollution Control 537
13.5 Radical Social Policies Leading to Pollution Control 542
13.6 Case History: Automobiles and Highways 543
13.7 Economic Problems of Pollution Control in Developing Nations 545
13.8 Conclusion 546

APPENDIX

A.1 The Metric System (International System of Units) 553
A.2 Chemical Symbols, Formulas, and Equations 554
A.3 Abbreviations 556
Logarithms to the Base 10 (Three Places) 556
Table of Relative Atomic Weights 557

INDEX 559

1

THE ECOLOGY OF NATURAL SYSTEMS

1.1 WHAT IS ECOLOGY?

No machine can perform as many diverse functions as a living organism. Animals and plants, unlike machines, can feed and repair themselves, adjust to new external influences, and reproduce. These abilities depend on intricate interrelationships among the separate parts of the organism. Each mammal, for instance, is far more than an independent collection of brain, heart, liver, stomach and other organs. What affects one part of the body affects its entirety.

Even with all its built-in mechanisms of life, however, an individual plant or animal cannot exist as an isolated entity; it is dependent upon its environment, including other organisms to which it relates. It must convert some source of energy (food or light), ingest water and minerals, dispose of wastes, and maintain a favorable temperature.

Plants and animals occurring together, plus that part of their physical environment with which they interact, constitute an **ecosystem.** An ecosystem is defined to be nearly self-contained, so that the matter which flows into and out of it is small compared to the quantities which are internally recycled in a continuous exchange of the essentials of life. The dynamics of the flow of energy and materials in a given geological environment, as well as the adaptation made by the individual and the species to find a place within the environment, constitute the subject matter of the ecology of natural systems.

Ecosystems differ widely with respect to size, location, weather patterns, and the types of animals and plants included. A watershed in New Hampshire, a Syrian desert, the Arctic ice cap, and Lake Michigan are all distinct ecosystems. Common to them all is a set of processes. In each there are plants which use energy from the sun to convert simple chemicals from the environment into complex, energy-rich tissues. Each houses various forms of plant-eaters, predators who eat the plant-eaters, predators who eat the predators, and organisms that cause decay.

Throughout this book we shall sometimes refer to ecosystems as geographically distinct areas with unique characteristics. But ecosystems are of course interconnected, so we shall also discuss the nature of their linkages.

In the box on page 48 at the end of this chapter, the definitions of ecology and ecosystem are discussed in more detail.

1.2 ENERGY

(Problems 2, 4)

After reviewing some general concepts of energy, we shall apply them to the study of ecosystems.

Energy is the capacity to do work or to transfer heat. In ordinary speech we refer to "physical work" or "mental work" to describe a variety of activities that we think of as energetic. To the physicist, however, "work" has a very specific meaning: work is done on a body when a body is *forced to move.* Merely holding a heavy weight requires force but is not work, because the weight is not being moved. Lifting a weight, however, is work. Climbing a mountain or a flight of stairs is work. So is stretching a spring, or compressing a gas in a cylinder or in a balloon, because all of these activities move something.

When heat is transferred to a body, its temperature rises or its composition or state is changed, or both. For example, when heat is transferred to air, the air gets warmer; when ice is heated enough, it melts; when sugar is heated enough, it decomposes. All of these changes require energy.

Now imagine that you see an object lying on the ground. Does it have energy? If it can do work or transfer heat, the answer is yes. If it is a rock lying on a hillside, then you could nudge it with your toe and it would tumble down, doing work by hitting other objects during its descent and forcing them into motion. Therefore, it must have had energy by virtue of the potential to do work that was inherent in its hillside position. If it is a lump of coal, it could burn and be used to heat water, or cook food, and therefore it too has energy. If it is an apple, you could eat it and it would enable you to do work and to keep your body warm, and so an apple has energy (see Table 1.1).

Of course, if you roll an apple down a hill, it has energy as a falling body. It is proper to assign a quantity of energy to an apple as a food (one small apple contains 64 Calories) rather than as a falling body because people eat apples; they don't build power plants in orchards to generate electricity from the falling fruit. This means that the energy you assign to a body depends on the process you have in mind. A person who looks at a warm lake and says, "There is energy in all that water," may be thinking of its ability to spill over a dam and force turbines to generate electricity, or its ability to melt ice, or its potential for use in a fusion reacter (see Chapter 8), where some of its hydrogen would be converted to helium with accompanying release of energy. Therefore, we refer to the energy of a body only in relation to a specific process. The potential energy of the water to be used in the hydroelectric plant is the *difference* between its energy at the foot of the dam and its energy at the top of the dam. It doesn't matter what absolute value we assign to any one of the two states—it is the energy difference between them that counts.

The energy released when an apple falls 15 feet would be only about 1/1000 Calorie, Sir Isaac Newton's head notwithstanding.

Let us return to the apple. The apple has 64 Calories because that is the amount of energy released when it is converted by oxidation to carbon dioxide and water. It does not matter whether this oxidation is performed by metabolism in a human being, or in a worm, or by combustion in a fire. It is only the difference between the final and initial states that matters. But if the apple is left in a closed vessel where there is insufficient oxygen to convert it all to carbon dioxide, it may ferment to yield alcohol. It will thus have reached a different state and the amount of energy transferred will be different—only about two Calories. Of course, the alcohol thus produced may also be considered to have energy, because it could be burned completely to carbon dioxide and water, releasing another 62 Calories. So, either path yields 64 Calories (Fig. 1.1).

TABLE 1.1 ENERGY: WHAT IT CAN DO

	HEAT A BODY ABOVE THE SURROUNDING TEMPERATURE	*WORK*
Apple	Eating it helps you to maintain your body temperature at 98.6°F, even when you are surrounded by air at 70°F.	Eating it helps you to be able to pull a loaded wagon up a hill.
Coal	Burning it keeps the inside of the house warm in winter.	Burning it produces heat that boils water that makes steam that drives a piston that turns a wheel that pulls a freight train up a hill.

Energy is necessary to all living organisms because many biochemical reactions are energy-requiring. In complex ecosystems such as those that

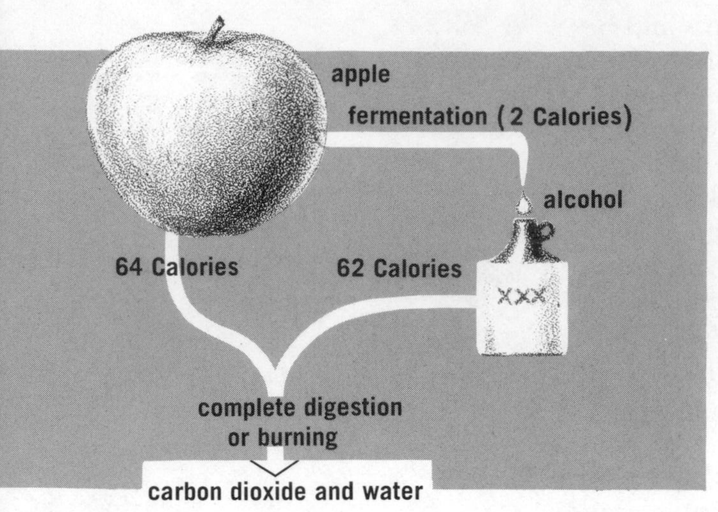

FIGURE 1.1 Energy from an apple.

exist today, all organisms are interrelated with other organisms by the food- (energy-) gathering process. Therefore, the study of the energy relationships within an ecosystem are of primary interest to the ecologist.

1.3 ENERGY RELATIONSHIPS WITHIN AN ECOSYSTEM

(Problems 3, 5, 6, 7, 8, 9, 36, 37)

Plants are able to trap the sun's energy and transform it into chemical energy to build molecular structures such as those of sugars, starches, proteins, fats, and vitamins. For this reason plants are called **autotrophs,** meaning self-nourishing. All other organisms obtain their nourishment (energy) from other sources and are called **heterotrophs** (other-nourishing). This broad classification includes such widely diverse species as cows, grasshoppers, mountain lions, sharks, maggots, and amoebas. There are many ways to classify the heterotrophs, depending on the type of study one wishes to conduct. A comparative anatomist would make separate categories based on evolutionary and morphological similarities. An ecologist wishes to focus attention on function, specifically on the position of the organism in the energy flow. He is therefore interested in levels of nourishment, or **trophic levels.** Autotrophs occupy the *first trophic level.* All heterotrophs that obtain their energy directly from autotrophs are known as **primary consumers** and are said to occupy the *second trophic level.* They may be as different from each other as a grasshopper is from a cow, but both have similar ecological functions: they are grazers. Praying mantises eat grasshoppers, and owls eat field mice; therefore, both of these predators are secondary consumers. That is, they obtain energy from the plants only indirectly, in two steps, and are said to occupy the *third trophic level.* Let's take another step: shrews eat praying mantises, and martens eat owls; therefore, both are tertiary consumers. Owls who eat shrews are quaternary consumers, and martens who eat the owls who have eaten shrews are still another step removed from the original plant. Now

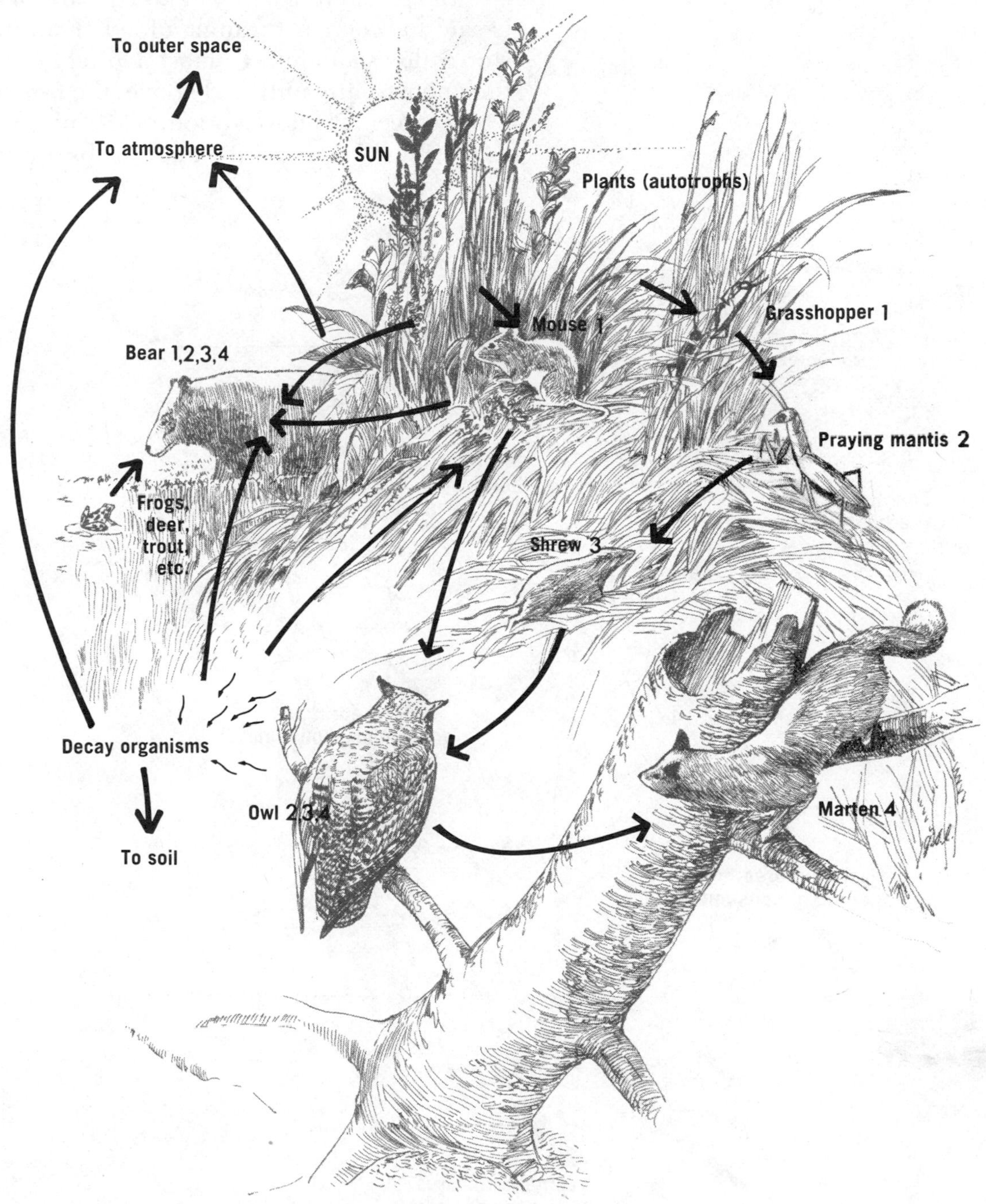

FIGURE 1.2 Land-based food web. Arrows are in the direction of progressive loss of energy available for life processes. Numerals refer to trophic levels.

just a minute—how can an owl be both a secondary and a quaternary consumer? The answer is that these categories are not mutually exclusive; a hungry owl doesn't care about human definitions, but dives at whatever is little and running about, regardless of how we classify him.

In very simple ecosystems the flow of food energy is said to progress through a food chain in which one step follows another. In most natural systems, such as the one partially outlined above, the term **food web** is a more accurate description of the observed interactions (Figs. 1.2 and 1.3). Food webs are further com-

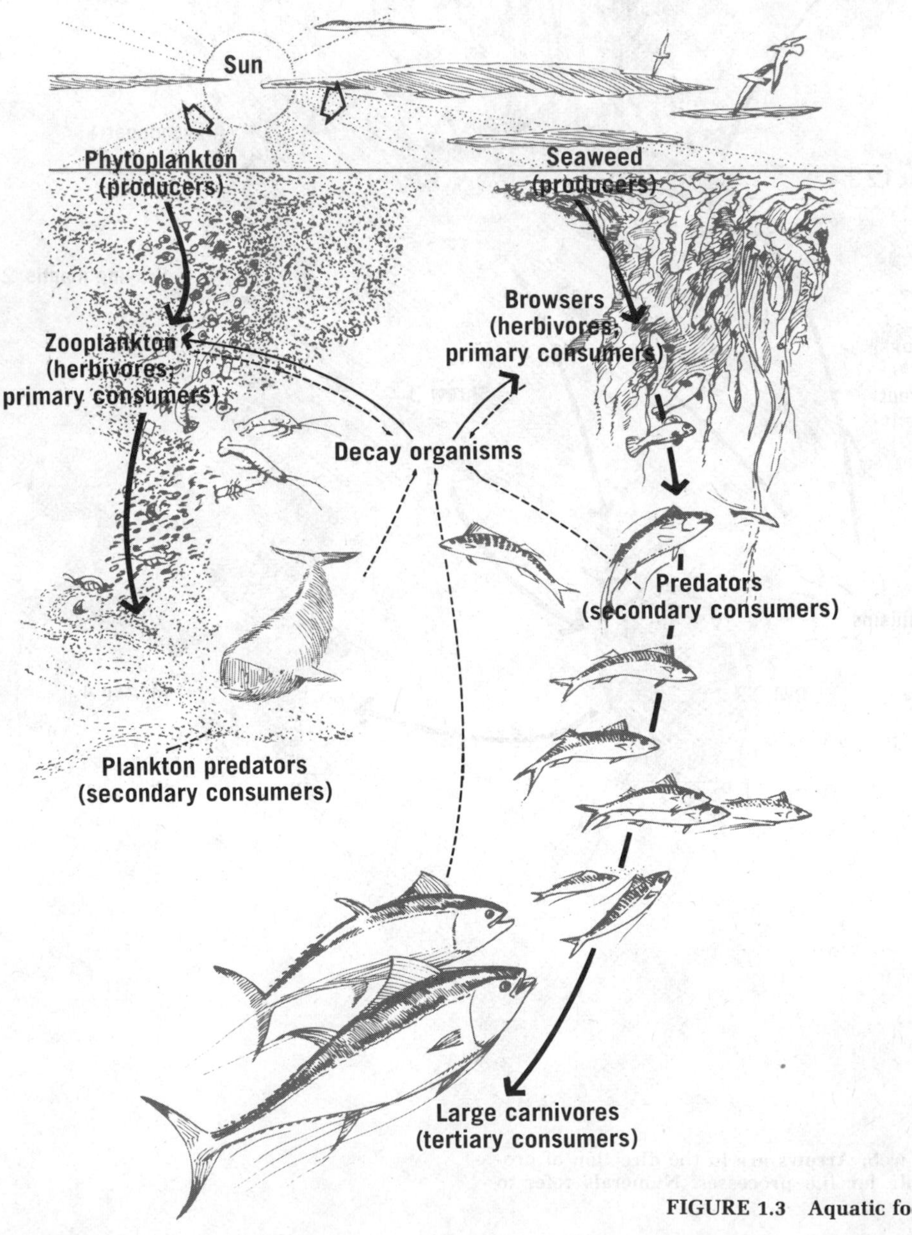

FIGURE 1.3 Aquatic food web.

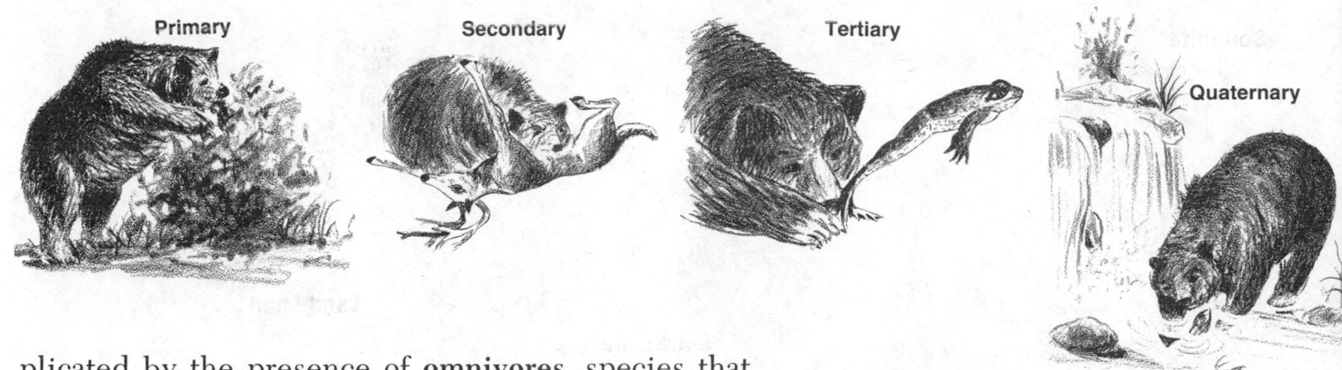

plicated by the presence of **omnivores,** species that eat both plant and animal matter. Bears, rats, pigs, man, chickens, crows, and grouse are all omnivores. These species are often very difficult to place within any simple scheme. When a grizzly bear eats roots and berries, he is a primary consumer; when he eats a deer, he is a secondary consumer; eating a frog makes him a tertiary consumer; and eating a trout qualifies him as a quaternary consumer.

Of course individuals often die without being immediately consumed, but instead remain to decay. The decay process is performed by many widely varying types of species, such as fungi, bacteria, soil mites, nematodes, ostracodes, and snails. These organisms, both plants and animals, are also classified by function and are called **saprophytes** (Fig. 1.4). Since saprophytes are also subject to predation, and some are omnivorous, the food web within a forest floor or on a lake bottom is quite complex. All the non-living organic matter in an ecosystem is known collectively as **detritus.** This debris is consumed by a group of organisms known as **detritus feeders.** Thousands of different species of plants and animals, and billions of individuals, feed on detritus. Although the complex interactions among the organisms are poorly understood, we do know that all stable ecosystems depend on the detritus food web to maintain stability.

One characteristic of ecosystems must be reemphasized: Subclassifications are helpful in focusing attention on a particular aspect of study, but the lines are not sharply drawn in nature. Consider the bear (clearly not a decay organism) who rips open a rotting log in search of termites. In consuming them, he derives energy. Some energy has now been removed from the detritus food web and has entered the food web of the larger animals.

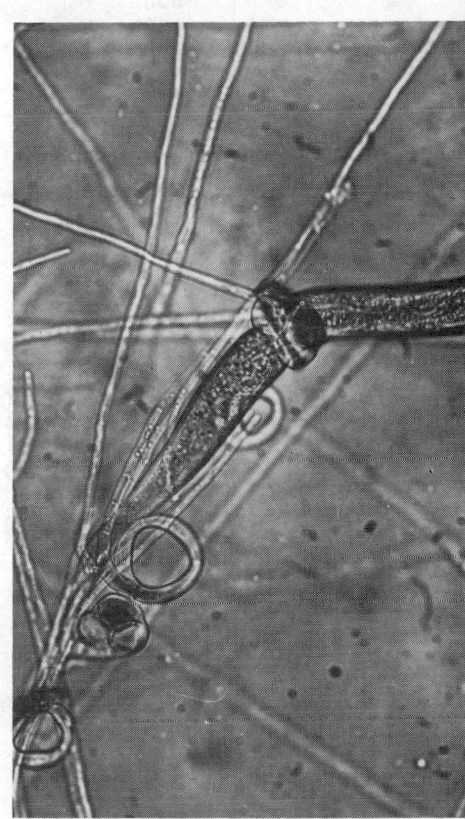

Nematode captured by the predaceous fungus. *Dactylella dreschleri* **(× approx. 560). (Courtesy of Dr. David Pramer.)**

Energy is continuously received from the sun at a constant rate. Ultimately this heat is radiated off into

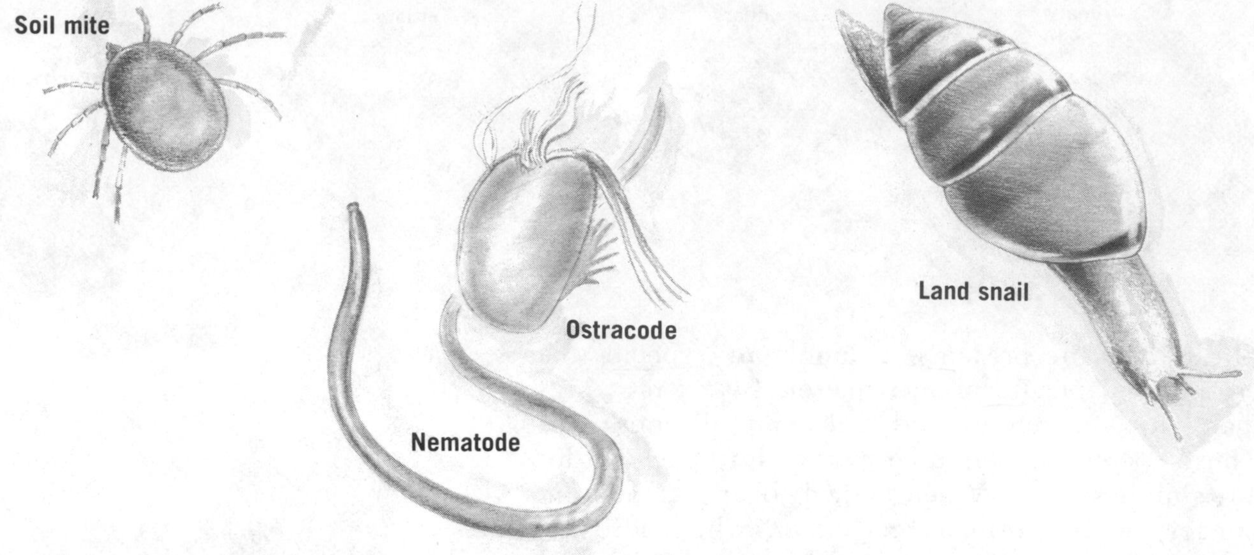

FIGURE 1.4 Saprophytes.

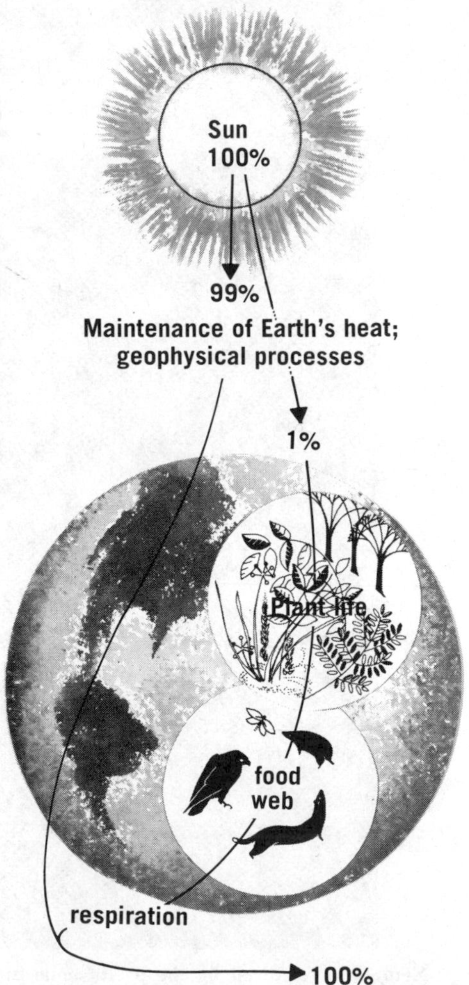

FIGURE 1.5 The flow of energy on Earth.

space. Because we observe that the average annual temperature of the earth remains relatively constant from year to year, we know that the energy gain and loss of the Earth must be in balance. The condition in which the inflow equals the outflow is called a **steady state.** The flow and ultimate fate of energy from the sun is represented schematically in Figure 1.5.

The chemical energy available for life processes constantly decreases through the food chain.* To understand the workings of an ecosystem, one must unravel the energy flow through the food web. Therefore, the ecologist is interested in the total production of organic tissue and the total consumption, which can be measured as respiration. The ratio of total (gross) production to total respiration is an indicator of the energy balance of the system; it tells us whether the system is growing, aging, or maintaining a steady state. Another important variable that is related to both the present and past production/respiration ratios is the **biomass**—the total mass of organic matter present at one time in an ecosystem. Finally, the *biomass at each trophic level* is of fundamental importance. For terrestrial ecosystems, a large primary production is required to support a small weight of predators, so that a bar graph of biomass of a self-contained ecosystem usually resembles a pyramid (Fig. 1.6).

*See Chapter 5 for a discussion of the First and Second Laws of Thermodynamics.

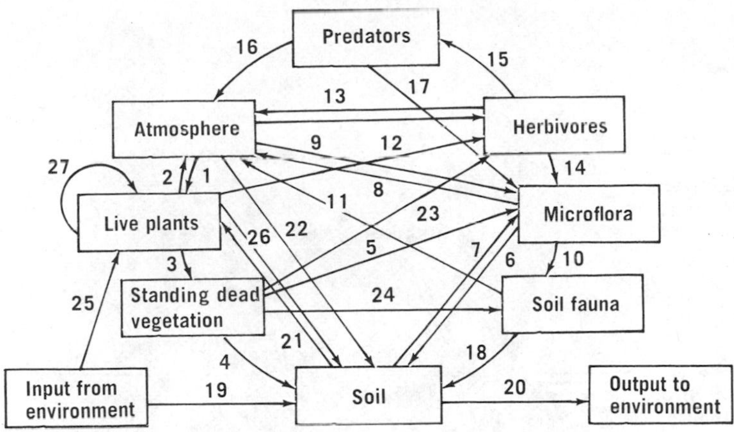

Food web (see Problem 6). From George M. Van Dyne, *The Ecosystem Concept in Natural Resource Management,* **Academic Press, New York, 1969.**

FIGURE 1.6 Food pyramid. The mass shown in each box is the amount required to produce the mass of tissue in the box above it. The areas in the boxes are proportional to the masses.

We shall now follow the energy flow through a simple food chain. Consider a plant that receives 1000 Calories of light energy from the sun in a given day (Fig. 1.7). The efficiency of conversion of the sun's energy to chemical energy depends on the species of plant and the conditions of growth, but, in any case, most of the energy is not absorbed; instead, it is reflected or transmitted through the tissue. Of the energy that is absorbed, most is stored as heat and used for evaporation of water from leaf surfaces and other types of physical processes. Of the energy that is used in the life processes of the plant, most is expended in respiration. The remainder, about five calories, is stored in the plant tissue as energy-rich material, suitable as food for animals.

Now suppose a herbivore, say a deer, eats a plant containing five calories of food energy. She would dissipate about 90 per cent of the energy received to maintain her own metabolism as well as to retain some energy for muscle action to move around. As a result, the deer would convert only about ½ calorie to weight gain. A carnivore eating the deer is likewise inefficient in converting food to body weight, so the energy available to each succeeding trophic level is progressively diminished. The energy advantage of the herbivores is one important reason why there are so many more herbivores than carnivores. It is obvious that man, who can occupy primary, secondary, or tertiary positions in the food chain, uses the sun's energy most efficiently when he is a primary consumer, that is, when he eats plants.

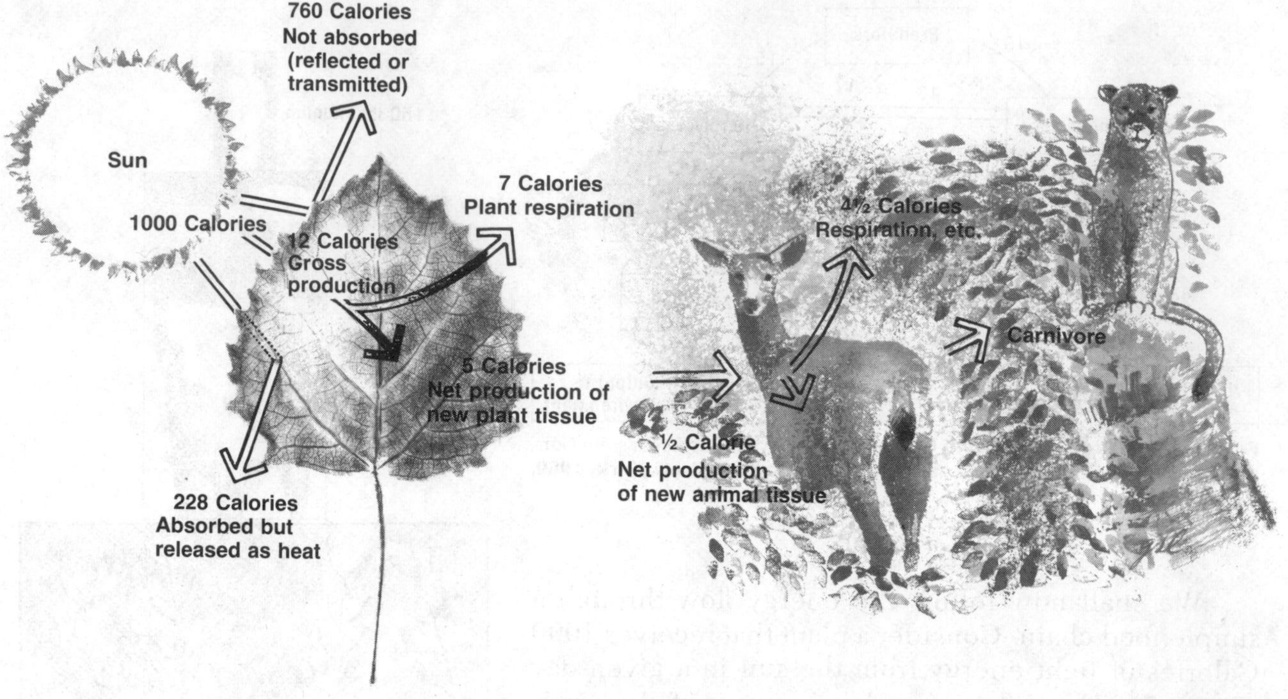

FIGURE 1.7 Energy flow through a simple food chain.

1.4 NUTRIENT CYCLES

(Problems 10, 11, 12, 13)

Energy alone is insufficient to support life. Imagine, for example, that some aquatic plants such as algae were sealed into a sterilized jar of pure water and exposed to adequate sunlight. The process of photosynthesis which utilizes the atmospheric carbon dioxide and releases oxygen would not be balanced by the plants' own respiration, and soon the plants would starve for lack of carbon dioxide. If the experimenter maintained the proper atmosphere through some gas-supply and exhaust lines, the plants still would not survive. They would starve for lack of the chemicals necessary for life. Suppose, then, that the experimenter fertilized the jar with the proper inorganic chemicals, which had been sterilized to insure that no additional living organisms entered the system. The plants would grow, and the cells which died in the normal processes of life would accumulate. Soon there would be no more room to introduce new fertilizer, and, unless the jar were enlarged, the fertilization would have to be stopped and the algae would die. The jar might be stabilized indefinitely, however, if some plant-consumer, a snail for example, were introduced, and the pure water were replaced by pond water to supply inorganic nutrients and saprophytes. Assuming an initially balanced ratio of snails to algae, the jar would now perhaps become a balanced, stable ecosystem in which the nutrients would be continuously recycled. The algae use water, carbon dioxide, sunlight, and the dissolved nutrients to support life and build tissue. Oxygen is one major waste product of algae and at the same time is an essential requirement of snails and other heterotrophs.

In general, plants produce more food than they need. This overproduction allows animals to eat parts of the plants, thus obtaining energy-rich sugars and other compounds. These compounds, in turn, are broken down during respiration, a process that uses oxygen and releases carbon dioxide. Other animal waste products, such as urine, though relatively energy-poor, are consumed by microorganisms in the pond water. The waste products of the microorganisms are even simpler molecules which can be reused by the algae as a source of raw materials. The consumption of dead tissue by aquatic microorganisms is essential both as a means of recycling raw materials and for waste disposal. The interaction among the various living species permits the community of the jar to survive indefinitely. In fact, biology laboratories have sealed aquariums in which life has survived for a decade or more. We would like to stress that the picture presented of this cycle has been greatly simplified. Cycles which occur in nature are so complicated that they are not fully understood today.

The biosphere itself can be considered to be a sealed jar, for although it receives a continuous supply of energy from outside, it exchanges very little matter with the rest of the universe. Thus, for all practical purposes, life started on this planet with a fixed supply of raw materials. There are finite quantities of each of the known elements. The chemical form and physical location of each element can be changed, but the quantity cannot.

This statement does not apply to radioactive elements which are discussed separately in Chapter 6.

A generalized nutrient cycle can be visualized as follows (Fig. 1.8): Suppose substance A is a vital raw material for a certain species of organism 1, and substance B is one of the organism's products. Substance B must then be a vital raw material for a second organism, which then produces a new product. This alternation of products and the organisms that consume them must continue until the cycle is complete; that is, until some organism produces our original substance A. When averaged over a long period of time, each prod-

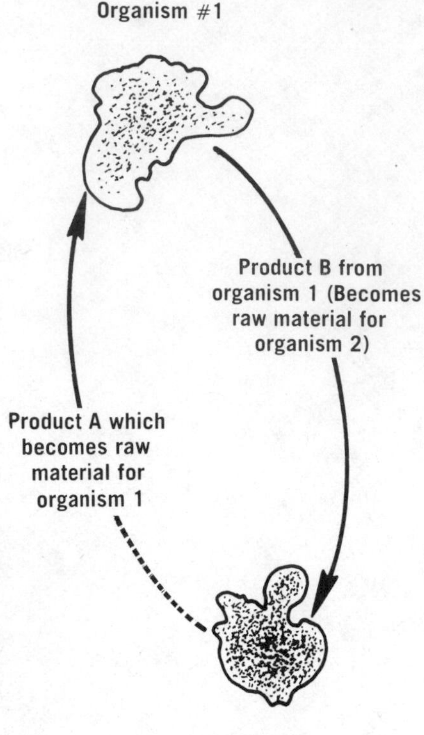

FIGURE 1.8 Generalized food cycle.

uct must be produced and consumed at the same rate; consumption must balance production. If any substance in the cycle, for example, substance B, were produced faster than it was consumed, it would begin to accumulate. The population of organism 2 would then increase in response to its greater food supply, and it would begin to consume more of substance B. Eventually the consumption of B would be increased enough to match its production, and the cycle would thus have regulated itself. Conversely, if B were produced at a rate *less* than the demand for it, the population of organism 2 would decline because of this insufficiency. The resulting decreased demand would enable production to catch up with consumption, and balance would be reestablished. Any permanent interruption of the cycle, for example by extinction of one of the species of organisms and the failure of any other organism to fulfill its function, would necessarily lead to the death of the entire system.

The substances produced by living organisms are widely varied. First, there are wastes associated with the act of living. Plants produce more oxygen than their life processes require; animals exhale carbon dioxide. However, organisms themselves can be viewed as products. A tree produces leaves, twigs, a trunk; a lion produces bones, a mane, a tail. All these parts of an organism, indeed the entire organism, when it is either living or dead, is a product which can be consumed by other organisms.

When studying nutrient cycles, it is helpful to separate those elements that can be found in large quantities in the air from those which are dissolved in water or located in the soil or the earth's crust. The continuous state of turbulence of the atmosphere insures a constant composition (omitting water vapor and air pollutants) throughout the earth. Moreover, gases in the air are relatively quickly assimilated and released by living organisms. The net result is the existence of a large, constant, readily available pool of nutrients. This does not mean an infinite, unchanging pool, for atmospheric composition has, in fact, changed considerably in the course of time.

The elements that occur naturally on Earth range from hydrogen (atomic number 1) to uranium (atomic number 92), but four atomic numbers are missing: technetium (43), promethium (61), astatine (85), and francium (87).

Of the 88 natural elements, about 40 are needed by living systems to maintain life. Many of these 40 are needed in only trace amounts, while others such as carbon, oxygen, hydrogen, and nitrogen constitute a large proportion of the mass of a system.

Oxygen Cycle

Oxygen is present in the earth as gaseous molecular oxygen (O_2), as a chemical part of gaseous carbon dioxide (CO_2), as a component of water (H_2O), as an important element in many organic compounds such as sugars, starches, and proteins, as a constituent of many dissolved ions such as nitrate (NO_3^-) or carbonate ($CO_3^=$), and as a major component of the earth's mineral crust. Geochemical exchange of oxygen between rock and gaseous or liquid form does occur but is slow compared to other forms of exchange. Presently, the most common cycling of oxygen is initiated by organic processes (see Fig. 1.9). A plant synthesizes carbohydrates by combining carbon dioxide with water in the presence of sunlight and discharging oxygen as a byproduct:

$$6CO_2 + 6H_2O \xrightarrow{\text{sunlight}} \underset{\text{sugar, a carbohydrate}}{C_6H_{12}O_6} + 6O_2 \qquad (1)$$

This process is called **photosynthesis.** Most molecules in plant tissue contain the element oxygen. Heterotrophs, which are unable to build sugars from carbon dioxide and water, consume plants and assimilate their complex chemicals, including the oxygen contained therein. As these chemicals move through the food web to consumers of other orders, they are all ultimately converted to carbon dioxide and water by respiration or other oxidative processes. The organic oxygen participates in these conversions and so it, too, finds its way into CO_2 and H_2O molecules:

$$6O_2 + C_6H_{12}O_6 \xrightarrow{\text{respiration}} 6\ CO_2 + 6H_2O \qquad (2)$$

This completes the cycle, for the carbon dioxide and water may then be reused as raw materials for new synthesis.

Carbon Cycle

Carbon is present in the atmosphere primarily as carbon dioxide, which constitutes only about 0.03 per cent of the atmosphere by volume. It can be readily seen from Equations 1 and 2 that the *biological* carbon cycle is intimately related to the oxygen cycle, for whenever oxygen is cycled by life processes, carbon must accompany it (see Fig. 1.9). Carbon is incorporated into organic tissue by photosynthesis and released by respiration and decay. However, less than half of the total carbon cycling occurs through biological or other organic pathways. Atmospheric carbon dioxide dissolves readily in water, and some dissolved molecules escape from the sea to the air. This exchange occurs as a dynamic equilibrium, that is, atmospheric carbon dioxide molecules are constantly being dissolved into the ocean, while those that were previously dissolved are constantly escaping from the

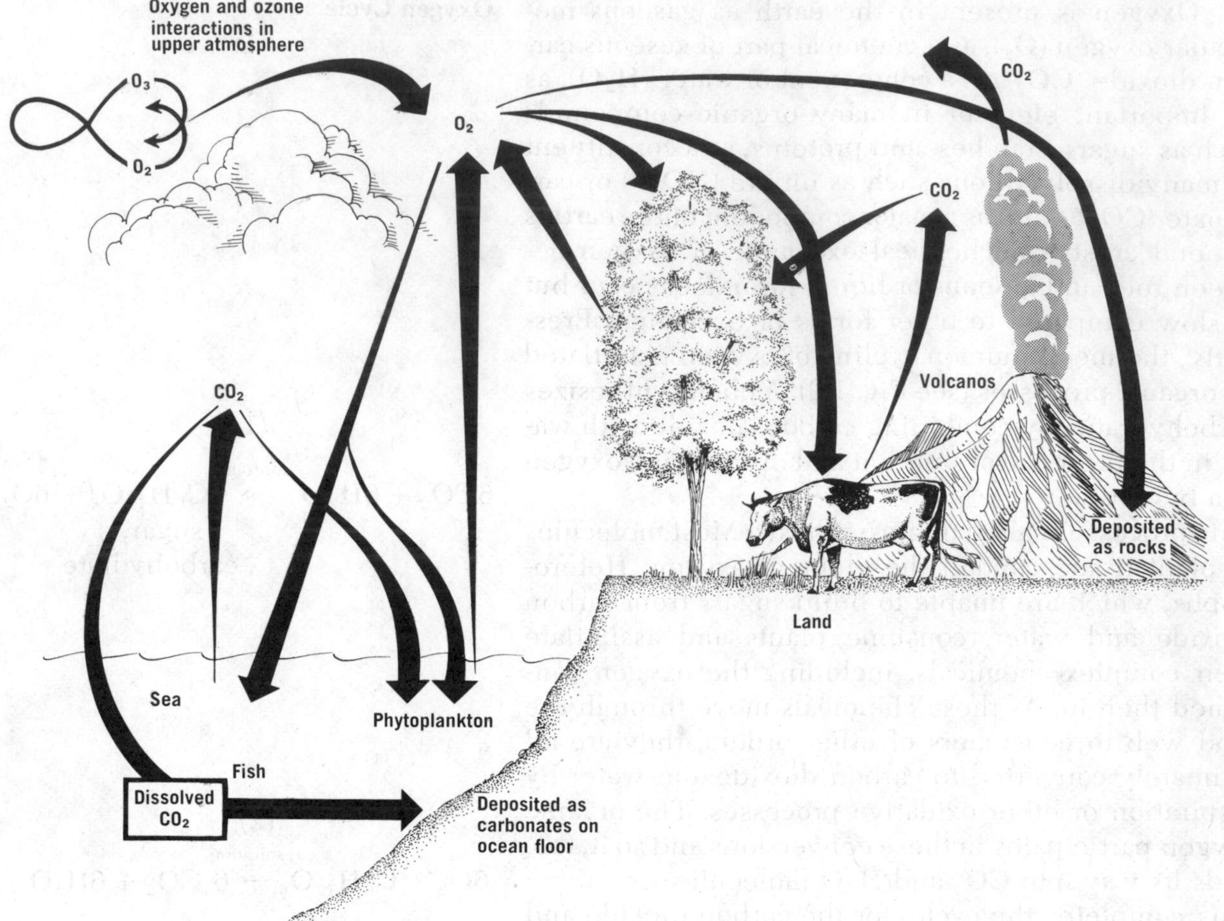

FIGURE 1.9 Oxygen-carbon cycle.

sea to the air. There is little net change in the ratio of dissolved carbon compounds to atmospheric carbon compounds. This geochemical cycling is independent of any living process; it is an inherent property of the chemistry of carbon dioxide and water.

Some of the dissolved carbon dioxide reacts with sea water to form carbonates, which settle to the ocean floor as calcium carbonate either in the form of inorganic precipitates (limestone) or as skeletons of various forms of sea organisms. This loss is partially balanced by the action of inland water which slowly dissolves limestone deposits on land and carries the carbonates to sea.

Currently there is about 15 times as much carbon locked into fossil fuel deposits as there is present in the atmosphere. These deposits were caused by past imbalances in the carbon cycle. The burning of coal and oil since the Industrial Revolution has released much carbon, permitting it to reenter biotic cycles. In playing an active role in the long-term recycling processes of the biosphere, industry is measurably changing man's environment. In Chapter 9 we shall discuss possible harmful effects of the introduction of large quantities of carbon dioxide into the air.

Nitrogen Cycle

Nitrogen is an important constituent of proteins, and therefore necessary to both plants and animals. Although nitrogen is roughly four times as plentiful in the atmosphere as oxygen, it is chemically less accessible to most organisms. Almost every plant and animal can utilize atmospheric oxygen, but relatively few organisms can utilize atmospheric nitrogen (N_2) directly. The nitrogen cycle (Fig. 1.10) must therefore provide various bridges between the atmospheric reservoir and the biological community. Lightning, photochemical reactions (see Ch. 9), and specialized bacteria and algae (called nitrogen-fixers) transform molecular nitrogen into forms usable by living organisms. Usually, fixed nitrogen is assimilated by plants and then travels into the heterotrophic chain. It may now cycle for a considerable time within the food web. Organic decay might return it to the soil,

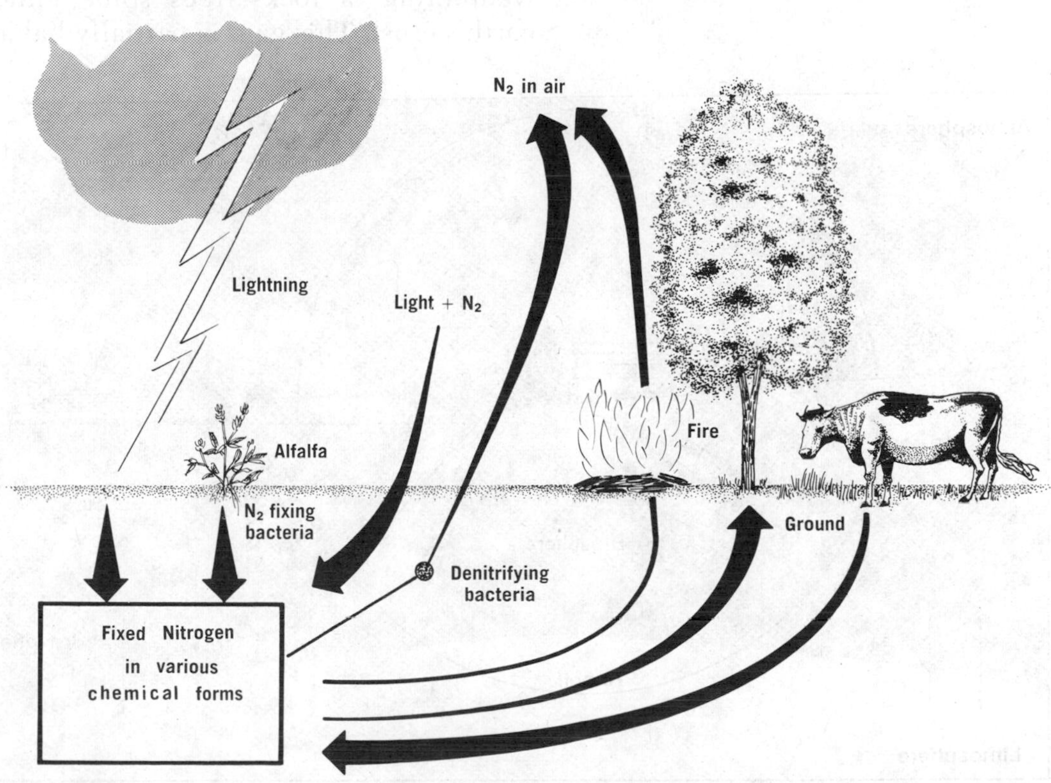

FIGURE 1.10 Nitrogen cycle.

or it might be digested and returned to the soil through feces or urine. Once in the soil, it may reenter plant systems. Denitrifying bacteria and fire provide mechanisms for the return of nitrogen to the atmosphere.

Despite the greater chemical complexity of the nitrogen cycle as compared to the oxygen cycle, the large available reservoir permits relatively facile transport of nitrogen in natural systems.

Mineral Cycles

The cycling of minerals such as phosphorus, calcium, sodium, potassium, magnesium, or iron is much more fragile than the cycling of oxygen, nitrogen, or carbon because of the absence of a large mineral reservoir in a readily available form (see Fig. 1.11). Let us consider the mineral cycles in a particular watershed in the mountains of New Hampshire. This area includes a ring of mountains and the enclosed valley. Rain is the area's only significant source of water; the only important exit for flowing water in this watershed is through a single stream, because the geology of the area is such that seepage is negligible. In such a system the total input of minerals is limited to two processes: the rain deposits some mineral matter, and weathering of rocks frees some minerals from the earth's crust. This gain is partially balanced by losses.

For the details of this example, refer to Chapter 4 of *The Ecosystem Concept in Natural Resource Management.* (See references at the end of this chapter.)

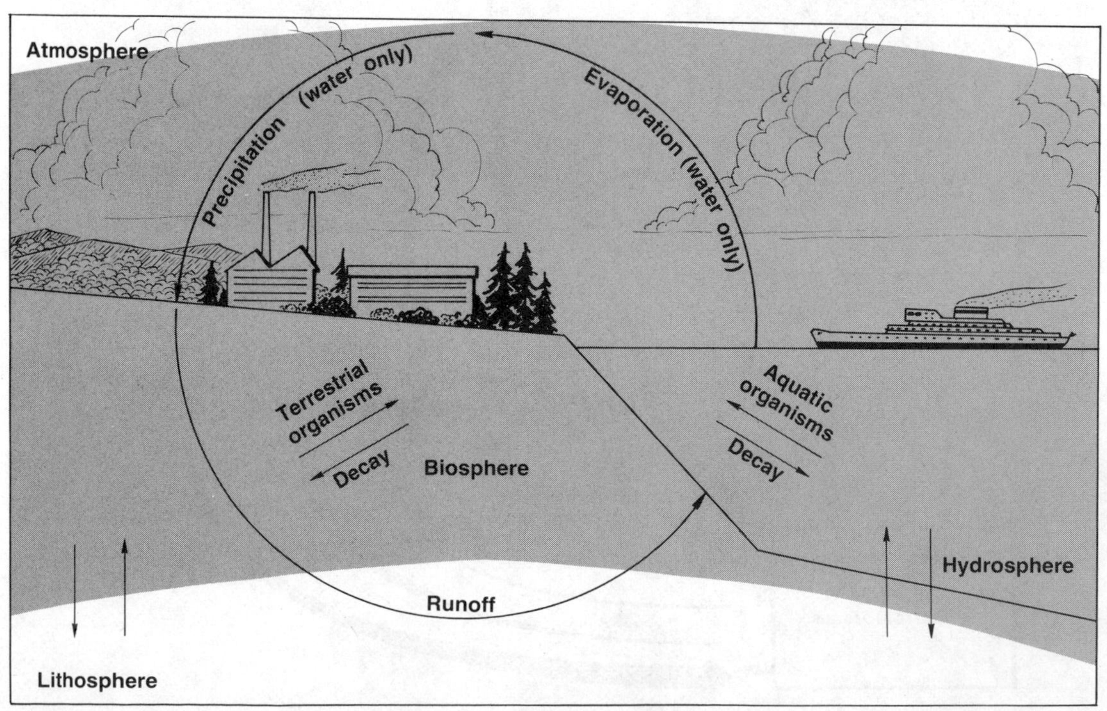

FIGURE 1.11 Mineral cycle.

In the particular watershed that was studied, it appeared that the stream outflow removed fewer minerals than weathering and rainwater were bringing in. However, the net gain for the ecosystem was very small compared to the requirements of all the life forms in the valley. This means that the valley life was operating on mineral capital, not mineral income. In fact, the entire biosphere depends for its supply of mineral matter upon the ability of each individual incoming inorganic ion to undergo countless transformations from soil to plant tissue, from plant tissue into the organic detritus or on to animal tissue, and then to the detritus, back to the soil, and around again and again before being washed out of the system. Therefore, input and output rates bear little relationship to the quantity of inorganic matter in the private pool of any given ecosystem.

1.5 ECOSYSTEMS AND NATURAL BALANCE

(Problems 14, 15, 28, 29, 30)

We have seen that food webs perpetuate the flow of energy and the cycle of nutrients. Naturally, all processes are being carried on simultaneously. A diagram illustrating all the energy and nutrient transfers of even a simple system would be a blur of arrows showing sugars, mineral ions, nitrogens, oxygens, and carbons going this way and that. How is the ensemble regulated? How can a natural meadow exist in an orderly and relatively unchanged state for many years?

Many opposing forces operate within a natural ecosystem (see Fig. 1.12). Organisms are born and die. Moisture and nutrients travel out of the soil and they are transferred back into the soil. Furthermore, many of these oppositions are exquisitely protected against disruption. During a dry season, when the mice in a grassland have less food, their birth rate decreases. But their behavioral response to lack of food is to retreat to their burrows, and thus their death rate also decreases. Their behavior protects their own population balance as well as that of the grasses, which are not consumed by hibernating mice. Such a tendency is called **ecosystem homeostasis.**

The "balance of nature" thus refers to the tendency of natural ecosystems to maintain their existence by appropriate opposition of processes and by regulatory mechanisms which protect these processes against disruptions.

How does this dynamic balance operate; how is it

Constancy of a system does not, of course, imply constancy for each individual, for balance is achieved not by stagnancy but by an equality of oppositions. The carbon cycle is balanced when the absorption of CO_2 matches the release of CO_2; a scale is balanced when the weights or forces on each pan are equal; population size is considered to be balanced when the opposing processes, the birth rate and the death rate, are equal. What happens when a balance is disturbed? That depends on the nature of the system. Some systems go out of balance easily; others resist change. As an example of the latter type, your blood is balanced between acidity and alkalinity; that is, opposing chemical processes maintain a constant level (one that is very slightly alkaline). Now suppose a certain quantity of acid is added to your blood. The chemical composition of blood is such that it will tend to oppose this disturbance of its acid-alkali balance and to return toward its normal condition. Such resistance to disturbance of acid-alkali balance is called buffering. **Buffering is thus a protective action. An organism tends to maintain a balance of various life processes by feeding itself, keeping itself in repair (healing itself), and adjusting to external changes. This tendency to maintain a stable internal environment is called** homeostasis.

FIGURE 1.12 Moist coniferous forest, a stable ecosystem, in the Olympic National Forest, Washington. (U. S. Forest Service photo. From Odum: *Fundamentals of Ecology*. 3rd Ed. Philadelphia: W. B. Saunders Co., 1971.)

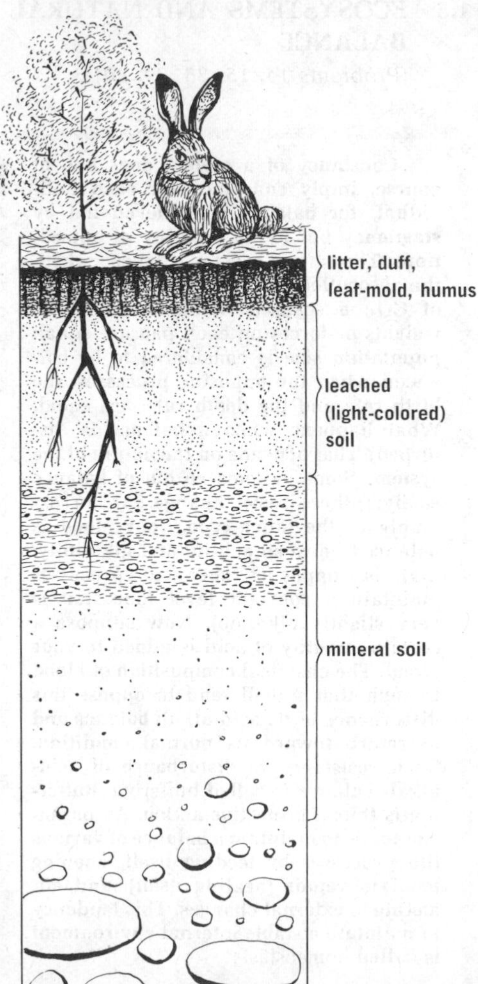

FIGURE 1.13 Cross section of forest soil.

controlled? One clue comes from the observation that in natural, stable ecosystems the biomass is quite large compared with the biomass of less stable systems. Some areas are incapable of supporting a large biomass; in deserts or in tundra the rate of growth and of accumulation of litter is slow, and massive plants such as trees cannot survive. However, the systems of greatest stability (the climax structures, see p. 43) in these harsh areas have a large biomass compared with emergent systems in the same climatic region.

Stable ecosystems are not necessarily in balance all the time, but if they are imbalanced in one direction, they must become imbalanced in the opposite direction at some time in the future if they are to survive with no essential change in character. In fact, all ecosystems naturally fluctuate. For example, climate varies from year to year. Other disruptions, such as migration, drought, flood, fire, or unseasonable frost can cause imbalance in an ecosystem. The ability of an ecosystem to survive depends on its ability to adjust to an imbalance.

Let us consider as an example the relative populations of tigers, grazing animals, and grass in a valley in Nepal. Assume that the mountains surrounding the valley are so high that no animals can enter or leave the valley. One year rainfall is limited and mild drought conditions exist. Because water is naturally stored in pools, the drinking water supply is adequate. However, the grasslands suffer from the lack of rainfall. Therefore, food is scarce for the grazing animals, and many are hungry and weak. Such a situation is actually beneficial to the tigers because hunting becomes easy, and the tiger population thrives.

The following spring, when rains finally arrive, the grazing population is low. This permits the grass to grow back strongly because not much of it will be eaten. However, because the food supply for the tigers is low (and the grazers that are left are the strongest of the original herd), the tigers suffer a difficult time the summer after the drought. The third year there are fewer tigers, so the herd resumes its full strength and balance is achieved again.

Stable ecosystems, if examined superficially, do not seem to go out of balance at all. Actually, their homeostatic mechanisms function so well that slight imbalances are corrected before they become severe. Such sensitive responses are, in fact, the essence of stability, because if severe imbalances are prevented, it is very unlikely that the system will be destroyed and, as a result, it lasts for a long time. Usually, when we refer to a stable ecosystem, we mean one which regenerates itself again and again with little overall change. Sometimes, however, an ecosystem is stable not because it regenerates often without change, but because the dominant plants are long-lived. A wonderful example is a redwood forest. To the hiker entering a redwood forest on a hot summer day, two characteristics of the environment are immediately apparent. It is cool, and there is a thick floor of spongy matter. The coolness, caused by the extensive shade of the tall trees, reduces water loss due to evaporation. The spongy floor is formed by a large bed of decomposed organic matter known as **humus,** as well as a liberal supply of fallen needles and other as yet undecomposed litter that catches and holds the rainwater, and thus serves as a water and nutrient reservoir, and as a home for the members of the detritus food chain (Fig. 1.13). Thus, the effect of a drought does not have to be balanced by the death of trees. In addition, the large amount of organic material available ensures a steady functioning of the recycling system. Thick mats of partially decomposed organic matter also retain the heat produced by saprophytes and help maintain a warm environment known as a **microclimate,** which is often more favorable for metabolism than the climate on the surface. Anyone who has ever operated a compost heap has observed this warming effect.

The coolness of the forest offers still another advantage. Clouds passing overhead are more likely to discharge their moisture where it is cool, for cold enhances the tendency of water to agglomerate into raindrops. Thus, the redwood forest tends to cause rain, to restrict evaporation, and to retain a substantial supply of water. Furthermore, decay organisms require moist environments to be effective, so the redwood forest insures the proper recycling of materials by environmental control.

This unusually stable ecosystem also exhibits other highly effective homeostatic abilities. One of these can be observed readily by walking through the forest on the west side of Route 1 as it runs through the town of Big Sur, California. Some time ago a tremendous fire swept through the forest. However, because the bark and the wood of the redwood tree are fire-resistant, the big trees were scarred but not killed. Because the forest environment was maintained, the forest community was quickly regenerated. If a fire of equal intensity had burned through a stand of pine, the trees would have died and many more years would have been needed for the forest to be rebuilt.

1.6 MAJOR ECOSYSTEMS OF THE EARTH

(Problems 1, 16, 17)

The **biosphere** is the region including all the life-supporting portions of our planet and its atmosphere. This section briefly describes some of the major ecosystems of the biosphere.

The ocean. (Courtesy of H. Armstrong Roberts.)

The ocean holds a varied and intricately interwoven set of ecosystems.

Despite the fact that the ocean is known to be more than 19,000 feet (6000 meters) deep in places, light does not penetrate more than about 650 feet (200 meters) in sufficient quantities for photosynthesis to occur. The depth of this illuminated section, or **euphotic zone**, varies considerably with the turbidity of the water, reaching its maximum limits in the central oceans and narrowing to 100 feet (30 meters) in coastal regions. Although the depth of the photosynthetic zone is greatest in the central ocean, the rate of photosynthesis is not greatest there, for the concentration of plant nutrients is low. Indeed, the central ocean has often been likened to a great desert.

The primary autotrophs in all marine ecosystems are phytoplankton, the one-celled chlorophyll-bearing organisms suspended in the water (Fig. 1.14). In areas near shore, various species of algae, commonly known as seaweed, are abundant and account for a significant portion of the net production of living matter.

A large portion of the primary consumption in the sea is carried out by myriads of species of small grazers called zooplankton (Fig. 1.14). These animals range in size from roughly 0.2 mm (.008 in.) to 20 mm (0.8 in.) in diameter and consist both of permanent zooplankton and larvae of larger forms of life. The composition by species of a sample of plankton varies with the climate, and plankton occupies several trophic levels. There are few large grazers in the sea analogous to land-based bison, deer, or cattle.

Many omnivores and predators live in the sea. These species represent various biological classifications, including fish, mammals, invertebrates, and reptiles. Organic debris from the life processes on the surface and in the body of the sea are continuously raining down toward the ocean floor. In the deep ocean, some of this food is intercepted, some decomposes, and some falls to the bottom, where **benthic species** (organisms that live in and on the bottom sediments) receive their nourishment (Fig. 1.15). Some of these are semimobile or immobile animals, while others crawl or swim actively in search of food.

While gravity is constantly pulling nutrients to the bottom of the sea, the mechanisms for upward recycling of nutrients are comparatively feeble. Cur-

Ocean Systems

FIGURE 1.14 (Courtesy of Professor John Lee, Dept. of Biology, The City College of the City University of New York.)

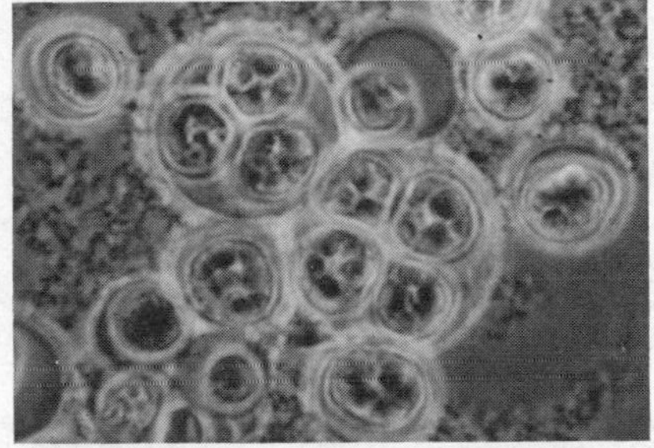

A

A nonmotile marine phytoplankter.

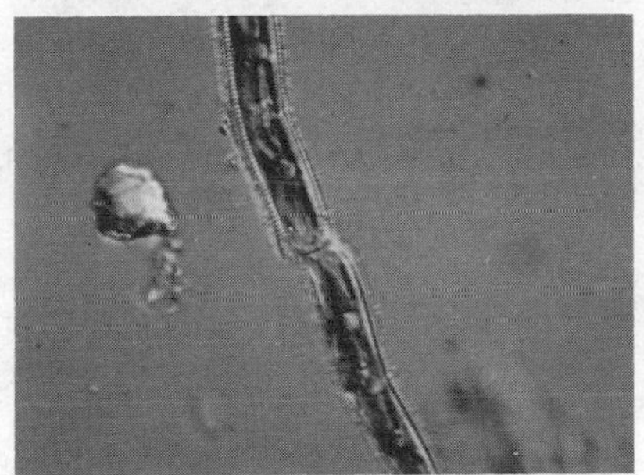

B

A microscopic marine plant.

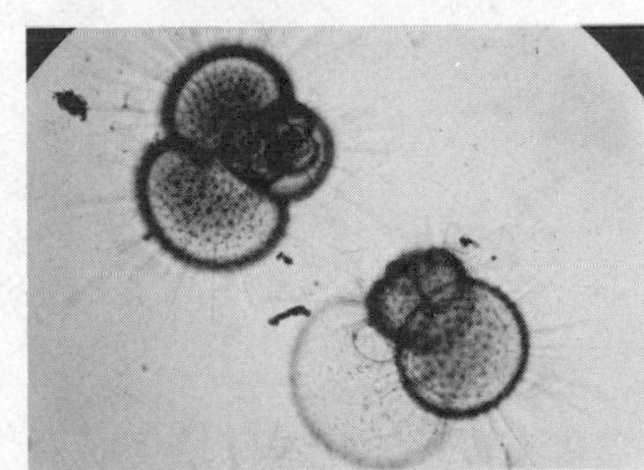

C

A planktonic marine animal *Globigerina bulloides*. Large areas at the bottom of the sea are covered by their remains.

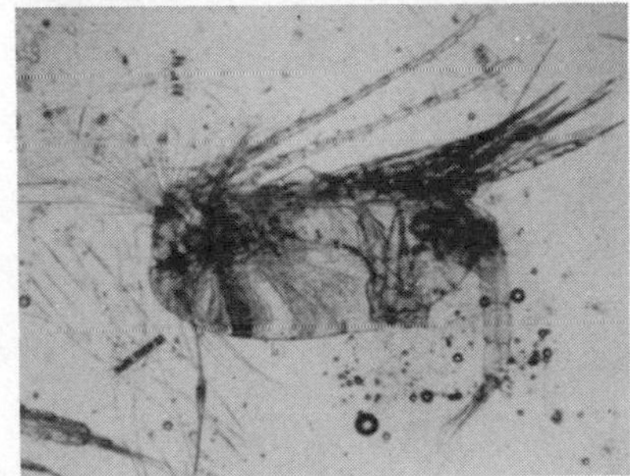

D

Another planktonic animal: marine copepod.

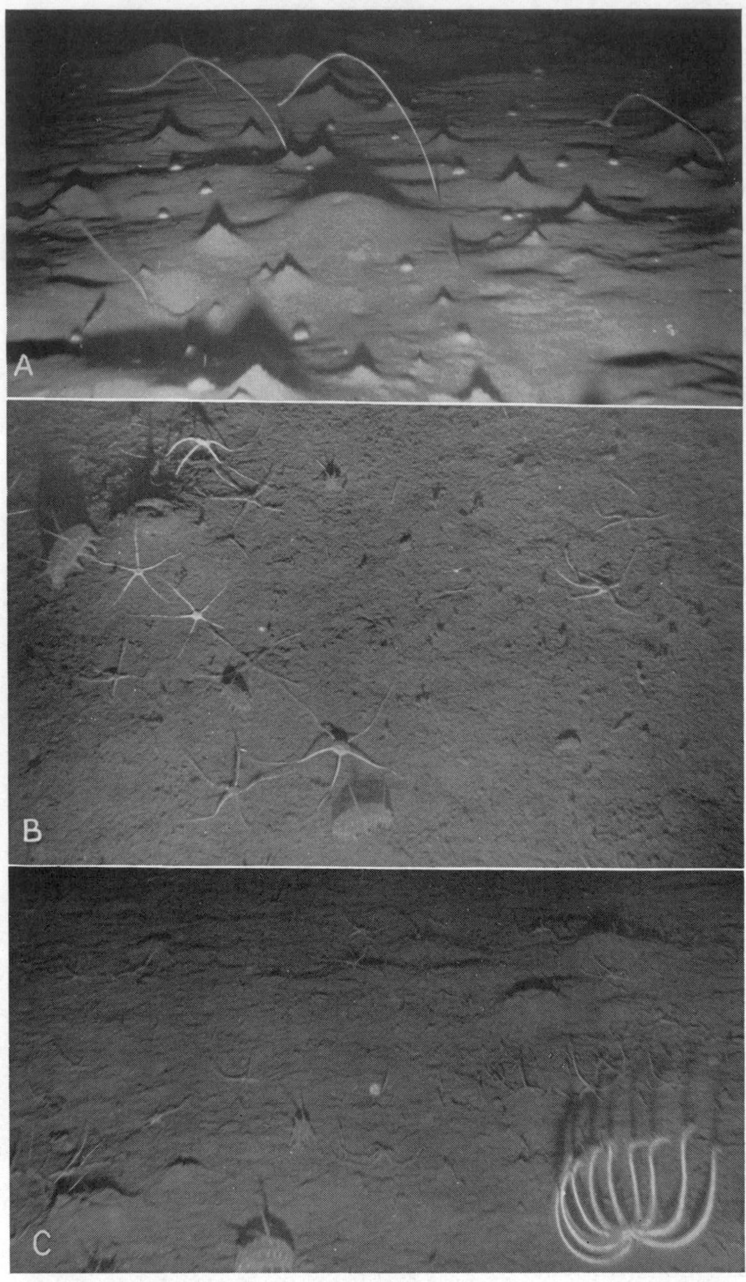

FIGURE 1.15 Benthic species. (From Odum: *Fundamentals of Ecology.* **3rd Ed. Philadelphia, W. B. Saunders Co., 1971.)**

rents and water turbulence serve to move some nutrients back to the surface, but a net loss does occur to the deep sea floor. These chemicals may eventually be recycled by geological upheavals. One net result of poor recycling in the ocean is that the concentration of nutrients dissolved in the bulk of our ocean waters is low and imposes limits on the growth rate of organisms. By contrast, in the food-rich coastal areas, wave actions, nutrient inflow from rivers, and a relative shallowness all combine to support a large biomass.

Estuary Systems

Coastal bays, river mouths, and tidal marshes are all physically contiguous to the open ocean, yet their proximity to land and fresh water affects the salinity and nutrient composition to such a large extent that these areas, known as **estuaries,** are characteristic neither of fresh nor of salt water (Figs. 1.16 and 1.17). The combination of such factors as (a) easy access to the deep sea, (b) a high concentration and retention of nutrients originating from land and sea, (c) protective

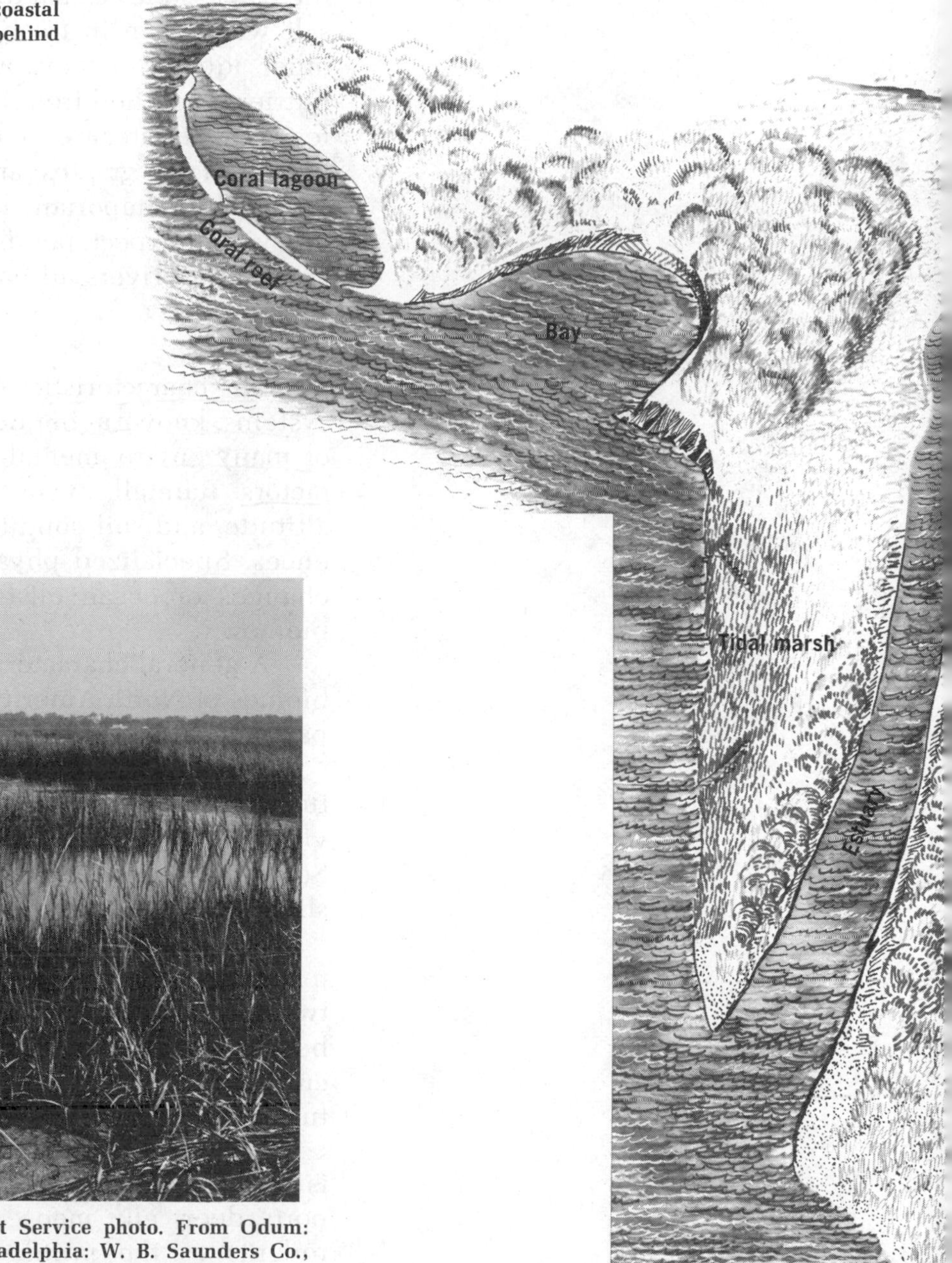

FIGURE 1.16 Shoreline features: coastal bay; estuary; tidal marsh; coral lagoon behind coral reef.

FIGURE 1.17 Salt marsh. (U.S. Forest Service photo. From Odum: *Fundamentals of Ecology.* **3rd Ed. Philadelphia: W. B. Saunders Co., 1971.)**

shelter, and (d) rooted or attached plants supported in shallow water makes estuaries very productive areas indeed. Estuaries provide nurseries for many deep-water fish which could not produce viable young in the harsher environment of the open sea.

Freshwater Systems

Freshwater systems involve ecological relationships similar to those which operate in the oceans. Again, the food web starts with plankton and culminates in large predators. A major difference between the two systems is that there are more trophic levels in salt water than in fresh. Freshwater systems, unlike other aquatic systems, are continuously fertilized by nutrients leached from the nearby soil. Because bodies of fresh water are shallower than the oceans, rooted plants, marsh grasses, and lilies, as well as algae, are much more important in the food webs. Finally, as one would expect, ponds, lakes, mudpuddles, springs, creeks, and rivers all have unique and characteristic species.

Terrestrial Systems

The characteristics of large, stable, terrestrial ecosystems, known as **biomes**, result from the interactions of many environmental, biological, and evolutionary factors. Rainfall, average and seasonal temperatures, altitude, and soil conditions all have profound influences. Specialized physical factors such as seasonal changes or ocean mists can help to create unique biomes.

A general characterization of the major terrestrial biomes of North America is outlined in the following paragraphs.

In the coldest regions, where summer temperatures average under 10°C (50°F) and trees cannot survive, stable plant cover is known as **tundra** (Fig. 1.18). Some of the grazer-plant and predator-prey relationships in this area will be discussed later.

The region between the tundra and the northern forest is characterized by tundra grasses growing between dwarfed, gaunt, windblown conifers. (Such border ecosystems, known as **ecotones**, will be discussed further on page 47.) Animals from both the tundra and the forest systems live here.

Still farther south, where the weather is warmer, is the **taiga**, or northern evergreen forest, which supports deer, elk, moose, caribou, wood buffalo, and rodents as the primary grazers and wolves, lynx, mountain lions, and man as the largest predators.

PLATE 1

The most favorable life conditions for different species of plants and animals are clearly displayed by the zonation on rocks along the Atlantic coast. (Courtesy of William H. Amos.)

PLATE 2

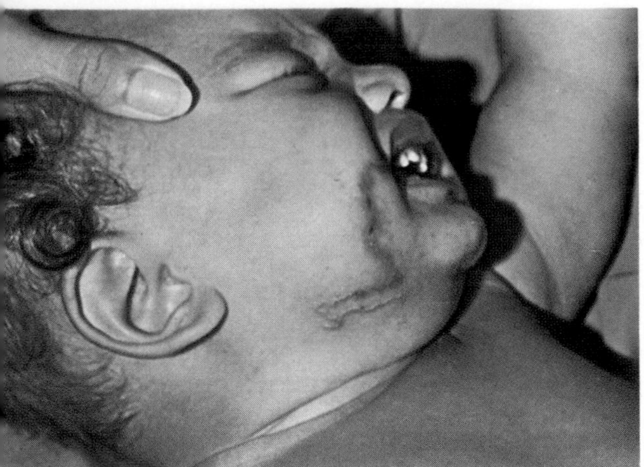

(A) Migration of a roundworm parasite through the skin. (Courtesy of Dr. Herman Zaiman.)

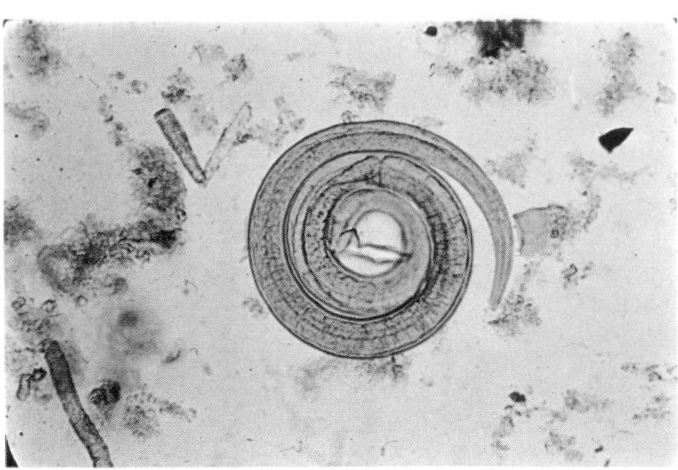

(B) Larva of the pork roundworm, *Trichinella spiralis*. (Courtesy of Dr. Herman Zaiman.)

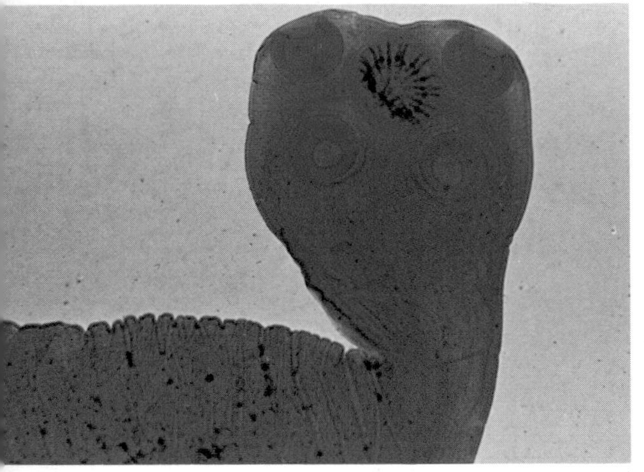

(C) The anterior end of the pork tapeworm, *Taenia solium*, as it appears in the intestine. (Courtesy of Dr. Herman Zaiman.)

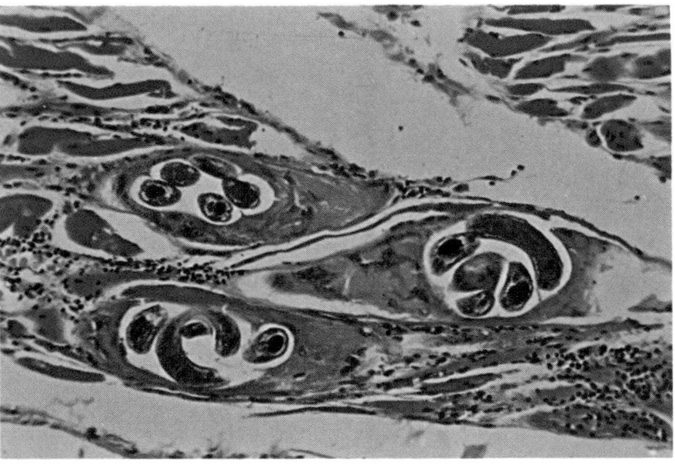

(D) Young *Trichinella* roundworms in muscle. (Courtesy of Dr. Herman Zaiman.)

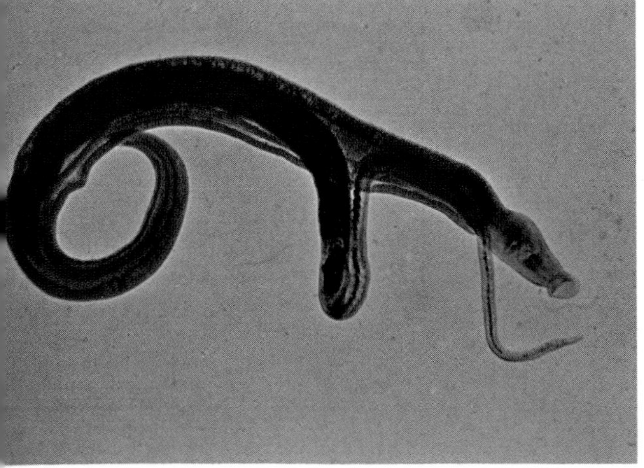

(E) *Schistosoma japonicum*, male and female flatworm parasites which infect the blood vessels of millions of people. The female lies within a canal of the larger male. (Courtesy of Dr. Herman Zaiman.)

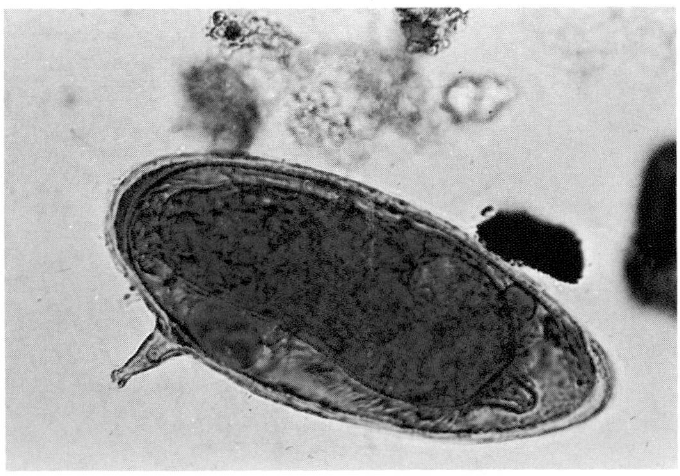

(F) An egg of the flatworm parasite, *Schistosoma mansoni*, from human stool. Note the characteristic spine on the shell. (Courtesy of Dr. Herman Zaiman.)

FIGURE 1.18 Tundra in July on the coastal plain near the Arctic Research Laboratory, Point Barrow, Alaska. (Photos by the late Royal E. Shanks, E. E. Clebsch, and John Koranda. From Odum: *Fundamentals of Ecology.* **3rd Ed. Philadelphia: W. B. Saunders Co., 1971.)**

South of the taiga are forests consisting largely of deciduous trees. Such forests need 30 to 60 inches (75 and 150 cm) of rain per year and an average temperature during the growing season of about 60° to 65°F (15° to 18°C). Therefore, they do not exist farther north than central Maine or southwestern Quebec or farther south than central Georgia or southern Louisiana.

In the temperate areas in which the rainfall is too low to support forests, the stable ecosystems are

prairies or **grasslands** (Fig. 1.19). In wet areas, the virgin grasses of North America were as tall as a man; elsewhere, they were short enough that the early scouts could "see the horizon under the bellies of millions of buffalo."

In very dry areas, where rainfall is less than 10 inches (25 cm) per year, the climax is a **desert,** either barren or able to support only scrub brush and cactus (Fig. 1.20). Deserts can be hot, as in Nevada, or relatively cold, as in eastern Washington.

In addition to the above ecosystems are some not found in North America. Tropical **rain forests** are stable systems of the Amazon basin, parts of Africa, Central America, and parts of Southeast Asia, Malaysia, and New Guinea. Here the seasonal temperature changes are less pronounced than daily ones. Vegeta-

FIGURE 1.19 Natural temperate grassland (prairie) in central North America. *A.* **Lightly grazed grassland in the Red Rock Lakes National Wildlife Refuge, Montana, with a small herd of pronghorns.** *B.* **Short-grass grassland, Wainwright National Park, Alberta, Canada, with herd of bison. (From Odum:** *Fundamentals of Ecology.* **3rd Ed. Philadelphia: W. B. Saunders Co., 1971.)**

FIGURE 1.20 Two types of deserts in western North America. *A.* **A low altitude "hot" desert in southern Arizona.** *B.* **An Arizona desert at a somewhat higher altitude with several kinds of cacti and a greater variety of desert shrubs and small trees. (From Odum:** *Fundamentals of Ecology.* **3rd Ed. Philadelphia: W. B. Saunders Co., 1971.)**

tion is thick, and most of the available nutrients are found in the biomass. Decay and recycling are rapid.

Tropical **savannas** are fire-dependent systems occurring in areas with annual wet and dry seasons where the yearly rainfall is 40 to 60 inches (100 to 150 cm). Fires are common during the dry season. The flora consists of a few fire-resistant trees and annual vegetation **blooms,** or periods of rapid growth. Many of the animals are migratory.

When studying biomes of the world, it is of interest to compare similar climatic regions which have been separated for millennia. The classic comparison between North American plains and those of Australia shows similar trophic structure but dissimilarities between species. Thus, kangaroos and bison, both important grazers, are morphologically quite different.

1.7 THE ECOLOGICAL NICHE

(Problem 18)

There are roughly one and a half million different species of animals and one-half million species of plants on Earth. Each species performs unique functions and occupies specific habitats. The combination of function and habitat is called an **ecological niche.** To describe a niche fully, one would first have to describe all physical characteristics of a species' home. One might start with specifying the gross location (for example, the Rocky Mountains or the central floor of the Atlantic Ocean) and the type of living quarters (for example, a burrow under the roots of trees). For plants and the less mobile animals, one would have to describe the preferred micro-environment, such as the water salinity for species living at the interfaces of rivers and oceans, the soil acidity for plants, or the necessary turbulence for stream dwellers. An animal's trophic level, its exact diet within the trophic structure, and its major predators are also important in the description of its niche. Mobile animals generally have a more or less clearly defined food-gathering territory or **home range** which is another factor in establishing the physical niche.

A niche is not an inherent property of a species, for it is governed by factors other than genetic ones. Social and environmental factors contribute to the choice of niche. Suppose that we wished to use laboratory observations to study the niche of the sagebrush. We could plant several batches of seeds under conditions which differ only with respect to acidity of the soil. By measuring the rate of growth of the sage seedlings in each pot, we could draw a graph of growth as a function of soil acidity. The experiment could then be repeated under different conditions. Nutrients, light intensity, duration of darkness, moisture, and temperature could all be varied. For each factor, one might observe three important experimental points. There is a minimum value for each variable below which no seeds will sprout. There is also a maximum limit above which no seeds will sprout. Finally, there is an optimum growth condition. Since the variables will interact with each other, the optimum growth condition will be experimentally difficult to obtain, but a reasonable approximation can be established. Armed with the data collected in the laboratory, we can now search for sagebrush in natural ecosystems. We know from the experimental data not to look in marshes, in northern Alaska, or in old mineshafts. Searching through some Western rangeland, we find a sagebrush. If we measure all the physical properties of the immediate environment of this plant, we will most likely find that it is not growing under optimum laboratory conditions. If we find a sagebrush growing a thousand miles away from the first one, it probably will be living under slightly different conditions than the first. In an exact sense, the two observed field niches are different from each other and from the experimental optimum niche.

Squirrel in park, feral house cat as predator.

The difference may be seen more clearly by comparing a tree squirrel in the city park of Boulder, Colorado, with one in the Gore Wilderness area in the Rocky Mountains of Colorado. The diet, the daily activities, and the major predators affecting each are quite different.

Squirrel in wilderness, fox as predator.

The niche of a given species in a given ecosystem is not the set of conditions that would be best suited to the genetic make-up of the organism, but rather it is the best accommodation that the organism can make to the realities of its environment.

1.8 OVERLAP AND DIVERSITY OF NICHES

(Problems 19, 20, 21, 22, 23)

Imagine two tribes, X and Y, of nomadic hunters living in the area of the fictitious map in Figure 1.21. Suppose that the best hunting grounds are found along the flood plain of the river. Naturally, both tribes would compete for this prime hunting area, perhaps by sporadic warfare. One probable result of this competition is that tribe X and tribe Y would maintain different secure territories on the edge of the flood plain. Between these two home bases, some of the hunting area would be visited by the hunters of both tribes. If tribe X had sole possession of the best hunting areas, the people could be fed easily from the population of lowland animals, but in the face of competition from tribe Y, the hunters from tribe X must

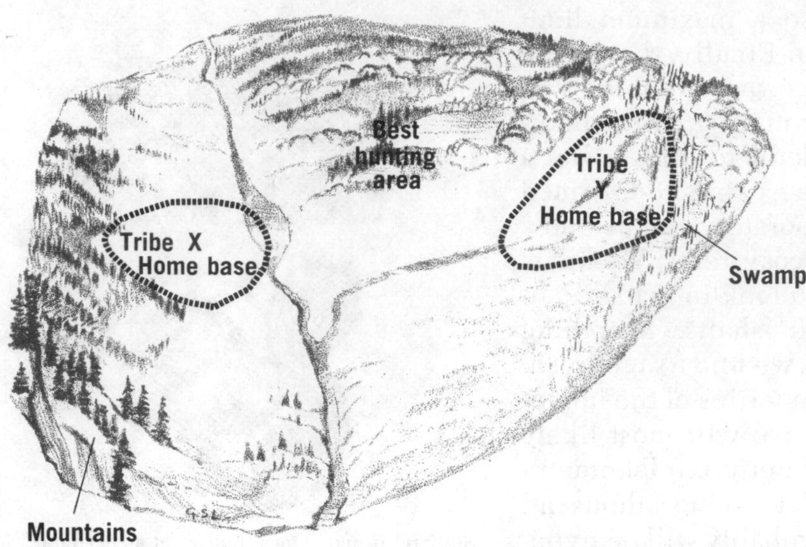

FIGURE 1.21 A model of competition.

obtain part of their game by hunting the mountainous area near their home base. Similarly, the hunters in tribe Y must search the swamp for a portion of their food. If some catastrophe, such as fire, destroys the lowland hunting areas, members of both tribes, experienced in hunting alternate sources of food, could probably feed their people.

Competition between animal species can be considered analogous to the competition between the two tribes. If the words "niche" and "species" are used in place of the words "home territory" and "tribe," the example just given illustrates a simplified case of competition between two animal groups.

These relationships can be clearly seen in a series of examples from natural systems. Consider two species of barnacles which live on the Scottish coast. The larger species dominates between the high- and low-tide mark where it can be assured of daily contact with the sea. The smaller barnacle species lives predominantly higher up along the beach where it must survive frequent and more prolonged periods of desiccation. Experiments have shown that the small barnacle can, in the absence of competition, survive anywhere in the intertidal zone, while the larger species can survive only in regions of plentiful moisture. However, the more adaptable species is unable to compete with the stronger one in the wetter environment. One species survives by virtue of its adaptability, the other by virtue of its competitiveness, and thus diversity is built into the system.

It is fair to ask whether or not two species of

plants living side by side, with their roots in the same soil and their tops touching, are not, in fact, occupying the same niche. Naturally, not every set of plants living near each other has been studied, but those studies that have been conducted have shown that if there isn't a difference in nutrient or light requirements, then there is usually some difference in root depth or in timing of life cycles. Of two species of clover that coexist, one was observed to grow faster and spread its leaves sooner than the other. The second clover species ultimately grows taller than the first. Thus, each has its period of peak sunlight, and they are both able to survive.

We can imagine that if the niches for two separate species were identical, competition for food and shelter would be particularly intense. In fact, both experiment and field observation have led to the generalization that two species in the same ecosystem cannot both survive in the same niche. If there are two species with the same preferred niche, competition will either lead to the elimination of one of them or to the adaptation of one of them to fit a new niche. A corollary to this rule is the generalization that in an ecosystem which houses many species, not only is the diversity of niches very large, but species are coadapted to overlapping niches so that elimination through competition is reduced.

One laboratory experiment showed that if two species of flour beetles were placed in a jar of flour, and an ample supply of food maintained, only one species would survive. Yet, in the absence of competition, either one could survive in the jar indefinitely. Further studies showed that if the jar were kept hot and wet, one of the species always won, while if the jar were cool and dry, the other species always won. It is easy to imagine what happens to these beetles in nature. For example, if grain is stored in a large elevator, it is likely that the conditions at the bottom of the pile are hot and wet while at the top the grain is drier and cooler. Each species of flour beetles then can fit into its own niche.

Niche competition and the resultant diversity are not merely a two-species interaction. Of the 10 principal mammalian predators living in the southern tundra and northern forests, no two species have identical hunting habits, yet significant overlap is apparent, as illustrated in Table 1.2. Let us see how this system

TABLE 1.2 NICHE COMPETITION IN THE TUNDRA*

	FOODS							
MAMMALIAN PREDATORS	***Rodents***	***Insects***	***Large Grazers***	***Eggs***	***Birds***	***Fish***	***Carrion***	***Vegetables and Berries***
Bear	X	X	X	X		X	X	X
Wolf	X		X	X	O		O	
Coyote	X		O	X	X		O	
Fox	X			X	X			
Wolverine[a]	X		X	X	X	X	X	
Otter	X			X	X	X		
Marten[b]	X			X	X			
Shrew		X						
Lynx[c]	X			X	X		O	
Primitive man	X		X	X	X	X		X

*Legend: X = regular or staple part of diet; O = occasional part of diet.
[a]Can gnaw bones unchewable by other species.
[b]Can climb trees and hunt tree-dwelling animals more effectively than other species.
[c]Migrates when food is scarce.

operates. All the predators but the shrew hunt the lemming, a rodent whose population is susceptible to periodic fluctuations (see page 40). During abundant lemming years, all the predators can eat lemmings. This prevents lemmings from becoming so populous that they destroy their food supply. Then, during years

Migrating herd in Africa. (From C. A. Spinage: *Animals of E. Africa.* Houghton Mifflin Co., Boston, 1963.)

with few lemmings, the wolves can eat more caribou, the foxes can subsist on birds, the lynx can move south, and so on. Thus the predator populations survive. If the diets of predators were limited to lemmings, some species might become extinct, and, in that case, there might not be sufficient control mechanism during peak lemming years.

We have shown some relationships among niches that occur when several species gather food in a single geographic area, or when several species of stationary organisms, such as barnacles, live along a gradient of climate. Sometimes niche interactions are defined by patterns of migrations. Of the numerous grazers in the Serengeti grasslands of Africa, the relationship between the zebra and the Thompson gazelle is illustrative. The former is a large, horse-like animal which does not chew its cud. Its physiology is therefore quite different from that of the cud-chewing gazelle. Since the zebra is able to eat large quantities of food and excrete the unnecessary carbohydrates, it can live on plants with a low protein content. The gazelle's digestive system is unable to handle such large quantities of food, and it must subsist on protein-rich grasses. During the abundant wet season both animals graze together on the hillsides, but when the yearly drought arrives, the zebra cannot subsist on the sparse growth and moves down into the floodplain. There it eats the tall grasses which are plentiful but nutritionally poor. The gazelle, meanwhile, remains to wander about the hillsides picking out the small, rich surface plants which the zebra did not consume. Because the gazelle is small and needs less total food than the zebra, he can afford the time to search for food. The zebra would have starved had he stayed. After some time, even the gazelle must migrate. The tall grasses of the floodplain have been cut down by this time by the zebras, leaving the lower, smaller, richer surface vegetation for the gazelle. The zebra continues to migrate in a path that eventually brings him back to the hillsides at the start of the next rainy season.

Top, **Zebras in Africa,** *middle and bottom,* **gazelles in Africa. (Courtesy of Dr. C. A. Spinage, College of African Wildlife Management, Mweka, Box 3031, Moshi, Tanzania.)**

1.9 INTERACTIONS AMONG SPECIES

(Problems 24, 25, 26, 27)

Food webs are complex. Ecological niches overlap in place and function. Many homeostatic equilibria exist simultaneously in the same system. It is just these complexities, the interplay of these checks and balances, the interactions between species, that maintain the system. Interactions among species are

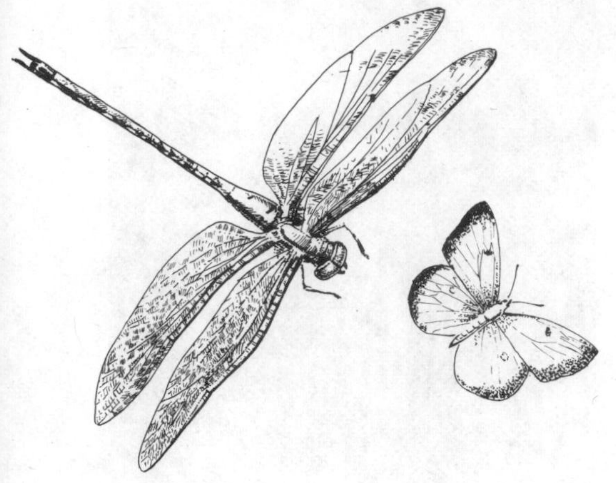

made up of separate events: an individual of one species interacts with one individual of another species at a given time. Of course, every isolated encounter has importance to the two interacting species. But when analyzing the system, we wish to focus attention upon the effects that the encounter has on the total community. In a larger sense, we are ultimately concerned with the effects of the sum of encounters on the total community. For example, knowing the relationship between deer and aspen saplings is necessary, but is only one prerequisite to the more comprehensive knowledge of the relationship of the community of grazers toward the community of plants.

Two-Species Interactions

Two-species interactions can be catalogued into eight major types: neutralism, competition, amensalism, predation, parasitism, commensalism, protocooperation, and mutualism.

Neutralism is the inconsequential case of very little interaction. Wild rose bushes and lynx have almost no direct dealing with each other.

In our study of niche selection we have already dealt with **competition.**

Amensalism is an interaction in which the growth of one species is inhibited, while the second is unaffected. As an example, the habitat of the small barnacles on the Scottish coast (p. 30) is reduced by the large barnacles, although the latter seem relatively unaffected.

Predation is an interaction in which some individuals eat others. Since all heterotrophs must eat to survive, predation is an integral part of the function of any ecosystem. In stable ecosystems, growth and predation are balanced in such a way that all species maintain viable populations. In the example on page 18, we saw that the grazer population in our hypothetical valley in Nepal was regulated by both the availability of its supply of food (the grass) and the size and vitality of its predator population (the tigers). In turn, the tiger population was regulated mostly by the size and health of the herds of grazers.

Of course, not all ecosystems are so finely regulated, and predators have destroyed certain prey populations in some regions. The homeostasis of an ecosystem is not dependent on individual restraint on the part of a particular predator, but rather on the statistical balance between offense and defense. Such natural controls do not always function smoothly, but ecosystems cannot remain stable unless there is an overall balance between consumption and growth.

Predation is a delicate and fascinating process. One of us (Jon) was recently fortunate enough to watch a lone timber wolf stalk a moose by a river in the northern Canadian forest. The wolf followed the moose along a river bank, always maintaining a separation of several hundred yards. While the moose frequently looked toward her predator, she never broke into a run, but continued to feed and move slowly downstream. After about five minutes the wolf turned abruptly, trotted over the hill, and disappeared from view. Both animals knew that a wolf was no match for a young, healthy cow moose. Empirical analyses of caribou killed by wolves have shown that the old, the crippled, the sick, and the very young are killed in disproportionately high numbers. A healthy adult caribou or moose is attacked only very rarely by a wolf or wolf pack. By selecting the less fit animals as victims, predation is a force in the genetic evolution of the hunted species.

This story is only a small part of the total predation picture, however. We have already mentioned that two species cannot occupy exactly the same niche because one always dominates or displaces the other. It has been shown that in the absence of predation, two species with different but similar niches will often be unable to coexist. Yellowstone Park has a food web that was once balanced and rich with diverse species. Man's initiation of the destruction of the predator population has led to a rapid rise in the number of elk. The abundance of elk has put so much competitive pressure on the white-tailed deer that they are threatened with elimination from the area. There is some evidence that predators generally prevent this type of imbalance by tending to put the greatest hunting pressure on the most abundant species.

Parasitism is a special case of predation in which the predator is much smaller than the victim and obtains nourishment by consuming the tissue or food supply of a living organism known as a **host.** Just as predator-prey interactions are balanced in healthy ecosystems, parasite-host relationships have also become part of the mechanism of homeostasis in nature. It must be stressed that this type of balance observed in old systems does not imply that a new parasite (or

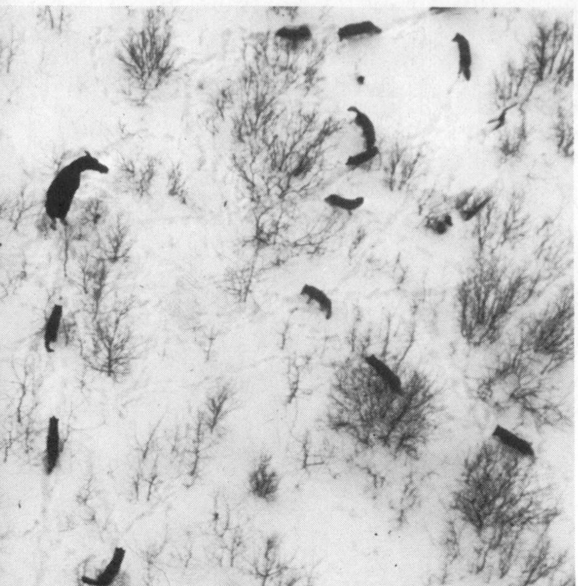

FIGURE 1.22 Wolves and moose on Isle Royale. (Courtesy of L. David Mech.)

Eagle—a predator of small game. (Courtesy of H. Armstrong Roberts.)

predator or grazer), artificially imported from another continent, will immediately establish itself as part of a stable system. On the contrary, a new species may find a new niche for itself and increase unchecked until, perhaps many years later, food supplies decline or another species migrates, is introduced, or evolves to control the rampant one. Violent oscillations caused by imported species are indications of an unstable ecosystem.

Commensalism is a relationship in which one species benefits from an unaffected host. Several species of fish, clams, worms, and crabs live in the burrows of large sea worms and shrimp. They gain shelter and often eat their host's excess food or waste products, but do not seem to affect their benefactors.

A relationship that is favorable to both species is called **protocooperation.** Crabs often carry coelenterates on their backs, and move them from one rich feeding ground to another. In turn, the crabs benefit from the camouflage and protective stingers of their guests. (Not all crabs and coelenterates are mutually cooperative.)

Mutualism is an interaction beneficial and necessary to both parties. Lichen, which grows on bare rock, resembles an extremely thin layer of vegetation. Actually, the lichen is a mixture of a fungus and an alga. The fungus, which does not contain chlorophyll and thus cannot produce its own food by photosynthesis, obtains all of its food energy from the alga. In turn, the alga cannot retain water and, in some harsh environments, would dehydrate and die if it were not surrounded by fungus. Here the dependence is direct because the organisms must grow together in order to survive.

Another example of a mutualistic interaction can be found within our own bodies. Millions of bacteria live in the digestive tracts of every person. These organisms depend on their host for food but, in return, aid in the digestive process and are necessary to the survival of all of us.

Community Interactions

Interactions within an ecological community are composed of a multitude of two-species encounters. The simplest type of community interaction is a chain reaction. Cats eat rats, rats attack beehives, bees pollinate flowers and produce honey. Thus, the popula-

Archer fish knocks ladybug off a leaf with accurately aimed drops of water. (From Roy Pinney – Globe Photos, Inc.)

tion of wild flowers, and the price of honey, is partly dependent on the population of cats. Because every animal has a place in a food web, the removal, addition, or exploitation of any species will necessarily cause reverberations, large or small, beneficial or harmful, throughout the system. This rule holds for destruction of "pests" as well as for the introduction of new game species.

We have previously suggested that the plant life of an area is generally healthier if grazed than if left ungrazed. Upon closer examination, we find that plant communities thrive best if they are grazed by a community of grazer species instead of by a single species. A single selective grazer is apt to put pressure on a few species of plants, allowing others to dominate.

Interactions between diverse species tend to promote stable ecosystems. A good example is afforded by the role of three species of grazers – bison, elk, and moose – on the ecological balance of the Elk Island National Park in Canada. Moose eat saplings and small bushes. With brush growth held in check, grasses find room to grow. Bison eat grass. Elk eat either leaves or

Moose, Elk and American Bison.

grasses. In this way, the community of grazers acts on the community of plants to ensure that an equilibrium *status quo* is maintained. Of course, there are many other interactions, such as the relationships between the community of predators and the community of grazers. The net result is a balanced, continuous, self-perpetuating ecosystem.

Diversity is also important in the plant kingdom. A single plant species does not constitute a stable system capable of buffering itself against changing weather. For example, both annual and perennial grasses grow in the prairie. The perennial grasses and some low bushes have deep roots, while the annual plants depend on much shorter and less extensive root systems. During dry years there is so little water that many annuals die. However, the perennials, which use water deep underground, are able to live; in doing so, they hold the soil and protect it from blowing away with the dry summer winds. In years of high rainfall, the annuals sprout quickly, fill in bare spots and, with their extensive surface root systems, prevent soil erosion from water runoff. Survival of both types of grasses is ensured by minimization of root competition because the different plants have root systems which reach different depths. In addition, not all species flower at the same time of the year, so that the seasons of maximum growth and consequent maximum water consumption differ.

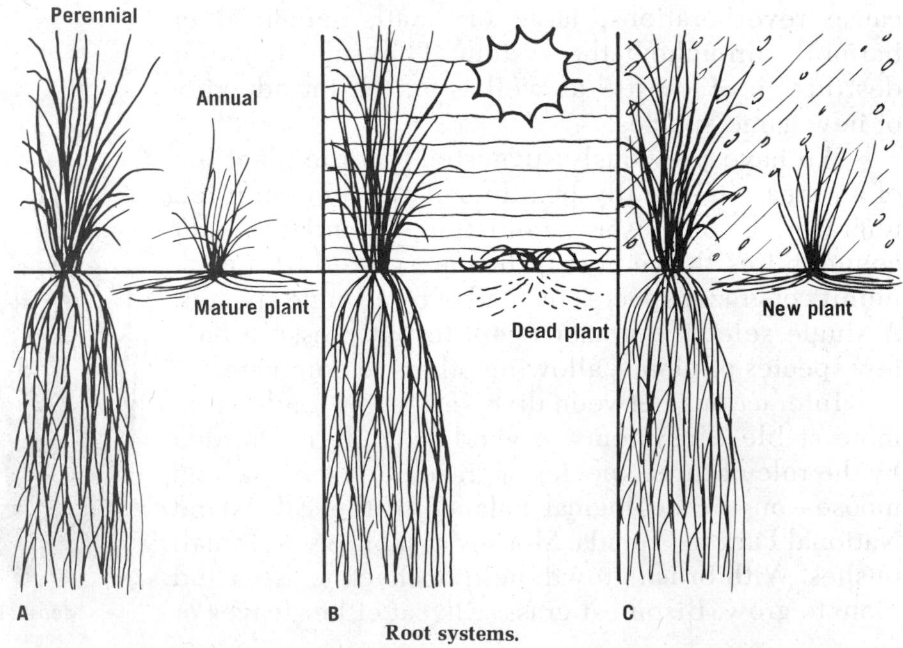

Root systems.

In canyons in the Colorado Rockies, Ponderosa pine, juniper, and small cacti predominate on the dry sunny side, and the blue spruce, Douglas fir, and flowering plants predominate on the wetter, shady side. Although both sides of the canyon contain representatives of all species, the niche of the pine is sufficiently different from that of the spruce to preserve the separation. If a prolonged drought were to strike the canyon, there would probably be some shift in the plant population in favor of those species that are more effective in controlling water. Thus, changing conditions often change the order of dominance of species. The advantage of community diversity is that the changes that do occur are relatively mild and the ecosystem is not disrupted.

1.10 SPORADIC AND CYCLIC DISRUPTIONS

While it is true that most old natural systems exhibit remarkable stability even in the face of changing environmental conditions, sporadic or cyclic disruptions occur. A dramatic example of a sporadic disruption is the red tide. Since recorded history, a species of red phytoplankton has been observed to periodically enter a period of rapid growth in coastal areas and turn the ocean red. Such a rapid growth of plants is called a **bloom.** There seems to be no predictable cyclic pattern associated with this phenomenon, but instead, the plankton grow very rapidly at unexpected times—seemingly in response to a new influx of nutrients. Traditionally, these "red tides" were often observed to occur after flood waters washed large quantities of soil nutrients into the sea. This influx created a plentiful food source. The red-colored plants were best able to take advantage of it. Unfortunately, these organisms discharge toxic substances into the sea, killing fish and aquatic mammals, and are there-

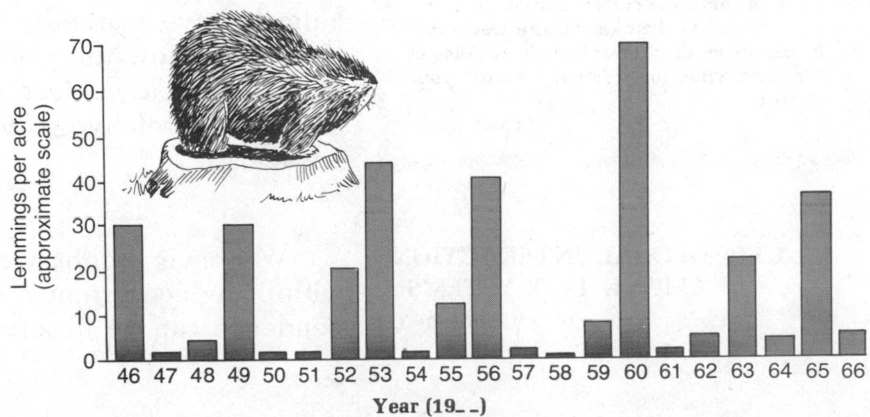

FIGURE 1.23 Lemming population cycles at Point Barrow, Alaska, from 1946 to 1966.

fore considered a great hazard. Recently, a series of red tides apparently unrelated to natural phenomena have been attributed to the nutrient content of pollutants discharged into the sea.

By contrast to the sporadic red tides, the very old and stable ecosystem of the arctic tundra exhibits predictable three- or four-year cycles of lemming population. One summer the population will be extremely high; the next year the population will rapidly decline, or crash. For another year or two, the population recovers slowly; then it skyrockets for a season and the cyclic pattern repeats itself.

Lemming abundance is associated with forage cycles (Fig. 1.23) During an abundant lemming year, the tundra plants are plentiful and healthy. Most of the available nutrients exist in plant tissue and there is little stored in the soil reservoir. In the spring following the overabundance of lemmings, the heavily overgrazed plants become scarce, and most of the nutrients in the ecosystem are locked in the dead and dying lemmings. The process of decay and return to the soil requires another year or two, during which time the population and health of both plants and rodents increase. Thus, it may be argued that the lemming cycles are caused by the cycling of forage supply. But no one is really sure which is the cause and which the effect. Perhaps both the lemming and forage cycles are due to other as yet undiscovered factors.

One might argue that because there are relatively few species in the arctic, the predator population is not itself stable enough to stabilize the lemmings. Data have not solidly supported this hypothesis. Foolproof answers are further complicated by the fact that similar population cycles occur in more hospitable climates. For example, locust plagues have occurred periodically in the region of the Mediterranean Sea since Biblical times. Many other species of plants and animals have markedly fluctuating populations. We just don't know the causes. It is interesting to note that many areas subject to natural cyclic blooms have survived in relative overall constancy since the last ice age.

That which the palmer worm hath left hath the locust eaten: and that which the locust hath left hath the cankerworm eaten; and that which the cankerworm hath left hath the caterpillar eaten.

Awake, ye drunkards, and weep; and howl, all ye drinkers of wine, because of the new wine, for it is cut off from your mouth.

(Joel 1:4–5)

1.11 GLOBAL INTERACTIONS AMONG ECOSYSTEMS

(Problem 31)

We have emphasized that the various species within an ecosystem are interdependent. This dependence can be observed within a radius of a few inches as in the case of lichen, or within several square miles, as is the case for many predator-prey relationships. It is also apparent that entire ecosystems depend on other ecosystems which may be located thousands of miles away. The life around a large river depends on the yearly cycles of river flow. In turn, the river flow depends on the water balance at the tributaries. This water balance is controlled by the forest systems. Thus, the overall state of the forests on banks of a small creek has a direct effect on the life cycles of organisms at the mouth of the river.

In a larger sense, the life forms on the continental and oceanic masses of the earth are linked together into a single interdependent system. Southerly winds and warm ocean currents bring heat to the northern tundra, an area which does not receive enough direct solar energy to support much life. The global balance of carbon dioxide and oxygen in the atmosphere is the classic example of biospheric homeostasis. Before life evolved, the primary constituents of Earth's atmosphere were nitrogen, ammonia, hydrogen, carbon monoxide, methane, and water vapor. Oxygen was present only in trace quantities. Although some theorists believe that geological processes altered atmospheric composition, most scientists feel that the excess oxygen released by the first autotrophs built up slowly over the millennia until its concentration reached about 0.6 per cent of the atmosphere. Multicellular organisms could have evolved only at this point, because aerobic respiration is a prerequisite for their development. The emergence of various organisms about 600 million years ago triggered an accelerated biological production of oxygen. The present oxygen level of about 20 per cent of the atmosphere was reached about 450 million years ago. While there have been some more or less severe oscillations since that time, an overall oxygen balance has always been maintained.

If the oxygen concentration in the atmosphere were to increase even by a few per cent, fires would burn uncontrollably across the planet; if the carbon dioxide concentration were to rise by a small amount, plant production would increase drastically. Since these apocalyptic events have not occurred, the atmospheric oxygen must have been balanced to the needs of the biosphere during the long span of life on Earth. By what mechanism has this gaseous atmos-

Nitrogen, $N \equiv N$

Ammonia, N (bonded to H, H, H)

Hydrogen, $H—H$

Carbon monoxide, $C \equiv O$

Methane, H—C—H with H above and H below C

Water, H—O—H

J. E. Lovelock: Gaia as seen through the atmosphere. Atmospheric Environment 6:579, August, 1972.

John E. Lovelock is an English chemist who has done important work on the analysis of trace gases, including atmospheric pollutants.

pheric balance been maintained? The answer appears to be that it is maintained by the living systems themselves. The existence of an effective homeostatic mechanism of the whole biosphere has led J. E. Lovelock to liken the biosphere to a living creature and call that creature *Gaia* (Greek for Earth). He believes that not only is the delicate oxygen-carbon dioxide balance biologically maintained but that the very presence of oxygen in large quantities in our atmosphere can be explained only by biological maintenance. If all life on Earth were to cease and the chemistry of our planet were to rely on abiological laws alone, oxygen would once again become a trace gas.

The concept of biological control over the physical environment warrants careful consideration. It says that, just as an organism is more than an independent collection of its organs and an ecosystem is more than an independent collection of its organisms, the biosphere is more than an independent collection of its ecosystems. The body chemistry of a human being can function only at or within a few degrees of 98.6°F, but we are normally not in mortal danger of very high or low body temperatures, because there are mechanisms by which our bodies maintain the proper temperature. Similarly, Lovelock believes that modern life can exist only in an atmosphere at or very close to 20 per cent oxygen, but we are not normally in mortal danger of conflagration or starvation because there are mechanisms by which we, the species of Earth, maintain the proper oxygen concentration.

An alternate theory claims that our physical environment has evolved through a series of inorganic reactions, and that biological and physical evolution were independent. The difference between these two beliefs is not trivial. If Lovelock is correct, then a large biological catastrophe such as the death of the oceans or the destruction of the rain forest in the Amazon Basin could cause reverberations throughout our physical world that might create an inhospitable environment for the rest of the biosphere. Alternatively, if the physical world did evolve independently of the biological and is presently controlled by inorganic processes, such a doomsday prediction concerning oxygen balance might be considered unnecessarily alarming.

1.12 NATURAL SUCCESSION

(Problems 32, 33, 34, 35)

We have discussed the mechanisms by which natural ecosystems maintain balance, but we must remember that these mechanisms do not prevent change; rather, they provide compensating processes that tend to drive the system back to its original state. These compensations do not always work.

Natural succession is defined as the sequence of changes through which an ecosystem passes as time goes on. The climax is the final stage, the stage that is "unchanging." Of course, the word "final" is used with reservation, because the slow process of evolution changes everything. The composition of the climax depends on temperature, altitude, seasonal changes, and patterns of rainfall and sunlight.

As succession continues, changes occur not only in the terrain and the types of species present, but also in the trophic structure of the system. In the final climax, photosynthesis and respiration have reached a balance. Much of the total respiration that does occur is initiated by decay organisms. Great quantities of organic matter are supported by a relatively low community energy consumption and it is as if metabolism has slowed down in old age. Thus, a climax would be a poor agricultural system, for agriculture must produce. But the climax structure is the type of system that has traditionally produced things of beauty—the redwood forests, the great plains, and the majestic hardwood stands that used to grow along the Eastern seaboard.

To a large extent, our perception of change in a

A marsh evolving into a meadow. (From North American Reference Encyclopedia)

system depends on how coarse or fine a time scale we choose for our measurements. A small lake is often used as an example of a stable ecological system. Plants, large animals, and microorganisms all exist in balanced relationships. Temporary imbalances occur and are adjusted as in all natural systems. However, for most lakes the incoming streams and rivers bring more mud into the environment than is removed by the outgoing streams. The effect is small; it may not be noticed in the lifetime of one man. A short-term study of the ecology of a lake would conclude that the ecosystem was in balance. But mass balance is disrupted by the steady addition of solid matter from incoming streams. In time, the lake will begin to fill up with mud. The vegetable and animal life will change. New plants that can root in the bottom mud and extend to the surface where light is available will appear. The trout give way to carp and catfish. If an ecologist studies the lake at this stage he might again say that it was a stable system. Again, minor imbalances and adjustments could be observed but the overall system appears stable. Yet mud continues to flow into the lake. In addition, since it is common for the plants in the lake at this stage to produce more food than is consumed by the herbivores, the bottom fills up with humus. Eventually the lake may become so shallow that marsh grass can grow.

$$6CO_2 + 6H_2O \underset{\text{respiration}}{\overset{\text{production}}{\rightleftarrows}} C_6H_{12}O_6 + 6O_2$$

(See Equations 1 & 2, page 13)

The marsh system is characterized by a net overproduction of organic matter, as indicated by imbalance in the equations shown at left.

The unequal lengths of the two arrows denote that more organic matter is produced than is oxidized by respiration. In biological terms, this means that there are not enough grazers to consume the large bloom of vegetation, resulting in a net accumulation of litter. Thus, the energy of photosynthesis is stored in plant tissue. This tissue remains largely uneaten and fills the marsh.

A marsh is considered a stable system over a short range of observation, but in most cases it is gradually evolving into a meadow. Litter accumulates because decay in the marsh system is slow. Slow decay thus implies poor nutrient recycling. Such a system is inherently unstable, as is typical of any early successional stage. The accumulated litter gradually fills in the marsh which slowly becomes a meadow.

The meadow, too, may change. If the climate is

right, trees can start to grow. First, shrubs appear, then quick-growing soft woods like birch, poplar, or aspen. The soft woods are replaced by pine; and finally, in what is called the climax, the pine is replaced by hardwoods.

During the succession of plant species, each species prepares the way for the next while contributing to its own extinction. Marsh grass could never grow without the rich soil of the partially decayed algae. However, by producing the rich soil, the algae helps to fill the lake and thereby destroy its own environment.

A "climax" originally referred to a gradually rising series, such as a gradation of more and more forceful expressions. More recently, the word has come to mean the highest point, or culmination, of such a series. It is in this latter sense that an ecological climax, such as a mature forest, means the end, or apex, of an ecological succession.

We cannot estimate the amount of time typically required to fill a lake, because that depends on many factors, such as the original volume of the lake and the net rate of accumulation of solids. For example, Lake Tahoe in California is so deep and clear that succession, if it occurs at all, would take geological time. The ecologist is generally not concerned with processes that are so slow. For all practical purposes, many lakes may be considered to be in a climax condition.

The time for progression from grassland to a climax forest has been measured. In the southwestern United States, grasslands give way to shrub thickets in one to 10 years, the shrubs become pine forests in 10 to 25 years, and the pine forests give way to hardwoods after about 100 years. We emphasize again that for an ecosystem to be indefinitely stable there must be a complete balance. This is never the case on Earth. Daily, seasonal, yearly, and long-term fluctuations occur in all natural systems, even in the most stable ones. The ultimate consideration of concern for the continuation of life on this planet is that the sum total of all the changes results in worldwide total balance.

The preceding discussion does not explain how a grassland, a marsh, or the tundra can be a climax ecosystem. We have presented situations which sometimes, but not always, represent reality. The great plains which once stretched unfenced from the eastern slope of the Rocky Mountains almost to the Mississippi River never became forests because the rainfall wasn't sufficient. Instead, the grasslands evolved into a climax system with all the trophic structure, energy balance, and stabilizing mechanisms characteristic of the most stable forest ecosystems.

Climax systems are often determined in large part by seasonal variations and geological cycles. A prairie plentiful with buffalo grass grew in Kansas because there wasn't enough yearly rain to support the forests, while the Everglades marsh has existed for a long time because the yearly rain comes in definite seasons. During the flood times, the Everglades behaves like an early successional marsh and is characterized by an overproduction of plant matter and a general silting of the streams and pools. If allowed to persist at this stage, the Everglades would soon follow the successional path toward a forest system. During the periodic droughts, the pools dry up and the litter decomposes, and sometimes burns rapidly. When the rainy season returns, the fallen leaves and logs have been recycled back to fertilizer, the pools have been cleared, and conditions are once again ripe for a bloom of marsh life.

The Everglades is the dominant biological community south of Lake Okeechobee in Florida. It includes extensive regions of sawgrass, cypress swamp, mangrove, and finally brackish waters leading to the Florida Bay and the Gulf of Mexico. It is a highly developed ecosystem with a large variety of plant and animal species and very complex biotic relationships, still not thoroughly understood.

The total cycle in the Everglades can't be classified neatly either as an early successional stage or as a climax system. During periods of bloom when net production greatly exceeds net consumption, we are reminded of a transient marsh, yet the overall stability and continuity of the system resembles that of a climax system.

The role played by fire in succession is also important. When the pioneers settled near the edge of the prairie in northern Wisconsin, plowed it, planted it, and controlled it, they discovered that any area left uncultivated soon became thick with new tree seedlings which sprouted and grew rapidly. A study of the area showed that the prairie that used to grow there had been periodically consumed by fire. Fast-growing grasses quickly regained control, while forests were never able to survive even though the soil and weather conditions were favorable. These grasslands exhibited a **fire climax,** a condition in which the continuance of a given system is maintained by fire. During growth after a fire, the ecology of the area resembled that of an early successional stage, for production was greater than consumption, and the standing quantity of organic matter was low. Eventually, evolution produced animals and plants which were able to accommodate to, or even depend on, the fires. Some grasses and trees developed seeds that sprouted only after being cracked open by fire. Those trees that did exist evolved thick, fire-resistant barks. Although fire returned fixed nitrogen to the atmosphere, this effect was counterbalanced because among the early successional plants were nitrogen-fixing legumes.

It is interesting to consider what determines natural boundaries between climax systems. In many places, local weather patterns such as those observed around a large river or a mountain range result in rather rapidly changing climates with the resultant change in climax systems. In other places, soil structure or local factors such as human work will affect the change. Even under these conditions, one wouldn't expect an abrupt transition between ecosystems but rather some sort of border area which harbors representative species of each neighboring region. If you live near a place that is part farmland and part woods, take a walk in the country some day and direct your attention to the boundary between plowed fields and woodlots. In most places you will find a band, sometimes not more than 10 or 20 yards wide, in which birch, poplar, or aspen saplings mingle with various grasses and maybe a corn stalk or two. Flora from both the field and the woods will be present. Additionally, many types of bushes which aren't characteristic of either major ecosystem will be found. If you have the eye of a naturalist, or of a hunter, you will discover that more rabbits and upland game birds live here, where they can take advantage of the rich plant life, the good cover, and easy access to crops or woods. Such a border area is known as an **ecotone** and varies greatly in width. Ecotones generally support more life and numbers of species than either bordering climax area. Recall that an estuary is an example of an aquatic ecotone.

1.13 THE ROLE OF PEOPLE

People, too, occupy a position in the flow of energy through the biosphere and must necessarily interact with thousands of other species of plants and animals. Why, then, do we consider humans separately? The answer is that today people have unprecedented power at their disposal, and with it the ability to increase the productivity or destroy the ecosystems of the earth. Man has been so successful at reducing competition from other species that his numbers have risen precipitously in a time span that is insignificant on the evolutionary clock. Before man, species and ecosystems evolved together over long stretches of time, but man's social and technological changes are orders of magnitude faster than evolution.

Nowhere is the rapid effect of human technological advancement more pronounced than on the

The Problems in Establishing Operational Definitions of Ecology and Ecosystem

Ecology, like economics, derives from the Greek stem eco-, or house. Economics, in its original sense, meant the management of the household, and, in the same spirit, ecology may be called "the science of the economy of animals and plants." When a physiologist studies an organism, he investigates the relationships of its component parts, while an ecologist delves into the relationships of that organism to others and to its physical environment.

Such an abstract definition gives no clue as to how to draw the boundary around an ecosystem. In some sense, every organism on Earth has some relationship, however small, with every other, and Earth itself depends on the sun for light and heat, the moon for tides, and the gravitational fields of the heavenly bodies for its course in space. Yet to call the universe the only ecosystem is frivolous. No ecologist studies every biological and physical phenomenon which impinges upon an organism. Operationally meaningful ecosystems must be smaller units with boundaries defined by the study being performed. If the ecology of an island is investigated, that entire island is regarded as an ecosystem, and the flow of material and energy into and away from the island is considered in the aggregate. If the ecology of a lake on the island is studied, the lake is called the ecosystem, and the relations between the lake and the rest of the island are regarded in the aggregate.

Of course, if the system to be studied is too small, the concept of an ecosystem is irrelevant. An individual organ or organism is not studied by an ecologist except as it relates to other species and to the physical world. A study of the mechanism by which a barnacle attaches to a rock would also not be considered an ecological investigation, even though it relates an organism to its physical environment. Even a study of the interrelationships among organisms is not necessarily ecological. Thus, a study of a virus in man, although it involves two organisms, is not called ecology.

What then, transforms an ordinary biological study into an ecological one? We can say that when a biologist studies an organism as it relates to at least one other or to its physical environment, and when that biologist says he is an ecologist, the study is ecological. Then the ecosystem is the smallest unit of environment that includes the study.

North American continent. Three hundred years ago, this land mass housed a set of diverse, balanced ecosystems where many types of plants and animals, as well as people, lived in a manner which had changed little since the last ice age. Today, some of the native ecosystems such as the tall-grass prairie and the Eastern hardwood forests are gone or altered almost beyond recognition. Those biomes that still exist, such as the Northern forest or the Rocky Mountain tundras are shrinking in the wake of progress, even as the original balances of species are being altered by predator control and hunting. In fact, every predator large enough to take a deer or a cow, with the exception of man himself, has been pushed into a few remote or bitter-cold areas. Even there the predators risk eradication. In place of the natural forests and ranges, millions of acres of intensely cultivated land produce large yields of foodstuffs to feed millions of non-farmers.

Such is our present condition. Moreover, it is completely unreasonable to believe that man will choose to return to the ways of the nomadic hunter or the Stone Age farmer. The problem for us is not to try to recapture the past but to make decisions for the future.

Many people believe that our present systems, controlled by man, are providing people with greater physical security and comfort than they have ever known. The major question of environmental policy is: what types of systems are viable and will continue to provide security and comfort far into the future?

Imagine a technology spectrum ranging from a Stone Age hunter to an automated city inside a huge environmentally controlled dome, where food is synthesized in large factories from sewage with the aid of sunlight. What levels of technology can survive for a long time, and of those that can, which are desirable? Before we attempt to find the answers to such grave questions, we must establish the criteria on which the answers must be based. If our environment is to be stable and hospitable to man for a long time, our decisions must be based on sound scientific reasoning by people whose goals go beyond personal gain or power, and whose interests extend beyond their own lifetimes. Public opinions on any given environmental problem, such as radioactive waste pollution, range from: "It doesn't present a problem," to "It is an annoyance but well worth the benefits gained," to "It's a debatable trade-off," to "It's an imminent threat to the existence of man." Of these, all statements must be given consideration except the first, for man-made pollution does indeed present problems.

Perhaps the main reason for the wide range of opinions from lay people as well as specialists is that the data can be interpreted in many different ways, depending on the individual's choice of priorities. We conclude this chapter on natural ecology with two examples of issues which are still to be decided. In both cases, the direct alteration of natural ecosystems is at stake.

CASE I

Every spring, birds of many species must fly north to their summer breeding grounds, and every fall they must return, or many would starve and freeze to death. Along the Eastern seaboard of North America there is still room in the northlands and southlands for the migrants to nest, but the area in between, from metropolitan New York south to Washington, D.C., is becoming increasingly inhospitable. One great and ancient swamp does remain in northern New Jersey, right in the heart of one of the most industrialized areas of the world. This marsh houses many species of its own, and supports millions of migrants who must stop there for rest and food. Without this haven, many birds could not survive the next migration.

It has been proposed that the marsh be converted into a model urban community complete with environmentally controlled shopping centers, comfortable apartment complexes, and spacious parks and recreational areas. The construction of the new city would provide jobs for many workers and, when complete, many needed new living units for our expanding population. These living units will undoubtedly provide modern conveniences for the new inhabitants and relieve some pressures from the already overcrowded cities. Let us assume that the planners of the project will install the latest in sewage recycling systems, solid waste disposal units, public transportation systems, and other pollution-limiting features. But what is the environmental cost? This particular project wouldn't exterminate all the remaining migratory bird populations along the Eastern seaboard, but it would destroy many birds. The lives of many species would end if projects of this sort proliferated.

Some questions the proposed development presents are esthetic and moral. Does a goose have a right, as a member of the biosphere, to live out its life according to its ability and wits, and to eat or be eaten as its ancestors were? Or, from a person's point of view, is it part of one's birthright to stand on the doorstep on that first cold day in October and watch the geese grouping into as yet poorly formed V's and honking their way south? Will the residents of the new housing projects want to go to Vermont each fall to hunt? How important is natural beauty to man? Are we willing to destroy some of the natural quality of our lives in order to provide more technology?

Moreover, threatening the survival of the geese presents concrete problems. Lumber and paper are harvested from the geese's summer homes; salt-water fish grow to adulthood in the southern

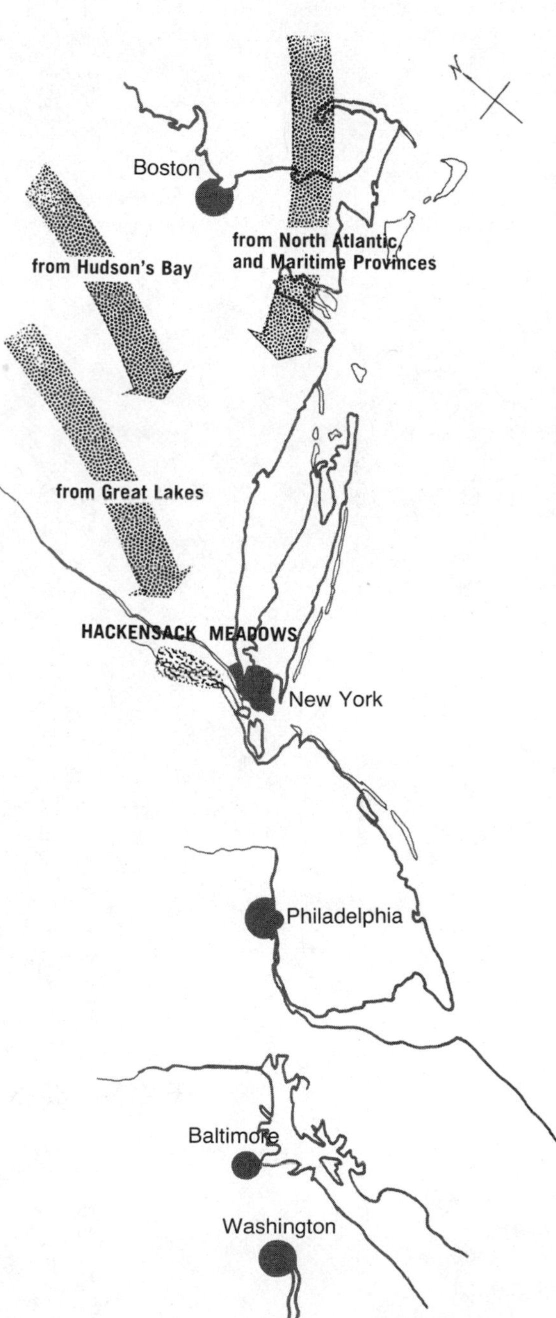

Migratory birds from northern Atlantic coastal and mid-west nesting areas use the Hackensack Meadows as a rest and feeding stop on their way south.

salt-water marshes where many geese winter. The removal of migrating birds would certainly affect the ecological relationships of both areas, but we are not sure of the effect on their ability to produce goods for man. Trees can be grown in monoculture farms, and perhaps the estuaries, without geese, could provide more food if we added more fertilizer, and perhaps both the forests and the estuaries could easily adjust to their new existence.

For the people who feel that nature must surrender to progress, the question remains: will monoculture forests and farmed estuaries be able to survive in time, or are we building our technology on a poor foundation? As technology improves, perhaps we will be able to construct plastic building materials and synthesize food from coal. But if the natural world is destroyed during the construction of a technological one, and the latter does not work, we will be able to re-create neither.

CASE II

Part of the shoreline of southern Long Island boasts some of the world's finest beaches. They provide recreation for millions of metropolitan New Yorkers. But the beaches are being eroded away by the sea. The wave action responsible for the sand removal isn't a factor initiated by some action of man. Rather, it is a natural geological process for oceans to form beaches and then displace them on land or move them out to sea. Shorelines and beaches have been moving and reshaping since the formation of the continents. Because of the recreational importance of this particular set of beaches, the Army Corps of Engineers has proposed to rebuild them at an estimated cost of 200 million dollars. The Corps plans to dredge some of the sand needed for the project from the inland bays and other estuary systems of the south shore. Conservationists argue that during the dredging process the ecological relationships in the estuaries would be severely disrupted. We know that many salt-water fish rely on estuaries as nursery grounds and that man relies on the fish as food. It is possible, then, that rebuilding the beaches would disrupt commercial offshore fishing.

Ultimately, a decision must be made: To what degree should man try to be the controller of his planet? Are humans wise enough to control the earth for their, and its, ultimate benefit or just powerful enough to change it for their own immediate gain?

PROBLEMS

1. **Vocabulary.** Define ecology, ecosystem and biosphere.
2. **Energy.** Classify each of the following substances as energy-rich or energy-poor with specific reference to its ability to serve as a food or fuel: sand, butter, paper, fur, ice, marble, and paraffin wax.
3. **Vocabulary.** Define autotroph, heterotroph, and trophic level.
4. **Energy.** Organisms have evolved to obtain energy from foods or from the process of photosynthesis. Would it have been possible for life forms to evolve

which obtain their energy by some other process, such as rolling down hills? Defend your answer.

5. **Food web.** What is a food web? Sketch a diagram of a food web that primarily involves life in the air, such as birds and insects. Will you need links to terrestrial or aquatic systems?

6. **Food web.** Consider the food web in the diagram on page 9. Arrow number 15 can be explained by the relationship: mountain lion eats deer. Write similar statements for each of the other 27 arrows.

7. **Food pyramid.** The discussion presented in Section 1.3 and illustrated in Figures 1.6 and 1.7 shows that more energy from the sun is used to nourish a human being who eats meat than a human being who eats vegetables. Explain why this fact, by itself, is *not* an argument for or against vegetarianism, taking into account the following questions: Does the choice of human diet affect the total energy flow through the food web? Does it affect the total biomass of plant matter on Earth? Does the choice between the alternatives of a large human population living largely on vegetable food or a smaller human population living largely on meat and fish significantly affect the total biomass of animal matter on Earth? In your answer do not take into account the possible destruction of species; this topic is treated separately in Chapter 3.

8. **Food pyramid.** An experiment has shown that the total weight of the consumers in the English Channel is five times the total weight of the plants. Does this information agree with the food pyramid shown in Figure 1.6? Can you offer some reasonable explanations for the findings?

9. **Food pyramid.** It is desired to establish a large but isolated area with an adequate supply of plant food, equal numbers of lion and antelope, and no other large animals. The antelope eat only plant matter, the lions, only antelope. Is it possible for the population of the two species to remain approximately equal if we start with equal numbers of each and then leave the system alone? Would you expect the final population ratio to be any different if we started with twice as many antelope? Twice as many lions? Explain your answers. (Assume that lions and antelope have the same body weight.)

10. **Nutrient cycles.** We speak about nutrient cycles and energy flow. Explain why the concepts of nutrient *flow* and energy *cycle* are not useful.

11. **Nutrient cycles.** Give three examples supporting the observation that nutrient cycling hasn't been 100 per cent effective over geological time.

12. **Oxygen cycle.** Trace an oxygen atom through a cycle that takes (a) days, (b) weeks, (c) years.

13. **Nutrient cycles.** Why don't farmers need to buy carbon at the fertilizer store? Why do they need to buy nitrogen?

14. **Ecosystem.** Could a large city be considered a balanced ecosystem? Defend your answer.

15. **Homeostasis.** Consider two outdoor swimming pools of the same size, each filled with water to the same level. The first pool has no drain and no supply of running water. The second pool is fed by a continuous supply of running water and has a drain from which water is flowing out at the same rate at which it is being supplied. Which pool is better protected against such disruptions of its water level as might be caused by rainfall or evaporation? What regulatory mechanisms supply such protection?

16. **Vocabulary.** Define euphotic zone, phytoplankton, and benthic species.

17. **Estuary systems.** Copy Figure 1.16 and indicate which bodies of water are salty, fresh, and brackish.

18. **Niche.** Define ecological niche. Are you living in your optimum niche? Defend your answer.

19. **Diversity.** Explain how niche competition promotes diversity.

20. **Stability.** Explain how ecosystem diversity promotes stability.

21. **Niche.** Decay organisms living in flowing streams often attach themselves to rocks. One species may predominate on the lee side and another on the current side. What factors might be involved in establishing such a relationship?

22. **Niche.** Can two individuals of the same species occupy exactly the same niche? Explain.

23. **Niche.** Discuss some differences between the home territory of the nomadic hunters and the niche concept of animals.

24. **Two-species interactions.** List the eight major types of two-species interactions. Include a brief explanation of each.

25. **Predation.** The ecologist Elton contends that predators live on capital, while parasites live on interest. Explain. Is this true from an individual or from a community viewpoint? Explain.

26. **Predation.** Would you expect a buffalo herd to be healthier after years of being hunted by men with bows and arrows or by men with guns? Explain.

27. **Community interaction.** Do you think that it might be economically profitable to raise deer along with cattle in some areas of North America? Explain.

28. **Stability.** The box in the margin of page 17 describes some balanced systems. List several examples of other such systems.

29. **Ecosystem stability.** Which do you expect to be better able to survive a drought: a cornfield or a natural prairie? Explain.

30. **Natural balance.** Discuss the statement, "Natural systems are perfect because they are always in harmonious balance."

31. **Ecosystems.** Would you say that the Earth includes many ecosystems that are relatively independent of each other, or that it contains only one ecosystem that occupies the entire biosphere, or that both statements are true? Present arguments in favor of your position.

32. **Climax systems.** It has been observed that large, complex plants and animals are more characteristic of climax situations than of early successional stages. (a) Name some organisms and their habitats that serve as examples of this observation. (b) To explain this observation, it has been suggested that species with long life cycles have evolved away from unstable environments. It has also been suggested that large, complex animal species cannot find proper sustenance from plants that grow in simple systems such as marshes. Repeat these hypotheses in your own words and argue for or against them.

33. **Succession.** Define natural succession. What factors bring about changes in an ecosystem? What is the climax of an ecosystem? Cite three examples of a climax ecosystem.

34. **Succession.** Imagine that a new island just arose in the South Pacific. Trace the succession that would be expected to occur. Estimate the time span required for the climax to be reached. Compare the energy balance of the early ecosystems of the island to the energy balance of a marsh. Compare the types of species present.

35. **Imbalance.** What do you think would happen to the Everglades if man initiated a set of dikes to ensure constant water levels all year round?

36. **Consumer levels.** Explain why a beer drinker is to some extent a secondary consumer.

The following question requires arithmetical computation:

37. **Energy.** Assume that a plant converts 1 per cent of the light energy it receives from the sun into plant material, and that an animal stores 10 per cent of the food energy that it eats in its own body. Starting with 10,000 calories of light energy, how much energy is

available to a man if he eats corn? If he eats beef? If he eats frogs that eat insects that eat leaves? Of the original 10,000 calories, how much is eventually lost to outer space? (*Answer*: 100 cal; 10 cal; 1 cal; 10,000 cal.)

BIBLIOGRAPHY

The basic textbook on ecology is:

Eugene P. Odum: *Fundamentals of Ecology.* 3rd Ed. Philadelphia, W. B. Saunders Co., 1971. 574 pp.

A periodical issue devoted in its entirety to "The Biosphere" is:

Scientific American. September, 1970. 267 pp.

Three books dealing with specific areas of natural ecology are:

R. Platt: *The Great American Forest.* Englewood Cliffs, N.J., Prentice-Hall, 1965. 271 pp.

B. Stonehouse: *Animals of the Arctic: the Ecology of The Far North.* New York, Holt, Rinehart and Winston, 1971. 172 pp.

G. M. Van Dyne, ed.: *The Ecosystem Concept in Natural Resource Management.* New York, Academic Press, 1969. 383 pp.

A classic study of ecology and conservation as seen through the eyes of a naturalist is:

A. Leopold: *A Sand County Almanac.* New York, Sierra Club/ Ballantine Books, 1966. 296 pp.

A very valuable and scholarly book that is oriented toward theoretical biology is:

Philip Handler, ed.: *Biology and the Future of Man.* New York, Oxford University Press, 1970. 936 pp.

There are several books that are not limited to ecology but are broadly concerned with environmental problems. Three are cited below; the first is oriented toward chemistry, the second and third are more general:

American Chemical Society: *Cleaning Our Environment: The Chemical Basis For Action.* Washington, D.C., 1969. 249 pp.

Paul R. and Anne M. Ehrlich: *Population, Resources, Environment.* San Francisco, W. H. Freeman & Co., 1970. 383 pp.

Ernest Flack and Margaret C. Shipley, eds.: *Man and the Quality of his Environment.* Boulder, University of Colorado Press, 1968. 251 pp.

A number of more popular books sound the alarm about the environmental dangers to man's survival. Five of these are:

Ron M. Linton: *Terracide.* Boston, Little, Brown, 1970. 376 pp.

J. Rose, ed.: *Technological Injury.* New York, Gordon & Breach, Science Publishers, 1969. 224 pp.

Robert and Leona Rienow: *Moment in the Sun.* New York, Ballantine Books, 1967. 365 pp.

Melvina A. Benarde: *Our Precarious Habitat.* New York, W. W. Norton & Co., 1970. 362 pp.

Osborn Segerberg, Jr.: *Where Have All the Flowers Fishes Birds Trees Water and Air Gone?* New York, David McKay Co., 1971. 303 pp.

2 HUMAN ADAPTATION TO ENVIRONMENTAL CHANGE

2.1 INTRODUCTION

The relationship between organisms and the environment is a reciprocal one. Fallen leaves, needles, and twigs maintain the composition of forest soil, which in turn promotes the growth of the trees themselves. The holes alligators dig in the Everglades regulate the yearly water flow and thus maintain the integrity of the swamp. Moreover, the whole biosphere maintains proper atmospheric conditions for itself. Such pictures imply homeostatic balance, and balance engenders stability. But we know that the Earth has not always provided an unchanging environment. If we could be transported back only about 50,000 years, the climatic conditions over much of North America, for example, would be unrecognizable.

We also know that some life forms have survived through many environmental changes; others have succumbed. The process of accommodation to change is called **adaptation.**

Certain forms of life have survived for millennia because they are well adapted to their biological and geological environment. There are basically two adaptive mechanisms. One of these is genetic variation. In every species the genes available (usually called the gene pool) are not constant in time but undergo continuous variation. As the environment changes, new genetic variants may be favored over previously well-adapted ones. In addition to this genetic mechanism of adaptation, most animals can also learn from their life experiences and alter their behavior accordingly. Thus, for example, a dog or cat learns to respond to the whims of its master. Compared to other species, however, man has developed this mechanism of adaptation by learning to an extraordinary degree. Of all animals, he possesses by far the greatest ability to reason in new situations, and the largest body of accumulated knowledge which he passes from generation to generation. His ability to accumulate and transmit knowledge allows him to deal with environmental challenges in ways that have no real parallel in any other species.

Most of this book deals with the various ways that man has affected the world about him. We shall see that his industry and inventiveness have often resulted in the injudicious use of natural resources, the fouling of water, land, and air, and the wholesale disruption of certain ecosystems. In addition, man has been so successful in reproducing that in some parts of the world his very numbers constitute a major obstacle to his own welfare.

But man's highly developed cultural adaptation does not prevent his genetic variation. The extent to which this type of variation takes place in man is not nearly so well studied as in the case of certain experimental animals, but, as we shall see, there is no doubt that such variation does occur.

2.2 THE BIOLOGICAL BASIS OF GENETIC ADAPTATION

Inheritance and the Experiments of Mendel

(Problems 1, 2, 3, 4, 5, 7)

Ancient peoples had undoubtedly observed that parental characteristics were often, but not always, passed on to members of succeeding generations. But beyond some simple observations, very little was known about the mechanism of inheritance until Gregor Mendel studied the problem carefully in the mid-nineteenth century. Mendel had observed that different strains of pea plants possessed certain clearly defined characteristics which remained invariant through the generations. He listed seven different characteristics which are shown in Table 2.1. A strain of yellow-seeded peas, for instance, produced only yellow-seeded offspring, while a green-seeded strain produced only green-seeded progeny. When Mendel crossed yellow-seeded peas with the green variety, all the second-generation plants had yellow seeds. Similarly, as we can see from Table 2.1, the smooth-seed characteristic **dominated** the wrinkled-seed characteristic, red flowers dominated white flowers, and so on. Mendel then allowed the second generation plants to pollinate themselves and found that in the third generation, the non-dominant, or recessive, characteristic reappeared. The ratio of dominant to recessive (for example, yellow seed to green seed) was always 3:1.

In order to explain the data, Mendel postulated the existence of discrete carriers of inheritance, which he called *Elementes.* In modern science these Elementes are called **genes,** and although Mendel never used the word "gene," we will use it throughout the discussion of his work to avoid shifting terms in the middle of the chapter. Mendel concluded that these genes occurred in pairs, with one member of each pair inherited from each parent. The data from the pea plants could then be explained in the following manner:

1. Both genes of the original yellow-seeded strain carried information for the production of yellow seeds.
2. Similarly, both genes of the original green-seeded strain carried information for the production of green seeds.
3. All members of the second generation, called

Johann Gregor Mendel (1822–1884) carried out his experiments in an Augustinian monastery in central Europe. He hybridized strains of peas, keeping numerical records of the numbers of individuals in each generation with various properties. He originated the concepts of dominant and recessive inherited traits.

TABLE 2.1 THE RESULTS OF MENDEL'S EXPERIMENTS

CHARACTERISTIC	*PARENTS*	*RESULTS OF FIRST CROSS* (F_1)	*RESULTS OF SELF-POLLINATION OF FIRST CROSS** (F_2)
1 Seed form	smooth × wrinkled	all smooth	2.9 smooth/1 wrinkled
2 Seed color	yellow × green	all yellow	3.0 yellow/1 green
3 Flower position	axial × terminal	all axial	3.0 axial/1 terminal
4 Flower color	red × white	all red	3.1 red/1 white
5 Pod form	inflated × constricted	all inflated	2.9 inflated/1 constricted
6 Pod color	green × yellow	all green	2.8 green/1 yellow
7 Stem length	tall × dwarf	all tall	2.8 tall/1 dwarf

*These numbers, within experimental error, equal 3:1.

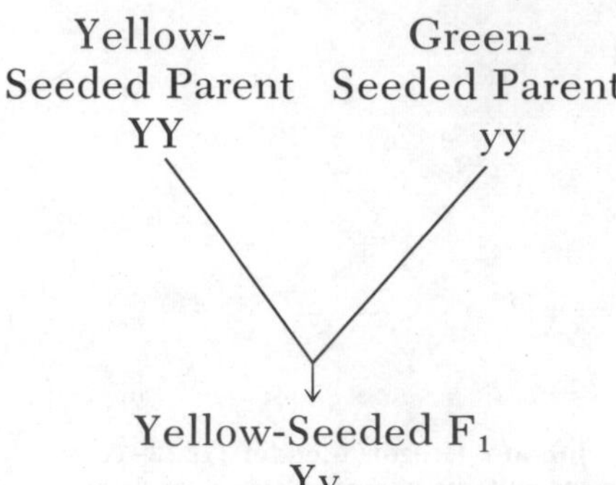

the F_1 generation in modern genetics, received one yellow-seeded gene from one parent, and one green-seeded gene from the other. Because yellow is dominant over green, the plants were all yellow. (At left, all the capital letters stand for dominant genes and small letters stand for recessive genes). We see that although both the yellow-seeded parent and the F_1 offspring contained yellow seeds, the two plants contained different genetic characteristics. In other words, two different genetic aggregates, called **genotypes,** produced the same observable character trait, called the **phenotype.** If both genes for a given characteristic are identical, as is the case of the yellow-seeded parent YY, the organism is said to be **homozygous** for that trait. On the other hand, if the genes for a given characteristic are different, as is the case for this F_1 generation Yy, the organism is said to be **heterozygous** for that character, and the phenotype is controlled by the dominant gene. The 3:1 ratio of yellow to green observed in the F_2 generation (produced by self-pollination of the F_1 plants) can be explained if random shuffling of the genotype Yy occurred during the mating process. This shuffling can be visualized in the following manner:

Dominance is not always as sharp as in the case of Mendel's pea characteristics and sometimes codominance or partial dominance occurs. See the description of the ABO blood group system on page 66.

Imagine that the genes of each of the F_1 plants segregate to give:

$$Yy \longrightarrow Y + y.$$

When fertilization occurs, the Y can join with either Y or y to give YY or Yy. Similarly, the y can join with either Y or y, leading to yY or yy. Thus, four combinations, all equally probable, arise for the F_2 generation: YY, Yy, yY, and yy.

Genotype	Color of Seed
YY (homozygous)	yellow
Yy, yY (heterozygous)	yellow
yy (homozygous)	green

The results from all these experiments can be summarized by the illustration, called a test cross or a Punnett cross. See diagram to the left.

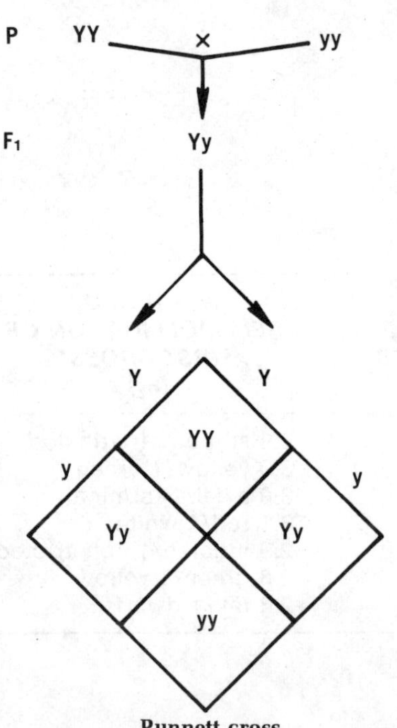

Punnett cross.

Modern science has added a great deal of sophistication to our understanding of inheritance, but Mendel's basic concept of discrete, paired *Elementes* remains a fundamental part of present-day genetic theory.

Biologists working in the late nineteenth century

Molecular Explanation of Mendel's Experiments

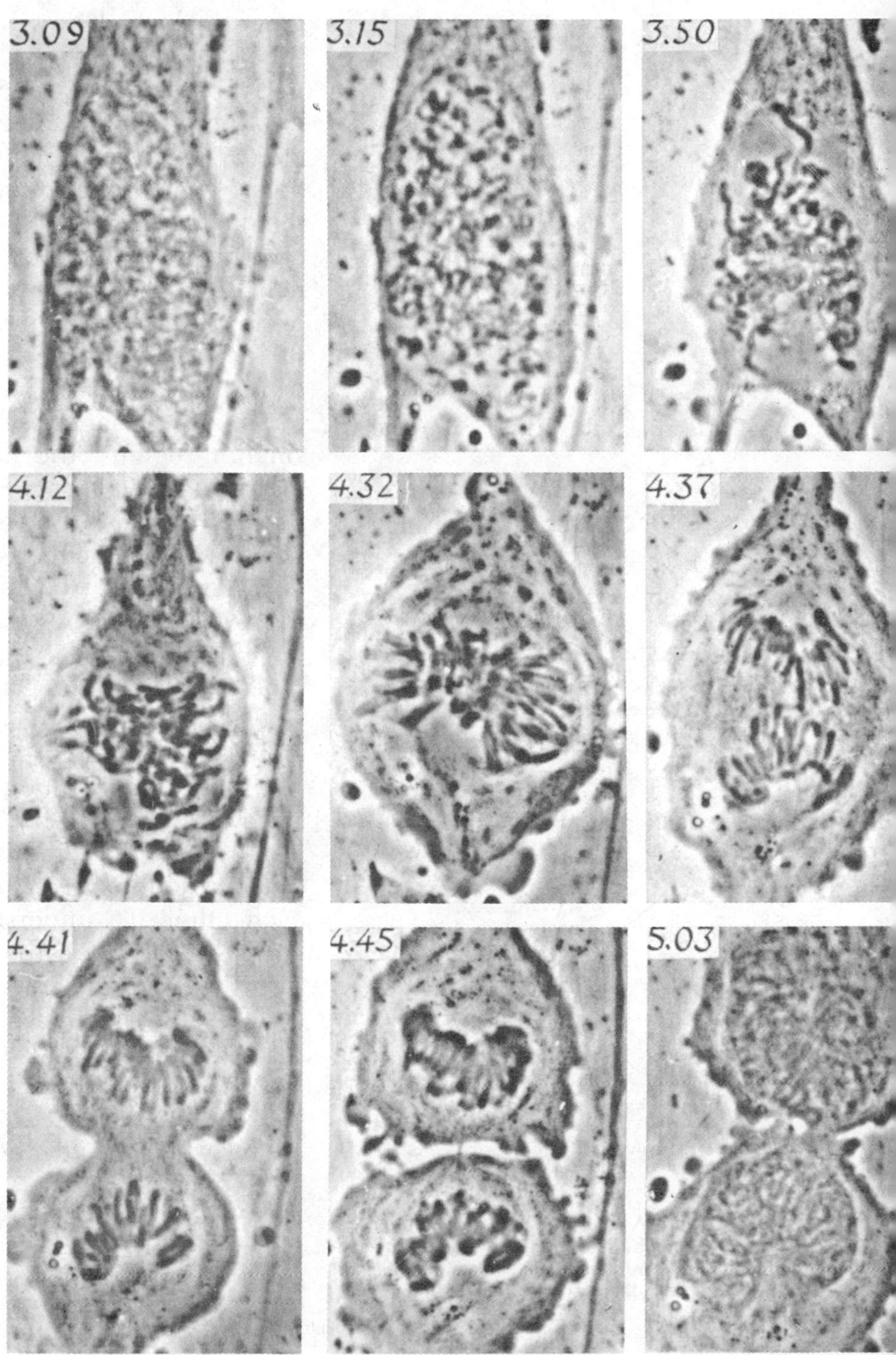

FIGURE 2.1 In this rapid sequence of photographs of a dividing cell, the chromosomal material condenses and becomes visible under the microscope and, at the time of cell division, is apportioned equally between the two daughter cells. (From Bloom and Fawcett: ***A Textbook of Histology,*** **8th Ed. Philadelphia, W. B. Saunders Co., 1962.)**

observed certain rod-like structures, occurring in pairs, in the nucleus of every cell. These structures, called **chromosomes,** were seen to separate from and recombine with each other in a characteristic manner during fertilization and cell division (see Figure 2.1). In 1902 the activities of the chromosomes were related to Mendel's paired genes. During the next 60 years, a great deal was learned about the molecular nature of inheritance.

Every animal species possesses a characteristic number of chromosomes; for example, fruit flies contain eight chromosomes grouped into four pairs, and every human cell except the sex cells (the sperms of the male and egg cells of the female) contains 46 chromosomes grouped into 23 pairs. In each of the pairs, one chromosome is a copy of a chromosome inherited from the father, and the other similarly from the mother. Thus, one half of any individual's genetic material is paternal in origin and the other half maternal. Each chromosome is subdivided into many smaller regions, the genes. As we have noted, for many years biologists considered genes as regions of the chromosome that controlled the appearance or non-appearance of certain observable character traits, for example, the shape or size of a pea, or eye color in the fruit fly. But not until about the last two decades have developments in biochemistry and molecular biology clarified the true chemical nature of the gene. The chromosome is composed chiefly of **deoxyribonucleic acid** (called **DNA** for short).

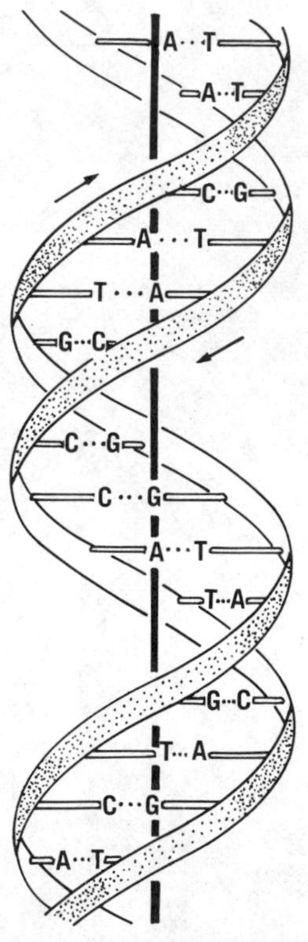

DNA is a very long and complex molecule whose structure is depicted diagrammatically in Figure 2.2. You can see that the overall shape of the molecule is that of a double helix, that is, two helices wound about each other. To understand this structure, let's construct it from its component parts. Each ribbon which forms the curving backbone of its respective helix is composed of molecules of a sugar called deoxyribose alternating with molecules of the ordinary phosphate ion ($PO_4^{\equiv}$), thus forming a very long chain:

–sugar–phosphate–sugar–phosphate–sugar–

FIGURE 2.2 A diagram of the manner in which the two helices of DNA wind about each other. (From Anfinson: *The Molecular Basis of Evolution.* **1st Ed. New York, John Wiley & Sons, 1959.)**

Attached to each of the sugar molecules in this chain is one of four nitrogen-containing bases. The names and chemical structures of these bases are shown in Figure 2.3. You do not need to know even the names of these bases, much less their structural formulas. We have included them so that you will appreciate how complex the final structure of DNA really is. Hence, with the bases inserted in their proper places, if each strand were stretched out along a straight line, we would have a structure that we can show schematically as follows:

```
-sugar-phosphate-sugar-phosphate-sugar-phosphate-sugar-
  |                |                |                |
  A                G                C                T
```

Reading from left to right, we see that this stretch of DNA gives the base sequence, -AGCT-. We could, of course, have chosen any of the bases, and arranged them in any order; this is simply an illustration of how the component parts of the molecule are arranged in a single strand. But, of course, in the whole molecule there are two strands; these are arranged spatially so that the bases from the two separate chains are brought into very close proximity. It turns out, however, that the structures of the individual bases are such that not all combinations of pairs are allowed; in fact the only pairs that are allowed are A-T and G-C (other combinations simply won't fit). We can therefore extend our schematic diagram one step further:

```
-sugar-phosphate-sugar-phosphate-sugar-phosphate-sugar-
  |                |                |                |
  T                C                G                A

  A                G                C                T
  |                |                |                |
-sugar-phosphate-sugar-phosphate-sugar-phosphate-sugar-
```

If we now take these long strands and twist them about each other, we have the double helix shown in Figure 2.2.

Now that you know the general structure of this monstrously complex molecule, you are undoubtedly curious to know what it does. The answer is that DNA is the molecule that makes up genes; that is, a gene is nothing more than a long stretch of DNA. But, you are asking, if genes determine something like eye color or skin color, or the shape and size of a pea, a gene must contain information of some sort, and therefore, if genes are DNA, then DNA must carry information. How can a molecule carry information? To understand how, we must realize what happens to DNA in

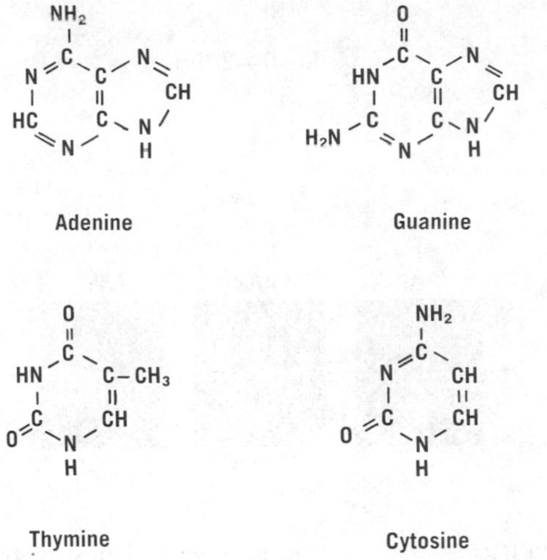

Adenine

Guanine

Thymine

Cytosine

FIGURE 2.3 The four nitrogenous bases used in DNA.

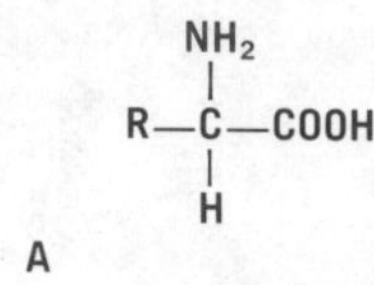

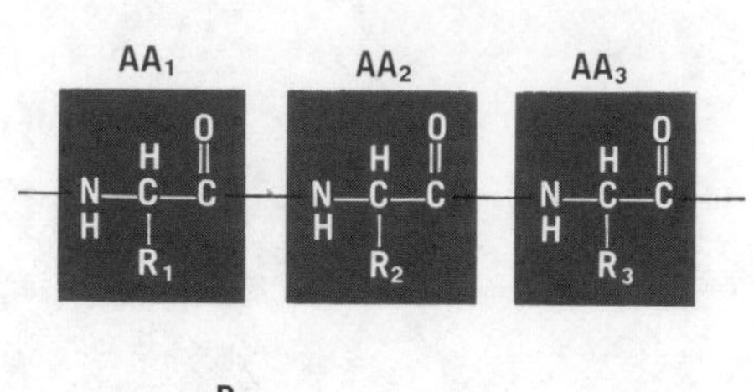

FIGURE 2.4 *A.* The general structure of the amino acids used in proteins. *R* stands for any one of many possible side chains. Each amino acid has its own specific side chain. *B.* The way in which these amino acids are bonded together in a protein chain. Here R_1, R_2 and R_3 denote the side chains of amino acids AA_1, AA_2, and AA_3, respectively. It should be emphasized that this sort of diagram is very schematic; in reality protein molecules are folded about each other in three-dimensional space, often in very complex configurations.

the normal course of cell function. DNA is used by the cell ultimately to direct the synthesis of proteins; that is, DNA is used in a series of biochemical reactions to help form another kind of molecule, similar to DNA, called **RNA**; the RNA is in turn utilized to direct the synthesis of protein molecules. Proteins are used by the cell in a variety of ways. Some, called **enzymes,** are responsible for catalyzing biochemical reactions inside the cell. Other proteins may serve structural functions such as the formation of connective tissue. Still others may transport different molecules from one place to another, as hemoglobin does with oxygen. Like DNA, protein molecules are long (not nearly so long as DNA), but their structure is totally different from DNA. They are composed of a series of organic molecules, called amino acids, linked together in a long chain (see Figure 2.4). The properties of a protein are determined largely by the specific sequence of amino acids that constitute it.

We are now ready to answer the original question: how does DNA carry information? The answer is that it is the specific *sequence* of bases in the DNA that determines the sequence of bases in the RNA, which in turn determines the specific sequence of amino acids in the protein molecules. For example, the sequence of bases –ATCCGTAACG– is different from the sequence –ATGGGTATCC–; the protein which is made from the information contained in this stretch of DNA will differ in its amino acid sequence in these two cases simply because the DNA base sequence is different. The information contained in the specific sequence of bases in the DNA is used by the cell to tell it what specific amino acid sequences its proteins should have. In effect, the "base language" of DNA and RNA is translated by the cell into the "amino acid language" of proteins.

We are now ready to give the term gene a modern definition and one that will help us in a later discussion. Very simply, a gene is a stretch of DNA that directs the synthesis (through an RNA intermediate) of one complete protein chain. We shall return later to the relationship of this "one gene– one protein chain" idea to the production of observable character traits.

Mutation

Since the sequence of bases in the DNA determines the sequence of amino acids in the protein made from it, you can see that if the DNA base sequence is changed in any way, the amino acid sequence of the protein will in general change, too. The DNA base sequence may change, for example, by deletion or addition of bases to the sequence, or by substitution of one base for another. In this way a different protein product will appear. We call any such change in the DNA base sequence a **mutation.**

Why do mutations occur? We probably do not know all the forces involved in the appearance of mutations inside living cells, but two of the most important are undoubtedly radiation and certain chemicals, both of which may result in a change in DNA base sequences by a complex series of reactions. Whether a mutation results in some kind of observable change in the individual animal, however, depends on two very important conditions. The first is simply whether the new mutant protein (that is, the protein formed from "translation" of the new mutant gene) can perform its assigned function in the organism as well as the old protein. If it can, no difference will be observed. If, however, it is defective and cannot perform its task, then the phenotype of the organism may change in some way; that is, the genetic change may become observable.

The second condition relates to whether the new mutant gene is dominant or recessive to the original gene in the organism that received the mutation. If the gene is dominant, the phenotype may be affected, but if the gene is recessive, no apparent change will be observed. However, if a recessive mutation occurs in a sex cell, the information will be passed on and may appear in the phenotype of subsequent generations.

Genetic Disorders

To make some of the previously described concepts clearer in the context of human biology, we cite a few examples.

Achondroplastic Dwarfism

This is a condition in which the affected individual is a dwarf. The arms and legs are short in relation to the trunk and head, which are of normal size. Genetic analysis of affected families has shown clearly that this condition is due to the presence of a dominant gene, and affected individuals are usually heterozygous for the dominant gene. It should be emphasized that a given individual may be such a heterozygote by one of two genetic mechanisms: (a) he may inherit the gene from a parent who is a dwarf, or (b) the parents may both be normal but a mutation may occur in the sex cells of either the mother or the father, giving rise to the gene for dwarfism which is then passed on to the child. The specific protein defect in this disease remains unknown.

Phenylketonuria (PKU)

In this disease the patient is homozygous for a gene that results in defective synthesis of the enzyme phenylalanine hydroxylase (see Fig. 2.5). This is the first enzyme in the pathway by which the amino acid phenylalanine is metabolized. Because this enzyme is defective or absent, the concentration of phenylalanine in the blood builds up to very high levels. Some of this large excess of phenylalanine is diverted into other metabolic pathways which lead to other compounds, some of which may be toxic to the infant's nervous system. As a result of one or more of these metabolic derangements, severe mental deficiency generally occurs. For an individual to have the phenylketonuric phenotype (that is, have the disease expressed), his genotype must be homozygous for the phenylketonuria gene. This means that each of his parents is usually heterozygous. Because the heterozygotes are indistinguishable from normal unaffected persons, we say that the gene for PKU is recessive to the normal, or simply, recessive.

Sickle Cell Anemia

Hemoglobin is the name of the class of proteins inside red blood cells that is responsible for transport-

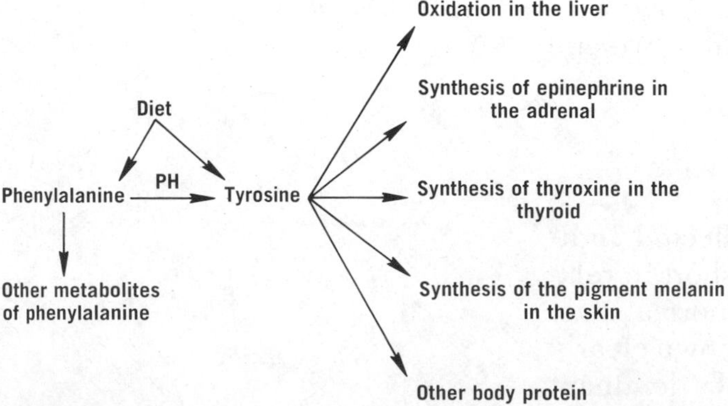

FIGURE 2.5 Phenylalanine and tyrosine are both amino acids that occur in any adequate diet. Once in the body, phenylalanine can be converted to tyrosine with the help of the enzyme phenylalanine hyroxylase (denoted by PH). Both amino acids are used in the synthesis of many normal body proteins; in addition, tyrosine is utilized in the several other ways shown. The phenylketonuric individual lacks the enzyme PH. Since phenylalanine can no longer be converted to tyrosine, the concentration of phenylalanine builds up to very high levels, as do the concentrations of other metabolites of phenylalanine, some of which may be toxic. (Modified from Stanburg, Wyngaarden, and Fredrickson: *The Metabolic Basis of Inherited Disease.* **2nd Ed. New York, McGraw-Hill, 1966.)**

ing oxygen from the lungs to the various tissues of the body. Most of the hemoglobin in the red cells of a healthy adult is composed of four protein chains in close association. Certain individuals carry a mutant gene which directs the synthesis of one protein chain with the amino acid valine substituted for the normal glutamic acid at one position of the chain. This may at first seem like a small difference, since each affected chain contains a total of 146 amino acids, but, in fact, this single substitution does indeed profoundly affect the ability of the whole hemoglobin molecule to function properly. In affected individuals, the hemoglobin tends to come out of solution when the red cells are in the capillaries of the circulatory system. This, in turn, causes the red cells to assume the shape of a sickle instead of their normal configuration (see Fig. 2.6). The result is that these individuals have a very severe anemia from early childhood and may die of the complications of their disease before reaching adulthood. As with PKU, individuals with sickle cell anemia have a double dose of the mutant gene and have usually inherited one copy from each heterozygous parent. Since the heterozygotes for the sickle cell gene usually display no symptoms of disease, we say the sickle cell gene is recessive.

The environment is very important in the expression of genetic information at the phenotypic level. If infants who are homozygous-recessive for PKU are fed from birth artificial diets containing no phenylalanine, they will develop normally and not suffer the mental retardation that would ensue if they ingested usual amounts of phenylalanine. An environmental effect is also seen with people who are heterozygous for the sickle cell gene. For if such individuals are placed in an environment in which the oxygen concentration is low, such as at a high altitude on a mountain or in a high-flying airplane that does not have an adequately pressurized cabin, then even *their* red cells may become sickled and massive death of cells in the spleen may occur. If such people happen, as most do, to stay out of such low-oxygen atmospheres, they may live a completely normal life unaware that they carry an abnormal hemoglobin. Thus we see that both dietary and atmospheric variation may mask or unmask genetic characteristics of the individual. Thus, the phenotype depends not only on the genotype, but on the environment as well.

Some amino acids:

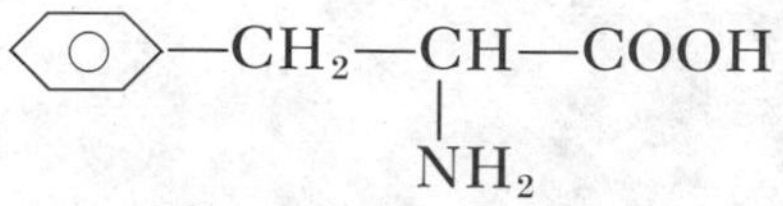

$$\langle\bigcirc\rangle\!-\!CH_2\!-\!\underset{\displaystyle NH_2}{\underset{|}{CH}}\!-\!COOH$$

phenylalanine

$$HO\!-\!\langle\bigcirc\rangle\!-\!CH_2\!-\!\underset{\displaystyle NH_2}{\underset{|}{CH}}\!-\!COOH$$

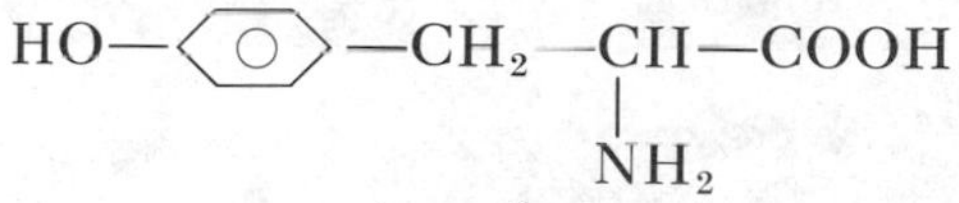

tyrosine

$$(CH_3)_2CH\underset{\displaystyle NH_2}{\underset{|}{C}}HCOOH$$

valine

$$HOOCCH_2CH_2\underset{\displaystyle NH_2}{\underset{|}{C}}HCOOH$$

glutamic acid

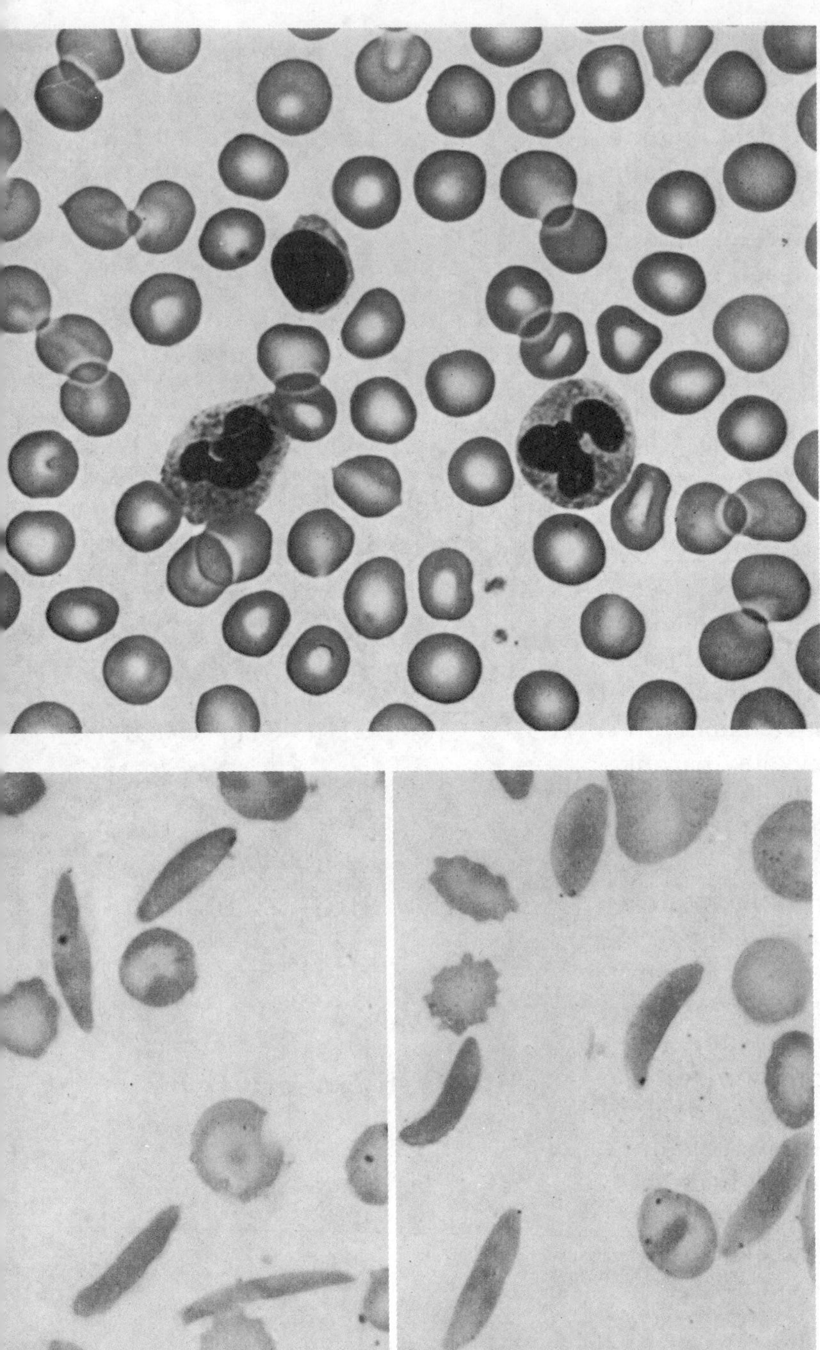

FIGURE 2.6 **Smears of normal blood and blood from a patient with sickle cell anemia. (From McDonald, Dodds, and Cruickshank:** *Atlas of Hematology.* **Baltimore, Williams and Wilkins, 1965.)**

The ABO Blood Group System

Human red cells contain a variety of substances on their surfaces which allow them to be distinguished by simple chemical tests. One such group of substances forms the so-called ABO blood group system. Human red cells may be type A, B, AB, or O, depending on whether the cells carry the A substance, the B substance, both A and B substance, or neither of the two.

The appearance of these substances on the red cell surface is controlled by a group of three genes, commonly denoted i^a, i^b, and i^o, as shown in the right column. Since i^ai^b results in a red cell with both the A and B substances on the surface, these genes are codominant.

Genotype	i^oi^o	i^oi^a, i^ai^a	i^oi^b, i^bi^b	i^ai^b
Phenotype	O	A	B	AB

2.3 THE INFLUENCE OF THE ENVIRONMENT ON THE GENE POOL—NATURAL SELECTION

(Problems 6, 8, 9, 10, 11)

We have seen that there are two mechanisms whereby genetic variation and hence individuality are introduced into a species. Existing chromosomes are shuffled into new combinations of old traits during sexual reproduction, and mutations introduce new genetic information. Now, if we have a genetically varied population subject to certain environmental stresses, some of the individuals may be more successful at reproducing than other members of the species. Obviously, individuals who produce the most progeny will make a greater contribution to the gene pool of the next generation than those who produce fewer offspring. Those individuals who enjoy a reproductive advantage over others are said to have greater fitness. This process by which environmental stresses give certain members of a species a reproductive advantage over others was termed **natural selection** by Charles Darwin, who was chiefly responsible for suggesting the overwhelming importance of this mechanism in evolution. Although Darwin himself did not know about genetic variation and mutation, he understood that somehow the characteristics of a species changed with time in response to environmental pressures.

What determines which individuals have greater fitness and hence which genotypes will be preserved? This question is a reworded form of the ecological question: what niche adaptations enhance the survival of a species? There is no single answer to this question, for there are many ways in which different species adapt. Some individuals, like antelopes, may survive predation because they can run fast, and others, like lions, because they are good fighters and can win territorial battles. But strength and speed are only two of many types of phenotypes which are selected for. Recall the example of the two species of Scottish barnacles from Chapter 1. One species sur-

A species is most satisfactorily defined as a group of animals or plants which exhibits reproductive isolation; that is, a group whose members form an interbreeding community but do not mate outside the group. The number of ways in which groups of living organisms can isolate themselves reproductively from other groups is unexpectedly large. For example, potential mates may have different breeding seasons or spawning grounds and hence may never meet during the time that they are interested in mating. Characteristic behavior patterns at the time of mating (for example, courtship and display behavior) may not be recognized. In some cases, mating may be physically impossible owing to incompatibility of the reproductive organs. Mating may take place but the egg may not fertilize; or if it fertilizes, it may die before many cell divisions. An egg may even develop into an adult organism but one which has greatly reduced fertility, as is the case with the mule (which is the result of a cross between a horse and a donkey). All these mechanisms are used in nature by various groups, singly or in combination, to ensure their identity as species.

vived because it was somehow better able to compete for the underwater niche; the other species gained fitness by its ability to grow in spite of periodic desiccation. The nature of this interspecies competition is quite complex and goes beyond conventional concepts of strength and speed. Ability to survive is also manifested in disease and parasite resistance, which is a powerful evolutionary force. Some species compensate for weaknesses by various forms of co-operation, such as the mutual relationship between algae and fungi that form lichens, or the herd and troop defenses of many animals.

Charles R. Darwin (1809–1882) was the great English naturalist who developed the concept that living species evolved by natural selection. His major work, published on November 24, 1859, was entitled "On the Origin of Species by Natural Selection."

We should emphasize at this point two important features of the concept of fitness. In the first place, it makes no sense in most cases to talk about an abstract concept of fitness. A phenotype advantageous in the jungle could be disastrous in the desert. One must almost always speak of fitness with reference to a particular environment. (We add the "almost" to the preceding sentence because some mutations, such as an animal without reproductive organs, would have a fitness of zero, independent of environmental variations. But most mutations which result in viable progeny do not involve such extreme changes as this.)

Secondly, the term fitness refers solely to reproductive advantage. Our lion mentioned above may be the largest and strongest lion in Africa and may have

FIGURE 2.7 Picture of moths having industrial melanism. (From Helena Curtis: *Biology*. New York, Worth, 1968. Courtesy of Dr. H. B. D. Kettlewell.)

won every territorial battle he ever fought, but his fitness will be zero if for any reason he is incapable of reproduction.

An example of natural selection will serve to tie some of these ideas together. In the mid-nineteenth century in England, rare black variants of the common pepper moth were noted for the first time. By the end of the nineteenth century, however, this once-rare color variant had become the predominating variety in most of the industrial sections of Britain. Such a phenomenon has since been noted in many other moth species in Europe and North America as well (Fig. 2.7). The most likely explanation for these striking observations was that the dark color protects the black moths from their natural predators (birds) in environments darkened by industrial pollution. In heavily industrial areas, then, one might predict that any individual moth with genes for black coloration will preferentially survive and thus have a reproductive advantage over more lightly colored individuals. Hence, these color genes will be enriched in subsequent generations.

Though this sounds like a reasonable explanation, it needs experimental verification, that is, a demonstration that color could be protective in different environments. This verification has been provided by experiments in which equal numbers of black and white moths were released together, both in woods whose trees were darkened by pollution and had no normal lichen on their bark, and in normal woods far from polluted areas. It turned out that there was great preferential survival of the light variants in the normal woods and of darkened variants in the polluted setting. This proves that the color of a moth may serve a protective function and strongly suggests a role for color in natural selection, at least for moths.

But the story of industrial melanism, as this phenomenon has been called, is even more interesting. At the beginning of this century, genetic analysis of the dark variants showed that the dark color was the result of a single dominant gene, which presumably arose from the original "normal" gene that gives moths their light coloration. Fifty years later, however, when the analyses were repeated, the results showed two interesting findings: (1) dark variants now are darker than they were at the beginning of the century, and (2) the darkness is no longer due solely to a single dominant

Melanism is the state or quality of being black.

gene, but rather to a complex of genes working in concert. In other words, we may reasonably suppose that dark color is of such selective advantage in certain environments that further modifying genes were selected to intensify this character. This means that moths with additional melanin survive to breeding age. As well as providing a beautiful example of the power of natural selection, this example also shows the frequently complex relationship between gene action and observable character traits. Frequently an observable, apparently simple character, such as the color of a moth, or indeed, skin color in man, is due to the interaction of many genes located at different places on the chromosome, or even on different chromosomes. One can, therefore, hardly speak of "a gene" for skin color.

We have seen that natural selection can result in the retention of those genes producing phenotypes with reproductive advantage in a certain environment. You might reasonably expect the converse to be true as well, that is, genes burdening their possessors with reproductive disadvantage ought to be eliminated from the population eventually. Though this seems reasonable, the issue is really rather complex. To see why, let us first assume that a certain dominant gene in a population gives a disadvantageous phenotype in a given environment. Then the more detrimental the phenotype, the stronger will be the tendency of natural selection to eliminate this gene. This tendency will be counteracted by random mutation, causing the disadvantageous gene to reappear in the sex cells of some individuals in the population. In other words, mutation alone would prevent the gene from disappearing entirely (unless, of course, the gene were lethal, in which case any of its possessors would die). With deleterious recessive genes, the power of mutation to cause accumulation of the gene in the population is even greater, since for a *true* recessive to be exposed to selection, it must be present in the homozygous state; that is, an individual bearing even a *lethal* recessive will not be selected against if he is heterozygous.

An even more interesting situation arises when the individual who is heterozygous at a given locus is for some reason *more* fit in a certain environment than either homozygote. When this happens, the mutant gene may occur in the population in rather high frequency, even if the mutant gene is lethal in the homozygous condition. The most famous example of this situation in man is the gene for sickle cell anemia mentioned above. This gene is present in about 10 per cent of black Americans.

But why should a gene which produces such devastating effects in the homozygous state be present in such a high proportion of the population? A probable answer to this question was provided by studies showing that *heterozygotes* for the gene (See p. 58.) have increased resistance to severe infection by the malaria parasite that is endemic in large parts of Africa. One can think of the unfortunate people with sickle cell anemia as the "price" the population pays for the increased resistance to a potentially lethal and disabling infectious disease.

From a genetic viewpoint, sickle cell anemia is a lethal illness, since very few afflicted patients live long enough to reproduce.

This type of situation, in which a single locus has two or more relatively common gene types occurring in the population, is called a **balanced polymorphism** (polymorphism because there are two or more types of genes). Other well-studied examples of genetic polymorphism in man include many of the blood group systems, of which the ABO system mentioned above is one. In this system a person's blood group is determined by the action of three different genes. Different populations have quite different frequencies of the various blood types. It is certainly possible that there are selective pressures favoring the observed frequencies of these blood group substances in the different populations, but no such forces have been identified to date.

2.4 ADAPTATION BY LEARNING

Thus far we have concentrated on the genetic means of adaptation which have played such a large role in the evolution of life. In addition to this most basic method of adaptation, many animals have evolved mechanisms for learning from their own prior experience. For example, in a part of Africa a fence was erected around a game park inhabited by many animals, including lions. Within one season the lions adapted their hunting habits to conform with, and even to utilize the fence. One of the lions would chase her prey toward the fence. On approaching the barricade, the prey would, of course, veer to one side or the other, whereupon it was seized by other lions who had stationed themselves there as part of the trap.

Lion chasing prey in a fenced game park.

Development of this scheme required reasoning, memory, and communication among the hunters and was a rapid response to an environmental change.

Of all the creatures ever to have lived, however, man has the most prodigious powers of reason, memory, and communication. Man alone seems to be able to accumulate large and diverse quantities of knowledge and pass increasingly complex thoughts on to future generations. For example, early man learned, like the lion, to herd game toward existing traps such as cliffs or tar pits. But then man went a step further by constructing traps in places where there were none. From traps he invented stockades and learned to retain and domesticate animals. Finally, from domesticated wild animals, he went on to breeding genetic variants of animals that suited his needs better.

The ability to add knowledge in a cumulative way to future generations is dependent on the capacity for symbolic expression of thought, that is, language. As we already mentioned, man is not the only species with the ability to communicate; group hunters, like the lion, must somehow "talk" to each other. Colonial animals, like bees, are also able to communicate. When a bee locates a source of food, it returns to the hive and, by a special dance on the combs of the hive, imparts this knowledge to the other members of the colony. Apparently, the speed of the dance reflects the distance of the source of the food from the hive, and the angle that the dancing animal makes with the vertical indicates the direction of the source from the hive. Birds do not sing for the benefit of poets or lovers; their songs serve very important functions as danger warnings or mating calls.

What non-human languages have in common, as far as we know, is that the individual animals are born not simply with the ability to learn to "speak" the language, but also with the specific vocabulary of the language already wired into the nervous system. After returning to the hive, the successful bee can tell its hive mates about the distance or location of a specific food source from the hive. It tells them nothing, for example, about the general principles of foraging for food. (It need not, of course, for these "principles" are wired in as well.)

When we compare what we know about the language of bees with the capability of human language, as exemplified, say, in the simple declarative statement, "The world is a sphere, and one can reach the East by sailing west," at least two differences become apparent. The most striking, of course, is that this statement is only one of an incalculably large number of sensible word combinations that man can make. Although there are clearly definable patterns to human speech, there is nothing stereotyped about it in the same sense as with our bees above. Secondly, human language, as well as allowing communication about specific bits of information, permits expression of generalities and abstractions.

It is easy to see the enormous selective advantage that the acquisition of this ability must have conferred on its possessors when it arose in human evolution. For a flexible symbolic language allows transmission of acquired experience to others and to subsequent generations. Even if it is possible to conceive of the human intellect without a symbolic language, it is hard to see what kind of progress could be made if the Pythagorean theorem, or the principles of visual perspective in art, or the equally tempered scale in music had to be rediscovered by each generation. But man's language has saved him from this task. From the time of the emergence of man until now, a period of at most a few tens of thousands of years, which is very short on the evolutionary time scale, man has learned to exist in environments ranging from the Congo jungle to the surface of the moon, from the tops of the Himalayas to the bottom of the ocean. The mechanism of adaptation in each of these cases has been cultural rather than genetic, that is, he can survive in these extreme situations because of the sum total of what he has been taught by others. The astronaut who orbits the moon surrounded by many of the comforts of home does so not because he is genetically different from a resident of ancient Athens, but because the values, beliefs, and scientific accomplishments of the whole Western cultural tradition have provided him with the means.

It has been said that insects are the most adaptable of animals and that of all animals, they can survive in the widest variety of environments. Certainly, one can find insects in virtually every type of terrain and climate. Remember, however, that with insects one is dealing with nearly three-quarters of a million distinct *species* of animals. Viewed in this way, the true extent of man's adaptability becomes clearer. For man, a single species, can live anywhere insects can. To do so, he simply lets his technology create the proper conditions, and by virtue of appropriate housing, clothing, and food, he remains alive and even comfortable.

In the previous discussion we have drawn a clear-cut distinction between genetic and cultural methods of adaptation. These general methods of adaptation differ in two important ways. The first is that the cultural is in many ways a more flexible and powerful mode of adaptation than the genetic. To understand why, we must first realize that it is not always possible for a species to adapt genetically to environmental changes. For if it were possible, obviously, no species would ever become extinct! The ability of a species to respond genetically to environmental challenge depends on his pre-existing genetic constitution and on whether it is possible to change, by either a new mutation or genetic recombination, in such a way that adaptive genetic variation will result. It is difficult, for example, to imagine that man could adapt by purely genetic means to all the climatologic conditions that his cultural methods allow him to master. The number of anatomic and physiologic adjustments that would permit survival, say, in the Antarctic without clothing or imported food are far too large and varied.

Undoubtedly, in the history of the biosphere there have occurred environmental changes which, though not so extreme as a drop to polar temperatures, have been radical enough so that the genetic constitution of a species did not have available the necessary background for flexible change that would have been required for continued survival. Quite obviously, man's cultural resources provide him with an enormous amount of additional flexibility, and have allowed him to extend his domination over the entire planet. As noted above, however, man's total lifetime as a species has been fairly short on the evolutionary time scale, and in his short lifetime, his cultural achievements have also included certain weapons that threaten his continued existence. In this capacity to destroy himself utterly, man is also unique among animals.

The second major difference between the two modes of adaptation is that the genetic method takes a long time for the human species. By contrast, cultural means may be far faster if the particular society already has methods to deal with the challenge. The response to malaria again provides a good example. It is uncertain how long it took for the sickle cell balanced polymorphism to become established where malaria is endemic, but probably several thousand years were required. (Recall that the average human generation time is 20 to 30 years.) The cultural methods of dealing with malaria involve both the treatment of infected individuals with anti-malarial drugs, and, more importantly, the use of pesticides to eradicate the mosquito that transmits the disease. From the introduction of DDT following World War II to the present, probably about 10 million lives have been saved by malaria eradication programs. There is, of course, no way to calculate the number of lives saved by the sickle cell balanced polymorphism. The only point that is important here is the enormous difference in time required by the two modes of adaptation.

2.5 THE INTERACTION BETWEEN CULTURAL AND GENETIC MECHANISMS

(Problems 12, 13)

Up to now we have been speaking of cultural and genetic adaptive mechanisms as quite unrelated entities. It has probably occurred to you that there must be a very close relationship between the two. First of all, the very existence of culture in human society is a reflection of the genetic foundation on which the anatomy and physiology of the human nervous system is based. In one sense, the brain is only another of man's organs, and like the heart, kidneys, and gonads, its development in fetal life and early childhood is in part determined by the organism's genes. We are indeed very far from even a remote grasp of how human genes, together with environmental influences, direct the formation of an intact nervous system. As we stated above, however, whatever the details of the mechanism, the selection for such a system must have been very powerful because of the enormous benefits that accrue to animals possessing such nervous systems.

Although genes can be said to influence culture, the reverse is also true, and we are only beginning to appreciate some of the subtler features of such a relationship. Unfortunately, the very term natural selection, when translated as "survival of the fittest," conjures up images of human beings locked in deadly struggles with sabre-tooth tigers and other now-extinct horrors. Perhaps at some time in the past man did face important challenges of this sort, and under these circumstances one might reasonably imagine that the largest, strongest men who could run fastest and fight best might survive preferentially. Although this is an oversimplification, one can at least identify some of the selective forces operating in this kind of life.

How has modern society changed this pattern? We must now be scrupulous in observing the distinction between "fitness" as ordinarily used, and fitness in the Darwinian sense. For the latter refers only to *reproductive* advantage, and, defined in this way, one's height, body build, and physical strength may be of secondary importance. In fact, the types of heritable characteristics which endow their possessors with reproductive advantage in modern societies cannot easily be determined. Nevertheless, a few statements can be made.

First of all, the aggregation of man in cities has probably influenced certain patterns of disease resistance which are most likely genetically inherited. European Jews, for example, have been a predominantly urban people for several centuries. Tuberculosis is an infectious disease that has been predominantly urban in its distribution. Present evidence indicates that Jews as a group now enjoy significantly greater resistance to serious infection by the tubercle bacillus than do members of other groups, such as blacks, who have not been exposed to the organism for so many generations. Presumably, generations of exposure to TB have allowed those to survive preferentially who tend not to be killed off or seriously disabled by the disease prior to their reproductive period.

Selective pressures are also modified by many of the scientific and technical advances of modern society, of which medicine is one of the most obvious. Consider diabetes, a common disease that is clearly inherited. Before the discovery and availability of insulin, many children who became ill with diabetes died before the reproductive period. Others might survive through young adulthood, but because of the severe manifestations of the illness, they were at a reproductive disadvantage with respect to the unaffected population. Insulin has of course changed this situation markedly, and while diabetes continues to be a severe illness for many of those affected, diabetics incontestably survive better than formerly. This means, of course, that they also tend to have more children than before, and hence we can expect the gene(s) for diabetes to increase in frequency as time goes on, at least in populations where reasonable medical care is available. That is, in several centuries a greater fraction of such a population will be either overt diabetics or carriers of the disease than is true today. Similarly, the children with PKU who used to languish in mental institutions will now grow up as phenotypically normal adults and have their own children, each one of which will be a heterozygote for the PKU trait.

Diabetes is a disease characterized by abnormally high concentrations of glucose (a sugar) in the blood. The blood sugar levels may be lowered by the administration of insulin, a hormone which is often deficient in these patients. Also present in diabetics is a propensity for diseases of the vessels of the eye, kidney, skin, extremities, heart, and brain; such vascular ailments are often the major cause of sickness in diabetic patients.

Another interesting example is the influence of dietary intake on the development of coronary heart disease. Although we do not yet know the specific biochemical steps that lead to the development of arteriosclerosis, research suggests that the dietary intake of certain kinds of fats may be important; that is, those with a high intake of such substances have an increased risk of developing vascular disease and heart attacks in later life, say from age 45 onward. But, you may ask, if natural selection is so powerful, why has human metabolism not evolved so that such high intakes of these substances are not harmful? The answer to this question is somewhat speculative. In the first place, since coronary heart disease causes illness and death largely in the post-reproductive period, selection for factors that reduce its occurrence might not be very powerful. Nevertheless, even if selective pressures are not very powerful, they will produce observable effects if allowed to act for a long enough time. A probably more important reason has to do with the fact that the general condition of the large bulk of mankind was, and continues to be, characterized by food scarcity rather than overabundance. Only for certain elements of the population of western Europe and America does the overfeeding problem exist at all. Viewed in this way, it is not surprising that physiological mechanisms have not yet evolved that can deal adequately with the glut of food in the affluent American diet.

Thus far all our examples have had to do with one disease or another. When we consider what effect modern culture has on inherited patterns having nothing to do with disease, the answers become even more speculative and sometimes even difficult to define. Intelligence, for example, is a human quality generally considered desirable, and one might naively think that a technologically oriented society such as ours would select for intelligence, if, in fact, such a quality is heritable. But again, we must be careful. For what we are asking can be better formulated in the question: Are intelligent people at a reproductive advantage in this society? If we define intelligence, admittedly circularly, as that quality which allows its possessors to do well on intelligence tests, then the answer is probably no. The economically disadvantaged have less access to formal education than the well-to-do, and this deprivation undoubtedly depresses their scores on so-called intelligence tests. Since these segments of U.S. society generally have the largest families, they can be considered to be at a reproductive advantage over those with higher test scores and fewer children.

But the fact that we cannot easily identify the kinds of selection that are taking place in human societies today does not mean that natural selection has ceased to exist as a force in determining man's future. For natural selection will always exist whenever genetic variation exists in a population and whenever individuals fail to reproduce equally because of genetically determined factors. For in this situation, those with a reproductive advantage will make a larger genetic contribution to future generations than those at a disadvantage. It is upon this single factor that the environment may exert its most telling influences on the future of man.

PROBLEMS

1. **Vocabulary.** Define the following terms: gene, homozygous, heterozygous, genotype, phenotype, dominant, recessive, fitness.

2. **Mendelian genetics.** What phenotypes would you expect from a cross of the F_1 pea plant Yy with the parent YY? With the parent yy? What genotypes? What would happen if you self-pollinated the children of the cross between Yy and YY and the cross between Yy and yy?

3. **Mendelian genetics.** Explain how a deleterious recessive gene is more likely to appear in the homozygous state, and thus be expressed in the phenotype, when mating occurs between close relatives than between unrelated individuals.

4. **Genetic variation.** What is the mechanism by which genetic variation is introduced into a population? Describe what this means with reference to the structure of DNA.

5. **Mutation.** What is the difference in the effect that selection has against a detrimental recessive and an equally detrimental dominant mutation?

6. **Mutation.** Do you think that the mutation causing the appearance of dark moths ever occurred before the Industrial Revolution? Explain.

7. **Mutation.** Does the presence of a dominant genetic disorder necessarily mean that one of the parents is affected with the disorder? Explain.

8. **Fitness.** The gene for sickle cell hemoglobin became prevalent among natives of the African continent because heterozygotes probably enjoyed greater fitness than either homozygote in the presence of malaria. What would you expect to happen to the frequency of this gene among blacks living in North America, where malaria is no longer a problem? Explain carefully.

9. **Fitness.** Examples of PKU and diabetes given in the text illustrate that one of the consequences of modern medicine is the accumulation of many genes in the gene pool of a technologically advanced society that would not accumulate so readily in a society where sophisticated medical care does not exist. In this sense it might be said that a technological society selects for the weak and sick. Discuss whether this is good or bad, with reference to: (a) the individual with the "bad" genes and (b) the society as a whole.

10. **Selection.** A skeptic who does not believe that people will stop smoking despite the data linking cigarette smoking to cancer and heart disease, replies, "What's the difference? If people continue to smoke as they are, pretty soon the human race will become genetically resistant to the bad effects of cigarette smoking just by survival of the fittest and the occurrence of lung

cancer will go down, even if nobody stops smoking." Discuss the validity of this assertion.

11. **Adaptation.** Some anthropologists feel that the dark skin of the black races is adaptive for the warm climates and hot sun in which these people have spent thousands of years. What simple experiment might you design to test the hypothesis that blacks tolerate warm climates better than whites?

12. **Fitness.** We have suggested in this chapter that in primitive societies, where physical strength and stamina may have counted for more than they do today, physical fitness may have endowed its possessors with Darwinian fitness. Can you imagine a type of social organization in which the best physical specimens were actually at a reproductive *disadvantage* with respect to others? Describe such a hypothetical situation.

13. **Fitness.** Suggest a way in which a disease which occurs principally in the *post*-reproductive period could have a selective effect on a population.

BIBLIOGRAPHY

A good summary of modern genetics is given in:

R. P. Levine: *Genetics.* New York, Holt, Rinehart and Winston, 1962.

Several excellent works dealing with various aspects of evolution are:

Theodosius Dobzhansky: *Genetics and the Evolutionary Process.* New York, Columbia University Press, 1970.

______: *Mankind Evolving.* New Haven, Yale University Press, 1962.

Ernst Mayr: *Populations, Species, and Evolution.* Cambridge, Mass., Harvard University Press, 1970.

George Gaylord Simpson: *The Meaning of Evolution.* New York, Bantam Books, 1971.

A summary of the research on industrial melanism is given in:

H. B. D. Kettlewell: The phenomenon of industrial melanism in the Lepidoptera. Ann. Rev. Entomol. 6:245, 1961.

A mine of information about all facts of human inheritable diseases is:

J. B. Stanbury, J. B. Wyngaarden, and D. S. Fredrickson: *The Metabolic Basis of Inherited Disease.* 3rd Ed. New York, McGraw-Hill, 1970.

3

THE EXTINCTION OF SPECIES

3.1 EVOLUTION AND THE EXTINCTION OF SPECIES IN PREHISTORIC TIMES

(Problems 1, 2, 3, 4, 5)

Individual plants and animals face many pressures, such as intraspecies competition, interspecies competition at a given trophic level, interspecies encounter in the form of predation or parasitism, and climatic variation. In the face of these pressures some individuals survive to reproductive age, and some do not. Although occasionally luck may bring long life for an individual, on a statistical basis better genetic adaptation to the environment leads to longer survival. Natural selection, acting through many generations of individuals, controls the eventual composition of a gene pool and thus the phenotypic characteristics of the population.

Genetic change explains how a species evolves but does not explain how a new species is formed. As long as there is sexual contact among members of a species, a new genetic trait, if beneficial, is likely to be dispersed in the population. For two species to diverge, however, populations must be isolated reproductively from each other for a long period of time, say 2000 to 100,000 generations. During the course of this isolation, genetic change in each population occurs independently, and as the specific mutations and selective pressures are in general different on opposite sides of some natural barrier, eventually two distinct species may arise (Fig. 3.1). While a large barrier such as an ocean, a jungle, or a high mountain range is

FIGURE 3.1 Different species of birds of paradise which evolved by isolation on New Guinea.

required to separate groups of birds, different species of rodents have evolved on opposite sides of a large river.

The extinction of species fits naturally into the evolutionary scheme. What appears in the fossil records as extinction can often be shown to be simply **development** of a species over a period of time. For example, the early horse, called the *Dawn Horse*, stood 10 to 20 inches high at the shoulder and lived in dense semitropical forests. As the climate and the vegetation of the world gradually changed, cooler temperatures and grassland ecosystems replaced hot jungles in many areas, and horses evolved along with the environment (Fig. 3.2). Over the years their tooth structure changed to accommodate to the new foods, and at the same time, the animals became larger and faster, presumably because size and fleetness were desirable on the plains. Figure 3.2 illustrates some of the stages in the gradual evolution of the modern horse. The point is that the dawn horse never died out but rather changed gradually in the course of millennia.

In other cases, extinction has occurred by ecological **replacement** of one species by another of similar size and niche requirements. In one period during the reign of dinosaurs, the giant stegosaurus, a herbivore, disappeared while a number of new herbivorous dinosaurs of similar size, such as the antasaurus and the triceratops evolved (Fig. 3.3). We cannot say that the change occurred through direct competition, but we do know that animals of one family replaced animals of another.

When we study the succession of ecosystems, we find that many extinctions occurred that can be classified neither as development nor as replacement. Throughout geological history, certain species have developed gradually, flourished, and then suddenly vanished, leaving behind them an empty niche. For example, approximately 500 million years ago, enormous numbers and varieties of primitive sponges populated the seas. Although for 30 or 40 million years these sponges dominated the seas from pole to pole, most species later disappeared, and different ecosystems developed. In more recent times, the rise and fall of the dinosaurs was certainly one of the most outstanding events in the history of the earth. Small dinosaurs first evolved some 225 to 250 million years

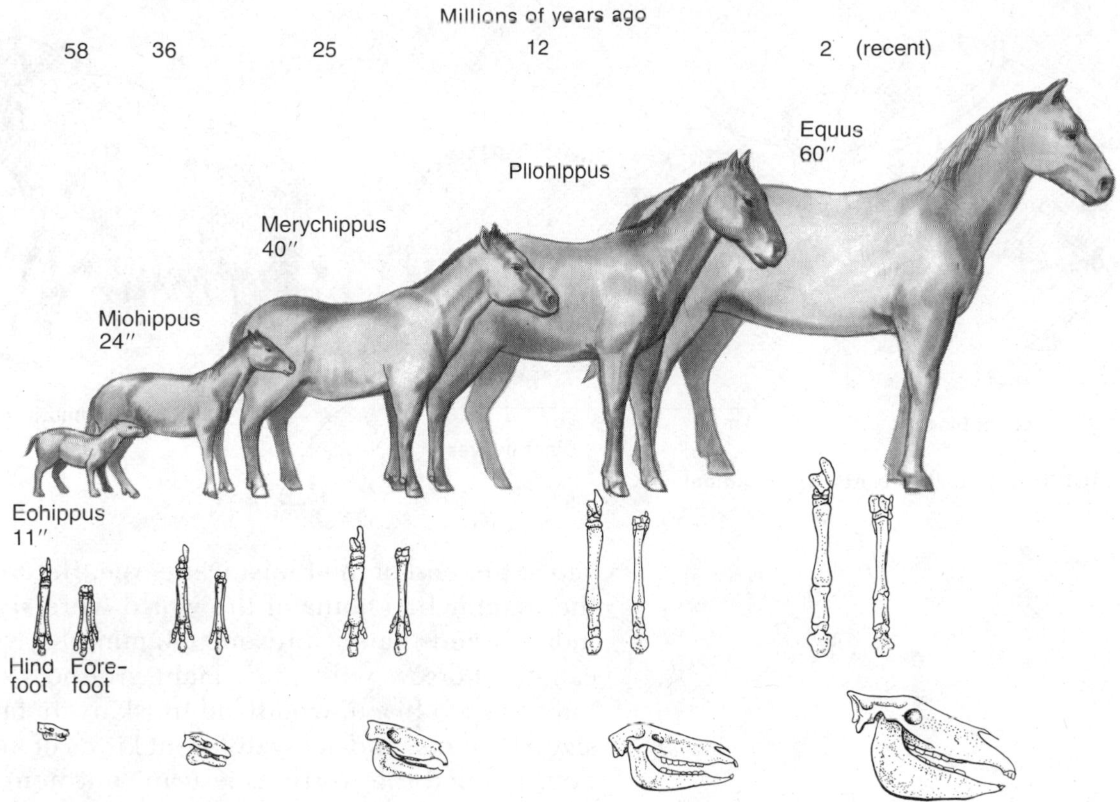

FIGURE 3.2 Some stages in the evolution of the horse.

ago. For perhaps 25 to 50 million years they slowly developed and grew in numbers and varieties until they dominated the earth. The dinosaurs reigned for 100 million years and then they all died, whereupon the established ecosystems and the balances of the food web collapsed. Dinosaurs have been often belittled in popular literature as being stupid and clumsy. But in fact the great reptiles were outstanding evolutionary successes; they dominated the earth longer than any other order has. The extinction of the dinosaurs reminds us that seemingly stable biological systems may someday collapse.

Of course, the period following the extinction of dinosaurs was not devoid of life, and during this time, the previously slow evolution of mammals accelerated. Many species, including man, arose. The next significant wave of extinctions occurred about 10,000 years

FIGURE 3.3 *Top,* **Triceratops and** *bottom,* **stegosaurus.**

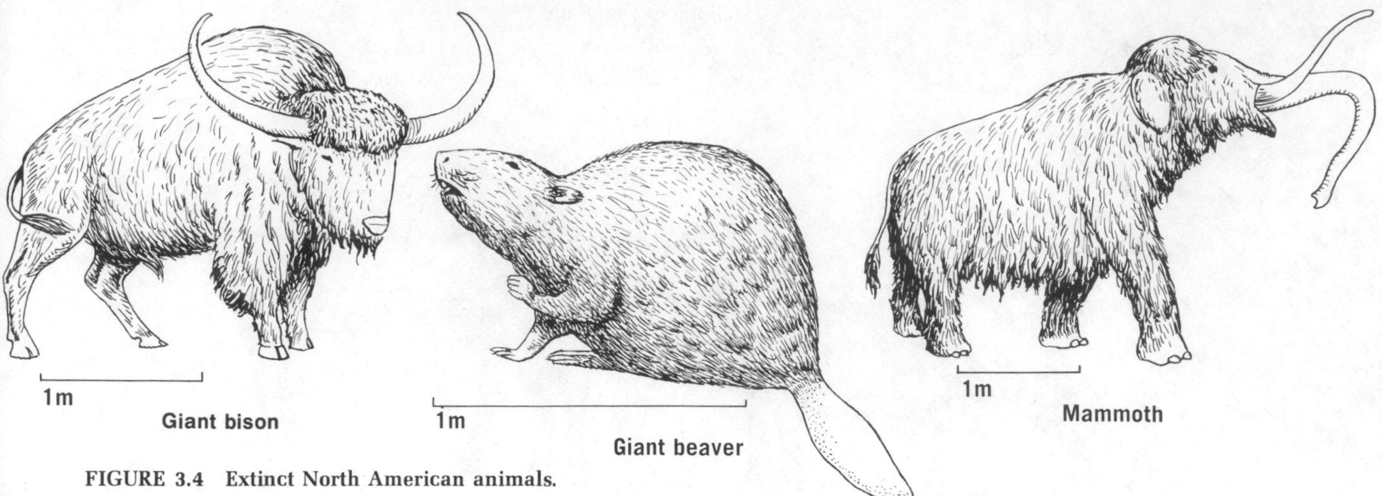

FIGURE 3.4 Extinct North American animals.

ago at the end of the Pleistocene Age. Before this time the mammalian fauna of the world were significantly more varied than at present. Mammoths, mastodons, camels, horses, wild pigs, giant ground sloths, giant long-horned bison, woodland musk oxen, tapirs, bear-sized beavers, and many different kinds of now-extinct deer roamed the North American continent (Fig. 3.4). They were hunted in part by sabre-toothed tigers, giant jaguars, and dire wolves. All these animals have become extinct, for the most part without ecological replacement. In fact, estimates are that 95 per cent of all large animal species in North America were lost, and mass extinctions occurred simultaneously in what are now South America, Northern Asia, Australia, and Africa.

What caused these extinctions? There are many uncertainties. Perhaps climatic changes had an effect. We know that rapid alterations in weather patterns occurred at this time, but we also know that the same animal species that suffered extinction had already lived through several advances and retreats of the glaciers. Also, few small animals followed their larger cousins into oblivion, and it is hard to explain why climatic variation affected large animals differently from small ones. It seems certain that the hunting activities of primitive people played some role in the drama. We know, for example, that all the herbivores that are now extinct were pursued by early humans. However, no major extinctions preceded the development of advanced hunting techniques such as throwing spears, shooting arrows, and setting fires to

"You're being recalled—He's going to try mammals."

(From Saturday Review, March 25, 1972. Copyright 1972 by Saturday Review Co. Used with permission.)

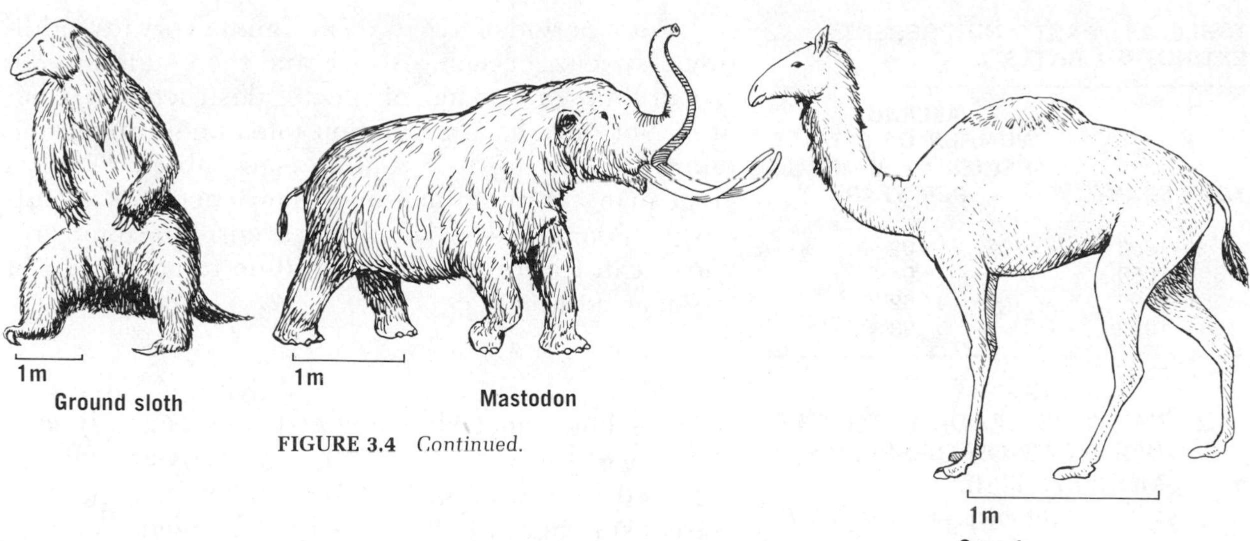

FIGURE 3.4 *Continued.*

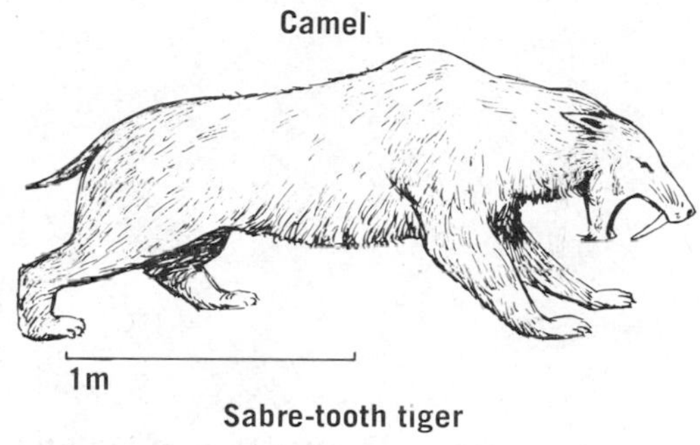

drive animals over a cliff or into a trap (Fig. 3.5). If large herbivores were hunted to extinction, the specialized carnivores like the sabre-toothed tiger must have perished soon after.

But now we come to an important question. How could man, the primitive hunter, have been even a small factor in the destruction of some 200 genera of animals across the face of the earth, and even if he were, how could he have survived after his prey had perished? One reasonable explanation lies in people's powerful ability to adapt culture to new situations. Imagine the relative plights of two great predators in post-glacial North America: the human being and the sabre-toothed tiger. Both depended at first on the mammals of the plains for their food. Of course, their hunting techniques differed. The tigers had evolved a specialized body structure to kill their prey. On the other hand, humans, though physically no match for their competitors, were extremely adaptive. When big game became scarce, people turned to other foods: freshwater mussels, insects, fruits, and berries. They could hunt larger prey only when it was convenient. But the tigers could not change their habits quickly—their hind legs were not structured for speed, their teeth and claws were not well adapted for opening mussel shells, their digestive system was not capable of utilizing berries—and they perished. The exact details of the extinction of the mammalian species in Pleistocene times will never be fully explained, but the fact is that man and a few big game animals adjusted and thrived.

FIGURE 3.5 How the Tule Springs site, Nevada, may have looked 12,000 to 10,000 years ago, with scattered sagebrush (*Artemisia*) and saltbush (*Atriplex*), and springs with ash trees, cattails, and other mesic species. (Courtesy of John Hackmaster.)

TABLE 3.1 PAST AND PRESENT EXTINCTION RATES

YEAR SPAN	*AVERAGE NUMBER OF EXTINCT SPECIES OF MAMMALS PER YEAR*
1–1800	.02
1801–1850	.04
1851–1900	.62
1901–1950	.93

For a period of five to six millennia very few additional species became extinct, and then suddenly, in recent times, a new age of species destruction has begun. Not only have extinction rates been increasing rapidly in the last half century (see Table 3.1), but a great many more animal and plant species are seriously endangered today. Unless current trends reverse, extinction rates will continue to accelerate in the near future.

3.2 FACTORS LEADING TO THE DESTRUCTION OF SPECIES IN MODERN TIMES

(Problems 6, 7, 11, 12, 14)

We know that climate has been relatively constant in recent years; certainly grizzly bears weren't expelled from California, or passenger pigeons from Wisconsin, because the weather changed. We also know that devastating disease epidemics have not recently occurred in wild animal populations, and that vegetation hasn't changed. We know, in short, that people are the major agents in the extinction of species today. Some of the mechanisms are discussed in the following paragraphs.

Destruction of Habitat

The relationship between human beings and nature is no longer a simple predator-in-the-forest system. Perhaps the greatest single cause for species extinction in the civilized world is man's destruction of natural ecosystems. This can occur in several ways: (1) The prodigious growth of human populations creates demands for living space. Accordingly, cities, suburbs, farms, and roads displace animal and plant communities. (2) Fuel consumption and energy utilization are multifaceted sources of environmental disruption, and the mining, transportation, and combustion of fuels (see Chapter 5) often destroy natural habitats. (3) Insecticides and herbicides have at times made water, air, and food supplies unhealthy for wild animals and plants.

When we think of the destruction of species, we usually visualize animals, but many plant species, too, are rapidly facing extermination, mostly from destruction of habitat. The prairie that once covered the midwestern United States and Canada was a beautiful sight, but it is gone and may never return. Small holdouts of true prairie plants still exist in a few old cemeteries and along old roadways, but grain fields and domestic grasses have displaced most of the old flora.

A

B

FIGURE 3.6 *A.* **Trailing arbutus. (Photo by Alvin E. Steffan, National Audubon Society).** *B.* **Giant white pine tree—a reminder of the primeval American forest. (From the book** *The Great American Forest* **by Rutherford Platt. © 1965 by Rutherford Platt. Published by Prentice-Hall, Inc., Englewood Cliffs, New Jersey.)**

An untold number of prairie plant species have become, or are in danger of becoming, extinct. But prairie plants are not the only species to suffer; forest plants, such as the trailing arbutus (Fig. 3.6*A*), can thrive only in the soils of climax systems, and as climax forests are being converted into stands of second-growth timber, the trailing arbutus and many other plants are becoming increasingly rare (Fig. 3.7*B*).

The Introduction of Foreign Species

In prehistoric times, migration of species from continent to continent, or even from watershed to watershed, often required thousands of years and in many instances did not occur at all. In the modern world, however, transportation is rapid. A foreign species introduced into one area often finds few natural enemies and may grow unchecked and unhindered until it has endangered or eliminated a native species.

For example, the sea lamprey is a predator-parasite that attaches itself to a living fish and sucks out the fluids and flesh until the prey dies. Lampreys have long been a part of aquatic ecosystems along the St. Lawrence River and the North Atlantic Ocean, and have been controlled by natural homeostatic mechanisms. In 1829, a ship canal was built to bypass Niagara Falls, thus allowing ships (and lampreys) access from the sea to the Great Lakes. For some reason, lampreys

FIGURE 3.7 Chestnut blight. (From Odum: *Fundamentals of Ecology*. 3rd Ed. Philadelphia: W. B. Saunders Co., 1971.)

failed to traverse the canal for 100 years, but when they finally did, they gained access to Lake Huron and Lake Michigan, where they multiplied rapidly and, in a 10-year period almost exterminated their major prey, the lake trout. This incident shows us that predators do have the potential to destroy a prey species, and may realize this potential if they find themselves in a new and hospitable environment.

The American chestnut tree used to forest much of the middle and southern Atlantic coastal areas of the United States, but now the chestnuts are seriously endangered because a parasitic tree fungus imported from China was lethal to the American trees (Fig. 3.7). Both American and Chinese trees had developed resistance to the parasites that grew in their own habitats but not to foreign parasites. As a result, all large American chestnut trees on the east coast died within 50 years of the arrival of the Chinese fungus. Perhaps some young trees will survive and provide a new breeding stock for the species, but the outcome is uncertain.

Hunting for Sport or Fashion

Perhaps one of the saddest and least excusable types of environmental disruption caused by man in the world today is the wholesale killing of endangered species to satisfy the sport or fashion demands of a few individuals (see Color Plate 3).

Conservationists estimate that between 500 and 700 wild Giant Sable Antelopes live today in west central Africa (Fig. 3.8). These animals are endowed with a set of magnificent curved horns, sometimes five feet long, which may easily be the animal's downfall, for hunters come from many parts of the world to shoot a big bull and bring home a trophy. Additionally, many hunters and tourists who are unable or unwilling to shoot a Giant Sable purchase their trophy from native hunters and return home with an expensive wall decoration. Hope for the survival of this species is slim.

FIGURE 3.8 Giant sable antelope. (Courtesy of Richard D. Estes.)

The snow leopard is one of the few large predators living in the high country of the Himalayas. Leopard skins are considered to be elegant, and snow leopard skins are warm as well. As a result, there are fewer than 500 snow leopards left alive. Snow leopard coats can still be purchased, even though some major furriers no longer sell them.

These two examples were chosen out of several hundred cases where sport and fashion hunters are responsible for large-scale killing of animals and the destruction of natural systems. Other well-known examples include Nile crocodiles and American alligators for handbags and shoes, many jungle cats for furs, and several species of birds for their elaborate feathers.

Hunting for Food

After the wave of Pleistocene extinctions, primitive hunting societies seemed to have existed in ecological balance with their prey. Yet today, many systems are so seriously disturbed that once again some species are being eliminated by food gatherers.

The pigmy hippopotamus has coexisted with tribal hunting societies in central Africa for a long time, but as industrialization reduces the animal's range and brings hunger to many people in the area, the hippo is being hunted more systematically today, and extinction may result.

Predator Extermination

Man has always shared his planet with other predators, and during the millennia of co-evolution, delicate relationships have developed. The most striking aspect of the interactions between man and his carnivorous competitors is that the large terrestrial predators of the earth habitually avoid attacking man. This does not mean that people are never killed by other predators, but just that confrontation has been minimal. For example, the grizzly bear is one of the largest and strongest carnivorous animals alive today (Fig. 3.9). These giants weigh from 450 to 900 pounds when mature and measure up to four feet high at the shoulder. They can run at about 25 to 30 miles per hour on flat ground and are strong enough to crush the skull of a large cow with one blow. Man's fear of this formidable animal is reflected by its scientific name, *Ursus horribilis.* Yet the bear's record is far less fearsome than its name. One of the last significant ranges of the grizzly south of Canada and Alaska is in the Rocky Mountains of northern Montana in and near Glacier National Park. In the past 50 years millions of hikers have toured the park and adjacent wilderness areas, thousands of unarmed people have sighted grizzlies, thousands of armed people have hunted and sometimes killed the bears, and on only two occasions have the giant animals killed persons.

The record of the relationship between man and wolves is even more remarkable. European literature is filled with horror stories of wolves, and English-speaking children learn the lesson early when they listen to the tale of Little Red Riding Hood. Yet, there have been *no* documented cases of wolves attacking man in North America during the past 50 years.

The list could continue on: American alligators do not attack unless cornered, lions generally avoid man, and cheetahs have *never* been known to attack man unless pursued. Other species, like the Bengal tiger, the polar bear, and the Nile crocodile attack man more often, but even for these animals, man is only a rare prey. Yet, though they do not constitute any real threat to life, predators have been hunted, sometimes vigorously, in the name of securing safety for humans (Fig. 3.10).

A second major source of conflict between predators and man is competition for food. Wild carnivores do kill game and domestic stock, and consequently hunters and farmers have often advocated the extermination of predators. But elimination of all predators does not mean more food for man. As an example, let us examine the relationships among ranchers, grass, sheep, and coyotes in the western United States and

FIGURE 3.9 Alaskan brown bear (grizzly) with salmon. (Photo by Leonard Lee Rue, National Audubon Society.)

FIGURE 3.10 Mountain lion cornered by bounty hunters in Colorado. (Photo by Carl Iwasaki, *LIFE* Magazine.)

Canada. A coyote's diet includes many different items, among which are field mice, rabbits, moles, pack rats, sage hens, carrion of lambs that have died of natural causes, and freshly killed lambs. If the coyotes were exterminated, then the rodents would flourish, and since rodents eat grass, the quantity of forage available for the sheep would decrease. On the other hand, a spiraling coyote population would prey on range lambs and inflict financial losses on the ranchers. Thus, though coyote control is defensible, extermination is ecologically unsound. Even limited control programs have been questioned by some study groups, who contend that the cost of the anti-predator campaigns does not warrant the meager livestock savings that are realized. Unfortunately, mass poisoning programs are being advocated by many ranchers who do not appreciate the true complexities of ecosystems.

The coyote is not now an endangered species, so why use this particular example here? Why not, instead, focus our attention on other animals that used to roam the rangelands of the West, such as the grizzly bear and the wolf, but that are now on the brink of extinction? The reason is that in the case of the coyote there is still time to initiate a balanced program of control and thus to avoid a crash extermination campaign that would later be followed by a crash conservation program to save this species.

The Zoo and Pet Trade

Zoos and pet dealers have been castigated for their role in species extinctions, for the populations of

FIGURE 3.11 Mother orangutan and young. (From LIFE March 28, 1969, p. 83. Photographer, C. Rentmeester.)

many species cannot support the pressures caused by trapping. Orangutans, for instance, have been severely affected by this trade (Fig. 3.11). After orangutan populations were severely reduced by the intrusion of agriculture, their already small community was invaded by hunters and trappers. Since orangutans are too smart to be trapped efficiently, and since adults are difficult to ship anyway, hunters often shoot a nursing mother and take the infant. As the survival rate of infants without maternal care is only about 15 per cent, for each orangutan that reaches the market in western cities about a dozen have died.

Zookeepers around the world have recognized the danger and in recent years have taken positive steps to reverse the trend toward extinction of many species. Reputable zoos will no longer purchase animals belonging to endangered species except in rare cases when an animal that is being overhunted can be saved by maintaining a viable breeding stock in captivity. Pet trading in endangered species, however, continues even in some areas in which it is prohibited by law.

Miscellaneous Mechanisms of Extinction

Some paths to extinction are strange. Powdered horns of the Javanese rhinoceros (Fig. 3.12) are reputed to be powerful aphrodisiacs by many Asians,

FIGURE 3.12 Indian Rhino—close cousin of the Javan Rhino. (From *LIFE* Feb. 7, 1969, pg 68. Photographer, Eliot Elisofon.)

and a single horn is worth $2,000 in Bangkok. The medicine may or may not work, but the species will probably not survive another decade.

The Florida Key deer was almost eliminated just after World War II. Since then the tiny population of 25 individuals has increased to 250, and perhaps the species will be saved. But a new and serious danger to the deer has arisen. They seem to have developed a taste for the tobacco in the cigarette butts thrown from passing cars. Unfortunately, roadways are hazardous places to graze, and many Key Deer are killed in automobile accidents.

3.3 HOW SPECIES BECOME EXTINCT – THE CRITICAL LEVEL

(Problems 8, 9, 10)

The six major types of pressures discussed in the last section reduce the populations of a number of species, but rarely eliminate the last few surviving individuals. What factors lead to the final extinction of a species, or, in other words, how does an endangered species become an extinct species?

Let us consider the case of the passenger pigeon (Fig. 3.13). In the late 1800's, approximately two billion birds flew over the North American continent in flocks that blackened the skies. Commercial hunters, shooting indiscriminately into the flocks, killed millions for food and many more for fun, for the species was thought to be indestructible. As hunting pressures increased, the pigeon populations naturally suffered, and by the early 1900's market hunting was no longer profitable. Yet thousands of pigeons survived. Then suddenly they all vanished. Poof! Ducks, geese, doves, and swans had all been hunted and they survived in reduced numbers; why did the passenger pigeons succumb?

The naturalist Aldo Leopold could offer none of the conventional ecological arguments and said of the pigeon,

> The pigeon was a biological storm. He was the lightning that played between two opposing potentials of intolerable intensity: the fat of the land and the oxygen of the air. Yearly the feathered tempest roared up, down, and across the continent, sucking up the laden fruits of forest and prairie, burning them in a traveling blast of life. Like any other chain reaction, the pigeon could survive no diminution of his own furious intensity. When pigeoners subtracted from his numbers, and the pioneers chopped gaps in the continuity of his fuel, his flame guttered out with hardly a sputter or even a wisp of smoke.

From Aldo Leopold: *A Sand County Almanac.* New York, Sierra Club/Ballantine Books, 1970.

In modern biological language, we say that the pigeon population was reduced below its **critical level**, and thus could not survive even though many mating pairs remained alive. The concept of the critical level is a general one and applies to nuclear reactors as well as to pigeons. However, it is often difficult to determine accurately the critical level for a given system. To understand how critical levels operate and why they differ quantitatively from species to species, the following aspects must be considered:

1. When animals face certain types of stress, many of them fail to reproduce normally. This phenomenon is fairly common in the animal kingdom; an abnormally high percentage of American males became impotent during the Depression of the 1930's; female rabbits reabsorb unborn fetuses into the blood stream during drought; and many animals, such as the Javanese rhinoceros, have never bred successfully in captivity. We just don't know enough about behavior to predict how stress will affect fertility of a given animal, but we do know that in certain cases destruction of range and harassment by hunters alter the reproductive potential of the remaining stock. A sudden decrease in the reproductive rates in the face of mounting pressures would certainly lead to a rapid population decline.

2. Often the pressures which act on one species of an ecosystem do not affect the entire ecosystem equally, and severe imbalance can result which might lead to extinction. The original North American population of a few billion passenger pigeons must have supported a large population of various predators. When commercial hunters slaughtered a significant number of pigeons, they did not shoot a proportional number of predators, and it is possible that the pigeon was exterminated because the ratio of predator to prey was so unfavorable.

3. The population density may decline to a point where members of the opposite sex have a hard time finding each other.

4. When the population of a species becomes low, that population is particularly subject to bad luck, and a few unfortunate events can lead to extinction. The situation is analogous to that of a gambler who is playing a game that is rigged so that the odds are in his favor. If he has a large cash reserve and can play for a long time, he will win, but if he starts off with a small

FIGURE 3.13 The passenger pigeon—a lesson learned too late.

stake, a few unlucky hands in the beginning can end the game. So it is with species. Any large population living in its own natural surroundings would be expected to survive. However, if only a small breeding stock exists, chance occurrences might lead to extinction.

An example is the Steller's Albatross. Despite decimation by hunters, a few viable flocks survived and breeded. Then, in 1933 as one flock was nesting peacefully on an offshore island, a volcano erupted, killing most of the adults and all of the young. Again, in 1941, another volcano erupted and destroyed a second nesting population. The species survived for another 20 years and reproductivity had just begun to accelerate when the last sizable flock was caught in a typhoon and destroyed. The future of the species is in question.

Sometimes, however, chance favors survival. The sea otter was extensively hunted for fur from about 1750 to 1900 (Fig. 3.14). By 1910 the otter was believed to be extinct. No one knows how many individuals actually survived the fur trade, or where they lived, but around 1930 this "extinct" animal reappeared and established itself in several locations. Obviously, a small breeding stock had survived and increased in spite of two decades of hard times.

Biological luck is more complicated than the presence or absence of a typhoon, volcano, landslide, avalanche, or other natural disaster. When breeding populations are very low, then genetic misfortunes can be disastrous. For example, appearance of a rare deleterious mutation in one offspring of a large population will have no effect on the survival of the population. But if the breeding population is tiny, such a mutation may threaten continuation of the species simply because the tiny community can ill afford to lose even one potentially reproductive member.

5. **Inbreeding** poses another danger to the existence of reduced populations. In general, unions between closely related animals, such as brother to sister and cousin to cousin, reduce the vitality of a population. As we saw in the last chapter, many potentially lethal characteristics survive in a gene pool because they are recessive and normally do not influence the phenotype. However, the probability is high that relatives have many similar recessive genes, and inbreeding raises the chance that an offspring will

FIGURE 3.14 Sea otter surfacing with sea urchin (top) and abalone (bottom). (Courtesy of James A. Mattison, Jr., Salinas, California.)

be homozygous for at least some deleterious recessives (see Chapter 2). When union between non-relatives occurs, the situation differs. The male may be recessive for deleterious characteristics *a* and *b*, for example, while the unrelated female may have a different collection of recessives, say *c* and *d*. If this were the case, none of the offspring of this mating could be homozygous recessive for any of these traits, and thus none of the deleterious traits would be expressed in the phenotype. In sum, small populations promote inbreeding, and inbreeding increases the probability that a species will weaken.

3.4 TWO CASE HISTORIES OF ENDANGERED SPECIES

The road to extinction is different for each individual species, and no single factor is responsible for the extinction of any species. Two case histories will be given here to illustrate how several factors combine to endanger a species.

The largest flying bird alive today is the California condor (Fig. 3.15). Spreading nine-foot wings and gliding along in search of food, these big birds still grace the mountains of southern California, but their days appear to be numbered. They are scavengers that have traditionally fed on the carcasses of elk, deer, bear, or other inland game, and on fish and an occasional whale washed ashore by the sea. The development of west coast lands for ranching and farming excluded game animals from these areas and thus reduced food available for condors. Their habitat preempted, they were forced to move further back into the mountains. At this point it is interesting to compare the condor with another avian scavenger, the magpie. Magpies have fared well in the twentieth century and are often seen along highways feeding on dead dogs, cats, and rodents. A condor could easily displace a flock of magpies, and the carrion on the California plains and beaches could support a sizable condor population. But the big birds do not come. Perhaps they are frightened away by the noise of cars on the nearby highways; we do not know.

FIGURE 3.15 Condor. (From LIFE June 13, 1969, p. 74. Photographer, John Dominis.)

But perhaps it is just as well that condors do not share space with man because they have been, and continue to be, hunted extensively. Many ranchers believe, quite incorrectly, that condors are predators and kill live lambs or chickens. Some hunters find sport in killing these magnificent birds. Shooting condors has been illegal since 1880, but in nearly 100 years only once (in 1908) has someone been punished for breaking the law. Condors are also inadvertently trapped or poisoned by programs to exterminate coyotes.

There are perhaps 50 to 75 birds still alive. Rapid population expansion is impossible because the condor's reproductive rate as compared to that of other birds is very low. A female does not breed until she is five or six years old. Then she lays a single egg, which she incubates for 42 days. The newborn chick is totally dependent on *both* parents for seven months, and partially dependent on them for another seven. Only about eight condor eggs are laid per year in the entire world, and since pressures that caused the original decline in condor populations are still operative, death rates appear to be higher than survival rates.

Apparently, condors are endangered because they are large, magnificent, unwieldy, unadaptive, and slow to reproduce. But don't be too quick to draw conclusions. If instead we shift our attention to the bandicoots of Australia, New Zealand, and Tasmania, we find that they are small, rodent-like, quick, and relatively fecund, yet four species have become extinct in recent times and three more are seriously endangered (Fig. 3.16). What has happened? At some early stage in evolutionary history, the islands of Oceania separated from the rest of the world and contact was not reestablished until modern times. The bandicoots that evolved on their islands were well adjusted to the life there; they faced and survived predation from monitor lizards and aborigines and did well eating insects, grubs, and some vegetables. When Europeans arrived, they brought a whole arsenal of new enemies. Cats, dogs, and rats preyed upon the bandicoots and proved to be effective hunters in this previously isolated world. Then man himself began hunting them, for valuable furs and tasty meat could be harvested with relative ease. Finally, after introducing the rabbit to Australia, man then decided that the rabbit was undesirable, and initiated a massive trapping, poisoning, and shooting campaign. Many bandicoots were inadvertently killed during the rabbit slaughters, and this type of predation added enough extra pressure to existing populations to exterminate some species and endanger others.

FIGURE 3.16 A bandicoot. (From *International Wildlife Encyclopedia.* **Vol. 1. Dr. Maurice Burton and Robert Burton (eds.) New York, Marshall Cavendish Corp., p. 133.)**

3.5 CURRENT TRENDS IN SPECIES EXTINCTION

What characteristics do the endangered species have in common? Highly specialized or immobile animals and plants are particularly vulnerable to pressures from loss of habitat. As an example, the ivory-billed woodpecker feeds only on insects that live in the decaying wood of standing virgin timber in the southeastern United States. It does not eat the small parasites of second-growth stands, nor does it approach the ground to search for food in fallen debris. As the large trees in the bird's range have fallen before loggers' saws, the ivory-billed woodpecker population has shrunk and the animal is now believed to be extinct. Disappearance of this magnificent bird has attracted much attention, but it is probably only one of many species that failed to survive the destruction of the climax forests. Undoubtedly, many insects and mites were also specialized to live in the large old trees of the southern forests, and uncounted numbers of these, too, are probably lost.

It is difficult to know how many invertebrate species are becoming extinct, but the number must be large. As an example, a species of freshwater crustacean was discovered about 30 years ago living only in small ponds in Ohio. Biologists, wishing to restudy this animal in recent years, found the original ponds drained, and the crustacean gone.

Among more mobile and adaptable creatures, large animals seem to be particularly at risk for several reasons. First, their habitat is most easily destroyed. Compare two major grazers of the Great Plains of North America—the buffalo and the field mouse. The bison population that now exists in Wyoming and Montana is not allowed to expand because the pastures and wheatfields of the West are not compatible with wild herds. Because it is easy to fence them out, the buffalo have few places left to roam free. In contrast, mice which cannot be barred by conventional fences, still range throughout the plains.

But destruction of habitat is often more subtle than is illustrated by this example. The white-tailed deer population has increased in the last quarter century in the United States. This increase has been hailed as a great achievement of conservation programs. Game management statistics often fail to mention that at the same time the larger grazers, such as moose, elk, and buffalo populations, have been dramatically reduced. Deer have flourished partly because they can thrive in small woodlots or woodlot-pasture ecotones. In contrast, the larger animals need more land to support themselves, and each one needs a sizable range.

Secondly, hunting pressures are generally more severe against large animals than small, for large game animals are more desirable trophies. Also, relatively small grazers like deer usually flee from their non-human predators, while a mature moose or bison is strong enough to stand his ground against wolves or mountain lions. Therefore, when men with modern firearms reached North America, they found that large game animals were far easier prey than the elusive deer. Consequently, the strong have fallen in proportionally greater numbers than the quick.

Thirdly, small animals like rodents and insects usually have a higher reproductive rate than larger animals. For example, a female bank vole is independent of her parents at 2½ weeks, reaches sexual maturity in 4 to 5 weeks, and in one year can give birth to five litters of five infants each. Insects lay several thousand eggs a year. These species have evolved amidst powerful predators and appear to adapt well to the addition of one more. By contrast, the mighty creatures have traditionally bred more slowly and maintained population growth through maternal care or defense by the herd. These protective measures are not effective against the repeating rifle.

Voles include various genera of rodents related to rats and mice. Voles are generally stouter, have smaller eyes, limbs, and tails, and move less briskly than rats and mice.

Many *predator* species are seriously endangered for reasons which should be clear by now. Large predators have traditionally known almost no external enemies and are even more poorly adapted to predation than are large herbivores. In addition, their niches are on the top of the food chain and they often need a large hunting range. Thus, as destruction of their habitat constricts their natural range so that the wild lands can no longer support them, the hunting animals often roam toward farmers' fields and prey on cattle. As a result, they are hunted extensively in extermination programs. Also, because environmental poisons tend to concentrate at the top of food chains, predators are often victims of pesticides. This phenomenon will be discussed in Chapter 5.

3.6 THE DISPLACEMENT OF SPECIES—EFFECTS ON ECOSYSTEMS

Although many species have not been able to compete successfully with the pressures of the twentieth century, many others have proliferated, expanding both in numbers and in range. Ecosystems have traditionally been separated from each other by barriers such as rivers, waterfalls, deserts, mountain ranges, and oceans. In recent times, however, bridges, canals, as well as land, air, and ocean trade have provided pathways through the natural barriers. Thousands of species have traveled along new routes, or have "hitchhiked" with cars, trains, planes, and boats to invade foreign ecosystems. At one extreme, many organisms have been unable to compete in foreign lands and migrations have been unsuccessful. At the other extreme, organisms such as the sea lamprey have succeeded so well in their new home that they have disrupted the ecological balance and have endangered other species.

Successful ecological invasions also present direct problems to man. In many cases, invaders quickly become pests and are responsible for large financial losses. The sea lamprey (Fig. 3.17) virtually eliminated a lake trout industry of 8.5 million pounds per year; the Japanese beetle, imported from the Orient, feeds on many crops, such as soybeans, clover, apples, and peaches; the American vine aphid, imported to France from the United States, was responsible for destroying three million acres of French

FIGURE 3.17 Young sea lampreys, *Petromyzon marinus*, attacking brook trout in an aquarium. (From Lennon, R. E.: "Feeding Mechanism of the Sea Lamprey and its Effect on Host Fishes." Fish Bulletin; U.S. Fish and Wildlife Service 56 #98; 247–293, 1954.)

vineyards. In fact, over half of the major insect pests in many areas are imports.

In addition to such direct disturbances, the invasion of species is responsible for more subtle perturbations. A few hundred years ago, individual continents possessed their own unique types of ecosystems. Thus, for example, temperate grasslands in America, Asia, and Australia all contain different groups of organisms. With the rapid immigration of some species and the destruction of others, the differences between ecosystems are becoming smaller.

3.7 THE NATURE OF THE LOSS

(Problem 3)

What will happen to the biosphere if these current trends continue? Ecosystems and food webs will become simplified, but the pattern of primary, secondary, and tertiary consumption will continue. If the deer becomes the largest grazer and the coyote becomes the largest non-human predator, grazing and predation will continue, so why worry about extinction?

There are several reasons to worry. In some ways the most compelling is the aesthetic and religious argument. Different individuals may express their feelings in different ways. To some, species and the wilderness should be preserved simply because they exist. Others might reformulate this in saying that man has no right to exterminate what God has created. To still others, an unobtrusive passage through an untouched wilderness area is a source of enormous aesthetic gratification, as valid and moving an experience as a great play or string quartet. As many of the greatest and noblest creatures of the earth fall prey to man's thoughtless acts, so too, the richness, variety, and fascination of life on this planet diminish with their passing.

A second reason is concerned with developments in medicine and the biological sciences, which have always been dependent on various plants and animals as experimental subjects. The range of species used in different experiments has traditionally been very wide. For example, when Thomas Hunt Morgan initiated his famous studies of genetics, he needed an animal that was easy to breed in large numbers and had few, large, and accessible chromosomes. He chose the fruit fly, *drosophila*, not because the life style of these insects was of particular interest in itself, but because the animals were easy to study and

Thomas Hunt Morgan (1866–1945) was an American geneticist whose experiments formed the basis for his theory that paired elements or "genes" within the chromosome are responsible for the inherited characteristics of the individual. He won the Nobel prize in medicine in 1933.

the lessons learned from them could be generalized to studies of other creatures. Another serious scientific loss in this category is that of wild populations of plants and animals that have traditionally been used in breeding hybrid species for agriculture. For example, domestic corns are often susceptible to various diseases. Geneticists have periodically crossed high-yielding susceptible corns with hardy wild maize in an effort to develop high-yielding resistant corn. However, natural maize growing along fence lines and roadways is often considered to be a weed and has been combated across North America with herbicides. The loss of this species would be a severe blow to modern agriculture.

A third example of species destruction that may prove detrimental is occurring among the herbivores on the African savannahs. A simplified picture of the savannah ecosystem, with its migrating species was discussed in Chapter 1, page 27. In many places natural grazers are being displaced to make way for cattle ranches. But cows are unable to utilize grass as efficiently as the heterogeneous African herds. Domestic stocks tend to overgraze certain species of plants selectively, thus upsetting the water table, and in addition they are susceptible to many tropical diseases, especially sleeping sickness. The result is that an untouched savannah is capable of an annual production of 24 to 37 tons of meat per square kilometer in the form of wild animals, while the best pasture-

Impalas in Africa. (In Vaughan, T. A.: Mammalogy, 1972, W. B. Saunders Co., p. 249. Courtesy of W. Leslie Robinette.)

cattle systems in Africa can yield only eight tons of beef per square kilometer per year. Yet in the name of agricultural progress, many ungulates are being threatened with extinction, and other herd sizes are being substantially reduced.

The final reason concerns the possible long-term effects of worldwide ecosystem alteration, which were mentioned on page 42. When 95 per cent of all large mammals died off in North America 8,000 to 10,000 years ago, climax plant systems and the human race survived intact. Does this mean that another wave of extinctions could occur without catastrophic side effects?

If the study of ecology and the environment teaches us only one lesson, it is that perturbations at one end of a delicately balanced web of interrelationships may shake the whole structure. In Chapter 2 we noted that the cell's metabolic machinery works so well precisely because the enzymes and structural proteins that constitute it have evolved together and are beautifully coadapted; indeed, this is the reason a mutation, a sudden change in only one of the components of the "well-oiled machine," is far more likely to be deleterious than beneficial. And so it is with ecosystems. The component parts here, of course, are the species, both plant and animal, which have also evolved coordinately. It is notoriously difficult to predict with any assurance what will happen to a given ecosystem when one or more of its species are extinguished; the system is far too complex and poorly understood. But it is a foolish person indeed who participates in, supports, condones, or ignores the needless destruction of species while maintaining steadfastly that no ill effects will supervene because none is obvious at present. For just as events at the cellular level generally require short times to become observable or measurable, events at the level of the ecosystem may take years, centuries, or millennia. What we have inherited from the evolutionary process is surely likely to be far more stable and rich than what we are likely to produce with aimless, hit-or-miss destruction.

PROBLEMS

1. **Evolution.** Outline briefly the types of environmental pressures to which an animal must adapt. Using this outline as a guide, discuss Darwin's concept of "survival of the fittest."

2. **Speciation.** Many zoos are finding it advantageous to breed specimens in captivity. If this practice continues, do you feel that there is a possibility that separate "zoo species" will evolve?

3. **Extinction of species.** Characterize each of the following as development, replacement, or extinction that leaves an ecological void: (a) the extinction of the Tasmanian wolf (a kangaroo-like carnivore) following the introduction of dogs into New Zealand and Tasmania; (b) the appearance of dark-colored moths and the disappearance of light-colored moths in England, which was discussed in Chapter 2, page 69; (c) the disappearance of the California grizzly bear following the arrival of the white man.

4. **Pleistocene extinctions.** Discuss the role of man in the Pleistocene extinctions. Do you feel that man, alone, caused these extinctions? Defend your answer.

5. **Pleistocene extinctions.** If the sabre-toothed tiger could be reintroduced into North America today, do you think that it would survive? Defend your answer.

6. **Habitat destruction.** List the major factors which lead to habitat destruction in the modern world. What factors would you list as being most disruptive to wild animals? Explain.

7. **Predators and man.** Considering the grizzly bears' record, do you feel that it is safe to feed park bears despite regulations forbidding it?

8. **Critical level.** Define critical level. Give two examples.

9. **Critical level.** Discuss how the critical level might differ greatly from species to species. In your answer, briefly explain how each of the following factors could affect different species differently: (a) reproductive failure, (b) ecosystem imbalance, (c) ability of males to find females, (d) luck, and (e) inbreeding.

10. **Effects of inbreeding.** Construct a test cross to show why a recessive characteristic can affect the phenotype only if both mother and father are recessive for the same trait. What is the probability that a recessive will appear in an individual if both parents carry it?

11. **Survival in the twentieth century.** Female polar bears give birth to one or two infants every three years. Discuss the survival potential of this low reproductive rate in past ages and during the present.

12. **Survival in the twentieth century.** Consider the following fictitious species and comment on the survival potential of each in the twentieth century: (a) A

mouse-sized rodent that gives birth to 40 young per year and cares for them well. This creature burrows deeply and eats the roots of mature hickory trees as its staple food. (b) An omnivore about the size of a pinhead. Females lay 100,000 eggs per year. This animal had evolved in a certain tropical area and can survive only in air temperatures ranging from 80° to 100° F (27° to 38° C). (c) A herbivore about twice the size of a cow adapted to northern temperate climates. This animal can either graze in open fields or browse in forests. It is a powerful jumper and can clear a 15-foot fence. Females give birth to twins every spring.

13. **Justification for the preservation of species.** Discuss the importance of wild grasses in the modern world.

The following problem requires arithmetical computation:

14. **Zoo and pet trade.** On page 92 it is stated that if orangutans are obtained for zoos by shooting nursing mothers and taking the infants, of which only 15 per cent survive, then a dozen orangutans have died for each one that reaches the market. Check this arithmetic.

BIBLIOGRAPHY

Four books which deal specifically with the destruction of animal species are listed below:

Roger A. Caras: *Last Chance on Earth.* New York, Schocken Books, 1972. 207 pp.

Kai Curry-Lindahl: *Let Them Live.* New York, William Morrow & Co., 1972. 394 pp.

H. R. H. Prince Philip, Duke of Edinburgh, and James Fisher: *Wildlife Crisis.* Chicago, Cowles Book Company, 1970. 256 pp.

Vinzenz Ziswiler: *Extinct and Vanishing Animals.* New York, Springer-Verlag, 1967.

Two valuable books which deal more specifically with ecosystem alterations are:

David W. Ehrenfeld: *Biological Conservation.* New York, Holt, Rinehart & Winston, 1970. 226 pp.

Charles Elton: *The Ecology of Invasions by Animals and Plants.* New York, John Wiley & Sons, 1958. 181 pp.

A detailed, fascinating, and advanced discussion of the Pleistocene extinctions is given in:

Paul S. Martin and H. E. Wright, Jr., eds.: *Pleistocene Extinctions.* New Haven, Yale University Press, 1967. 453 pp.

4

THE GROWTH OF HUMAN POPULATIONS

4.1 INTRODUCTION
(Problems 3, 4)

The 1970's promise to be a decade of unprecedented population growth. Not only will each year bring an increase in the total number of people on Earth, and not only is the rate of growth expected to become higher each year, but even the rate of the rate, the acceleration, will probably continue its current rapid ascent. Our forefathers did not live in times of such rapid change. Estimates of world populations before the twentieth century are very approximate. (Indeed, even current population data are sketchy for most of Africa, for all of China, and for various other areas.) Anthropological evidence suggests that modern man evolved one hundred thousand years ago. During prehistoric times, the total human population of the Earth must have fluctuated widely. In some years there were more deaths than births, causing human population to decrease temporarily. By the first century A.D., however, population had already established its present pattern of almost uninterrupted growth. Rates of growth were then very slow and extremely variable. The Black Plague of the fourteenth century is believed to have caused a temporary decrease in world population, while the European Renaissance marked the beginning of a rapid rise in world population. At the time of the discovery of America, there were about one-quarter billion people alive on Earth. In 1650, about a century and a half later, world population had doubled to one-half billion. In another 300 years, world population multiplied fivefold to 2.5 billion persons. During the 1950's, the population increased almost another one-half billion and, by 1970, world population was approximately 3.5 billion persons. In other words, the *increase* in world population from 1950 to 1970 was about twice the *size* of world population in 1650. Or consider another comparison. Today there are more people in China than there were people on Earth in 1650. Or yet another: two-thirds of all people born since 1500 are alive today.

Figure 4.1 presents a schematic graph of world population size since the emergence of modern man. Note that the smoothness of the curve reflects ignorance of details of population data rather than regularity of population increase. It is clear that the curve of growth is becoming steeper and steeper in time. In fact, a glance at Figure 4.1 may well cause you to panic. If the population continues to grow ever more steeply, or even if it continues to grow at its current rate, very soon there will be too many people for Earth to support. If there are poverty and starvation now, how can economic and agricultural development be expected to keep pace with an exploding population? Destruction of land, depletion of natural resources, production of waste, and pollution of the Earth can all be expected to increase with increased population. You have probably read dire predictions based on extrapolation of the growth curve of Figure 4.1. One projection which has gained a certain cur-

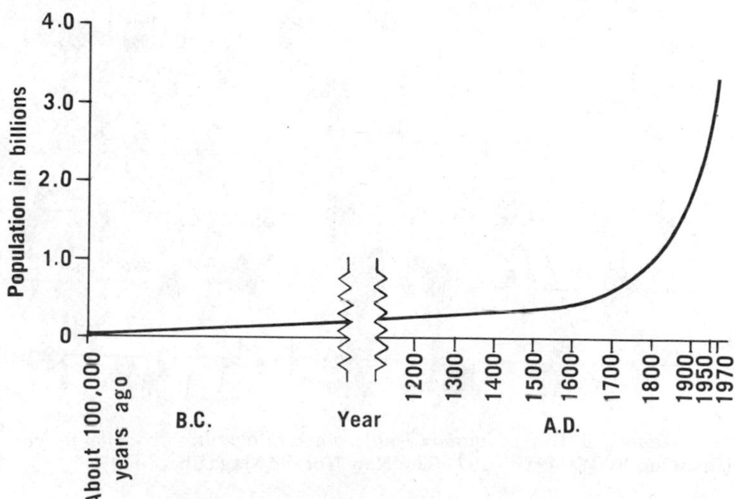

FIGURE 4.1 World population size from emergence of *Homo sapiens* **to 1970.**

rency in both lay magazines and scientific journals says that if the present rate of world population increase were to continue, there would be one person for every square foot of the Earth's surface in less than 700 years. If we accept the premise, the conclusion is necessary. But we know that the conclusion must be false. It is impossible for men to stand elbow to elbow on this planet. One square foot of Earth's surface cannot feed, clothe, and shelter a person. Thus the premise must be untenable. In other words, human population cannot continue to grow at current rates indefinitely. Indeed, similar reasoning should convince you that *population size cannot grow forever,* even very slowly. It will be checked by such factors as limits of space and food, by explicit

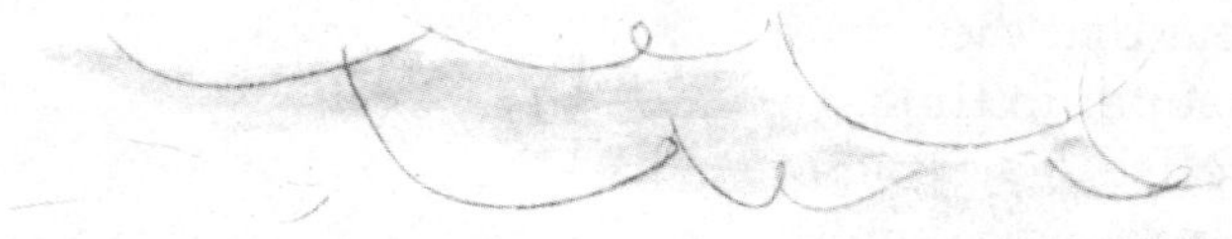

"Excuse me, sir. I am prepared to make you a rather attractive offer for your square."
(Drawing by Weber; © 1971 The New Yorker Magazine, Inc.)

decisions of families and nations, by famine and diseases, and by complicated interrelated social forces.

Clearly, though, it is of great importance to know where the curve in Figure 4.1 is going. One reason we need such estimates is to plan for the future. How much food must be produced during the next decades? How many schools should be built? Where should roads be placed? Parks? Power plants? Without some methods for projecting future population growth, planners would have insurmountable difficulties. Lest these questions appear to imply that population growth acts as an inexorable and independent force, we must emphasize that social and economic factors help determine population size just as population size is one determinant of social and economic situations. For example, a well-educated and well-fed society typically tends to grow slowly, for people are both aware of population control and motivated to practice it. Conversely, a society that grows slowly is more likely to be well fed. These are not firm rules. After World War II, for instance, American population growth was quite rapid.

Thomas Robert Malthus (1766–1834) was an English economist best known for his *Essay on Population,* **published originally in 1798 and republished in 1803, considerably revised and enlarged. In his first edition, Malthus asserted that war, famine, pestilence, misery, and vice prevent population from increasing beyond the limits of subsistence. In the second and later editions of his work, he suggested that "moral restraint" (postponement of marriage and strict sexual continence) acts as a further check on population growth. Throughout his life, Malthus remained pessimistic in his assessment of the future progress of mankind.**

A second reason for knowing how to predict population growth is of a political nature. If we can arrive at a reasonably accurate estimate of population size at some future date, and we can show it to be too large to be consistent with societal well being, we have a numerical weapon with which to fight for the implementation of population control measures. In order to evaluate for yourself the validity of statements, predictions, and proposals concerning population size, you must understand the mechanism of population growth and the terms used to express it.

4.2 EXTRAPOLATION OF POPULATION GROWTH CURVES

(Problems 2, 5, 6)

We noted in the previous section that the most obvious method for predicting population growth is to construct a graph that plots population size against time and to guess how the curve will continue. Guessing points on a curve outside the range of observation is called **extrapolation.** Extrapolation is a subtle art. Look at Figure 4.1. If you didn't know the labels of the axes and were asked to continue the curve, what would you do? Such an exercise is frivolous. One person might think the curve will continue to go up indefinitely; another might try to draw a curve that peaks and then falls below zero. Still

TABLE 4.1 COMPARISON BETWEEN GEOMETRIC AND ARITHMETIC GROWTH

YEAR	GEOMETRIC GROWTH
0	10 people
1	$2 \times 10 = 20$
2	$2^2 \times 10 = 40$
3	$2^3 \times 10 = 80$
4	$2^4 \times 10 = 160$
.	.
.	.
.	.

YEAR	ARITHMETIC GROWTH
0	100 pounds
1	$(2 \times 1) + 100 = 102$
2	$(2 \times 2) + 100 = 104$
3	$(2 \times 3) + 100 = 106$
4	$(2 \times 4) + 100 = 108$
.	.
.	.
.	.

another might finish it by continuing up for a while and then leveling off. And someone else might think that the curve will become wiggly and erratic. If, however, you knew you were graphing population, your historical knowledge would allow you to exclude several kinds of curves. Infinite and negative populations would be impossible, a zero population highly improbable for many years, and wide fluctuations unlikely. Exclusions, however, wouldn't construct your curve. On the other hand, if you had a theory of population growth you would have a basis for extrapolation.

A model for a mechanism of population growth was introduced in 1798 by the Reverend Thomas Robert Malthus. He noted that population, when unchecked, grew at a **geometric rate** of increase. For example, if there were x people in year 0 and ax in year 1 (where a is greater than 1), there would be $a(ax)$, or a^2x, in year 2, a^3x in year 3, and a^nx in year n. However, Malthus said that food supplies increase at an **arithmetic rate:** that is, if there were y pounds of food in year 0, there would be $y + a$ in year 1, $y + 2a$ in year 2, and $y + na$ in year n. Table 4.1 presents examples of both types of growth. In Figure 4.2A both types of growth curve are depicted graphically. Geometric growth is eventually much faster than arithmetic. Malthus therefore predicted that any uncontrolled population would eventually grow too large for its food supply. He continued:

> By that law of our nature which makes food necessary to the life of man, the effects of these two unequal powers must be kept equal.
>
> This implies a strong and constantly operating check on population from the difficulty of subsistence. This difficulty must fall somewhere and must necessarily be severely felt by a large portion of mankind.
>
> . . . The race of plants and the race of animals shrink under this great restrictive law. And the race of man cannot, by any efforts of reason, escape from it. Among plants and animals its effects are waste of seed, sickness, and premature death. Among mankind, misery and vice. The former, misery, is an absolutely necessary consequence of it. Vice is a highly probable consequence, and we therefore see it abundantly prevail, but it ought not, perhaps, to be called an absolutely necessary consequence. The ordeal of virtue is to resist all temptation to evil.

Thomas Robert Malthus: *An Essay on The Principle of Population.* Published originally in 1798. Republished in 1970 by Penguin Books Ltd., Harmondsworth, Middlesex, England, pp. 71–72.

The premises on which Malthusian theory is based have not proved true historically. First, because of rapid improvements in agricultural techniques, man's food supply has increased much faster than arithmetically; second, geometric growth curves do not adequately describe human population increases. However, the core of the Malthusian argument cannot be ignored, for there are limits to the number of persons Earth can support, and unless growth is checked rationally, something akin to the Malthusian's "misery and vice" *will* afflict mankind.

The inadequacy of geometric curves in predicting future population sizes has led to the search for other graphical representations of growth. Observing that population sizes of all species must have limits, ecologists often describe the growth of populations in the following terms. Consider some population which is initially very small. The very fact that it is small places it in danger of extinction, because it may not be able to recover from such setbacks as epidemics, famine, or poor breeding. Even if such factors do not totally destroy the population, they will limit its growth rate, and therefore the population will increase only slowly at first. However, once the population is established, its size will rise more rapidly as long as there are adequate food sources, relatively few predators and favorable living conditions. When the population becomes large with respect both to food supply and vulnerability to predators, and so dense that disease spreads rapidly, the rate of growth decreases. The whole curve of growth looks like an S and is called "**sigmoid**," or S-like. Figure 4.2*B* shows the shape of a sigmoid curve. However, the sigmoid curve, though useful for some species, does not describe all populations. Certainly, populations which become extinct exhibit a very different growth curve (see Fig. 4.2*C*). Sizes of many natural populations exhibit fluctuations seasonally over periods of many years.

Figure 4.3 plots population growth curves for mice and snowshoe hares. The data for the mice demonstrates a growth curve of a population declining to extinction; snowshoe hares exhibit a cyclical pattern of population growth. For another example of cyclical growth, refer back to the discussion of lemmings in Chapter 1, page 39.

Consider again Figure 4.1, the graph of total population size since man's arrival on earth. Even if we all agree that the curve will eventually become sigmoid, examination of Figure 4.1 gives no clue as to when the

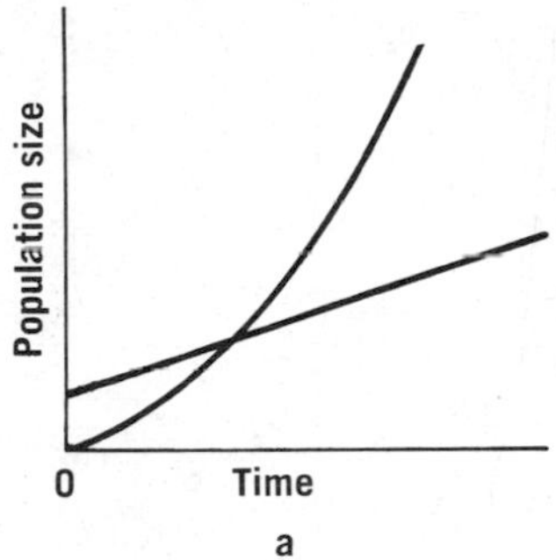

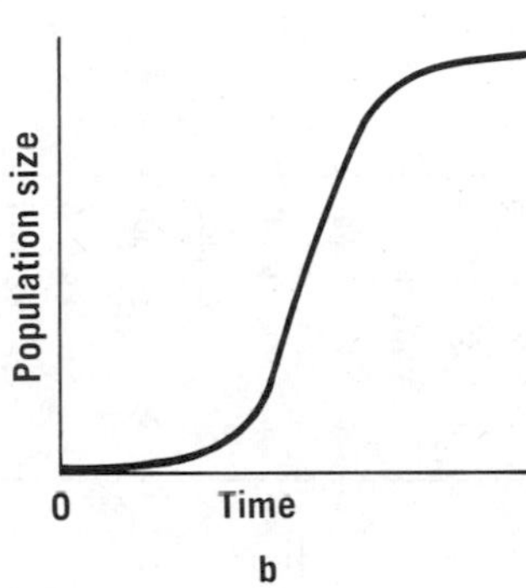

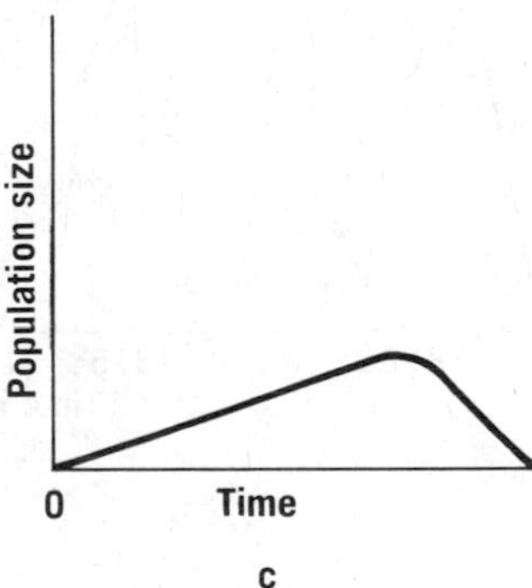

FIGURE 4.2 Schematic growth curves. *A.* **Arithmetic (straight-line) and geometric (curved-line) patterns of growth.** *B.* **Sigmoid curve of growth.** *C.* **Growth curve of a population that becomes extinct.**

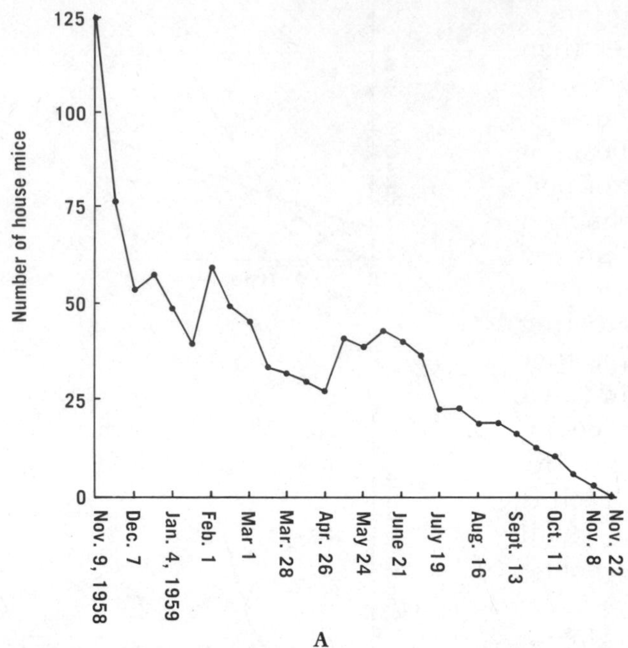

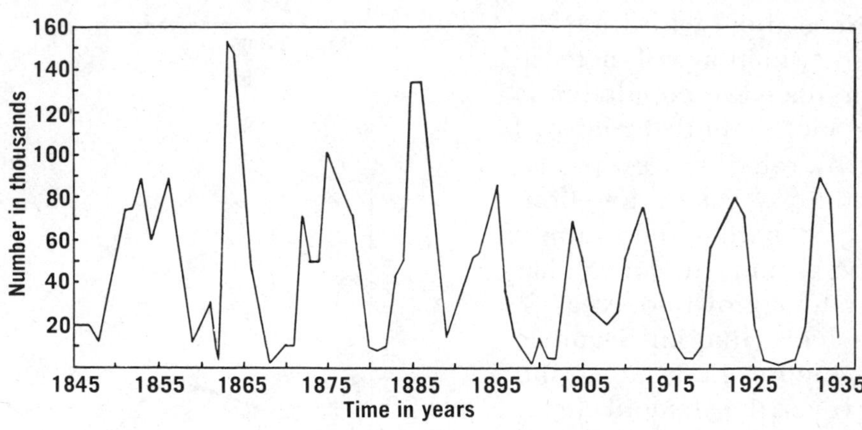

FIGURE 4.3 *A.* **Decline of a population of feral house mice on Brooks Island in San Francisco Bay. (Data from William Z. Lidicker, Jr.: Ecological observations on a feral house mouse population declining to extinction, Ecological Monographs** *36*:**27–50, 1966.)** *B.* **Changes in abundance of the snowshoe hare as indicated by the number of pelts received by the Hudson Bay Company. This example exhibits a 9- to 10-year cyclical oscillation in population size. (From Odum:** *Fundamentals of Ecology.* **3rd Ed. Philadelphia, W. B. Saunders Co., 1971.)**

top of the S will begin to form. Indeed, if we had looked at the curve depicting growth from the beginning of man until 1600, we would have had no idea when the population would start to increase rapidly. A much more reliable method of predicting population size is to look not at changes in total population size with time, but at patterns of changes in *rates of growth.* For this purpose, the reader needs to understand what rates of growth are and how birth and death rates affect the size of populations. He must be able to see the relationship between current and future growth and decline. These concepts are introduced in the following sections.

4.3 AN INTRODUCTION TO DEMOGRAPHY

In order to investigate the population dynamics, or the **population ecology**, of a species of animal or plant, the ecologist studies the geography of the region of interest, the food supplies, climatic fluctuations, predation by other species, and competition between and within species. For human populations, however, social and economic forces are more important determinants of population growth. Man has no important predators except, of course, other men, for intraspecies competition in the form of exploitation, subjugation, and war is at least as old as recorded history. If an animal species grows too plentiful for its habitat, its members often starve; human overpopulation and consequent food shortages, however, more frequently result in emigration, or importation of food, or even the development of more effective modes of agriculture. Because of the uniqueness of human patterns of growth, we shall consider it in demographic, rather than ecological, terms.

(Problems 1, 7, 8, 9, 10)

Demography is that branch of sociology or anthropology which deals with the statistical characteristics of human populations, with reference to total size, density, number of deaths, diseases and migrations, and so forth. The demographer attempts to construct a numerical profile of the population he is studying. He views populations as groups of people, but is not professionally concerned with what happens to any given individual. He wants to know facts concerning the size and composition of populations. For instance, he may want to know the number of males in a population or the number of infants born in a given year.

The subject may sound terribly dry to those of you who are uncomfortable with numbers and computations, but we emphasize that demographic data often reflect in fascinating ways the history of the country studied, the trends in medical care, and the occurrence of social changes.

In addition to studying the composition of populations, the demographer is interested in how populations change in time. He studies changes by counting the number of **vital events**—the births, deaths, marriages, and migrations. If he knew the composition of a population at any given time and the number of vital events occurring between that time and another, he would know the composition of the population at the end of the period. For example, suppose that in 1924 the population of some village were 732. In the next two years, if there were 28 births and 15 deaths and if

The word "vital" is from the Latin *vita*, meaning life. Vital means "pertaining to life" or "necessary for life," hence, "basic" or, even more loosely, "very important." When the term "vital" is used in "vital rates" or "vital statistics," the purpose is not to emphasize that the data are important, but rather that they pertain to lives.

four people moved in and one moved out, there would be $732 + 28 - 15 + 4 - 1 = 748$ people at the end of 1926.

Demographers often study **vital rates,** the number of vital events occurring to a population during a specified period of time divided by the size of the population. For example, the 1968 birth rate for the United States is the number of live births in 1968 divided by the midyear population in 1968. (Since the population itself changes during a given year, the population size is usually defined to be the number of people alive at midyear, June 30.)

Before we consider demographic rates, we must understand how a population grows. If we have \$100.00 in the bank accumulating 3 per cent annual interest, we know at the end of the year there will be \$103.00. We can think of the 3 per cent operating on each dollar within the population of dollars in the bank. Therefore, each individual dollar grows to \$1.03, or we can say that each dollar is growing at an annual rate of 3 per cent. Taking one dollar out of the bank will not affect the rate of growth of the remainder.

But what does it mean to say that a population is growing at 3 per cent? Certainly, if we know that 100 people were alive on January 1 and 103 were alive the following December 31, the annual growth rate has been:

$$\frac{(103 - 100)}{100} \times 100\% = 3\%.$$

However, if we thought of the 3 percent operating on each person, we would have to think of each person becoming 1.03 people. What does that mean? A 3 per cent growth rate means only that for every 100 people there were three more births and immigrations than deaths and emigrations. Even in very simple cases, removal of one person from the population will affect the rate of growth of the remainder. Suppose we have 100 people on January 1, and during the year there are no migrations, five births, and two deaths of people over 65. There would be 103 people on December 31, a 3 per cent increase. Now, suppose we looked at only 99 people, omitting one infant. There would still be five births and two deaths, because births do not occur to infants. So there would be $99 + 5 - 2 = 102$ persons at the end of the year. The annual rate of growth would be:

$$\frac{102 - 99}{99} \times 100\% = 3.03\%.$$

On the other hand, suppose we looked at the rate of growth of the population excluding, instead of an infant, one of the women who had a child. Now we would have four births occurring to 99 people and the rate of growth would be:

$$\frac{101 - 99}{99} \times 100\% = 2.02\%.$$

Finally, omission of one death would have produced a population of size 99 + 5 − 1 = 103, with a rate of growth of:

$$\frac{103 - 99}{100} \times 100\% = 4.04\%.$$

This very simple example has pointed out some of the difficulties confronting the student of population size, but it also leads to three important insights that are necessary for an effective approach to the investigation of growth:

1. An overall "rate of growth" is really the difference between a rate of addition (by birth or immigration) and a rate of subtraction (by death or emigration). The rate of growth is positive only when there are more additions than subtractions.

2. The probability of dying or of giving birth within any given year varies with age and sex.

3. The age-sex composition, or **distribution,** of the population has a profound effect upon a country's birth rate, its death rate, and hence its growth rate.

Three very basic measures of population growth are the **crude birth rate,** the **crude death rate,** and the **rate of natural increase.** For any geographical area, or ethnic group, being studied, we define:

$$\text{(a) Crude birth rate per thousand during year } X = \frac{\text{Number of live children born during year } X}{\text{Midyear population in year } X} \times 1000$$

and

$$\text{(b) Crude death rate per thousand during year } X = \frac{\text{Number of deaths during year } X}{\text{Midyear population in year } X} \times 1000.$$

Since the denominators of the birth and death rates are identical, we can define:

(c) Rate of natural increase per thousand during year X = Crude birth rate per thousand − Crude death rate per thousand

$$= \frac{\text{Number of live children born during year } X - \text{Number of deaths in year } X}{\text{Midyear population in year } X} \times 1000.$$

Another name for the rate of natural increase is the **crude reproductive rate.**

For simplicity, we shall assume that migration rates are negligible and therefore we shall not discuss data from Israel, Australia, or other countries where the assumption is patently invalid. Table 4.2 presents some data on international migration.

Patterns of migration do not greatly affect growth except when a large proportion of the population migrates, especially when the migrants are predominantly of one sex. For example, in the last century many Irish young men emigrated, leaving a surplus of

TABLE 4.2 LONG-TERM INTERNATIONAL MIGRATION DATA FROM SELECTED COUNTRIES*

CONTINENT AND COUNTRY	YEAR	TOTAL NUMBER OF IMMIGRANTS† IN THOUSANDS	RATIO OF IMMIGRANTS TO BIRTHS	TOTAL NUMBER OF EMIGRANTS‡ IN THOUSANDS	RATIO OF EMIGRANTS TO DEATHS
America, North					
Canada	1969	162	0.4	?	?
Mexico	1969	84	0.04	63	0.1
Panama	1969	0.9	0.02	?	?
United States	1971	370	0.10	?	?
America, South					
Colombia	1968	3	0.004	5	0.03
Venezuela	1967	0.3	0.0008	2	0.03
Asia					
Israel	1969	33	0.5	7	0.4
Japan	1969	756	0.4	747	1.1
Europe					
Bulgaria	1969	0	0	0.2	0.002
Czechoslovakia	1967	4	0.02	14	0.1
France	1968	93	0.1	198	0.4
Germany (West)	1969	701	0.8	440	0.6
Netherlands	1969	67	0.3	45	0.4
United Kingdom	1969	206	0.2	293	0.4
Oceania					
Australia	1969	249	1.0	108	1.0

*Data from *U.N. Demographic Yearbook* (1970) and *Statistical Abstract of the United States* (1972).

†Long-term immigrants are defined as persons entering a country with the intention of remaining at least one year.

‡Long-term emigrants are persons leaving a country with the intention of not returning for at least one year.

women behind. Consequently there were many childless women. Conversely, at the beginning of this century, there were many more men than women in the United States. The differential can be attributed in large part to the fact that the heavy immigration of the period was male-dominated. In general, groups of migrants are usually quite different with respect to age, sex, and vital rates from the inhabitants of the country either from which they leave or to which they go. In particular, migrants are often healthy males of reproductive age.

4.4 MEASUREMENTS OF MORTALITY

(Problems 11, 12, 13, 14, 33)

For long-term assessment of historical trends, the three crude rates introduced in the previous section are concise, useful, and graphic. Figure 4.4 plots birth and death rates for the developed and the underdeveloped parts of the world. The United Nations defines the developed regions as Europe, the United States, the Soviet Union, Japan, temperate South America, Australia, and New Zealand. The rest of the world is called underdeveloped. (See the map on pages 118–119.)

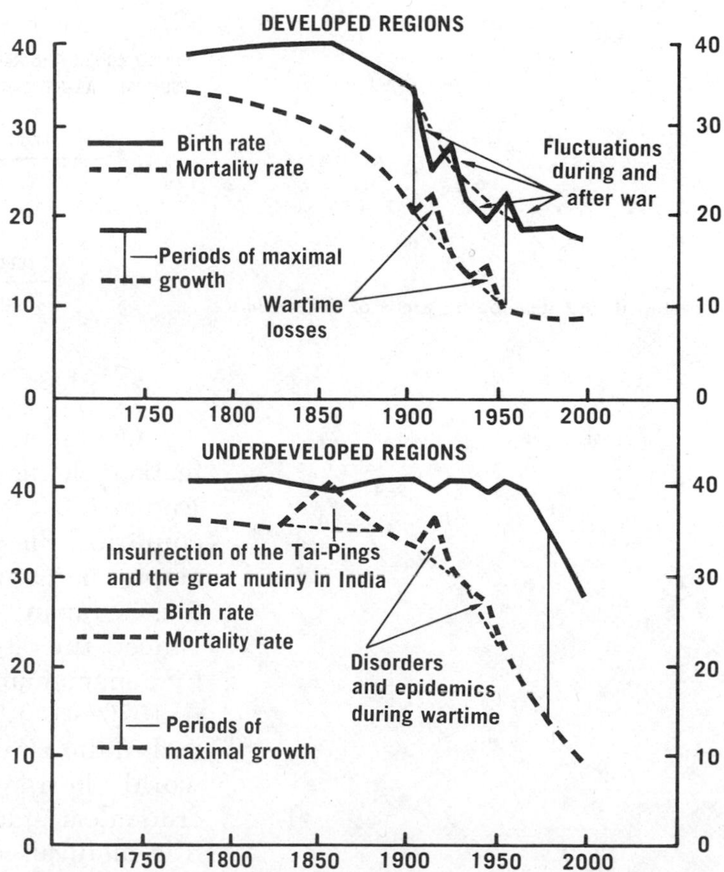

FIGURE 4.4 Estimated and predicted crude birth rates and crude death rates for the period 1750–2000 in the developed and underdeveloped regions of the world. (Adapted from *The Demographic Situation of the World in 1970,* **United Nations, 1971.)**

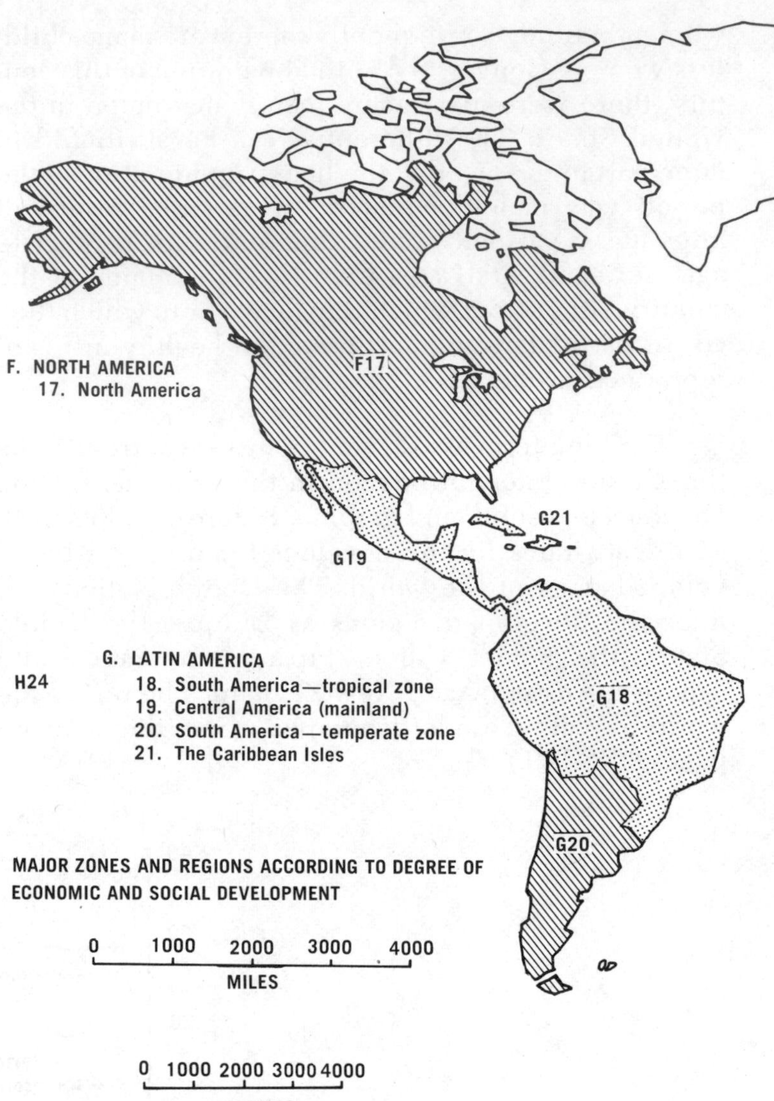

Developed and developing areas of the world.

Overall trends in birth and death rates, and the marked changes in times of war are clearly depicted in Figure 4.4. Yet for current data, analyzed by individual countries, these rates can be very deceptive. For example, in Taiwan, in 1970, the crude death rate was 5.1; in Japan, which is much more economically advanced, the rate was 6.9. In the Canal Zone, populated by Panamanians and Americans, the crude death rate in 1970 was 1.7; the rate in Panama itself was about 8 and in the United States 9.4. Sweden, with one of the world's best systems for delivering health care, had a crude death rate of 9.9 in 1970, and in Scotland, where free medical care is available to all, the rate in 1970

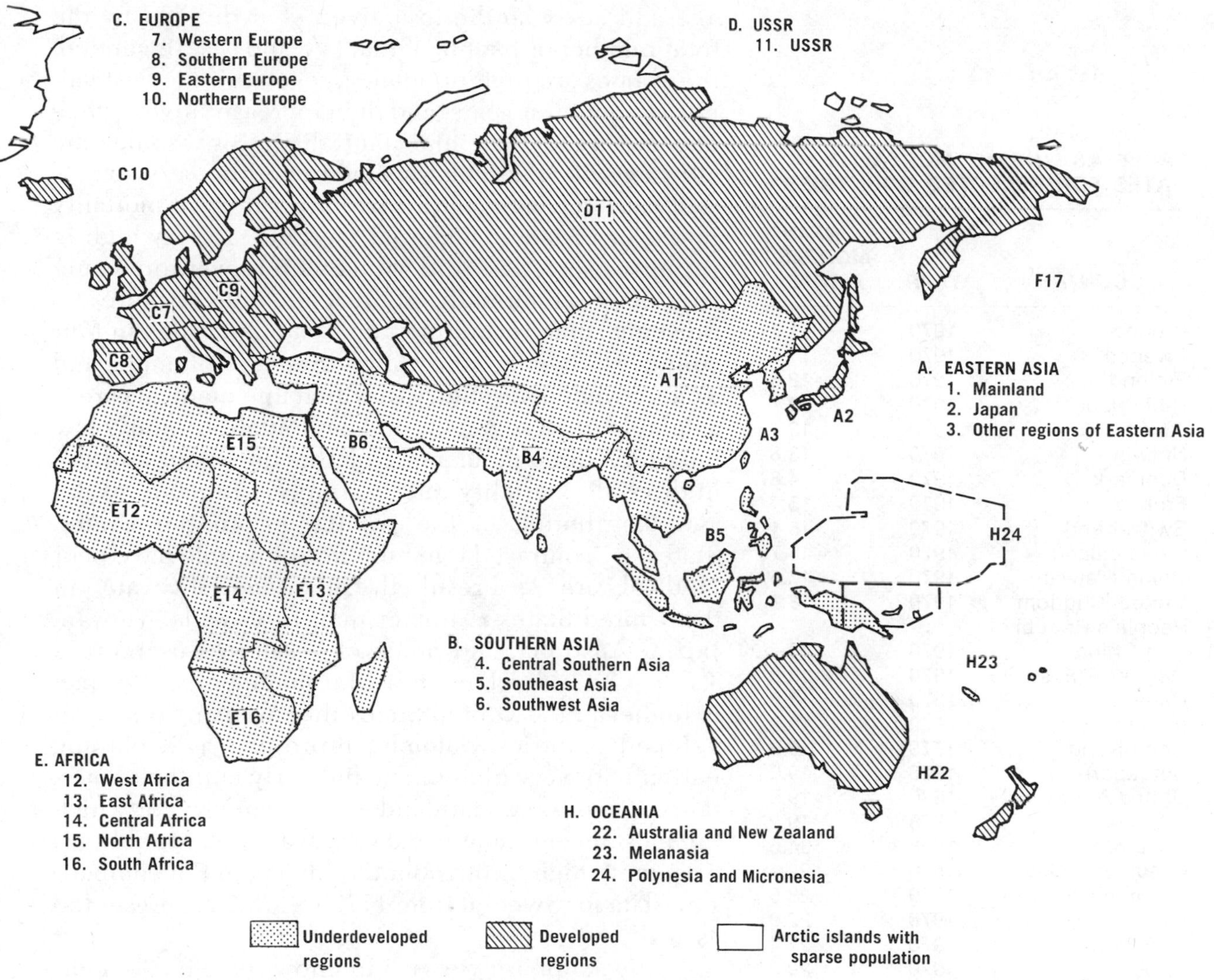

was 12.3. Why are there such apparent anomalies? The answer is that crude death rates are responsive not only to probabilities of dying, but also to the age distribution of the population. The populations of Sweden and Scotland are, on the average, much older than the population of Japan, which in turn is older than that of Taiwan. The Canal Zone is primarily a military installation, populated in large part by young Americans, and its very low death rate is simply a reflection of the fact that not many young men die each year.

For these reasons, serious studies of population growth use not crude death rates, but age- and sex-specific rates, that is, the number of people of a given

TABLE 4.3 INFANT MORTALITY RATES FOR SELECTED COUNTRIES*

COUNTRY	*YEAR*	*INFANT MORTALITY RATE*
Iceland	1970	11.7
Sweden	1970	11.7
Finland	1970	12.5
Netherlands	1970	12.7
Japan	1970	13.1
Norway	1970	13.8
Denmark	1970	14.8
France	1970	15.1
Switzerland	1970	15.1
New Zealand	1970	16.7
China (Taiwan)	1970	17.5
United Kingdom	1970	18.3
People's Republic of China	1970	17–20
Germany (East)	1970	18.8
Canada	1970	18.8
Ireland	1970	19.2
Hong Kong	1970	19.6
Singapore	1970	19.8
United States	1970	19.8
White	1968	19.2
Other	1968	34.5
Czechoslovakia	1970	22.1
Germany (West)	1970	23.6
Israel	1970	22.9
USSR	1970	24.4
Italy	1970	29.2
Greece	1970	29.3
Poland	1970	33.2
Yugoslavia	1970	55.2
Argentina	1967	58.3
Costa Rica	1970	67.1
Mexico	1970	68.5
Ecuador	1970	76.6
Guatemala	1970	88.4
Egypt	1969	119

*Data from *UN Demographic Yearbook* (1971) and *Statistical Abstract of the United States* (1972). Figure for People's Republic of China from *Studies in Family Planning,* Vol. 3, Number 7. Supplement.

†Rates are number of deaths of infants under one year of age per corresponding 1000 live births. These rates are difficult to compare because the definitions of live births and the effectiveness of birth and death registration vary from country to country.

age and sex who die in a given year divided by the total number of people of that age and sex. In general, these rates are high in infancy, reach their lowest values at about ten years, and then increase slowly. They become higher than infant mortality rates at some time after age 60. Particular societal conditions or events will cause changes in this overall pattern of mortality rate. For instance, maternal mortality is very high in primitive societies. And young men in war-torn countries have abnormally high death rates. Figure 4.5 presents some recent mortality data for males in four countries: Sweden, the United States, Colombia, and Togo. Sweden, a country in which fine medical care is available to all segments of the population, enjoys low mortality throughout most of the life span, and the high crude mortality rate is simply a function of the age distribution of the population. In the United States, by contrast, large numbers of people have poor medical care. As a result, the **infant mortality rate*** in the United States is higher than in Canada, Taiwan, Japan, Australia, and much of Western Europe (see Table 4.3). Indeed, mortality for much of the life span is higher in the United States than in many other developed countries. Colombia represents a developing nation with very high infant and early childhood mortality. Togo is the mainland African country for which the most recent reliable data are available. Mortality is extremely high throughout the life span. For comparison, data for Sweden from 1778 to 1782 are presented as well.

Now, suppose we wish to know the effect of mortality on a population; for example, how many of the infants born in a given year (called a **birth cohort,** since a cohort is a group of people with some common starting point) will live to celebrate their tenth birthday. Such information can be computed from the

*Infant mortality rate is usually defined as:

$$\frac{\text{Number of deaths of infants less than one year old in a given time period}}{\text{Number of live births in the same given time period}} \times 1000.$$

The definition of "live birth" varies from country to country, rendering international comparisons of infant mortality rates rather tricky. In particular, in some countries, infants who die within the first 28 days of life are not considered to be live births. Since the United States includes all live births in its definitions of infant mortality, the high infant mortality in the U.S., as compared with that of many other nations, may be in part an artifact of data collection and not an indictment of U.S. systems of delivering medical care. Even if this is the case, however, the differences in rates between whites and non-whites in the U.S. seem largely due to differences in medical care.

age- and sex-specific death rates of Figure 4.5 by constructing the **survivorship function,** defined as the proportion of the birth cohort surviving a given length of time after birth. Suppose, for example, that infant mortality were 30 per 1000 live births. Then 97,000 members of a hypothetical birth cohort of 100,000 would reach age one, or the probability of survival to age one would be 0.97. If, further, the death rate for one-year-olds were 10 per thousand, only 96,030 would reach their second birthday, for 97,000 × (1 − 10/1000) = 96,030. The probability of survival to age two would therefore be approximately 0.96. Continuing in this manner with successive age-specific mortality rates produces the entire curve. The calculations are slightly more complicated when the rates are given for five-year age groups instead of singly.

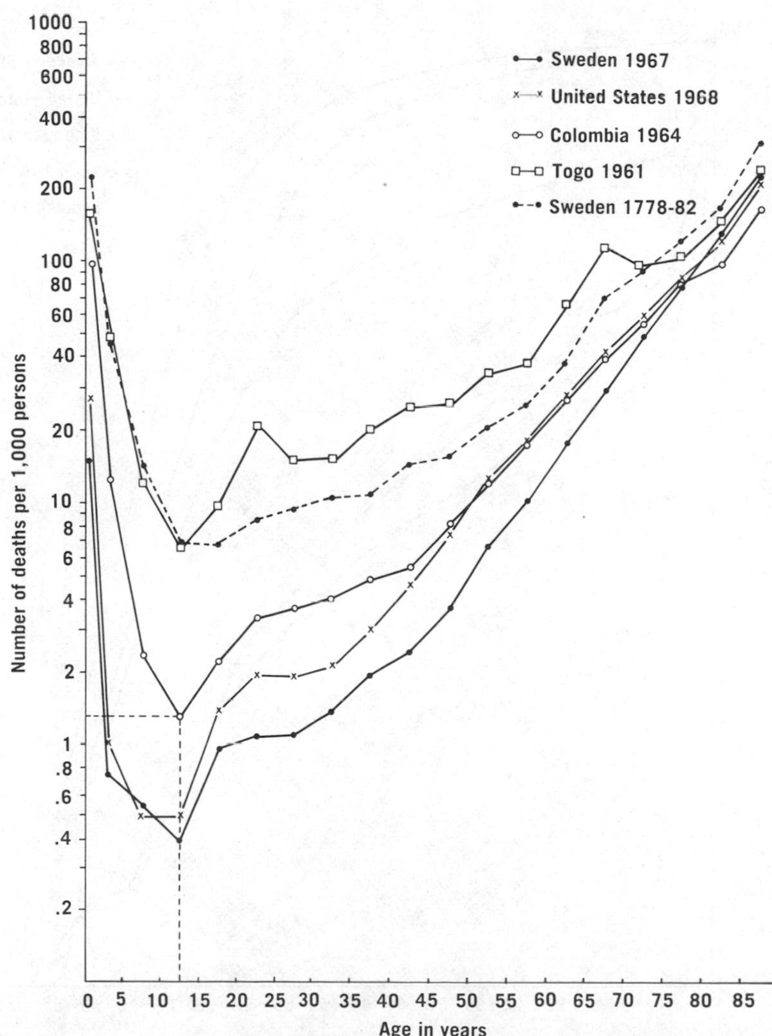

FIGURE 4.5 Age-specific death rates of males in Sweden (1967), the United States (1968), Colombia (1964), and Togo (1961). Example: The point shown at the intersection of the dashed lines tells us that in Colombia in 1964, the male death rate for 10- to 14-year-olds was 13 per thousand. (Data from Keyfitz and Flieger: *World Population.* **Chicago, University of Chicago Press, 1968, and from** *U.N. Demographic Yearbook.)*

The age-specific rates used to construct a survivorship curve are usually the rates that prevail at any point in time. They are examples of so-called **period measures,** and do not reflect the mortality experience of any current or historical cohort. For instance, in 1970, the age-specific death rate for 70-year-olds is the death rate for people born in 1900, while the infant mortality rate refers to people born in 1970. Period measures present a profile of many cohorts at a point in time. A true cohort survivorship cannot be completed until all members of the cohort have died, but even incomplete cohort data may provide insights often lacking in period measures. Unless otherwise stated, reported demographic indices are period measures.

The survivorship functions for Sweden, the

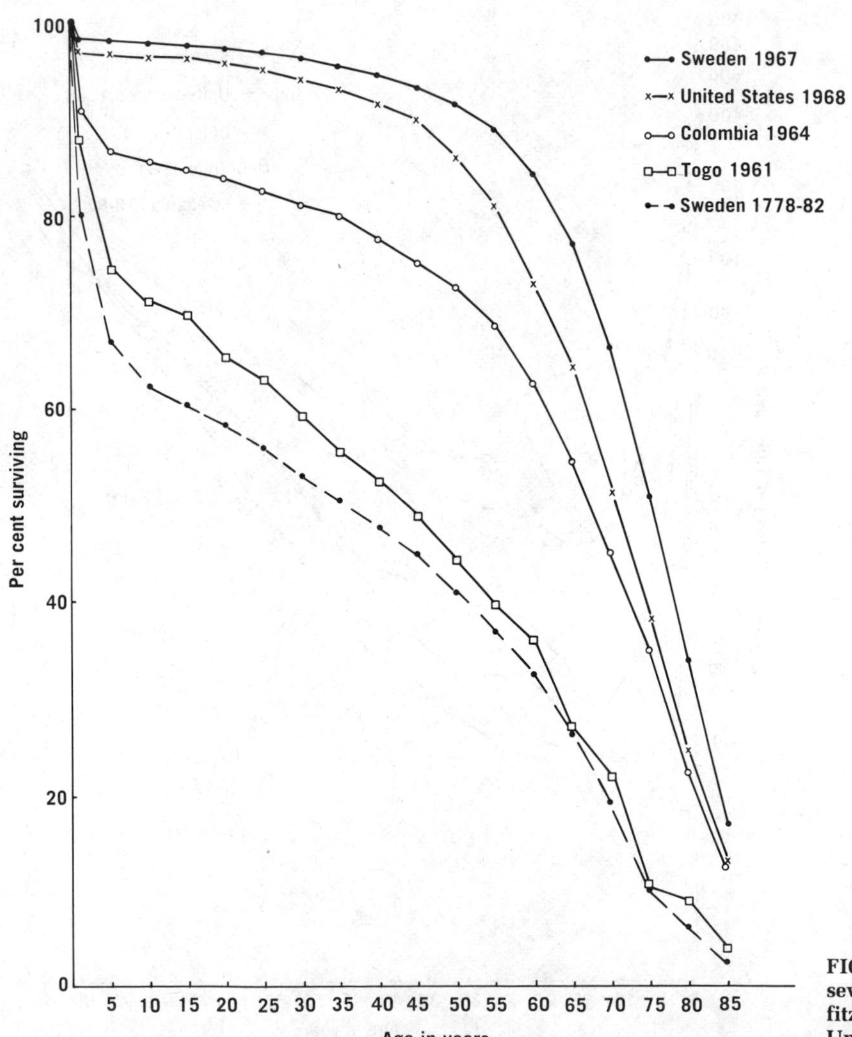

FIGURE 4.6 Male survivorship curves for several selected countries. (Data from Keyfitz and Flieger: *World Population.* **Chicago, University of Chicago Press, 1968.)**

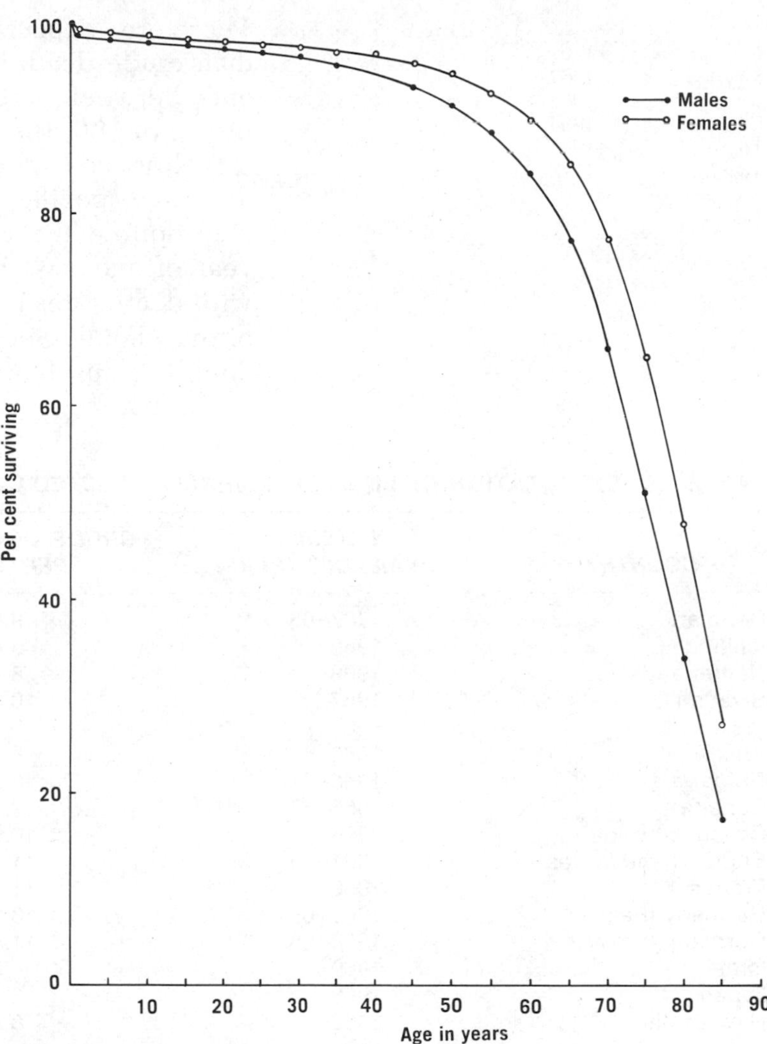

FIGURE 4.7 Male and female survivorship curves in Sweden (1965).

United States, Colombia, and Togo are plotted in Figure 4.6. Notice how smooth the curves are. Figure 4.7 plots male and female survivorship curves for Sweden in 1965. The women have lower infant mortality rates and generally higher survivorship. (Where is infant mortality plotted on the survivorship functions?) Later we shall see how very useful the age-specific death rates are for projections of population size. They are, however, very cumbersome for a static picture of the general health of a population as reflected by its mortality experience. For this, we use what is called **expectation of life.** The expectation of life is the average number of years a newborn infant will live given a set of age-specific rates. The measure depends only on age-specific death rates and not on the prevailing age-sex distribution of the country.

Life table methodology is used to obtain average life span from the survivorship function. See any introductory text in demography.

Country	Year	Expectation of life Male	Female
Sweden	1967	71.9	76.5
United States	1968	66.6	74.0
Colombia	1964	58.2	61.6
Togo	1961	33.4	40.2
Sweden	1778–82	36.0	38.5

Therefore, expectation of life is much more useful than crude death rate in comparing mortality experience between nations. For the countries represented in Figure 4.6, the expectations are shown to the left.

Now consider a hypothetical population with fixed age-specific vital rates. Further, pretend that we may double the number who live through any one year of age we choose. (This assumes a population with death rates no lower than 50 per cent in the year of our choice. See Table 4.5 for a demonstration that doubling survivors is not the same as halving the

TABLE 4.4 EXPECTATION OF LIFE FOR SEVERAL SELECTED COUNTRIES

COUNTRY	*LATEST AVAILABLE YEAR*	*CRUDE DEATH RATE PER 1000*	*EXPECTATION OF LIFE* Male	Female
Denmark	1967–68	9.8	70.6	75.4
Iceland*	1962	6.8	71.4	76.5
Netherlands	1968	8.3	71.0	76.4
Sweden	1967	10.1	71.9	76.5
Australia*	1965	8.8	67.7	74.1
Bulgaria	1965–67	8.5	68.8	72.7
Canada	1965–67	7.5	68.8	75.2
Czechoslovakia	1966	10.0	67.3	75.6
England and Wales	1967–69	11.5	68.7	74.9
France	1968	11.1	68.0	75.5
Germany (East)	1965–66	13.0	68.7	74.9
Germany (West)	1966–68	11.5	67.6	73.6
Israel	1969	7.0	69.2	72.8
Japan	1968	6.7	69.0	74.3
New Zealand*	1965	8.7	68.2	74.2
Austria	1969	13.4	66.5	73.3
Hong Kong	1968	5.0	66.7	73.3
Poland	1965–66	7.3	66.8	72.8
United States	1968	9.7	66.6	74.0
Soviet Union	1967–68	7.6	65.0	74.0
Taiwan	1965	5.5	65.8	70.4
Uruguay	1963–64	10	65.5	71.6
Yugoslavia	1966–67	8.5	64.7	69.0
Mexico	1965–67	9.5	61.0	63.7
Guatemala	1963–65	16	48.3	49.7
Pakistan	1962	20	53.7	48.8
India	1951–60	20	41.9	40.6
Burundi	1965	25	35.0	38.5
Chad	1963–64	25	29.0	35.0
Nigeria	1965–66	25	37.2	36.7
Togo*	1961	29	33.4	40.2

*Data without asterisks from *U.N. Demographic Yearbook* (1970).
Data with asterisks from Nathan Keyfitz and Wilhelm Flieger: *World Population, 1968.*

death rate.) What age should we choose to double if we desire the greatest increase in population? Clearly, we will be most effective if we reduce infant mortality (deaths from age 0 to age 1), for if in the long run there were no other changes in rates, two times as many viable infants would lead, in just a century, to a doubling of population size. On the other hand, if we were to double the number of surviving 80-year-olds, the population size would hardly change. Why? First, few people reach 80; second, even healthy 80-year-olds cannot survive very much longer; finally, the aged do not produce babies.

To gain insights into the numerical effects of reduction of infant mortality, we shall compare Chile in 1934 to Sweden in 1970. In Chile in 1934 the infant mortality rate was extremely high—262 deaths for every 1000 live births! By contrast, the infant mortality rate in Sweden in 1970 was about 13 deaths per 1000 live births. What would have happened to the population of Chile after 1934 if all birth rates and death rates had remained constant but the infant mortality had suddenly dropped to levels approaching Sweden's? It is easy to see that the sudden reduction in infant mortality with no concomitant change in birth rate would have quickly led to a rapid increase in total population. In 1934, the probability that any infant who was born in Chile would reach the age of 1 was about 75%. If he reached 1 year, the probability that he would reach 15 years was about 90%. Thus, for every 1000 babies born, only about 650 reached reproductive age. On the other hand, if infant mortality had been reduced to about 13 per thousand per year, and if all other mortality rates had remained unchanged, about 900 of every 1000 infants born would have reached 15 years of age. This would have been about a 30% increase in the number of people who would have reached the reproductive age. Furthermore, if mortality rates in the entire pre-reproductive age had been reduced to the level of Sweden's, 950 out of every 1000 people would have reached 15, causing a small further increase in population growth. Thus, lowering rates of death in the pre-reproductive years, without changing birth rates, can induce a very rapid population increase.

TABLE 4.5 WHY IS HALVING THE DEATH RATE NOT THE SAME AS DOUBLING THE SURVIVAL RATE?

EXAMPLE 1.* HALVING THE DEATH RATE

(1) *Death Rate/1000*	*(2)* *Survival Rate/1000*
1000	0
500	500
250	750
125	875
↓	↓
Successive halving	Decreasing change
↓	↓
8	992
4	996
2	998
1	999

EXAMPLE 2. DOUBLING THE SURVIVAL RATE

Survival Rate/1000	*Death Rate/1000*
1	999
2	998
4	996
8	992
↓	↓
Successive doubling	Increasing change
↓	↓
100	900
200	800
400	600
800	200

*In both examples, column (2) = 1000 − column (1).

4.5 MEASUREMENTS OF BIRTH RATES

Measurements of **natality** are data concerning the number of births or the rate of birth in an area and time period of interest. The simplest measure of natality is the previously mentioned crude birth rate, the number of live births in a calendar year divided by the total population of the area being studied. The rate is easily calculated, but, like the crude death rate, it reflects the current population distribution. Since the probability of a woman's giving birth is a function of her age, the most useful indications are age-specific birth rates which are the ratios of births to women in specified age groups. Figure 4.8 plots some age-specific birth rates for several selected countries.

The set of age-specific birth rates may be used to compute simple indices of birth rates independent of population distributions. The most widely used of these is the **total fertility rate,** or TFR, which can be interpreted as the average number of infants a woman would bear if she lived through age 49 and if the age-specific birth rates were to remain constant for her generation. To compute the TFR for Sweden in 1965, for instance, we need the seven age-specific rates shown at left.

Age-specific birth rate in Sweden 1965.

Age	Age-specific birth rate per thousand
15–19	49.3
20–24	136.4
25–29	152.1
30–34	90.0
35–39	39.6
40–44	10.1
45–49	0.7

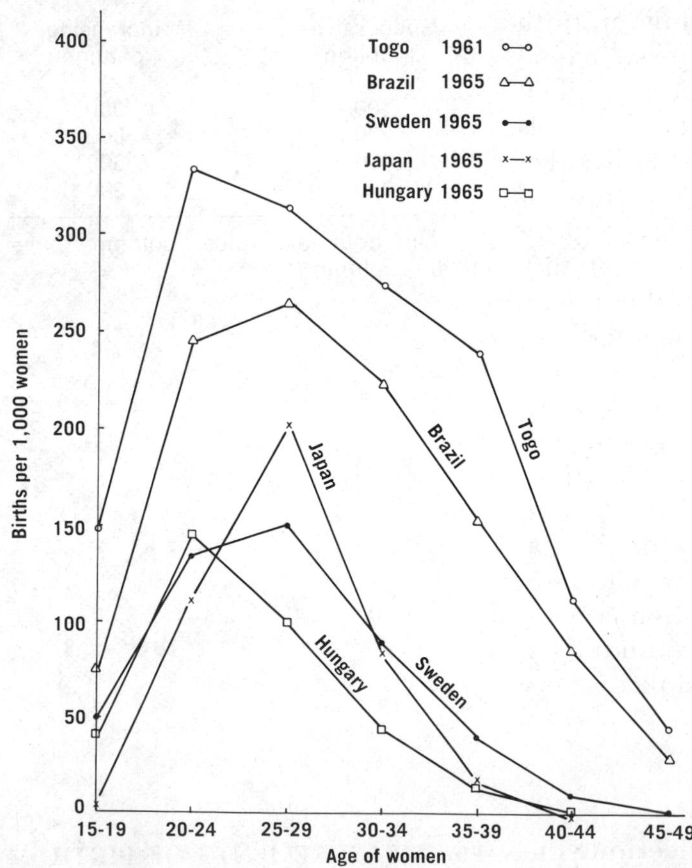

FIGURE 4.8 Age-specific birth rates in five selected countries.

TABLE 4.6 NATALITY MEASURES FOR SEVERAL SELECTED COUNTRIES*

CONTINENT	COUNTRY	YEAR	CRUDE BIRTH RATE/1000	TOTAL FERTILITY RATE
Africa	Togo	1961	54.5	7.1
North America	Canada	1965	24.3	3.2
	Costa Rica	1965	42.3	6.8
	Mexico	1965	44.2	6.7
	United States	1965	19.6	3.0
South America	Colombia	1965	36.8	6.6
Asia	Japan	1965	18.6	2.1
	Pakistan	1963–65	49.0	7.6
	Taiwan	1965	32.7	4.8
Europe	France	1965	17.6	2.8
	Hungary	1965	13.1	1.8
	Italy	1965	19.2	2.6
	Netherlands	1965	19.9	3.0
	Sweden	1965	15.9	2.4
Eurasia	Soviet Union	1966–67	18.2	2.4

*Data from Keyfitz and Fleiger: *World Population*, 1968 and from *Interim Reports of Conditions and Trends of Fertility in the World*, 1960–1965. United Nations, New York, 1972.

Adding the seven, multiplying by five (because the age intervals are five-year spans and the age-specific birth rates are expressed per year), and dividing by 1000, the total fertility rate was 2.4. (The comparable rate in the U.S. was 3.0; in Canada 3.2; in Hungary, 1.8; in Costa Rica, 6.8.)

Because the TFR is based only on age-specific rates and not on population distributions, it is useful for international comparisons of child-bearing. Table 4.6 presents crude birth rates and TFR's for several countries. In the developed countries, a TFR of roughly 2.1 is the so-called **replacement level**—that value of the TFR which corresponds to a population exactly replacing itself.

4.6 MEASUREMENTS OF GROWTH

(Problems 20, 21, 34, 35, 36)

Why do we want to measure the population growth of a country? Our purpose will in large part dictate the kind of measure we use. To learn how many more people were inhabitants at one time than at some other time, we need only subtract the two populations. For example, the population of India in 1950 was 358 million persons. By 1970, the population had reached 550 million. The total growth of population in the

20-year span from 1950 to 1970 was 192 million people, or slightly less than the current population of the United States. By contrast, in the United States, there were 152 million people in 1950 and 205 million in 1970, a difference of 53 million. Thus, the United States, with many times the wealth of India, and about three times the land area, had about one quarter the population increase.

It should be clear that these population differences give little feeling about how fast a country is growing relative to other countries. More useful are rates of growth, defined in Section 4.3. For the data from India, the average annual rate of growth from 1950 through 1970 was:

$$\text{Average annual growth rate (India, 1950-1970)} = \frac{(550 \times 10^6 \text{ persons} - 358 \times 10^6 \text{ persons}) \times 100\%}{358 \times 10^6 \text{ persons} \times 20 \text{ years}} = 2.4\%/\text{year}.$$

For the United States in the same period the average growth rate was:

$$\text{Average annual growth rate (U.S., 1950-1970)} = \frac{(205 \times 10^6 \text{ persons} - 152 \times 10^6 \text{ persons}) \times 100\%}{152 \times 10^6 \text{ persons} \times 20 \text{ years}} = 1.7\%/\text{year}.$$

These rates are averages, and do not imply, for example, that the population of India in fact grew by 2.4 per cent per year. For if it had, that percentage would be compounded. Thus, for the second year, the average growth would be 2.4% × (1 + .024) and so on.

A common popular measure of population growth is the **doubling time** (t_d) of the population. The doubling time is defined as the length of time a population would take to double if its annual growth rate (r) were to remain constant. The relation between doubling time and growth rate is:

$$t_d = 0.693/r.$$

To compute doubling time from rate of growth, think of an analogy with compound interest at the bank. If a rate of growth (interest rate) is applied once a year to a population of size P_o (capital in the bank), the population (capital) at the end of one year is:

$$P_1 = P_o + P_o r = P_o (1 + r).$$

More generally, if population growth is compounded n times each year, then:

$$P_t = P (1 + r/n)^{nt}.$$

It is reasonable to suppose that populations grow continuously, or that $n \to \infty$. From elementary calculus, $\lim_{n\to\infty} (1 + r/n)^{nt} = e^{rt}$. The doubling time is then the solution of the equation:

$$2P_o = P_o e^{rt_d}.$$

Or, taking logarithms on both sides of the equation:

$$t_d = 0.693/r.$$

Overall, such growth rates are interesting measures of long-term historical trends, but as our study of birth and death rates has shown, population composition affects overall rates too severely for them to be useful for reliable prediction. Note that although the population size and composition are irrelevant to calculating a theoretical doubling time, they are not irrelevant to the time a real population will take to double. Therefore, beware of conclusions made on the basis of projecting doubling times. Remember that although the term "doubling time" sounds like a projection, the measure uses only the crudest available population data.

To project future population size more accurately, the simplest effective approach is to apply current age-sex specific birth and death rates to the present population and calculate future size. However, even this type of prediction is subject to large error, for it assumes constant age-specific vital rates. Figure 4.9,

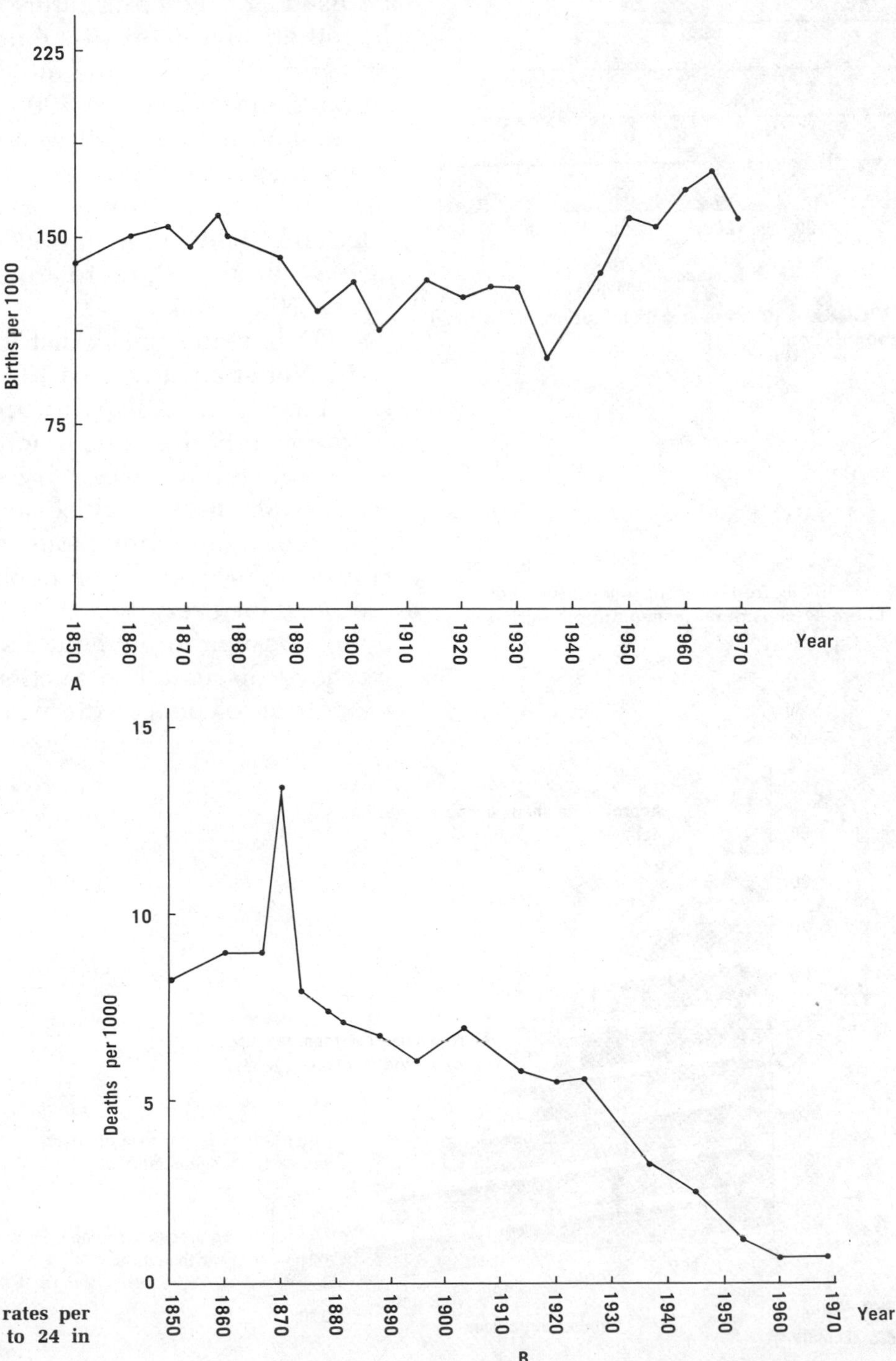

FIGURE 4.9 Birth and death rates per thousand for women aged 20 to 24 in France (1850–1968).

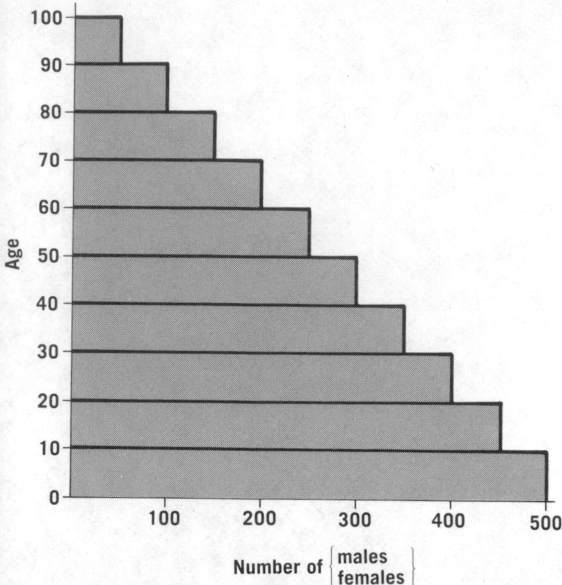

FIGURE 4.10 Age-sex distribution of an ideal population.

which shows birth and death rates for women aged 20 to 24 in France, demonstrates that these rates are in fact not always constant. (Certainly, the birth rates have been much more stable than the death rates.) Accurate forecasting of population size is impossible unless we know something about how vital rates are changing.

The current age-sex distribution of an area should be used in predicting future population sizes. In the hypothetical age-sex distribution of Figure 4.10, each age group has the same number of males as females. In particular, there are 500 boys and 500 girls under 10, and 50 men and 50 women between 90 and 100 years of age. Furthermore, there are exactly 50 fewer men and 50 fewer women at each succeeding age decade. Thus, while there are 300 of each sex between the ages of 40 and 50, there are only 250 of each sex between 50 and 60.

Do human age-sex distributions look like Figure 4.10? Not at all! Figure 4.10 would represent a population in which (a) boys and girls were born with equal frequency; (b) the same number of persons were born every year for over a century; (c) everyone died by the age of 100; and (d) any person, at birth, had an equal chance of dying throughout each year of his life span. However, in real human populations, about 106 boys are born for every 100 girls. Nor is the probability of dying constant throughout man's life span. Instead, as we have discussed in Section 4.4, a relatively large proportion of people die when they are very young,

This figure represents a worldwide average. There is considerable geographic variation.

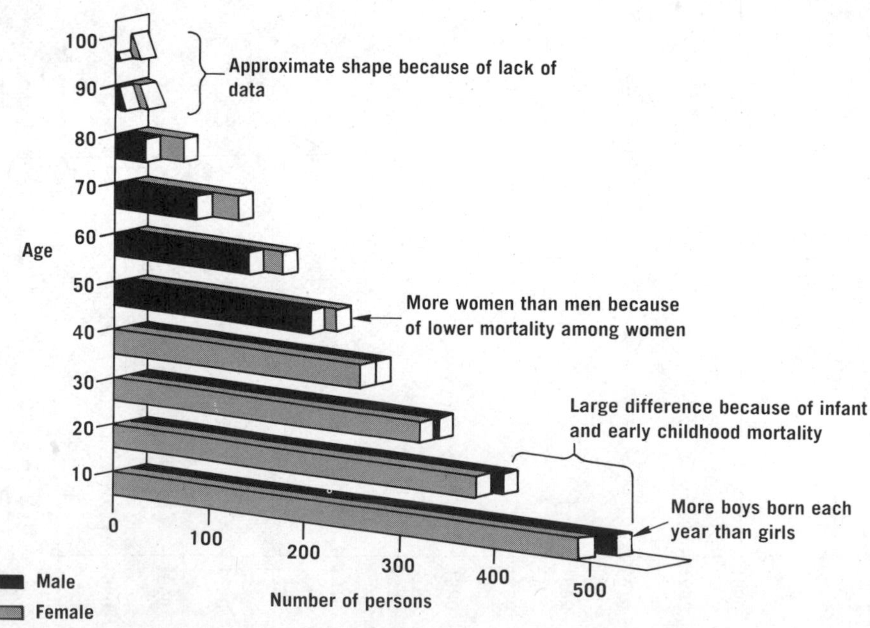

FIGURE 4.11 Typical age-sex distribution.

comparatively few die between the ages of 10 and 50, and the proportion of people dying each year after 50 increases rapidly. In addition, there are marked sex differences in mortality (the number of deaths which occur in a given period). Women have a higher probability of surviving from one year to the next throughout the life span except, in some societies, during the childbearing years.

Consider the effects of realistic patterns of vital

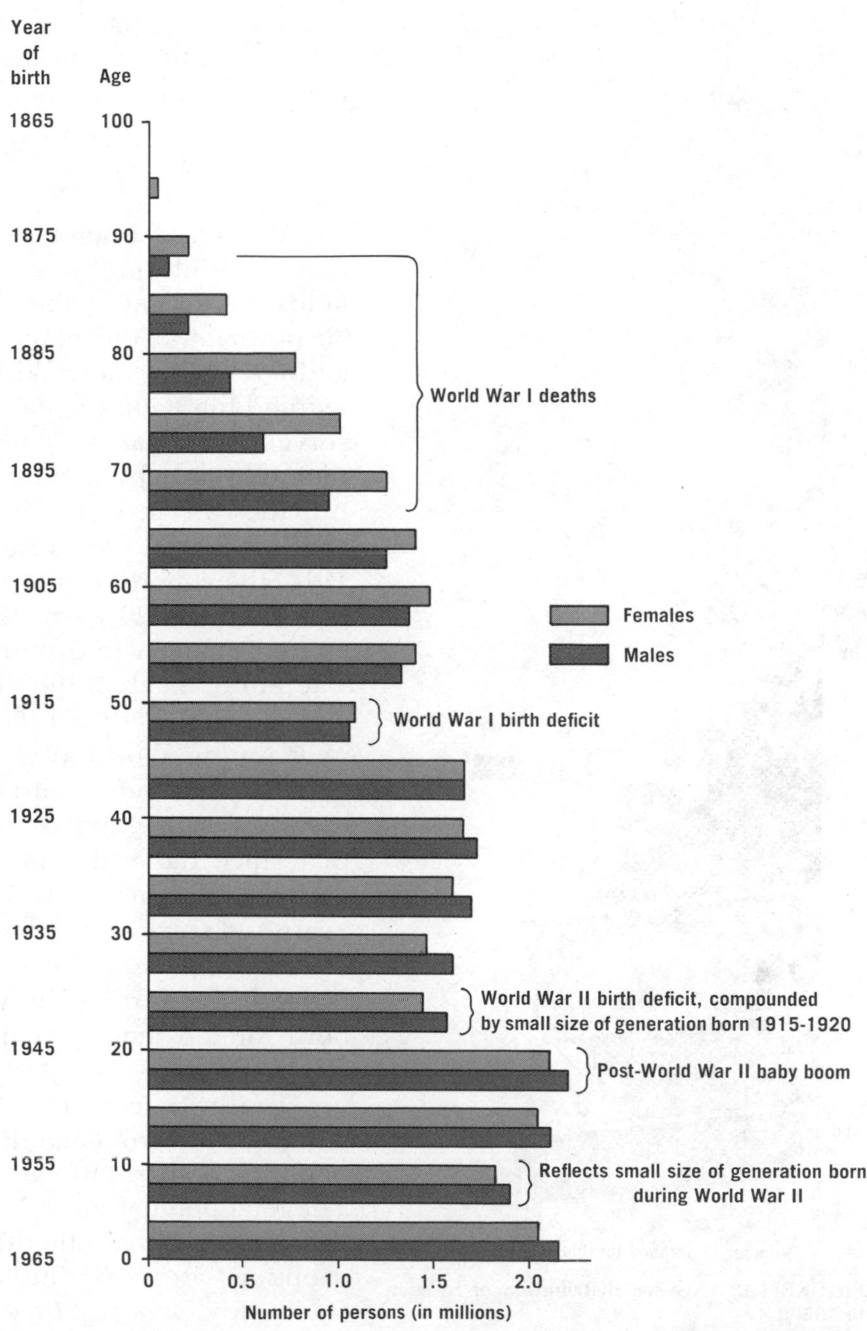

FIGURE 4.12 Population of France in 1965.

events on a birth cohort. The greater survivorship of women over men means that even though more boys are born than girls, the ratio of women to men increases as the cohort grows older. By the time the cohort is elderly, there are considerably more women than men. Also, data are usually collected in such a way that we know only the total population of each sex over 70, 80, or 85. Therefore, the graphs can be only approximate for the very old age groups. Figure 4.11 presents an age-sex distribution which more nearly reflects these demographic characteristics.

In addition to these reasonably predictable phenomena, many less predictable changes can occur. Population growth is affected by such events as war, famine, medical advances, and changes in social custom. For example, Figure 4.12 presents the age-sex distribution of France in 1965. This is a very bumpy curve. To interpret it we need knowledge from many fields. For example, the graph shows that in 1965 only 38 per cent of the population of France between 70 and 74 were males. Why so few? We have already learned that men die earlier than women, but the observed discrepancy is much larger than can be explained solely by natural differences in rates. History provides an answer. This cohort was born between 1890 and 1894. At the outbreak of World War I, in 1914, the cohort born in this five-year period was between 19 and 23 years of age. Its members, then, included many of the men who fought World War I. Therefore, much of the difference in numbers of men and women in 1965 reflects the mortality of French men during World War I. We can immediately see how the distribution curves can sometimes aid in predicting future population growth. We would expect that since the males of the 1890–1894 cohort were away from home during some of their reproductive years and since so many were killed, there should occur some very small cohort about one generation later. Indeed, the birth cohort of 1915–1920, the people born for the most part during World War I, is very small.

To illustrate another example of predicting population growth from age distributions, we examine the population distribution of Sweden in 1950 (Fig. 4.13). The total population in 1950 was nearly seven million persons. First, note that the male and female age distributions are very similar. Therefore, we need consider only one sex to examine population growth.

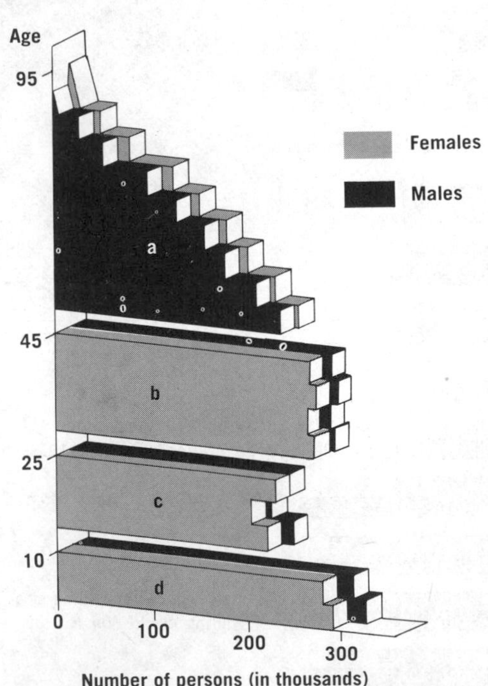

FIGURE 4.13 Age-sex distribution of Sweden in 1950.

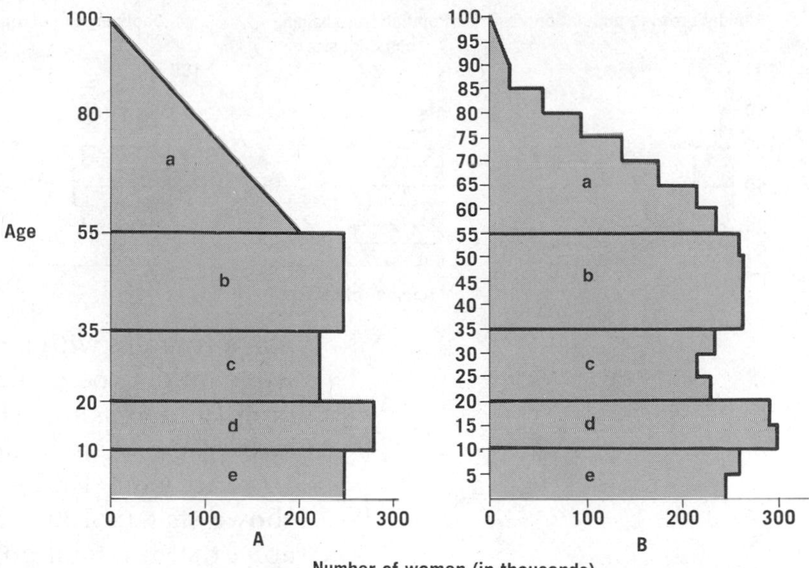

FIGURE 4.14 Female age distribution of Sweden in 1960. *A.* **Distribution guessed from age distribution of 1950.** *B.* **Distribution based on Swedish government population data (1960).**

Next, we see that the distribution is separated into four distinct parts. Since the distribution is very narrow for persons of reproductive age, we would expect the age distribution for women 10 years later to look something like Figure 4.14*A*. Each of the distinct sections of Figure 4.13 would move up the age axis 10 years, with each decreasing in size because of mortality, and the small reproductive class would produce a small number of births. In fact, the shape of the actual distribution curve in 1960, shown in Figure 4.14*B*, is remarkably similar to Figure 4.14*A*. The male age distribution is nearly the same shape.

A population curve need not be bumpy to be a useful predictor of growth. The age distribution for women in India in 1951, shown in Figure 4.15, looks

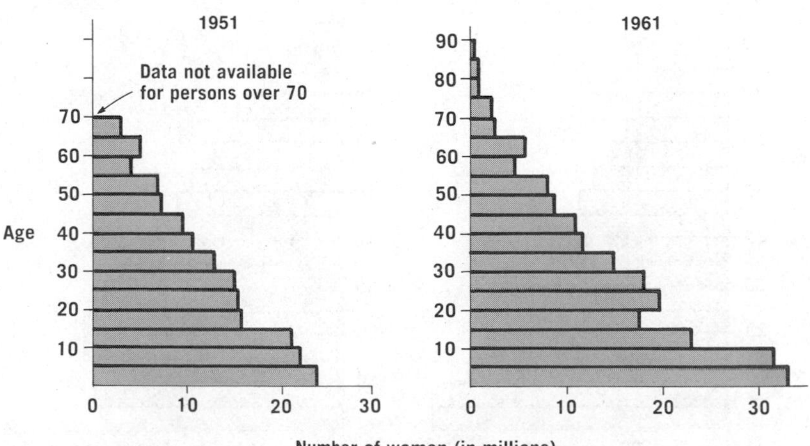

FIGURE 4.15 Female age distributions of India in 1951 and 1961.

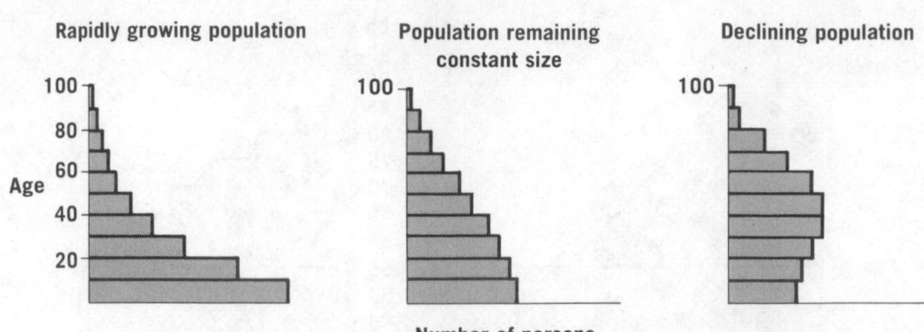

FIGURE 4.16 Schematic population age distributions.

like a triangle with a very wide base. From this graph, we would expect that such a population would be rapidly increasing. Hence, we would guess that the base of the triangle described by the age distribution 10 years later (1961) would be very wide. Figure 4.15 shows that in 1961 the base had widened considerably, that the total population had increased by about 25 per cent, and that the proportion of the population under 10 had increased from about 25 per cent to about 30 per cent. The population distribution of 1961 predicts an even more rapidly increasing population during the next decades.

The deficit of 15- to 20-year-olds could not have been predicted from examining the 1951 distribution alone.

The shapes of smooth population curves tell much about the growth of a population. Rapidly growing populations have very large bases; populations that are remaining constant have relatively narrow bases, and declining populations have pinched bases. (See Figure 4.16.)

Predictions about future population growth from the age of the distribution alone are not sufficient if there is considerable change in mortality and fertility. For example, Figure 4.17 shows that the distribution

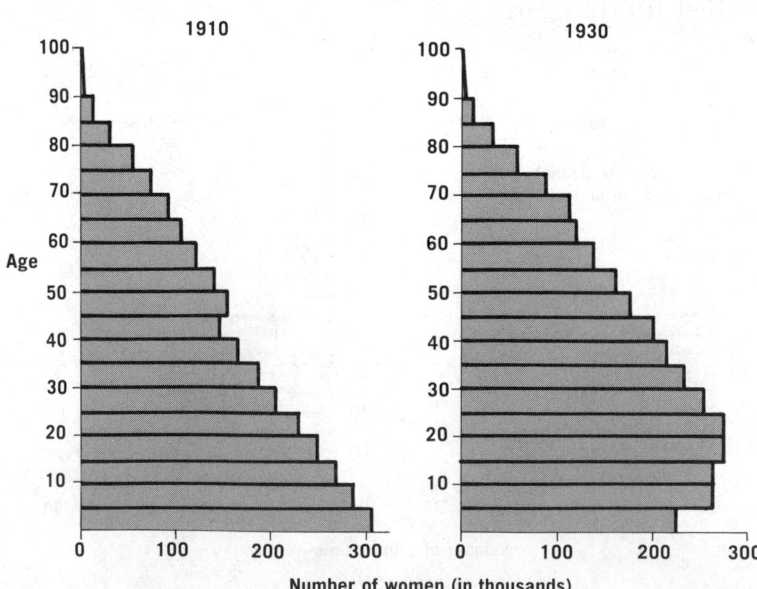

FIGURE 4.17 Female age distributions of Sweden in 1910 and 1930.

for Sweden looked like a triangle in 1910. Not knowing anything about changes in vital rates, we would predict a triangle in 1930. However, in 1930 the base of the age distribution was pinched. Reexamination of Figure 4.12 shows that the pinched base persisted through the birth cohort of 1940. A demographer would have had to predict World War I and its profound economic and social effects on all of Europe to have projected accurately the population of Sweden from 1910 to 1930.

4.7 ACCURACY OF THE DATA

(Problems 17, 18)

Some readers may have found the choice of examples inexplicable. Why cite data for Togo, an African country with roughly one and a half million persons, and totally ignore Nigeria with close to 60 million persons? Why use Taiwan as an Asian example and not the People's Republic of China? Why, for South America, no mention of Brazil? And why not use 1970 data for all countries? The reason for the choices is practical. We have used Togo as an example of a continental African country not because it necessarily represents social, political, and demographic trends in Africa, but because Togo is the only continental African country with reliable data. We cannot use mainland China, or Brazil, for data are unavailable. The choice of years was dictated by availability of information.

The more refined the measure we wish to report, the more difficult it is to collect the data. Knowledge of the current population size of a country requires only that one number. Crude birth rate needs two pieces of information—total population and number of births within a year.

Data collection for the purpose of computing age-specific rates is much more complicated. Registration systems would require that the person reporting a birth know the mother's age, while a person reporting a death would have to know the deceased person's birthday. Many opportunities for error arise. The reported information may be incorrect, and the procedures used for tabulating the data may introduce serious inaccuracies. Data-gathering techniques other than registration systems, such as sample surveys, are expensive and require considerable expertise.

Even when data appear reliable, the cautious demographer must question their validity. Figure 4.18

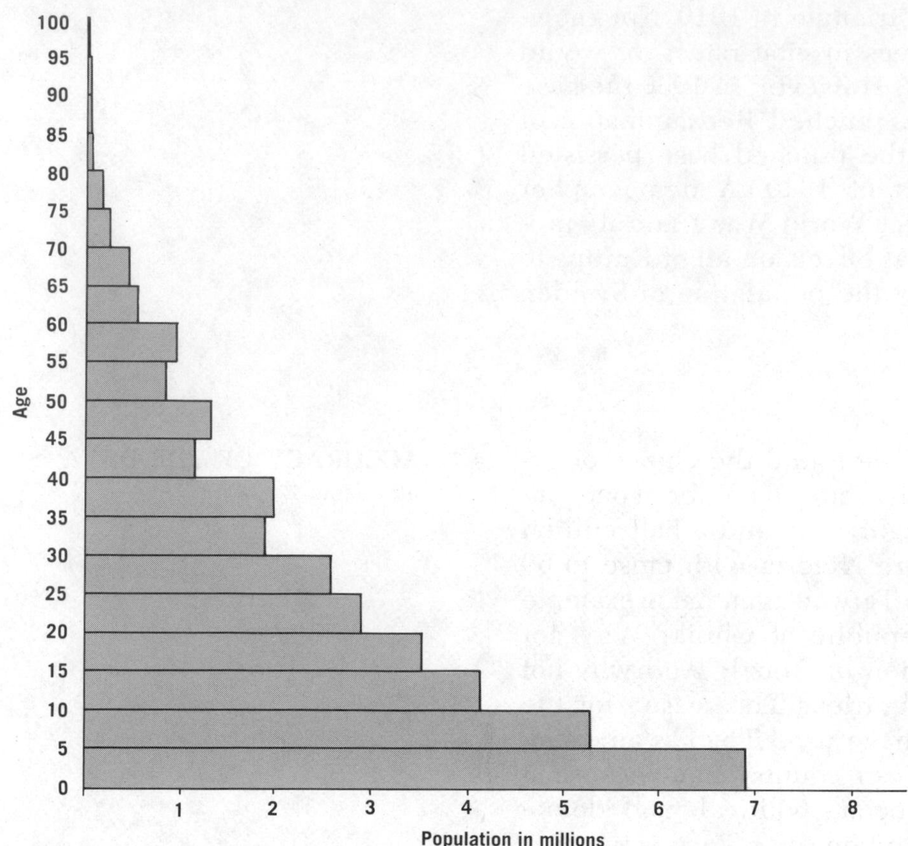

FIGURE 4.18 Population distribution of Mexico by age categories in the 1960 Census. (Data from *U.N. Demographic Yearbook,* **1962.)**

depicts the age distribution by five-year age groups for Mexico in 1960. The data were collected in the decennial (tenth annual) **census,** or count, of the population. Each person was asked to report his age. The graph appears odd, especially between the ages of 30 and 60. There are more 36- to 40-year-olds than 31- to 35-year-olds; more 46- to 50-year-olds than 41- to 45-year-olds; more 56- to 60-year-olds than 51- to 55-year-olds. The population dips cannot be attributed to successive birth deficits because they are too close together. If we replot the data by single year of age (Fig. 4.19), the results are shocking. According to the 1960 census of Mexico, people were much more likely to report an age divisible by 10 than to report the next year. If we accept the reported data as true, we would have to believe that there were three times as many 30-year-olds as 31-year-olds; five and a half times as many 40-year-olds as 41-year-olds; six times as many 50-year-olds as 51-year-olds; 10 times as many 60-, 70-, 80-, and 90-year-olds as 61-, 71-, 81-, and 91-year-olds, respectively. We would have to postulate a demographic process which could explain the overabundance of people whose age ended in five, and the lack of people whose age ended in seven. The answer, of course, is that we are dealing here with a tendency of people to report their ages in some round form, for instance, to the nearest five, or to some even number close to the true age. Such marked **digit preference** is not uncommon in underdeveloped countries. In such cases, use of data as reported leads to unreliable age-specific measures.

Other sources of error abound. For instance, the definitions of live birth, immigrants, and emigrants vary from country to country. Another problem in population enumeration arises because conflicting political claims over territory cause some people to be counted twice—once as subjects of one country, once as subjects of another. Other difficulties arise because

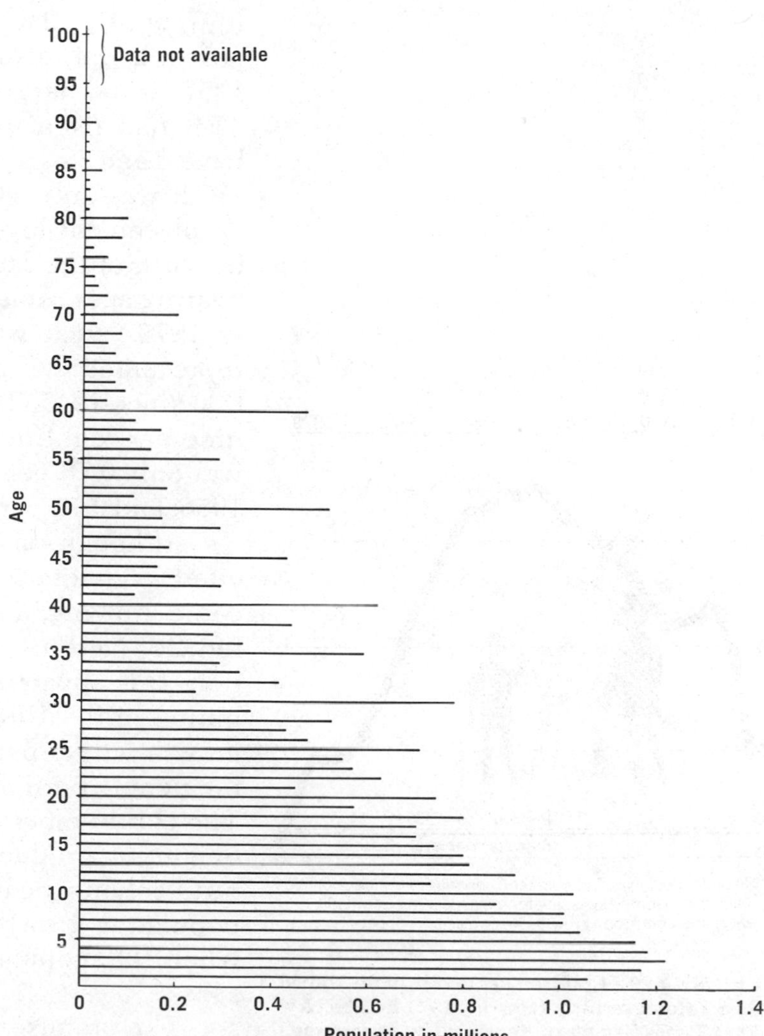

FIGURE 4.19 Population distribution of Mexico by single year of age in the 1960 census. (Data from *U.N. Demographic Yearbook,* **1962.)**

of the habits of the group being counted. Nomadic bands, for example, are notoriously hard to count.

All the foregoing is to warn against hasty uncritical acceptance of a demographic "fact." It is not a plea to ignore data, for even if data are imprecise or sketchy, general trends in birth or death rates will almost always appear. In using demographic data, always consider what sources of error might be present, and that such error might change your interpretation of the subject under investigation.

4.8 DEMOGRAPHIC TRENDS IN THE UNITED STATES

(Problems 22, 28, 29)

American demographic history after World War II has been characterized by sharp changes in fertility patterns. In 1945, the total fertility rate was about 2.5. That is, on a period basis, the average American woman was giving birth to 2.5 children during her lifespan. By 1946, the TFR had risen to 3.2, reflecting post-war optimism and the reuniting of families. In 1957, at the height of the so-called "baby boom," the TFR had risen to 3.7. The 1960's and early 1970's have been years of rapid decline in TFR which, by 1972, was only 2.03 children, significantly below the "replacement level" of 2.1 (see Fig. 4.20). However, because of the large number of women in the child-bearing ages, American population continued to grow in 1972. What was remarkable was that the growth represented the smallest annual increment in population since 1945. The actual number of births declined nine per cent from 1971 to 1972. The total growth rate was only 0.78 per cent as compared to 1.83 per cent in 1956, and 0.71 per cent in 1945 (see Fig. 4.21).

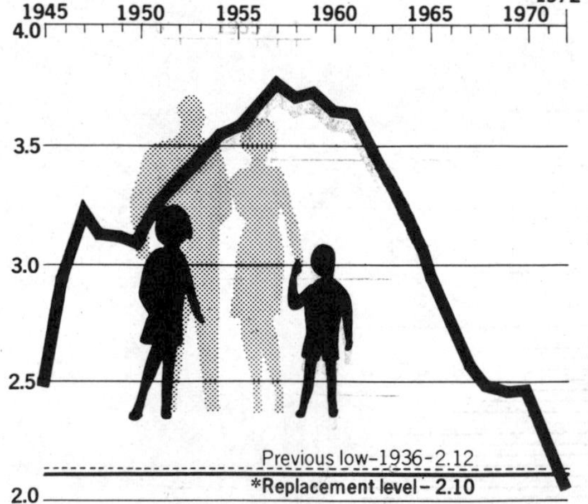

*Replacement level: The number of average births per woman over her lifetime necessary for the population eventually to reach zero population growth. This would take about 70 years.

Sources: National Center for Health Statistics, Census Bureau

FIGURE 4.20 Children per family in the United States (total fertility rate). (© 1973 by The New York Times Company. Reprinted by permission.)

Although this rapid decline in the TFR has been cited as evidence that the American population will cease growing, the actual growth depends not only on fertility patterns but also on the population distribution. It is apparent from the female population distribution of the United States that the future American growth will depend largely on the fertility patterns of the people born during the baby boom (see Fig. 4.22). The children born during the baby boom have not yet begun to produce babies. If they, and successive birth cohorts, bear children at replacement level, zero population growth will be realized in about 70 years, when U.S. population will level off at 320 million persons.

The premise that fertility rates will continue to be

quite low is only an assumption. A rapid rise in population will result from even a small increase in the TFR for the baby-boom children. Accurate prediction of the behavior of fertility patterns depends on accurate assessment of future social patterns. Increasing age at marriage, increasing rates of divorce, the current rapid increase in the proportion of women at work, and the increasing popularity of the two-child family have all contributed to the decline in fertility in the U.S. If these trends continue, U.S. fertility rates will undoubtedly remain low. On the other hand, if economic and social pressures begin once more to encourage the values of homemaking, fertility rates may again rise steeply. The increased use of more effective contraceptive methods, the steady liberalization of abortion laws, and the increasing social acceptability of abortion are also important factors in limiting population growth. These factors, however, are probably responses to social trends discouraging births. In the U.S. as elsewhere, contraceptive devices are most widely used and are most effective among married, motivated women. Indeed, recent studies show that about 53 per cent of unwed women do not use contraceptives when they have sexual relations.

If Americans continue to reproduce at replacement levels, the population will grow older. In 1970, 10 per cent of the population of the United States was aged 65 or older. By 2000, if the average family size continues to be approximately two per family, and if

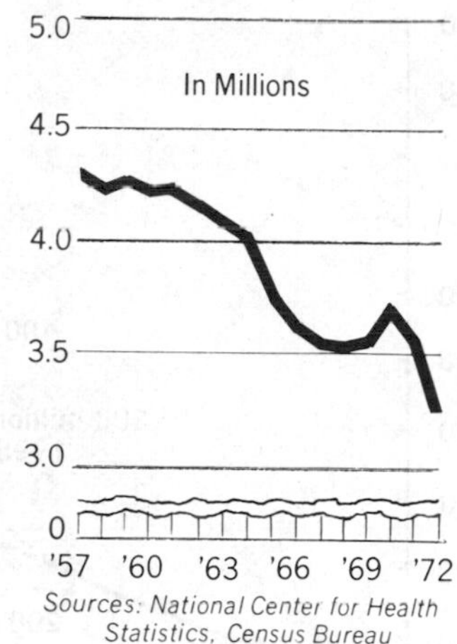

FIGURE 4.21 Births in the United States since 1957, the peak year. (© 1973 by The New York Times Company. Reprinted by permission.)

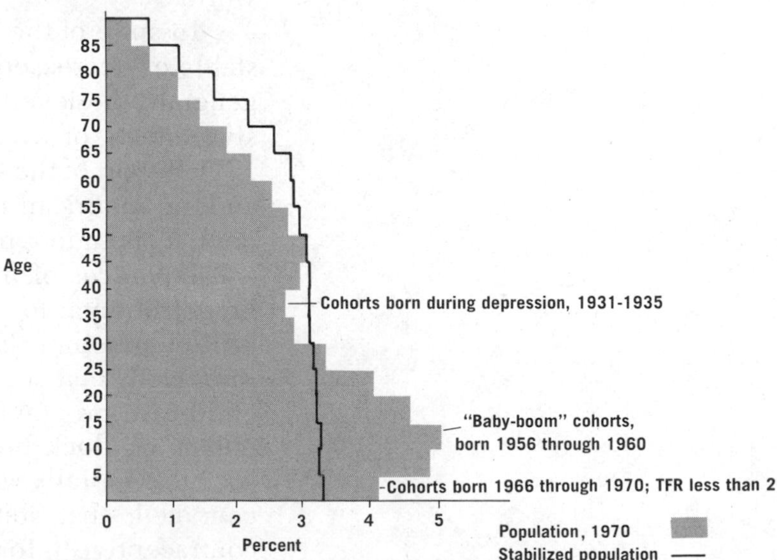

FIGURE 4.22 Per cent female age distribution, United States.

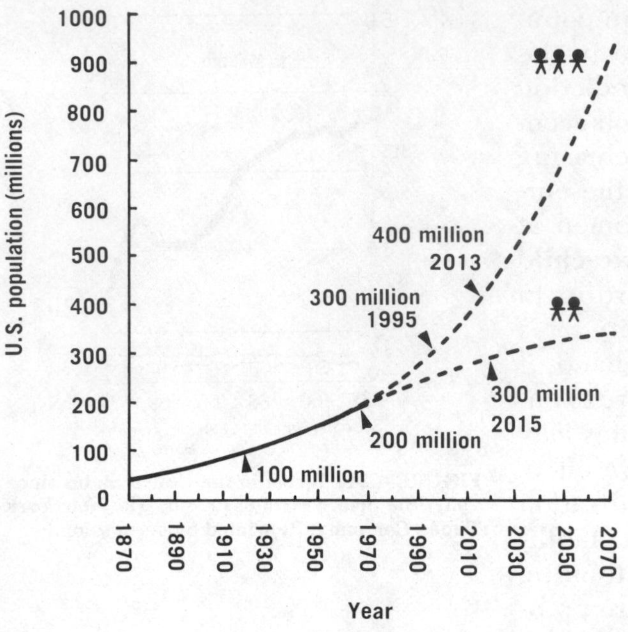

FIGURE 4.23 Projected population size of United States if average family size remains constant at two children or at three children.

there is no migration, the figure will rise to 11 per cent. When population is stabilized, that is, when it stops growing, 16 per cent of Americans will be over 65. The implications of such a dramatic change in population will be profound. (Refer again to Figure 4.22). Unless the age at retirement is raised to, say, 70, a large proportion of the population will depend on Social Security and other forms of old-age assistance. Priorities in medical care will have to change as more geriatric and less pediatric skills are demanded. Repercussions will be felt throughout the business world.

In spite of the economic and social adjustments a stable or decreased population will demand, they are generally believed to be small compared to the massive effects of a rapidly expanding population. The 1971 Report of the Commission on Population Growth and the American Future, commissioned by the President, concluded that the social and environmental consequences of increased population growth would be detrimental to the United States, and recommended, among other things, that population education be improved; that sex education be available to all; that child-care programs be improved; that children born out of wedlock be accorded "fair and equal status, socially, morally and legally"; that adoption be encouraged; that states eliminate legal restrictions to contraceptive information, procedures and supplies;

that states adopt laws permitting contraceptive and prophylactic information to teenagers; that laws be written affirming that only the patient and the doctor should be involved in the decision to undergo voluntary sterilization; that abortion be permitted by the states; that employers of illegal aliens be punished; that immigration levels not be increased; and that comprehensive land-use planning be implemented. The President rejected the recommendations of his population commission, especially those which would legalize abortions and make contraceptive information available to minors. The importance of the President's position is that during the next few years the executive branch of the government will probably neither propose nor support measures encouraging population stabilization.

4.9 CONSEQUENCES OF POPULATION DENSITY

(Problems 19, 24)

As the population density in a given area increases, each person's share of the available supplies of land, water, fuels, wood, metals, and other resources must necessarily decrease. In the past, people in many parts of the world have raised their standard of living despite a rising population by using the available resources with increased efficiency. However, there are eventual limits to population growth. Since we cannot predict the future advances of technology accurately, we cannot predict the maximum possible human population. The total world food and energy supplies will be discussed in more detail in Chapters 5 and 6.

Before humans are eliminated by death through starvation and thirst, it is certain that the quality of life on earth will change. Many forests and wild places will disappear and be replaced by cities and indoor environments. Some individuals will welcome the change, some will be adversely affected, but what will happen to society as a whole as populations continue to increase?

Violence, disunity, political upheavals, and personal unhappiness are attributed by some to population density. In a series of experiments with strains of rodents, John Calhoun studied the effects of extreme crowding. He constructed cells supplied with enough food and water for many more rodents than the space would normally hold. A few animals were placed in each cell and allowed to breed. The population and

See John B. Calhoun in Scientific American, *206*:139, February, 1962, and "Environment and Society in Transition," *Annals of the New York Academy of Sciences*, June, 1971.

the density grew quickly, and the animals began to act bizarrely. The females lost their ability to build proper nests or to care for their infants. Some of the males became sexually aggressive; most retreated from communication with others. In short, the normal processes of socialization were destroyed. Other literature presents evidence of decreased fertility and strange behavior in many animal species under conditions of overcrowding. However, we do not know in what way conclusions from animal experiments apply to human populations.

POPULATION DENSITY (1971)

REGION	*PERSONS/km²*
Africa	12
North America	11
Latin America	14
Asia	76
Europe	94
Oceania	2
U.S.S.R.	11
World average	27

Available data do not support the hypothesis that high population density of a nation is necessarily associated with social upheavals, violence, or poverty. Research into the nature of the relationship between population density and social problems has yielded a morass of conflicting conclusions. No consistent patterns emerge when national population densities are defined as population size divided by total national area. Such densities are grossly misleading, for they do not measure population densities in populated areas. For example, the Netherlands, one of the most densely populated areas of the world, had 319 persons per square kilometer in 1970. By contrast, the population density in India was only 168/km² and in Algeria, only 6/km². Since, however, nearly all of the Netherlands is inhabitable, while much of India is jungle and most of Algeria is desert, the fact that the density of the Netherlands is so high and its society so stable does not by itself disprove a hypothesis that high population densities are socially detrimental. These examples are not isolated instances, for the 62 per cent of the earth's surface that is semi-arid, taiga, tropical jungle, arctic, tundra, or desert holds only one per cent of its population.

The relation between population density on arable land and social problems is also confusing. Japan, which supports 1700 persons per square kilometer of arable land, is an example of a very densely populated country which maintains a prosperous and relatively crime-free society.

Some studies of population density contend that the violence-producing effect of density is most apparent in urban areas, such as cities in the United States. However, the more densely populated cities are not necessarily those with higher crime rates. For example, New York City has a lower violent crime rate than Los Angeles, even though New York is five times as densely populated. Crime rates are probably more fundamentally related to poverty than to density. Table 4.7 compares murder rates and density in the 15 largest American cities.

The degree to which density is an important factor in breeding antisocial behavior is still unknown. One difficulty in studying the effects of population density on humans is that spatial requirements may in part be culturally determined. Much sociological research is needed to gain an understanding of the factors causing members of different societies to feel crowded.

We do know that density is associated with forms of government and with patterns of social life. In primitive societies, for example, population density is related to the kind of animals hunted. If the prey is large and strong, many men are needed for attack. This leads to concentrated bands of hunters. An example of such a society was the bison-hunting Plains Indians. On the other hand, if the prey is small enough to be caught by a single hunter, families tend to live isolated from each other. The Australian bushman is an example of such a pattern of isolation. Of course, the type of prey is only one of many factors which determine living patterns. In more modern cultures, very rural societies are often too dispersed for economical government services. Schooling, health care, and postal service, for example, are extremely expensive to deliver in sparsely populated areas. Some societal responses to injustices, manifested in labor unions or protest movements, are rare without a population density high enough so that large groups of people are regularly in contact with each other. On the other hand, in very densely populated areas, government cannot be immediately responsive to individual needs, and the civic role of most citizens is severely limited.

Change in density is not usually a useful measure in examining effects of population growth. A country with rich soil and advanced agricultural techniques can support many more persons per unit area than an agriculturally backward land; industrial nations which can afford importation of food can sustain a very dense population. The economic and social effects of population growth are far more important than the effects attributable to density itself.

TABLE 4.7 MURDER RATES AND DENSITY IN THE TEN LARGEST AMERICAN CITIES (1970)*

CITY	*PERSONS PER SQ. MILE*	*MURDERS PER 100,000*
New York	26,343	14.2
San Francisco	15,764	15.5
Philadelphia	15,164	18.1
Chicago	15,126	24.1
Washington, D.C.	12,321	29.2
Baltimore	11,568	25.5
Detroit	10,953	32.7
Cleveland	9,893	36.1
Milwaukee	7,548	7.0
Los Angeles	6,073	14.0

*Data from *Statistical Abstract of the U.S.* (1972).

4.10 THE DEMOGRAPHIC TRANSITION

(Problem 23)

Historically, much of the current population explosion is due to a rapid decrease in infant mortality without a decrease in birth rate. In Western Europe, United States, Canada, Russia, and Japan, both birth rates and death rates have decreased rather slowly over the past century. In many countries of Asia, Africa, and Latin America, introduction of modern medicine has led to changes in mortality patterns, while birth rates have remained relatively constant. Typically, medically primitive societies maintain a population balance by having both high birth and high death rates. When infant mortality drops, birth rates remain high for a while, leading to a period of time characterized by rapid population growth arising from continued high birth rates and newly achieved low death rates. This period is called the time of **demographic transition**. So-called developed societies are characterized by both low birth and low death rates. They have completed the transition by reducing birth rates. Sometimes, changes in the population age distribution cause an increase in the crude death rate after transition, even though the age-specific death rates may be continuing to decline.

The reader should understand that declines in birth rates have not usually occurred simply because of propaganda, statements of public policy, or even the availability of birth control. Rather, small families have been the result of complicated social and economic forces. The developed countries have been lowering their death rates for about a century. Thus, they have had many years to allow fertility patterns to change and have avoided explosive rates of growth. Furthermore, the technological and economic development of these countries has led to societal patterns which encourage, at least implicitly, a reduction in the number of births. One important example of such a social pattern is an economic structure encouraging women to work.

The underdeveloped countries remain today in a transition period. With the assistance of aid programs, missionaries, and others, these countries have been able to drop their death rates significantly in a few decades. Unfortunately, they have had only limited time and meager resources with which to develop the technological and social patterns which, in other countries, have preceded or coincided with a drop in

TABLE 4.8 SOME TYPICAL VITAL RATES BEFORE, DURING, AND AFTER DEMOGRAPHIC TRANSITION*

		CRUDE BIRTH RATE/1000	CRUDE DEATH RATE/1000
Before transition	*Very high birth and death rates*		
	Afghanistan (1965–70)	50.5	26.5
	Angola (1965–70)	50.1	30.2
Transition period	*High birth and death rates*		
	Bolivia (1965–70)	44.0	19.1
	Indonesia (1965–70)	48.3	19.4
	High birth rates, moderate death rates		
	India (1965–70)	42.8	16.7
	Morocco (1965–70)	49.5	16.5
	Moderate birth rate, low death rate		
	Argentina (1967)	20.7	8.7
	Spain (1970)	19.6	8.5
	United States (1970)	18.2	9.4
After transition	*Low birth rate, low death rate*		
	Hungary (1970)	14.7	11.7
	Sweden (1970)	13.6	9.9

*Source: *U.N. Demographic Yearbook* (1970).

birth rates. This lack of preparation is one reason why their current growth rates are generally viewed with alarm. Table 4.8 shows some typical birth and death rates for countries at various stages of demographic development.

4.11 WORLDWIDE DEMOGRAPHIC TRENDS

(Problem 30)

Because of the marked differences in mortality and fertility patterns between the developed and the underdeveloped nations, and because of the lack of reliable demographic data for much of the underdeveloped world, future projections of world population size are much more difficult to construct than are projections of the growth of an individual country. The developed nations are characterized by low mortality and fertility. Since these patterns are likely to persist for some years into the future, population growth is expected to be slow during the next century. The total population of the developed nations was 1.1 billion persons in 1970. If the TFR were to drop to 2.1 by 1975, the population would stabilize at 1.3 billion by 2050. Such a projection represents a minimum likely population. A maximum likely population is projected by assuming that the TFR will stabilize at 3.3 by 2000. In such a case, the total population of the developed nations would be 1.8 billion persons by 2050. Since the range between the likely maximum and likely minimum is small, if we assume no nuclear war or other catastrophe, the range of estimated population size, 1.3 to 1.8 billion persons, is a useful guide.

For the underdeveloped nations, the patterns are clouded. In the highly unlikely event that the TFR were to drop to replacement level by 1975, the population of the underdeveloped countries would rise sharply from 2.5 billion in 1970 to 4.0 billion in 2050 and continue to rise rapidly thereafter. Such striking population growth, even under conditions of replacement fertility, is a legacy of the second stage of the demographic transition, for the recent decades of low mortality and high fertility have left the underdeveloped countries with large numbers of young people. Barring catastrophe, we can regard 4.0 billion persons as the minimum number of inhabitants of the underdeveloped world in 2050. If however, the underdeveloped nations were to take as much time to complete the demographic transition as did Europe, the TFR will slowly drop until it reaches 3.3 in 2000. Such a condition would lead to a population of over 11 billion persons by 2050. The range between four billion and 11 billion is so large that accurate prediction is unrealistic. When we consider, moreover, that current data for the People's Republic of China and for continental Africa are lacking, the value of predictions of population size in the underdeveloped world is even more suspect.

We can conclude that, barring widespread famine or war, the proportion of people living in what are now called underdeveloped nations will rise rapidly during the next century, unless massive emigration occurs. Today, about 70 per cent of the world's population lives in underdeveloped nations. The figure will certainly rise to at least 80 per cent by 2050, and will continue to increase thereafter. This shifting balance of population will probably bring more abject poverty in many areas of the world, for the economic corollary to our demographic predictions is that the poor nations will become poorer as they become much more populous.

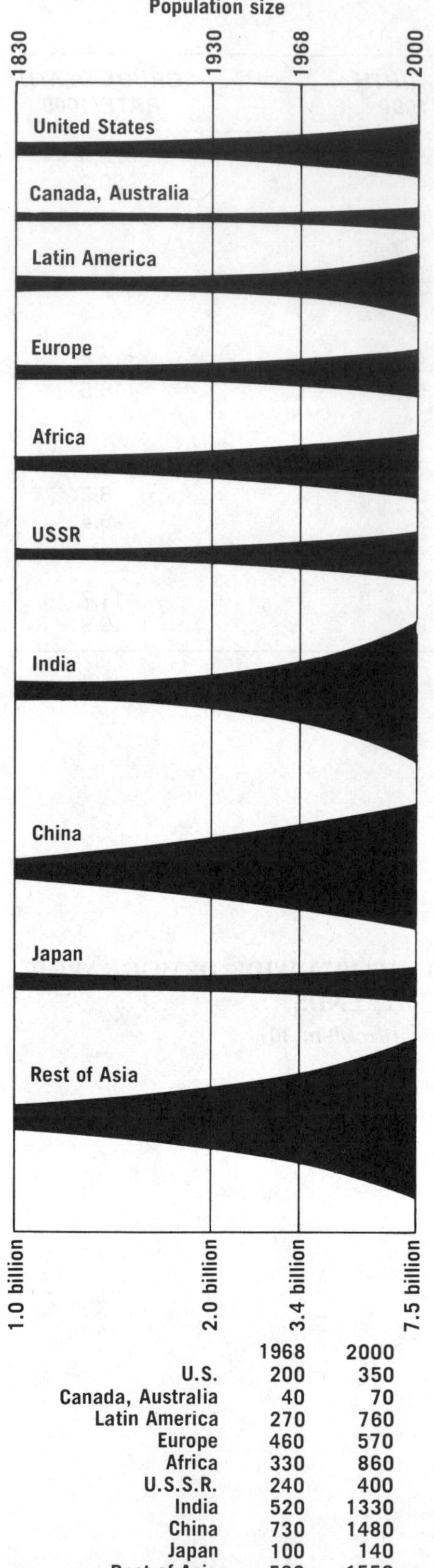

	1968	2000
U.S.	200	350
Canada, Australia	40	70
Latin America	270	760
Europe	460	570
Africa	330	860
U.S.S.R.	240	400
India	520	1330
China	730	1480
Japan	100	140
Rest of Asia	590	1550

FIGURE 4.24 Worldwide distribution of population: estimates and projections from 1830 to 2000.

4.12 CONSEQUENCES OF RAPID POPULATION GROWTH IN UNDERDEVELOPED COUNTRIES

(Problem 25)

Rapid population growth, especially when it leads to increasing urbanization and rising unemployment, may be detrimental to the economic and social aspirations of an underdeveloped country. An underdeveloped country is often defined, arbitrarily, as one in which the average per capita annual income is under $600. Birth rates in such countries are well above 30 per thousand, and often more than 40 per thousand. If an underdeveloped country is growing rapidly, much of the economic effort which could have been expended in development to improve the per capita income is instead necessarily diverted to efforts for increasing agricultural production. Many schools are needed, because a very large proportion of the population is under age 15. The development of better schools, improved health care, greater industrialization, more modern housing, and pollution control can only be undertaken after an initial effort is made to assure minimal standards of nutrition. Although an adequate diet is often not attained, rapid population growth precludes raising the per capita caloric or protein content of the diet.

Problems caused by rapid urbanization are especially acute in the underdeveloped world. In the West, urbanization has often been associated with

Houses in Brazil. (Photo courtesy of Planned Parenthood Federation of America.)

economic, industrial, artistic, and intellectual productivity. Migration in the West from rural areas to the cities has, in the past, been accompanied by fundamental changes in styles of life. Cities have been wealthier and more modern than the rural areas and have been characterized by a considerable occupational differentiation. By contrast, the patterns of urbanization in many underdeveloped countries are not associated with productivity. The modern Indian, African, and South American city is often culturally rural. Migrants to the city may not be the upwardly mobile, but rather the farmers whose crops have failed. Cities in underdeveloped nations are often poor and agricultural; the literacy rate may be very low; housing is often shoddy; food is scarce. Dakarta, India, for example, is a coastal city which has five times more people now than it did before World War II. For food, it depends upon imported grain. Since India has not dispensed food efficiently in the past, the hungry flock to the seacoast cities to be near sources of food. To transform these poor cities to modern ones would require assured supplies of food, the introduction of profitable industry, and population control.

The economic success of some developing nations, even at great odds, is truly remarkable. An example of such success is Ghana. If there were no external investment, and if fertility were constant, the per capita income would probably fall eight per cent between 1960 and 1985. If, however, the fertility were to decline only one per cent annually, per capita income would rise nine per cent; a two per cent decline in fertility should lead to a 24 per cent rise in income. These figures mean that the per capita growth in productivity is very high. In most underdeveloped countries, the rapid rise in productivity rates has not been sufficient to prevent per capita incomes from diverging farther and farther from incomes in the developed nations. The race between population growth and increased agricultural productivity is discussed in Chapter 6.

As well as having large-scale societal and economic effects, rapid population growth has implications for individual family units. Since fertility patterns change slowly, the sudden recent decline in death rates has meant larger family sizes in many underdeveloped regions. In the Matlab region of Pakistan, for instance, the average woman of age 45 has already borne seven or eight children of whom two or three have died. If the current age-specific death rates had prevailed for the past 30 years, only one or two would have died. In other words, the average 45-year-old woman in Matlab has five living children. If the current generation of mothers were to bear babies at the same rate as their mothers did, an average woman of the current generation at 45 years would have six living children. The consequences within the family are multiple. Maternal and child health, nutrition, education, and even the intelligence of the child may all be adversely affected by large family sizes.

4.13 ARGUMENTS AGAINST STABILIZATION OF POPULATION

(Problem 31)

The governments of several developed nations have expressed alarm when population growth has fallen to or below the replacement level. The fears have centered on possible adverse economic effects and the problems associated with an aging population. In 1970, responding to several years of population growth below replacement level, the Japanese government instituted a program to encourage births.

Another example is Rumania. In 1957, when the birth rate was 24.2 births per thousand, the government officially approved induced abortions for fetuses under 12 weeks; the fee was the equivalent of less than $3.00. In 1958 there were 29 legal abortions for every 100 live births; by 1966 there were 400 abortions per 100 live births. The birth rate in 1966 was 14.3. At that time, abortion was the only important method of birth control. On October 1, 1966 the liberal abortion laws were repealed. Nine months later, in July 1968, the birth rate had reached 38.7 per thousand. (By 1970, it had declined to 21.1; presumably contraceptive methods were being used.)

This figure, though officially cited, has been attacked as unrealistically high. The figure for Hungary, which is demographically similar to Rumania, in that year was 136 abortions for every 100 live births.

Clearly, the poorer the nation or ethnic group, the more it can gain by population stabilization. A wealthy group can absorb large population increases without a large or significant decrease in standard of living. A poor nation, on the other hand, has great difficulty maintaining subsistence levels of support for all its citizens, and increasing size leads to further economic pressures. Yet many spokesmen for the underdeveloped nations, and several leaders of American blacks, view the popular call for population stabilization as a fundamentally racist or imperialist position.

They claim that wealthy societies urge population limitation as a device for maintaining their own dominance. For example, if black reproductive rates were to continue, and white rates were to remain unchanged, or decrease, the proportion of blacks in the U.S. would increase from the current 12 per cent. Some American black spokesmen, feeling that increased per cent representation will have political value, suspect that the current emphasis on population planning is a white ruse designed to prevent increased black influence. Similar arguments are expressed by some representatives of poor countries. The accusation is difficult to answer, for although population stabilization will certainly tend to maintain existing political inequities, allowing continued population growth will lead to all the problems mentioned throughout this chapter.

See, for example, the testimony of Rev. Jesse Jackson as quoted in *Population and the American Future*, pp. 72–73.

4.14 METHODS OF POPULATION CONTROL

(Problems 26, 27, 32)

Social trends leading to decreased fertility, such as the increasing role of the woman outside the home, do not in themselves prevent births. Instead, such patterns lead to the desire to control actively the number of births. Three important approaches to birth prevention are available—**contraception, sterilization,** and **induced abortion.** Sterilization procedures render a person incapable of fathering or conceiving a child. Examples of contraceptives, devices or techniques which prevent conception, are the condom, the intrauterine device (IUD), contraceptive pills, the diaphragm, and spermicidal agents. Induced abortion is the artificial termination of pregnancy. Sterilization and induced abortion are, by their very nature, effective methods of population control, while the effectiveness of various contraceptive methods is a function of both the theoretical efficacy of the method and the motivation of those who practice it. For example, the rapid decline in birth rate in nineteenth-century Europe has been credited in large part to the widespread use of *coitus interruptus,* intercourse with withdrawal before ejaculation. The historical success of this error-prone method has been attributed to the strong motivation of its users. By contrast, vaginal foam should theoretically be a highly effective method of birth control, yet in studies of contraceptives, many babies are born to women using foam. A likely explanation for its relative ineffectiveness is that highly motivated women chose other methods of conception control, and therefore, the less highly motivated women relying on foam use it improperly or often neglect to use it at all.

In addition to modern methods of contraception, many traditional practices tend to limit births. Perhaps the most important is breast-feeding, for fertility declines during lactation.

Most countries have laws concerning the teaching, advertising, dissemination, and use of various birth-control methods. The laws are often contradictory and widely disobeyed. In many countries which forbid birth-control methods, the well-to-do members of the society are able to circumvent the law by leaving their country or state for an abortion or sterilization, by purchasing smuggled contraceptives, or by going to a private physician who prescribes some contraceptive method as "therapy" or as a "disease preventive." The poor in any country have fewer opportunities to violate population-control laws. Until the recent Supreme Court decision allowing abortions throughout the United States, an unmarried pregnant American girl of moderate means could have gone to New York to have a safe, legal abortion, while many of her poor counterparts who wanted an abortion were forced to choose an illegal abortionist, where the operating conditions were often quite unsanitary. Out of every 100,000 American women who are pregnant and do not have an abortion, about 20 die of direct complications of pregnancy. Of every 100,000 who have a legal abortion in a hospital, only three die. By contrast, there are 100 deaths for every 100,000 women who undergo an illegal abortion; moreover, many of the survivors are rendered infertile.

Similar problems arise in other countries. In Japan, the laws are contradictory. The criminal code forbids abortion, but the Eugenic Protection Law permits it. In Eastern Europe, abortion is often prohibited except under special circumstances. In Bulgaria, for instance, women with at least three children and women over 45 are allowed abortions, others not necessarily. Again, there are many illegal abortions. Many countries allow abortion for socio-economic or humanitarian reasons, such as after rape.

Laws concerning contraceptives are varied. They are forbidden in Ireland and Spain but are required to be stocked by pharmacies in Sweden, and by pharmacies, department stores, and rural cooperatives in the People's Republic of China. Until July 1970, when the law was declared unconstitutional by the U.S. Court of Appeals of the First Circuit, Massachusetts forbade the sale of contraceptives and the dissemination of information about birth control to unmarried women. In Poland, information about contraception is mandatory before an abortion; in Denmark it is required after childbirth and abortions. In some North African countries, the manufacture and importation of contraceptive devices are illegal.

Sterilization laws are perhaps the most ambiguous. In the United States, 26 states permit compulsory sterilization to be performed on mentally infirm patients maintained at state institutions. Only five states specifically allow voluntary sterilization for therapeutic or socio-economic reasons. In the many states without any law concerning sterilization, the resulting legal ambiguity causes many doctors to be reluctant to perform sterilizations.

In 1968, the United Nations Conference on Human Rights unanimously proclaimed that family planning is a basic human right and declared that governments should abolish laws conflicting with the implementation of such rights and adopt new laws to further such rights. The legalization of contraception, abortion, and sterilization throughout the world would help reduce population growth and curtail maternal deaths from unsanitary, illegal abortions. It should be noted, however, that this proclamation of the United Nations is a recommendation which carries no legal weight anywhere in the world.

Among people of the developed countries, the choice of various methods of birth control is largely personal. Several methods may be tried and the most satisfactory chosen. In underdeveloped countries, the problem is quite different. Usually, a team of health workers makes a decision about the type of birth-control techniques to introduce. The wrong choice can lead to a rejection of family planning. Family planners in underdeveloped countries have learned that introduction of a highly reliable method of birth control to a relatively small group of couples is often more effective in reducing births than a moderately effective method introduced to many. Methods such as diaphragms, foams, and other techniques which require careful use are often not effective in the underdeveloped countries.

The acceptability of contraceptives varies by culture. The highly successful birth-control program in Puerto Rico is due to the widespread acceptability of sterilization there. The program in Taiwan has used the IUD, whereas Japan and Eastern Europe depend on abortion for limiting population size. However, no family planning program can be successful unless the families involved are motivated to practice birth control.

4.15 THE ULTIMATE HUMAN POPULATION

We cannot know what ultimately will limit the growth of human population. Pessimists predict mass starvation, total war, irremediable destruction of drinking water, a lethal upset in the oxygen-carbon dioxide balance. They cite evidence of cannibalism in overcrowded rats, suicide in lemmings. Optimists speak of rational self-limitation. However, it is reasonable to guess that limits, either cataclysmic or rational, will not occur on a worldwide basis but rather will depend on factors within smaller geographic units. Countries or areas with decreasing birth rates may achieve self-limitation with little disruption of life and environment. In countries with rapid rates of growth, where food shortages threaten, the mechanisms that will limit population will not be pleasant ones. Various setbacks are likely to precede significant decreases in birth rate and perhaps thereby avert mass death from actual starvation. For example, there may gradually occur a general state of undernutrition, leading both to increased susceptibility to disease and to lower fertility. Contamination of drinking waters may lead to widespread epidemics of contagious diseases. Technological development, if it causes increased air pollution, will lower the general health of a population. Moreover, these factors affect other forms of life as well. For example, if pure water becomes scarce, populations of many species decrease, lowering man's food supply.

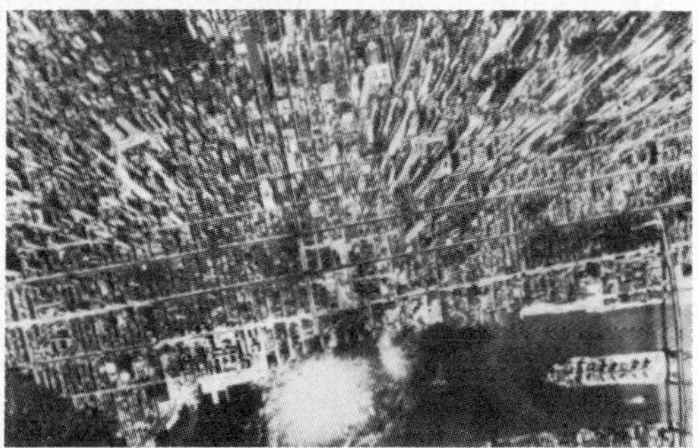

The exploding metropolis. (Courtesy of U.S. Department of Commerce. Coast and Geodetic Survey.)

The contrasts between the developed and the underdeveloped nations will undoubtedly become more marked. The most highly regarded projected growth rates predict that the less-developed regions of the world will nearly double in size by the end of the century, while the developed nations will grow to less than one and one-half times their present size. Economic and military disparities will magnify.

Traffic jam.

Predictions about man's potential growth must take into account the interrelationship of man and his environment. When we predict the amount of growth in some small period of time, it is reasonable to ignore the populations of other species. However, when we predict far into the future, we must account for other living things. We must not think of world population growing wildly until the time when there is not enough food, air, water, or room in which to stand. Rather, we would expect increasing population size to cause a gradual deterioration of levels of health and nutrition, and to lead to increases in the number of areas of the world beset by poverty and overcrowding.

PROBLEMS

1. **Population ecology.** Discuss some important factors relating to the population growth of species other than man. Contrast these factors with those important to man's population growth.
2. **Vocabulary.** Define the following terms: rate of growth; rate of change.
3. **Exponential growth.** Explain why world population size cannot grow forever.
4. **Population prediction.** Discuss some uses of predictions of population size.
5. **Growth curves.** Outline Malthus' theory of population growth. What is the distinction between arithmetic and geometric growth? Does the growth of money in a bank proceed by arithmetic or geometric progression? What about the growth in the age of a person?
6. **Growth curves.** What is a sigmoid curve? Discuss its relevance as a model for human population growth.
7. **Vocabulary.** Define demography; vital event; vital rate.
8. **Migration.** Immigration and emigration data are defined in terms of the stated intention of the migrant at the time he applies for the proper visa. Do you think such data are reliable measures of long-term migration? Define migration in terms that would be more demographically useful. Why is your definition not employed by data-collection agencies?
9. **Migration.** For which of the countries in Table 4.2 might we safely ignore migration in investigating population growth? In Japan, there were almost as many emigrants as immigrants. Can we therefore say that since net migration is low, we can ignore migration? Why or why not?
10. **Migration.** "Since there is no migration to and from

earth, to predict population size we need only consider birth and death rates." Argue for or against this statement.

11. **Age-specific mortality rates.** Discuss possible explanations for the fact that the graphs of mortality rates in Figure 4.5 cross each other.

12. **Survivorship curves.** Compare the survivorship curves for males in the U.S. and Sweden (Fig. 4.6). The probability that a person will survive *to any age* is higher for populations in Sweden than for populations in the United States. What does this information tell you about the relative average ages of men in the two countries? What about average life span? Explain.

13. **Infant mortality.** Explain the relationship between infant mortality and expectation of life.

14. **Population explosion.** Comment on the assertion that modern medicine, by prolonging the lives of elderly persons, has contributed greatly to the population explosion.

15. **Replacement value.** Under what conditions would the replacement value for the TFR be exactly 2? Can you think of a population with this value? Why is India's replacement value higher than Canada's?

16. **Family size.** What would happen to human population size if no family had more than two children? If the average family size were two children?

17. **Accuracy of data.** In Table 16 of the United Nations Demographic Yearbook (1970), the infant mortality rate in Angola for 1968 is cited as 15.9. What are the units of infant mortality rates? A footnote states, "Rates computed on number of baptisms recorded in Roman Catholic Church records." Comment on the accuracy of the quoted rate.

18. **Accuracy of data.** The U.N. cites that the crude birth rate in the Netherlands Antilles in 1968 was 23.0 and that the infant mortality rate was 20.2. These data exclude most live-born infants dying before registration of birth. How do these definitions affect international comparisons?

19. **Population density.** Population density is usually defined in terms of people per unit of land area. Density is often cited as an index of poverty. How useful is such a definition in a city with many luxury high-rise buildings, and large areas of decaying tenements? What alternative definitions would you suggest?

20. **Population distribution.** Explain why some declining animal populations have age distributions with large bases. (Consider populations where there is heavy predation of adults.)

21. **Population distribution.** How are age-sex distribution

curves useful in predicting future growth? What are some of their limitations?

22. **Population distribution.** How would you expect each of the following to affect population growth? Consider which age groups are most likely to be affected by each event and the interrelationships between the event and the population change: (a) famine; (b) war; (c) lowering of marital age; (d) development of an effective method of birth control; (e) outbreak of a cholera epidemic; (f) severe and chronic air pollution; (g) lowering of infant mortality; (h) institution of a social security system; (i) economic depression; (j) economic boom; (k) institution of child labor laws; (l) expansion of employment opportunities for women.

23. **Demographic transition.** Describe the demographic transition. In Bolivia (1964–66), the average food consumption included 1,760 Calories per person per day. In India (1968–69) the comparable rate was 1,940 Cal/day; in Spain (1970) 2,750 Cal/day; in Hungary (1969) 3,180 Cal/day. How do these caloric values relate to the stages of demographic transition (see Table 4.8)? Consider social and economic factors as well as those relating to population structures. (For example, what segments of the population eat most? Least? How would you explain Argentina's average of 3,170 Cal/day? Sweden's 2,750 Cal/day?

24. **Urbanization.** New York, Chicago, and London have been called cities ecologically dependent upon the railway and the steamship; Los Angeles and the Boston-New York-Washington megalopolis are said to be ecological consequences of the automobile. Discuss.

25. **Rapid population growth.** In times of rapid population growth from excesses of births over deaths, what types of professions and services must become increasingly available very rapidly?

26. **Birth control.** The following measures have been proposed to measure contraceptive effectiveness: (a) number of failures (pregnancies) to 1,000 women on contraceptives; (b) number of failures per year to 1,000 women on contraceptives; (c) birth rate to population on specific contraceptives; (d) completed family size to women on specific contraceptives. Discuss their relative merits.

27. **Birth control.** What factors are most important in introducing birth-control methods to an underdeveloped nation?

28. **Trends in the United States.** What factors will tend to raise American birth rates? lower them?

29. **Social policies.** In the United States, families are allowed an income-tax deduction for each child. In most

developed countries, each family is allotted an annual grant for each child. How do you think these tax laws affect family size?

30. **Social policy.** If you had the responsibility of discouraging population growth, would you consider curtailing income tax deductions if a family has more than four children? curtailing health benefits? Whom would such policies harm?

31. **Social policy.** What arguments and programs might allay the fears that population limitation programs are designed to preserve economic inequalities?

32. **Population control.** Joan P. Mencher, in an article entitled, "Socioeconomic constraints to development: The case of south India" (Transactions of the New York Academy of Sciences, 1973, pp. 155–167), points out that poor people are often treated shabbily in medical clinics. In south India, she says, "To have a baby does not require contact with hospital people, but to *avoid* having a baby requires contact with maternity assistants, doctors, etc., all of whom tend to treat the poor and low-caste people as 'animals.'" How do you think such treatment affects population control among the poor of south India?

The following problems require arithmetical reasoning and computations.

33. **Infant mortality.** Plot infant mortality rates (Table 4.3) against male or female expectation of life (Table 4.4) for Iceland, Sweden, the Netherlands, Japan, Denmark, France, New Zealand, East Germany, Canada, the United States, Czechoslovakia, West Germany, Israel, the Soviet Union, Poland, Yugoslavia, Mexico, and Guatemala. Describe and explain the relationships.

34. **Family size.** This problem investigates family size as it relates to population growth: Consider a population where 10 per cent of all women are childless. Roughly 10 per cent have only one child. Assume that all others have exactly two children and there there is no migration.

(a) How many children will a cohort of one thousand women produce?

(b) Suppose that 90 per cent of babies reach reproductive age. How many children reaching reproductive age will the cohort produce?

(c) How many children who reach reproductive age must a cohort of 1000 women produce in order to exactly replace itself? (Assume equal numbers of boy and girl babies.)

(d) How many children must be born to the cohort in order for it to replace itself, that is, to produce an eventual zero population growth?

(e) What must be the average number of babies produced per woman in order to achieve an eventual zero population growth?

(f) What must be the average number of babies produced by the 80 per cent of women who have more than one baby in order to achieve an eventual zero population growth?

(g) If women were to produce the average number of babies computed in part (e), why would a zero population growth not necessarily be achieved immediately?

35. **Population distribution.** In this problem we shall study vital rates and changes in population distribution. Let us consider a population on January 1, 1970 with the age distribution and vital rates shown at left. Assume there is no migration.

Age	Population, both sexes, in millions	Death rate/1000	Birth rate/1000
<1	4	150	0
1–14	40	15	0
15–44	44	10	100
45–64	10	20	0
65+	2	100	0

(a) What is the total population represented by the table?

(b) What is the overall (crude) birth rate per thousand per year?

(c) What is the crude death rate?

(d) What is the rate of natural increase, i.e., the difference in per cent, between crude birth and crude death rate?

(e) Suppose there are 3,000,000 14-year-olds, 1,000,000 44-year-olds, and 500,000 64-year-olds. Assume the death rates are the same for each age class. (For example, the mortality rate for 20-year-olds and for 44-year-olds is 10 deaths per thousand per year.) Finally, assume that all births occur on January 1 and all deaths occur on December 31. Show that the age distribution on January 1, 1971 is as shown at left.

Population Distribution: Jan. 1, 1971

Age	Population, both sexes in millions
<1	4.400
1–14	39.845
15–44	45.525
45–64	10.300
65+	2.290

(f) What is the total population on January 1, 1971?

(g) What has been the rate of growth during 1970? How is this answer related to your answer to part (d)? Explain.

(h) How has the population distribution curve changed shape during 1970? Discuss the implications of the decline in 1- to 14-year-olds. Which age group had the largest per cent of increase?

(i) In order to compute a population distribution for 1971, we made many simplifying assumptions in part (e). What were they? What would be the effects on your conclusions had more realistic assumptions been made?

(j) If there had been migration, how would the overall rate of population growth compare with the rate of natural increase, computed in part (d)?

36. **Stable population.** If a given combination of birth and death rates were to remain in effect for many years, and if there were no migration, a characteristic percentage composition would eventually develop. It would remain unchanged indefinitely so long as the vital rates remained unchanged. This population is the so-called **stable population.** The crude birth and death rates that would be computed if stability were achieved are called the **intrinsic birth rate** and **intrinsic death rate.** Their difference is the **intrinsic reproduction rate.** For

Colombia, Hungary, and the United States, (a) compute all reproductive rates. (b) Which crude rates represent growing countries? (c) Which countries will continue to grow if the age-specific birth and death rates were to remain unchanged? (d) Discuss reasons for the difference between the crude and the intrinsic rates. (e) Why is the intrinsic death rate for Hungary so much higher than the crude death rate? (f) Discuss the demographic differences between the United States rates for 1935 and 1965.

Vital Rates per Thousand (Females)

		Crude Rates		Intrinsic Rates	
Country	Year	Birth	Death	Birth	Death
Hungary	1965	12.3	10.0	10.5	17.6
United	1935	17.2	10.0	13.4	18.1
States	1965	18.8	8.0	21.3	8.6
Colombia	1964	36.9	9.4	38.4	10.1

BIBLIOGRAPHY

This chapter has introduced demographic techniques for analyzing population growth. There are several valuable texts available for those interested in further study of demography.

Two excellent introductory texts requiring no calculus are:

Peter R. Cox: *Demography.* 4th Ed., Cambridge, England, Cambridge University Press, 1970. 469 pp.

Donald J. Bogue: *Principles of Demography.* New York, John Wiley & Sons, 1969. 899 pp.

More mathematical introductions are:

Mortimer Spiegelman: *Introduction to Demography,* Rev. Ed. Cambridge, Mass., Harvard University Press, 1968. 514 pp. (Spiegelman includes an extremely large bibliography covering a wide range of topics related to population size, control, measurement, and so forth.)

Nathan Keyfitz: *Introduction to the Mathematics of Population.* Reading, Mass., Addison-Wesley Publishing Co., 1968. 450 pp. (This highly technical and mathematical text is especially careful in its presentation of interrelationships among various measures of population composition and vital rates.)

Sociological factors, as we have noted, are of crucial importance to the study of population growth. A useful introductory text which combines sociology and demography is:

William Petersen: *Population.* 2nd Ed. New York, Macmillan Co., 1969. 735 pp. (Petersen includes a fine annotated bibliography at the end of each chapter.)

A more advanced sociological discussion is presented by:

James M. Beshers: *Population Processes in Social Systems.* New York, Macmillan Co., The Free Press, 1967. 207 pp. (This book is useful for learning the interrelationships between social systems and patterns of demographic transition, fertility, migration, and mortality.)

Several volumes of collected papers afford most interesting reading in many areas of importance to the student of population. A fascinating collection of essays is found in:

Stuart Mudd, ed.: *The Population Crisis and the Use of World Resources.* The Hague, Dr. W. Junk, Publishers, 1964. 562 pp.

Another recommended reader is:

Garrett Hardin, ed.: *Population, Evolution, and Birth Control.* 2nd Ed. San Francisco, W. H. Freeman & Co., 1969. 386 pp.

Several books sound a tocsin for our crowded planet. One of the most popular of these is:

Paul R. Ehrlich. *The Population Bomb.* New York, Ballantine Books, 1968. 201 pp. (Ehrlich includes a bibliography of similar discussions.)

On the other hand, there is an important argument for encouraging moderate population growth expressed in a very provocative work:

Alfred Sauvy: *General Theory of Population.* (Translated by Christophe Compos.) New York, Basic Books, 1969. 550 pp.

Arguments pointing to an implicit elitist attitude among people advocating birth control are cogently presented in:

Richard Neuhaus: *In Defense of People.* New York, The Macmillan Co., 1971. 315 pp.

For the student interested in data sources, two works are highly recommended. The most useful and complete source of world population data is the *United Nations Demographic Yearbook,* published annually since 1948. For many nations and areas of the world the *Yearbook* includes the most recent available information on population sizes, vital rates, and many more specialized demographic statistics.

Another interesting source of data is:

Nathan Keyfitz and Wilhelm Flieger: *World Population: An Analysis of Vital Data.* Chicago, University of Chicago Press, 1968. 672 pp. (The authors summarize data for several countries, extrapolate various demographic measures into the future, and perform many calculations useful to the student of population.)

For up-to-date trends in population studies, refer to *Studies in Family Planning,* a monthly bulletin published by *The Population Council,* New York City.

Finally, two recent reports of commissions are especially valuable as summaries of current trends in population. The first is *Population and the American Future* (1972), the Report of the Presidential Commission on Population Growth and the American Future, and is available through the U.S. Superintendent of Documents. The second is a two-volume work entitled *Rapid Population Growth,* published for the National Academy of Sciences by Johns Hopkins Press, Baltimore, Maryland, 1971. (Vol. 1, 105 pp.; Vol. 2, 690 pp.)

5

ENERGY: RESOURCES, CONSUMPTION, AND POLLUTION

5.1 ENERGY CONSUMPTION
(Problems 40, 41)

In a natural ecosystem, radiant energy that is received from the sun and trapped in the form of potential energy in plant tissues flows through the food web which consists of various animals and decay organisms, until it is degraded completely into heat and radiated back into space. We showed in Chapter 1 how the system depends upon a continuous inflow of energy from the sun. Moreover, the total amount of sunlight puts a limit on the total metabolism of a system, for the biotic community cannot utilize more energy than it receives. The ecosystems that have existed in the past evolved to survive within these constraints and were very long-lasting.

It is clear that technological man no longer lives within this ancient energy flow pattern. Before the advent of fire, our ancestors needed only 2000 kcals of energy per day per person. The energy used was in the form of food. Later, people domesticated animals, engaged in agriculture, and used fuel for cooking and heating. Per capita energy requirements rose by a factor of roughly six, to 12,000 kcals per day. By 1860, small amounts of coal were being mined, heat engines

Remember, 1 kcal = 1 kilocalorie = 1000 calories = 1 Calorie (Capital C). The energy values in food tables are given in Calories. See Appendix.

Energy consumption by man.

had been invented, and a resident of London used about 70,000 kcals per day. In Western Europe at this time, the total population and the per capita energy requirements were so high that man needed more energy than could be simultaneously replenished by the sun. He began to use reserves of energy stored as fossil fuels. Such a situation is inherently unstable, for consumption cannot be forever higher than production. In the United States in 1970, man's per capita utilization was 230,000 kcals per day, a rate which greatly accelerates the exhaustion of fossil fuels. This prodigious rate is unique in the history of the world. At no other time, and in no other place, have people utilized energy faster than Americans do today.

TABLE 5.1 ENERGY CONSUMPTION FOR SELECTED NATIONS*

	PER CAPITA DAILY ENERGY CONSUMPTION IN THOUSANDS OF KCALS	*PROPORTION OF WORLD'S ENERGY CONSUMPTION*	*PROPORTION OF WORLD'S POPULATION*
United States	230	35%	6%
Canada	165	3%	0.6%
United Kingdom	145	6%	1.5%
Germany	110	5%	1.5%
U.S.S.R.	85	16%	7%
Japan	40	3%	3%
Mexico	30	1%	1.3%
Brazil	15	1%	3%
India	6	2%	15%

*Data from Scientific American, September 1971, p. 142 and U.N. Demographic Year Book, 1970.

The United States, home for six per cent of the population of the world, is responsible for 35 per cent of the world's energy consumption (see Table 5.1). What does it mean to say that the average U.S. citizen uses 230,000 kcals of energy daily? Recall from Chapter 1 that energy isn't a material, like steel, that you can hold in your hand, but rather it is the ability to do work or to heat a body above the temperature of its surroundings. (See Table 5.2 for units of energy.) The

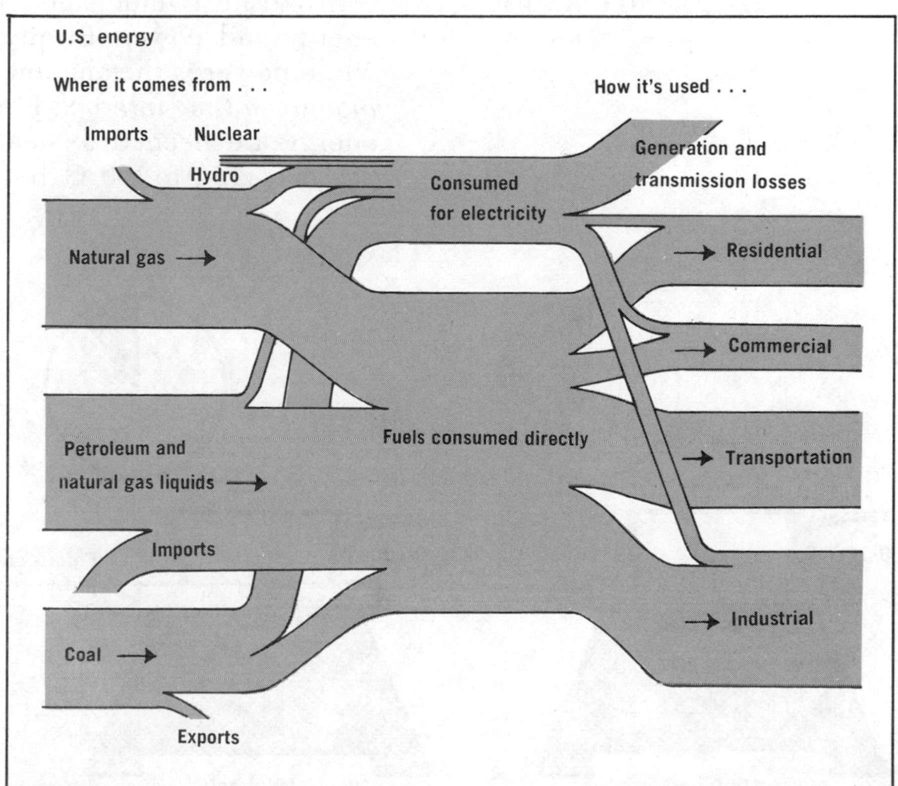

NOTE: Based on graphical scheme by Earl Cook, Texas A&M University. Relative scale is based on 1972 study by Office of Science and Technology of energy consumption in 1968 in B.t.u.'s. (From *Chemical & Engineering News*, November 13, 1972, p. 21.)

calorie was originally defined as a unit of heat. Translated into American units, 230,000 kcals is the energy needed to heat 10,000 gallons of water approximately 11 Fahrenheit degrees. But technology is primarily concerned with doing work, not heating water. In these terms 230,000 kcals is also equal to the work of lifting 710,000,000 pounds one foot. Since people are more versatile in their daily activities, the 230,000 kcals of energy are therefore spread out in a variety of ways: a man heats his home, drives his car, bears responsibility for the fuel used by the trains, trucks, and tractors that serve him; he also burns lights and runs electric appliances; indirectly he uses the energy consumed by the industries that supply him with goods. This utilization therefore seems less personal than the 2000 daily kilocalories of food for primitive people, or even the 12,000 kcal rate for early agriculturalists. In fact, industry consumes about 41 per cent of the total energy used in the United States. Furthermore, 230,000 kcals is the *average* utilization per person. The poor in America consume only a fraction of this amount daily; the wealthy consume much more.

5.2 ENERGY AND POWER

(Problems 1, 2, 3, 4)

We must understand the relationship between energy and power. Briefly, energy is heat or work, while **power** is the amount of heat or work delivered *in a given time interval.* For example, 1000 calories of energy are needed to heat 10 grams of water from 0°C (freezing) to 100°C (boiling). The correspondence

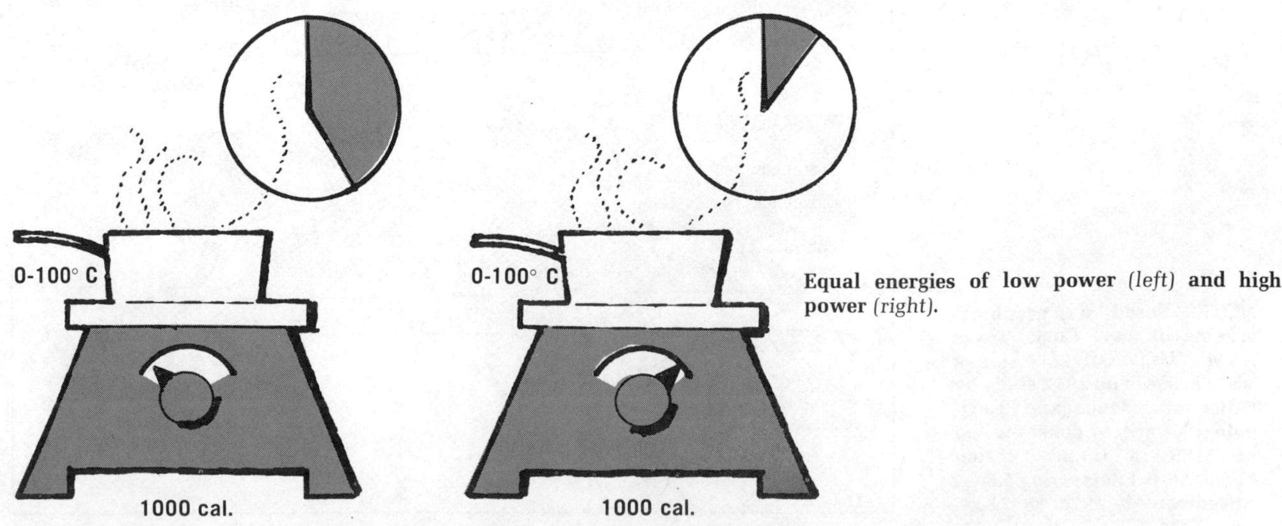

Equal energies of low power *(left)* and high power *(right)*.

between energy and the temperature rise of a quantity of water is an inherent property of water and is independent of how we do the operation. We could put the water on a hot plate, or rub two sticks together under the water, or spill the water down a long pipe and allow friction to heat it, or do any other conceivable operation involving heat or work. The energy required to heat 10 grams of water from freezing to boiling would always be 1000 calories. If we used a hot plate, we could choose to turn the switch to "low" and wait a while, or turn it to "high," and accelerate the process. When turned to "low," the hot plate uses small quantities of energy per unit time, but more time is required to deliver the 1000 calories. In the "high" position, large quantities of energy per unit time are delivered for a short while. The total energy delivered to the water is the same, but the power delivered in the two operations is different, simply because power is defined as the energy per unit time (see Table 5.2).

Where does energy go after it has been used? A lump of coal turns to ash, but what becomes of its energy? This non-substance, called "the ability to do work," is elusive, hard to define, and hard to keep track of. A piece of steel may wear out with use and get bent up a bit, but at least it remains recognizable. Could we not find used energy and reuse it? Or better yet, if energy isn't really matter, could we find some for nothing? These two questions plagued scientists for a long time, and the search for the answers led to the development of the science of heat-motion, or **thermodynamics.**

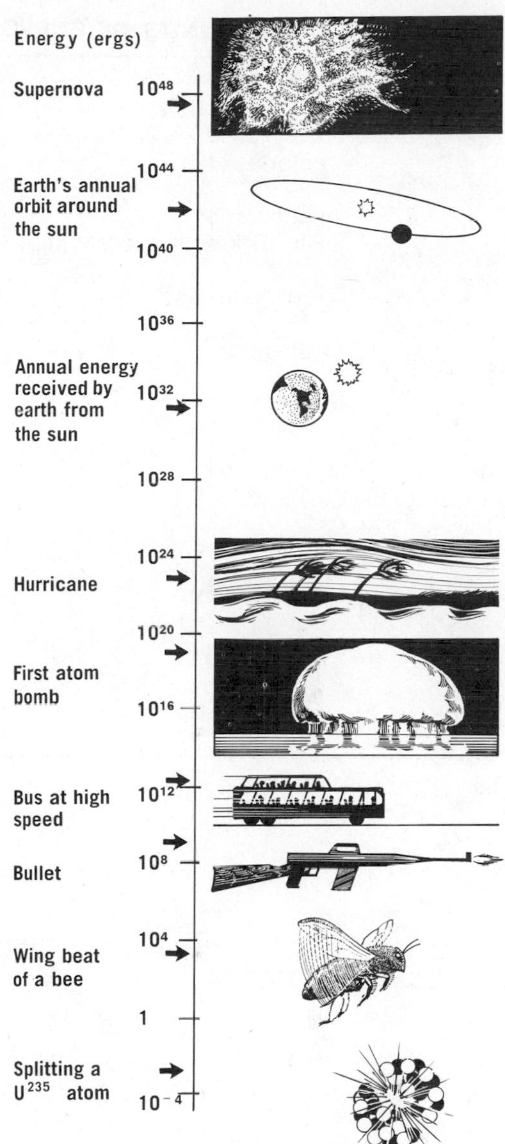

5.3 WORK BY MAN, BEAST, AND MACHINE

To understand the first two Laws of Thermodynamics and the answers to the hopeful questions posed above, let us search for solutions to the following imaginary work problem:

A landslide causes a large boulder to roll downhill; it comes to rest in a position where it blocks the entrance to a cave, and a man wishes to remove the obstruction. What can he do? Primitive man might have tried to push it, roll it, or drag it. If he lacked the strength to get it out of the way, he would have been forced to search for a new cave. However, if he could cooperate with his fellows, perhaps a group of men, working together, could have been able to push or roll

TABLE 5.2 COMMON UNITS OF ENERGY AND POWER

Unit	Meaning
	ENERGY
calorie	The heat required to raise the temperature of one gram of water one degree Celsius.
kilocalorie	1000 calories.
Btu (British thermal unit)	The heat required to raise the temperature of one pound of water one degree Fahrenheit.
foot-pound	The work required to raise one pound one foot in the air.
watt-hour	The energy released when one watt of power is delivered for one hour.
erg	A fundamental metric unit, the work required to accelerate one gram of mass at a rate of one centimeter per second per second through a distance of one centimeter
	POWER
horsepower	The power required to lift 550 pounds one foot in the air in one second.
watt	The power produced when ten million ergs are supplied in one second (equal to 0.0013 horsepower).
Btu/hr.	The rate of power at which one Btu will be delivered in one hour.

or drag it out of the way. The situation was still further improved after people had learned to domesticate large animals and use them as beasts of burden. But a much more far-reaching development for doing work was the invention of machines. The lever, the wheel, the roller, the screw, the inclined plane, and, later, the gear and the block and tackle were all devices that could increase the force that a man or his animals could exert.

These three techniques—cooperation among people, the use of animal power, and the application of machines—enabled man to build (and sometimes to destroy) such wonders as the pyramids of Egypt, the Great Wall of China, and the cities of Athens, Rome, and Carthage.

We will now discuss other ways to get work done. But do not forget the cave man and his boulder; we shall return to him.

5.4 THE FIRST LAW OF THERMODYNAMICS (OR "YOU CAN'T WIN")

As machines became more and more effective in extending the force of living muscle, inventors saw no reason to doubt that, by providing machines that would produce work indefinitely just because their mechanisms were so clever, continued improvement in design would eventually free man and his animals entirely from their labor. If our cave man had such a device (which we call a "perpetual motion machine of the first kind"), he could move all the boulders he wished at his leisure. It may be difficult for the modern reader to appreciate the fact that this objective seemed entirely reasonable, apparently requiring only continued progress along the lines that had already been so successful. However, all attempts failed. The failures have been so consistent that we are now convinced that the effort is hopeless. This conviction has been stated as a law of nature and is called the **First Law of Thermodynamics.**

(Problems 5, 7. We recommend especially that you try Problem 7 before you laugh at what you think was the naiveté of the attempts at "perpetual motion.")

This law can also be expressed in terms of conservation of energy by the statement "energy cannot be created or destroyed."

Don't ask for a proof of the First Law. There is none. The First Law is simply a concise statement about man's experience with energy. If it is impossible to create energy, then it is hopeless to try to invent a perpetual motion machine, and we may as well turn to some other method of doing work.

5.5 HEAT ENGINES AND THE SECOND LAW OF THERMODYNAMICS (OR "YOU CAN'T BREAK EVEN")

(Problems 6, 8)

Modern technology demands far more energy than can be supplied by men and beasts. Instead, it makes use of heat engines, which consume fuel to produce heat, and convert the heat into work.* The idea that heat could be converted into work was far from obvious. In fact, heat engines have been used successfully only during the past 200 years or so. (James Watt developed his steam engine in 1769.) The first experimental proof that energy can be converted, without gain or loss, from one form to another was supplied by James Joule in 1849. Heat and work are two forms of energy, and it is therefore possible to convert heat into work or work into heat. Thus one can create heat by rubbing two sticks together, and the heat produced is exactly equivalent in energy to the work required to rub the sticks. It was also discovered

*Of course, a person (or an ox) is also a heat engine. People consume food, which is their fuel, and convert its energy into the work of muscle contraction. Mechanical heat engines, however, can be much larger than animals, and can consume fuels such as gas, oil and coal that animals do not eat. The result is that the total amount of work done and heat produced is increased to an extent that has new effects on the environment.

that fuels contain potential heat. Thus, a pound of coal contains stored energy which can be released by combustion. But heat is also stored in substances that are not fuels—in such ordinary substances as water. Water loses energy when it turns to ice; therefore, water must *have* energy. Why not, then, use this energy to drive a machine to do work? Such a machine, although seemingly not quite so miraculous as the perpetual motion machine of the first kind, would extract energy from its surroundings (for example from the air or from the ocean) and convert it into useful work. The air or water would then be cooled by the extraction of energy from it, and could be returned to the environment. Automobiles could then run on air, and the exhaust would be cool air. A power plant located on a river would cool the river while it lighted the city. Such a machine would *not* violate the First Law, because energy would be conserved. The work would come from the energy extracted from the air or water, not from an impossibly profitable creation of energy. Such a device is called a "perpetual motion machine of the second kind"; alas, it too has never been made and never will be.

Let us return now from the impossible to heat engines that use fuel, where the situation continues to be discouraging. It was learned through experiment that the potential energy inherent in a fuel could never be completely converted into work; some was always lost to the surroundings. We say "lost" only in the sense that this energy was no longer available to do work; what it did, instead, was warm the environment. Ingenious men did try to invent heat engines that would convert *all* the energy of a fuel into work, but they always failed. It was found, instead, that a heat engine could be made to work *only* by the following two sets of processes: (a) Heat must be absorbed by the working parts from some hot source. The hot source is generally provided when some substance such as water or air (called the "working substance") is heated by the energy obtained from a fuel, such as wood, coal, oil, or uranium. (b) Waste heat must be rejected to an external reservoir at a lower temperature.

A heat engine cannot work any other way. The original form of this negative statement, as made by Lord Kelvin (1824–1907) is, "It is impossible by means of inanimate material agency to derive mechan-

William Thomson (later Lord Kelvin) was a British physicist who proposed the absolute scale of temperature (the "Kelvin" scale) in 1848. His major contributions to science were in the field of thermodynamics.

ical effect from any portion of matter by cooling it below the temperature of the coldest of the surrounding objects." This is an expression of the **Second Law of Thermodynamics.**

To help us gain further insight into this very fundamental concept, we return to our cave man. Imagine that he discovers he can move his boulder by wedging a bar of copper between it and the cliff and then heating the bar with a flame. Because heated metal expands, and the cliff is stationary, the boulder will move. He has constructed a basic heat engine. Now let us assume the following circumstances:

(a) The cave man has several copper bars, each 10 meters long.

(b) It is very awkward to build a fire under the copper bar between the boulder and the cliff, but it is convenient to have a fire nearby to heat the copper bars to a temperature of 100°C.

(c) The outside temperature is 20° C.

Refer to the Appendix for a discussion of temperature.

(d) The bar expands 0.17 millimeter for every degree of temperature rise.

The cave man, thinking he has discovered that hot copper bars can do work, heats one of his bars on the fire, wedges the hot bar between the cliff and the rock, then waits and watches. The bar cools and contracts, and no work is done. This failure teaches him his first lesson in the design of heat engines: Work is done only by heating materials and allowing them to expand. Therefore, temperature differences rather than absolute temperatures are important. Having learned this, our cave man now decides to wedge a *cold* bar between the rock and the cliff and to bring hot bars to the cold one. Let us say that he heats three bars to 100°C (we will ignore heat losses to the air). He places the first hot bar on the cold one, as shown in Figure 5.1. The cold bar gets warmer and expands, while the hot one cools and contracts. When both reach the same temperature, all heat transfer (and hence all work) stops. The temperature of the working bar, which was originally 20°C, rises 40 degrees while the temperature of the heating bar drops 40 degrees. The final temperature of both bars is 60°C. In order to move the rock further, he places the second hot bar on top of the working bar and observes that the working bar warms from 60° to 80°, and the new hot bar cools from 100° to 80°. Repeating the process a third time, the working bar heats from 80° to 90°. Recall that for

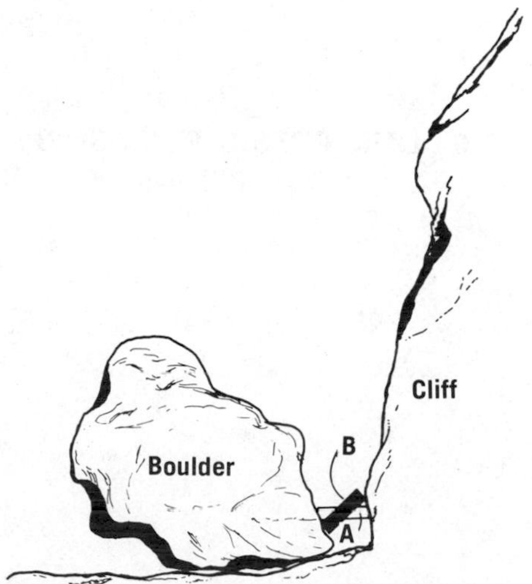

FIGURE 5.1 Bar A, originally at 20°C, warms up to 60°C. Bar B, originally at 100°C, cools down to 60°C.

every degree rise, the working bar expands 0.17 mm. Thus, the first time he heats the bar the rock moves 6.8 mm, the second time it moves 3.4 mm, and the last time it moves only 1.7 mm. Therefore, the amount of work that can be extracted from a given quantity of heat depends on the temperature difference between the hot and the cold bars. This limitation is *not* imposed by the First Law; if the hot bar were to cool to 20°C and the cold one warm to 100°C, their total energies would be conserved. But such an event is contrary to all experience. In fact, the universal observation that two bodies in thermal contact eventually reach a common temperature can be taken as another expression of the Second Law.

Of course, if our cave man were clever, he would not need to tolerate a diminished output of work after the use of each successive hot bar. Instead, after each heating he would cool his working bar back to 20° in some nearby stream, replace it on the cliff (with an additional stone wedge to make it tight), and move the boulder a full 6.8 mm for every hot bar he used. The excess energy would be dissipated as heat and the stream would become warmer. But because the stream would be heated only a small amount, no useful work could be obtained from it, and, as we shall see in Section 5.18, it would become thermally polluted. Thus, only some of the energy from the firewood could be used to produce work; the rest would be lost to technological processes. This observation leads to a restatement of the Second Law: "Energy cannot be completely recycled."

5.6 OUR FOSSIL FUEL SUPPLY

(Problems 9, 10, 11)

In 1970, 95 per cent of the total energy used in the United States was derived from fossil fuels—coal, oil, and natural gas. Only four per cent was from a renewable supply—mostly power from falling water, plus some wood and solar energy. The remaining one per cent was from nuclear fission. The Second Law assures us that someday our reserves will be depleted. How much time do we have? For a realistic estimate of the number of years before man has used all the earth's energy reserves, we must forecast man's population growth rate (see Chapter 4), estimate the quantity of the remaining reserves, and predict accurately man's future rate of consumption. All such forecasts are subject to large errors.

Geologists have used several methods to estimate

the remaining reserves of fossil fuels. Fuels that have been positively located and identified are called *proved* reserves. Moreover, from seismic recordings and other data, the location and size of *probable* sources have been recorded. Educated hunches and preliminary data give us a third category, *possible* deposits. By adding to the proved sources a reasonable fraction of the probable ones, and assuming that new, as yet unthought-of sources will balance disappointments among the possible deposits, we can get a rough guess at the world's supply of fuel.

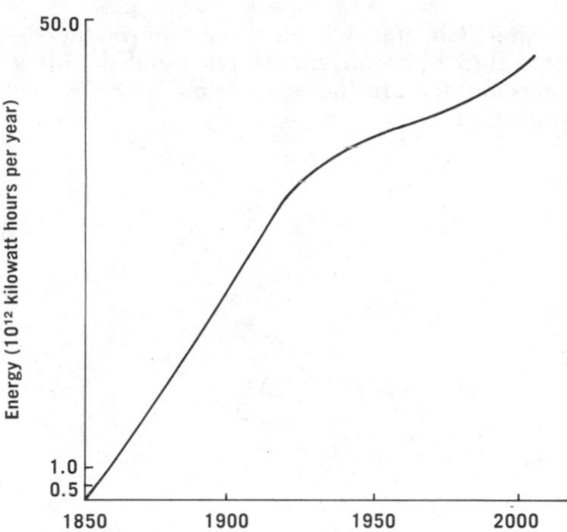

FIGURE 5.2 Growth of U.S. energy production. (Adapted from The Energy Resources of the Earth, by M. King Hubbert.)

To estimate the energy requirements from the 1970's into the early twenty-first century, we start by graphing the past energy consumption, and then try to guess how the curve will continue in the future. Figure 5.2 shows the production of energy in the United States from 1850 to the present. From 1850 to about 1915, the average growth in energy production was 6.9 per cent per year, compared to a population growth of only four per cent. During this period, then, the quantity of energy used doubled every ten years. Between 1915 and 1955 the consumption rate increased only 1.85 per cent per year, corresponding to a doubling in use roughly every forty years. In the period of time between 1961 and 1970, consumption has been increasing about 4.5 per cent per year, corresponding to a 15- to 16-year doubling period. During the same period the population was rising at a rate of 1.1 per cent. Total world use of energy is increasing at a rate of roughly seven per cent per year. If that rate remains constant, world energy requirements will double every 10 years.

It is obvious that the world cannot double its consumption of a depletable resource every ten years for very long, for each doubling corresponds to an increasingly large increase in the magnitude of energy. Think of it this way: If energy consumption is doubling every 10 years, then in the past 10 years we have used as much energy as in our whole previous history. Or another way: If energy consumption is doubling every 10 years, then by the time we have used up half our total reserves, we will have only enough left for another 10 years. Even if there were a very large reserve of fuel, the environmental side effects of energy production would eventually limit expansion. In that case, a reasonable prediction for growth with time is depicted in Figure 5.3. This resulting curve, called a

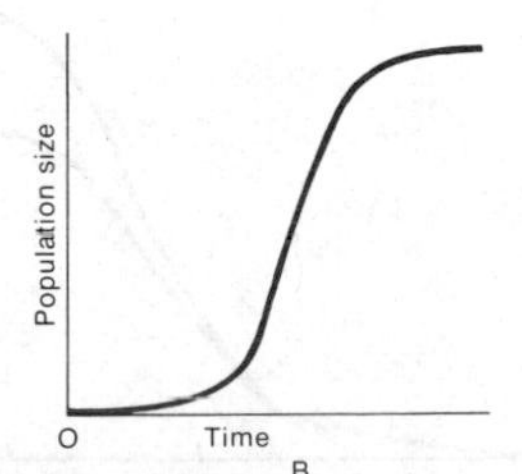

FIGURE 5.3 Sigmoid growth curve.

See Chapter 4 for a discussion of the relationship between growth rates and doubling times and for a further discussion of curves and projection.

sigmoid function, shows an initial growth rate, followed by a rapid rise, and finally a leveling off. This level portion would correspond to a condition in which the environmental cost of increased use would be higher than the expected gain.

In the real world it is reasonable to suspect that the combination of environmental side effects, and depletion of resources will lead to the leveling off and eventual decline of consumption of a depletable fuel. We can expect that for some time consumption of coal, gas, and oil will increase rapidly as technology improves the methods of exploration, mining, and transmission. As poorer reserves must be tapped and pollution control becomes more expensive, the cost of power will rise and people will consume less.

It is generally believed that natural gas is our least abundant fossil fuel. Moreover, government price regulations in the United States have had the effect of discouraging exploration. The peak consumption rate will probably be reached about 1985, or before many of the readers of this book have reached middle age. Already, contractors in New Jersey and other areas are unable to supply gas lines for heating new houses because the industry claims it will not have enough fuel for new customers. The scarcity of natural gas is environmentally unfortunate since it burns cleaner than any other widely used fuel.

Figures 5.4 and 5.5 combine estimates of world supply with the predicted rates of use to display past and predicted coal and oil production. We have barely begun to utilize our coal reserves. We do not mean to imply that there will be sufficient energy available from coal in the year 2300 to supply all of our needs. Figure 5.4 tells us simply that world production of

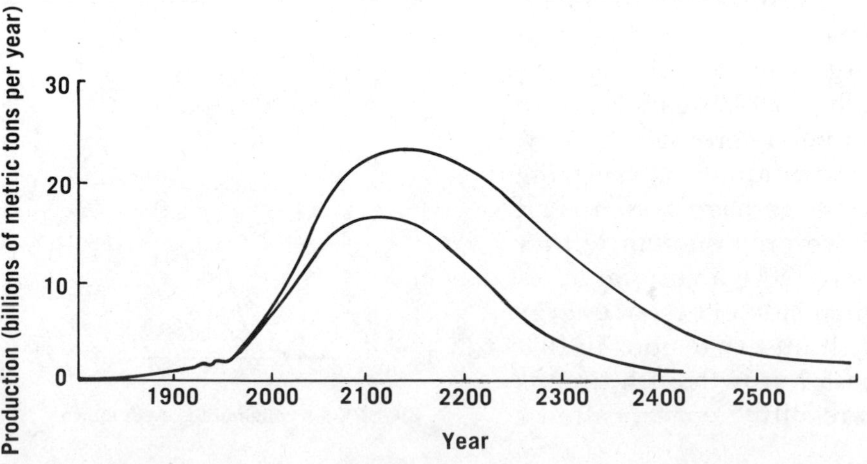

FIGURE 5.4 **Past and predicted world coal production based on two different estimates of initial supply. (From** *Scientific American,* **September, 1971, p. 69. Copyright © by Scientific American, Inc. All rights reserved. Adapted from "The Energy Resources of the Earth," by M. King Hubbert.)**

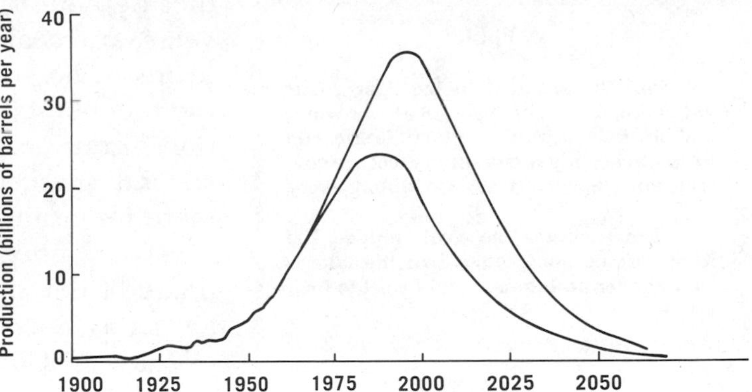

FIGURE 5.5 Past and predicted world oil production based on two different estimates of initial supply. (From *Scientific American,* **September, 1971, p. 69. Copyright © by Scientific American, Inc. All rights reserved. Adapted from The Energy Resources of the Earth, by M. King Hubbert.)**

coal will be around 10 billion metric tons per year at that time. Probably the world will require even more fuel and people will be exploiting other sources. The fossil fuels cannot be the primary source of energy for a rapidly expanding population for more than a few generations. When 90 per cent of the reserves have been removed, the energy needed to mine and extract the fuel approaches the energy released when it is burned, so that further exploitation is uneconomical. According to reliable estimates, natural gas in the United States will be 90 per cent depleted around 2015; the world supply of oil will be 90 per cent depleted between 2020 and 2030, and world coal between 2300 and 2400.

Articles in the popular press often present estimates quite different from these, for they often use two other, much less accurate ways of projecting the world's supply of fuel. One is to predict the length of time the reservoirs would last at the current rate of consumption. However, since it is unreasonable to assume that both population and per capita consumption will not increase above present levels, this viewpoint is unrealistically optimistic. A survey in 1966 showed that at present rates of consumption coal reserves would last 1440 years. On the other hand, some authors present alarming predictions of depletion based on the current rate of acceleration in use. Just as population cannot be expected to grow at a rapid pace until one fine day we wake up to find that each person has one square foot of land area allocated to him, so the world society will not continue to accelerate utilization of a resource right up to the day that it is all gone. Using the alarmist prediction, coal reserves would reach 90 per cent depletion by the year 2088. The more realistic arguments already presented pre-

"Be practical — people always need coal."

(From: The Energy Crisis. Science & Public Affairs Book Bulletin of the Atomic Scientists, p. 63.)

A Fable

Said the scientist to the King, "Our only lake, from which we get all our water and our fish, is growing a terrible slime on its surface. Only a few square feet are covered now, but the area is doubling every day."

"Don't bother me yet," replied the King. "Come back when half the lake is slimed over and then we will start to look into it."

dict depletion of gas reserves in 35 years, oil in 50 years, and coal in 350 to 400 years. Will we be driving teams of horses to town when the oil runs out? Will the technological age die completely with the depletion of our coal? The answer is almost surely no, for several stopgap measures are available, and a few more long range solutions may be found.

Until now, most of our petroleum has been extracted from underground wells or "lakes" made up of a viscous, but pumpable liquid. Underground deposits of shale in Colorado, Wyoming, and Utah have been found to contain large quantities of petroleum impregnated in the pores of the rock, and sand fields, laden with oil, have been discovered in Alberta, Canada. The quantity of oil that can be obtained from oil shales and tar sands is unknown. The richest deposits, containing about 100 gallons of oil per ton of ore, are believed to add another 25 per cent to the world's initial supply of liquid petroleum. At present, importation of oil is usually more economical than mining the shales, although a Canadian firm is currently extracting oil from tar sands, and a large pilot plant for handling oil shale is in operation in western Colorado. As imported oil becomes more costly and as the technology needed for extraction becomes more highly developed, shale deposits will be used more. Many shales have smaller quantities of fuel per ton of rock. If all the shales containing from five to 100 gallons per ton are considered to be recoverable for fuel, the total reserves are equal to *100 times* the initial estimated supply of coal. We do not know how much of this can in fact be used. Present technology would require more than five gallons of fuel to mine, transport, extract, and refine a ton of shale. Thus, it is clearly uneconomical to mine some shales. But what about the shale with 50 gallons of oil per ton? Twenty-five? As the cost of fuel from other sources increases and as the technology of retrieving oil from shale advances, oil shale exploitation will probably increase, for lower-grade ores will become relatively more economical. The break-even point, and hence the available reserves, cannot now be predicted.

Another method of increasing the supply of oil and gas is to convert coal to liquid fuel. Today, chemists can convert organic chemicals from one form to another to synthesize new chemical products. The state of this art is so highly advanced that many medicines and plastics are presently synthesized from coal. Attempts have been made to convert coal, which is relatively plentiful, into methane, which is scarce. Methane is a very desirable fuel because it burns cleanly and is simple to transport, while coal is dirty and imposes handling problems that are particularly difficult for small users. The theory of the conversion of coal to methane is well understood, and the practice has been worked out in the laboratory. The United States Bureau of Mines has been searching for a contractor to construct a pilot plant to convert coal to gas. The only major problem now is cost. Synthetic methane can be expected to cost two to three times as much to produce as the natural fuels now cost to extract. But that does not mean that the cost of the gas to the consumer would be increased proportionately. Initial cost is only a small fraction of the price of natural gas. If transportation costs, distribution costs, and profits were held constant, the total proportional price rise would be much less. Moreover, the present price base for natural gas is rising.

During World War II, the Germans converted coal to gasoline and, surprisingly, the cost was not unreasonable. Coal can also be converted to kerosene, motor oil, and other petroleum products. Unfortunately, research in the United States in this area has not been pursued vigorously enough in previous years. However, the cost of refining crude oil is growing while new technical improvements are reducing the cost of converting coal to oil. By the end of the century we shall probably see the conversion of solid to liquid fuels with little change in the prevailing market prices of the liquid products.

$$\text{coal} \xrightarrow[\text{high pressure}]{\text{hydrogen}} \text{liquid fuels}$$

5.7 ENVIRONMENTAL PROBLEMS CAUSED BY COAL MINING AND OIL DRILLING

(Problem 12)

Coal can be mined either in the traditional underground tunnels or in open pits. The former method is more expensive and less efficient, for coal seams lying between tunnels are often left untouched. Compared to open pits, underground coal mines are more prone to fire, and thus more dangerous to the miners. However, the surface contours of the land are left undisturbed and the topsoil is not removed, although occasionally buildings situated above old coal mines have collapsed as the land above the tunnels settles.

To improve the efficiency of mining, coal com-

Strip mining operation. (Photograph by Arthur Sirdofsky.)

panies are switching an increasingly large portion of their operations to open pits, otherwise called **strip mines.** In the operation of a strip mine, the surface layers of topsoil, subsoil, and rock are first stripped off by huge power shovels or draglines, exposing the underlying coal seam. The soil is piled alongside the cut or in an adjacent cut which has already been thoroughly mined. The coal seam is then mined and a new cut is started. By 1965, 1.3 million acres of land in the United States had been mined in this unsightly way. Less than half of this land has been reclaimed to its former agricultural productivity. The credit for this rejuvenation must be shared about equally between natural processes of succession and remedial actions by man. The least harm is done to flat, fertile land, which can be mined, reclaimed, and farmed again within five years. On the other hand, hillsides in West Virginia have been ravaged so badly that they cannot be expected to be useful for farming or recreation within our lifetimes. Some of the land that has been reclaimed by green plants has not been returned to its original contours and its former productivity, but instead low-quality trees or grasses have replaced hardwoods or corn.

Unreclaimed strip mines are a multifaceted environmental insult. Neither facts nor figures are needed to convince a discerning person that open, rootless, dirt piles are uglier, less useful, and more liable to water erosion than the natural forest or prairie. The uglification is an esthetic loss to all of us. Erosion silts streams and reservoirs, and kills fish. Furthermore, sulfur deposits are often associated with coal seams. This sulfur, generally present as iron sulfide, FeS_2, reacts with water in the presence of air to produce sulfuric acid, which runs off, together with the silt, into the stream below.

Strip mined land.

The United States has roughly 40 million acres of coal deposits which can be strip mined in the near future. Such vast areas magnify the problems outlined above. A much greater effort at land reclamation should be made. If the true environmental cost, the ruined streams and land, are levied against the price of coal, an incentive would be provided to rework and replant the land. Naturally, the cost will be reflected in higher fuel bills, but it is a question of direct cost in dollars and cents or indirect cost in terms of our recreation and food reserves. The price must be paid in some form of currency.

There are various possible reclamation procedures available. The simplest technique is to regrade the old mine to produce pleasing contours, and then cover it with topsoil and replant. Land reclaimed in this manner will necessarily be lower than the original surface, and this change may upset the water table of the area.

Strip-mined areas can be refilled with various solid wastes, such as ashes. A one-million-watt electric generator fueled by coal produces 850 to 1000 tons of ashes every day, and thus provides an abundant source of filler. Unfortunately, rainwater reacts with coal ash to produce lye, which is a strong caustic and therefore a water pollutant. However, lye seepage can be prevented by sealing the pit with an impervious material.

Lye is any strongly alkaline (basic) solution, especially that which is formed by extraction of ashes. A caustic substance is one that is destructive of living tissue; chemically the term refers to a strong base, particularly NaOH (caustic soda, or sodium hydroxide) and KOH (caustic potash, or potassium hydroxide).

Another procedure is practiced by the Aloe Coal Company of Pennsylvania, which fills its pits with compacted garbage from the city of Pittsburgh, thus reducing two environmental problems in one economical operation.

Alternatively, we could return to tunnel mining, but, aside from the waste and increased cost, more men would die to provide a cleaner environment for the rest of us.

In contrast to the disruption caused by coal mining, oil drilling in an area of flat land, such as the plains of Texas, is relatively innocuous. As our easily accessible oil fields are being pumped to depletion, however, man has invaded the delicate ecosystems of the continental shelves and the Arctic for more petroleum.

Despite careful precautions, accidents seem to occur periodically in all industrial operations, and drilling for oil is no exception. Broken drill pipes, excess pressure, or difficulties in capping a new well have repeatedly led to blowouts, spills, and oil fires. When these have occurred on land, the problems have been locally contained and the environmental disruptions have been minimized. However, when they occur on offshore rigs such as those currently in operation in the coastal waters of Louisiana and Southern California, the result is disastrous. Refer to Chapter 10 for a more detailed discussion of the possible ecological consequences of oil spills.

5.8 TRANSPORTATION OF FUELS: A CASE STUDY OF THE ALASKA PIPELINE

(Problems 13, 14)

After fuels are extracted from the earth, they must be transported to the refinery or the power plant. Thus far, the shipment of coal or uranium by train, and oil or gas by pipeline, has produced a minimum of environmental disruption. As the more accessible sources become exhausted, however, fuels are increasingly sought in more inhospitable places, where transportation may cause environmental problems. Our example concerns the large quantities of oil and gas that have been discovered in the north slope of Alaska. Because fossil fuels are needed by our civilization, these deposits will eventually be exploited. However, the Arctic tundra is such a delicate ecosystem that transporting the oil may be very disrupting to the environment. The Alaskan tundra and forests are very much larger than any wilderness in the contiguous United States, and any pipeline across this region will necessarily affect its wilderness quality.

The proposed pipe would run from Prudhoe Bay on the north slope to the ice-free port of Valdez in southern Alaska (see map), a distance of 789 miles. Much of the route travels through ecosystems of the northern tundra, while the southern section travels through forests. Each day the pipeline would carry about two million barrels of oil, which would be heated to 145°F (63°C) for practically all of the journey. Heating the line would be necessary because at the prevailing frigid temperatures of the surroundings the cold oil would be too thick to flow easily.

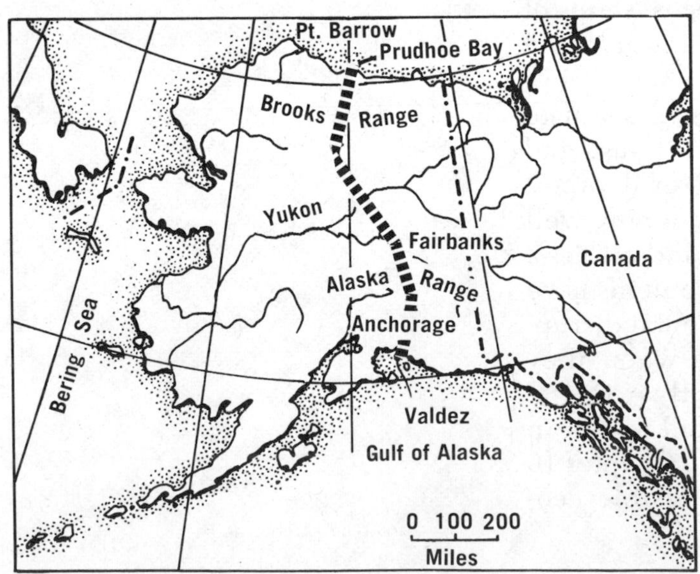

Proposed pipeline route.

The following are some of the important environmental side effects of the pipeline:

1. During pipeline construction, new roads would be constructed in areas previously inaccessible to motor vehicles. These roads would provide easy access for hunters and tourists, necessarily leading to deterioration of the wilderness.

2. Construction workers riding in jeeps, skimobiles, or airplanes have run caribou to death. Even though these practices are illegal, importation of thousands of additional workers for pipeline construction would virtually assure continuation of such violations of the wilderness.

3. The roadbuilding and pipeline projects would require roughly 80 million cubic yards of sand, gravel, and rock. The strip mining of hillsides and stream beds near the construction route to obtain this fill would almost certainly cause silting of previously pristine wild rivers.

4. The land of the tundra consists of a thin layer of topsoil over a layer of permanent frozen subsoil, or **permafrost.** During the summer, the hot line would melt the ground ice and kill vegetation close to the line. The resulting wastelands, though very small in area, would be in the shape of a continuous band. In turn, removal of the plant cover would alter the insulation properties of the soil adjacent to nearby permafrost, causing uneven freezing and thawing. When mud is in contact with ice, soil slippage is probable, and slipping soil can cause a break in the pipeline. Even with emergency shutdown procedures, a break could easily cause spillage of about two and a half million gallons of oil.

5. Although some southern areas of the tundra have been successfully replanted after being disrupted by heavy equipment, experimenters have thus far been unable to replant tundra vegetation in many northern areas.

6. The caribou is a migratory animal that spends its summers near the northern seas and its winters in the southern forests. Herds numbering about 150,000 individuals still group in the tundra every spring at their calving grounds, and many of the animals live a life that is, as yet, largely undisturbed by humans. The effect of a narrow band of soft, unfrozen, barren earth on the migration of caribou is unknown. They might be able to cross it, or they might falter when their feet

Picture of caribou on tundra. (From: Bernard Stonehouse: *Animals of the Artic.* **Holt, Rinehart & Winston, 1971, p. 19.)**

sink into unknown turf, and not be able to complete their journey.

7. To reduce the probability of a break caused by ground slippage, it is proposed to elevate about half of the pipe either on wooden stilts or on a bed of gravel. Such structures might impose another serious barrier to the migrating caribou.

8. The proposed route travels across an area prone to earthquakes, and the probability is high that a large quake would occur sometime during the operation of the pipeline.

9. If, during the operation of the pipeline, a break does occur, the consequences may be severe. At the very least, hundreds of acres of delicate tundra would be destroyed and rehabilitation would be quite slow, but widespread economic or environmental perturbations would be minimal. At worst, however, a break in the pipeline could spill millions of gallons of oil onto the Gulkana River, a commercially valuable salmon spawning ground. The Gulkana flows into the Copper River, which presently supports heavy summer populations of wild ducks and geese.

10. Because of routine losses and possible accidents large quantities of oil would probably spill into the ocean at Valdez. The pipeline corporation has announced that it will build special ballast treatment tanks to reduce the amount of oil discharged. But we do not now know how much oil would be spilled, how much reduction would be realized, or how much oil could be tolerated by the aquatic life without severe damage to the northern oceans.

The Alaskan Pipeline presents complicated environmental problems. Conservationists have proposed alternate routes which they feel would reduce the destruction to the tundra. One suggestion is to run the oil pipeline adjacent to a natural gas pipe which is proposed to run across the Canadian Yukon and along the Mackenzie River. Since the same roads would serve both lines, this plan would reduce migratory disruptions and other environmental stresses. No tankers would be needed at Valdez, since the pipe would go directly to the central United States. However, no solution will leave the environment unscathed.

Before the pipeline is built, people should consider how much the existence of a remote area matters. Americans today want both a clean environment and the benefits of a sophisticated but cheap technology. To achieve both, compromises must be made, and the Alaska pipeline project is a case in point. If the project were delayed for a few years to allow time to study the environmental impact of the pipeline, the cost to the consumer might be a small rise in gasoline prices, or a similar increase for fuel oil or for plastics.

Recent legislation passed by the U.S. Congress in the summer of 1973 has greatly reduced the chance for a delay in the construction of the pipeline.

There are no people alive today who can remember seeing large buffalo herds travel across the North American prairie. Very few have paddled a canoe along the Mackenzie, the Gulkana, or the Perth Rivers and seen moose in the northern forests, or geese nesting in the muskeg country, or caribou darkening the tundra. The construction of the pipeline is not likely, by itself, to bring about the extinction of any species, but it will certainly reduce the herd and flock sizes; it will reduce the remoteness and desolation of the North American wilderness.

5.9 NUCLEAR FUELS

(Problem 16)

Oil shales, tar sands, and chemical conversions will undoubtedly extend the life of the Fossil Fuel Age, but such stopgaps cannot be the basis for a long-range continuation of our technological existence. Instead, we must turn either to renewable resources or to nuclear energy.

In 1970, only one per cent of the energy used in the United States was provided by nuclear fuels, but this fraction is expected to increase rapidly in the near future. By the year 2000, 25 per cent of the total energy of this country will probably be supplied by nuclear plants, and, of course, the percentage will continue to rise. We must therefore consider the availability of ^{235}U, the isotope now used as a fuel in nuclear reactors. Once again, we are concerned with both abundance and concentration in natural ores, since both factors determine cost. Presently, the uranium from which the fissionable ^{235}U is extracted (see Chap. 8) costs about eight dollars per pound, but the high-grade ores which make such a low price possible are scarce. Table 5.3 shows the relative amounts of ore available at various prices. Authorities predict that by the year 2000 the low-cost ores will be substantially consumed, and hence we will be paying higher electric bills. We will complain and perhaps curtail our consumption, but most of us will not change our lives much. Moreover, since expensive low-grade ores are

TABLE 5.3 AVAILABLE URANIUM RESERVES

COST IN DOLLARS PER POUND	TONS OF URANIUM AT THIS OR LOWER PRICES	INCREASE IN COST OF ENERGY, IN MILS/KWH, RESULTING FROM INCREASE IN FUEL COST
8	594,000	0
10	940,000	0.1
15	1,450,000	0.4
30	2,240,000	1.3
50	10,000,000	2.5
100	25,000,000	5.5

relatively plentiful, this high-priced energy will be available for a thousand years or more.

Within this century breeder reactors, which use plentiful non-fissionable ^{238}U or ^{232}Th to synthesize fissionable materials, are likely to be developed (see Chapter 8). These fertile fuels are present in the currently mined uranium ores, and additional reserves exist in the granites of the Sierra Nevada mountains, in the White Mountains of New Hampshire, and in many of our states just west of the Appalachian Mountains. Granite in the Sierra Nevada contains 50 p.p.m. of non-fissionable uranium. Converted into fuel, one pound of granite could be expected to produce a heat equivalent of 70 pounds (32 kg) of coal. If we used all the mountain granites and other sparsely concentrated reservoirs, we might obtain 100 to 1000 times the energy supplied by the original fossil fuel reserves.

Beyond the breeder lies the possibility that man will someday harness the fusion reaction as a power base (see Chapter 8). Some authorities expect fusion never to be functional, while others predict the first feasibility demonstration within 10 years. Most members of the scientific community feel that development of such a power plant is highly unlikely before the year 2000. If fusion is found to be practical, the power available to us will be enormous. Probably the first to be developed will be the deuterium-tritium reaction which requires lithium for the synthesis of tritium. We have only enough lithium to provide energy equal to our original supply of fossil fuel, but if the deuterium-deuterium fusion reaction can be used, fuel can be obtained from ocean waters, and power will be available to man for a million years or so.

5.10 RENEWABLE ENERGY SOURCES

At present, our efforts to explore various renewable energy sources are meager indeed. We have seen that the renewable sources on which we used to depend will no longer suffice. In the early nineteenth century, the power base for most of the world was wood, a renewable energy source. Forests cannot grow fast enough to provide fuel for people's varied needs today.

In this section,we shall discuss harnessing energy from other renewable sources: the sun, the rivers and streams, the tides, the wind, and the burning of garbage.

(Problem 17)

Solar Energy

Every 15 minutes, the solar energy incident on our planet is equivalent to man's power needs for a year at his 1970 consumption level. Thus, if we could trap, concentrate, and store solar energy, man's power needs would be fulfilled, but unfortunately the technology is not now available. At present, the cost involved in harnessing solar energy is huge. For example, in 1970 the investment in an electrical generating power plant amounted to about $500 per kw (kilowatt) for a fossil fuel station, $750 per kw for a nuclear facility, and $2,000,000 per kw for an electric solar cell collector. Experiments with new materials suggest that the cost may be reduced to $2500 per kw. This cost and the amount of land needed are prohibitively large. If research can provide new materials, perhaps the cost of energy from the sun can be greatly reduced and a virtually limitless supply of power will be available. The successful utilization of solar energy would solve all our energy needs during the life of the planet, for a renewed supply arrives daily, but many research workers feel that the full potential will never be realized. This prediction, of course, is based on the questionable assumption that new developments will merely refine our present techniques, but assumptions come and go, and never is a long time.

Top: Solar collectors for heating water. Bottom: Parabolic solar heaters oriented to receive maximum sunlight. (Courtesy Prof. Harry Lustig, Department of Physics, the City College of the City University of N.Y.)

Hydroelectric Power

Today, only four per cent of the power used in the United States is derived from hydroelectric sources. The sun evaporates water at sea level and the winds bring it to the mountains, where it collects into streams with potential to fall and release energy.

Energy in falling water.

Thus, hydroelectric energy is an indirect form of solar energy. As more sites are exploited, more power will be tapped by harnessing falling water. As the total power need of the country continues to rise, however, the relative importance of hydroelectric power will probably not increase much. There simply are not enough good locations for building large dams, and those that are available are often in recreational areas. Furthermore, damming the Grand Canyon or the Snake River Gorge would be a great aesthetic loss.

Geothermal Energy

Energy derived from the heat of the earth's crust, called *geothermal energy*, has gained increased attention in the past few years. In various places on the globe, such as in the hot springs and geysers of Yellowstone National Park in Wyoming, hot water is produced near the surface. Although no one is suggesting the harnessing of Old Faithful for generating electricity, there are several places where hot (500°F) underground steam is available for the price of a deep well. The Pacific Gas and Electric Company has connected a generator to well holes in central California and is presently producing electricity from them on a small commercial scale. This project is new in the United States, but an Italian facility at Larderello has been in operation since 1913 and is presently working at a capacity of 400,000 kw. Even optimistic supporters, however, do not expect a large portion of our power to be supplied in this manner, for there are not enough hot springs at or near the surface. Additionally, continuous exploitation for more than a century or two is expected to exhaust the water or heat content of these wet wells.

A more ambitious proposal to drill one to four miles through the earth's crust into its molten core has much greater potential. New types of drills have been devised which melt their way through bedrock. If the core could be reached, water could be poured down the hole and the resulting steam withdrawn. If successful, geothermal tappings of this type could provide man's requirements for energy far into the future. Great care must be exercised in such drillings, for the deep wells and large quantities of water flowing through the holes might cause slippage of rock layers, and hence earthquakes.

(A) Leopards mounted as trophies at a taxidermy shop outside Nairobi (Kenya), safari center of East Africa. (Photograph by Tom Nebbia, courtesy of DPI.)

PLATE 3

(B) Baby giraffe being taken for the zoo trade. (Photograph by Tom Nebbia, courtesy of DPI.)

PLATE 4

Pipeline installation on Arctic tundra. (Courtesy of Carlson, Rockey & Assoc., Inc., New York, N.Y.)

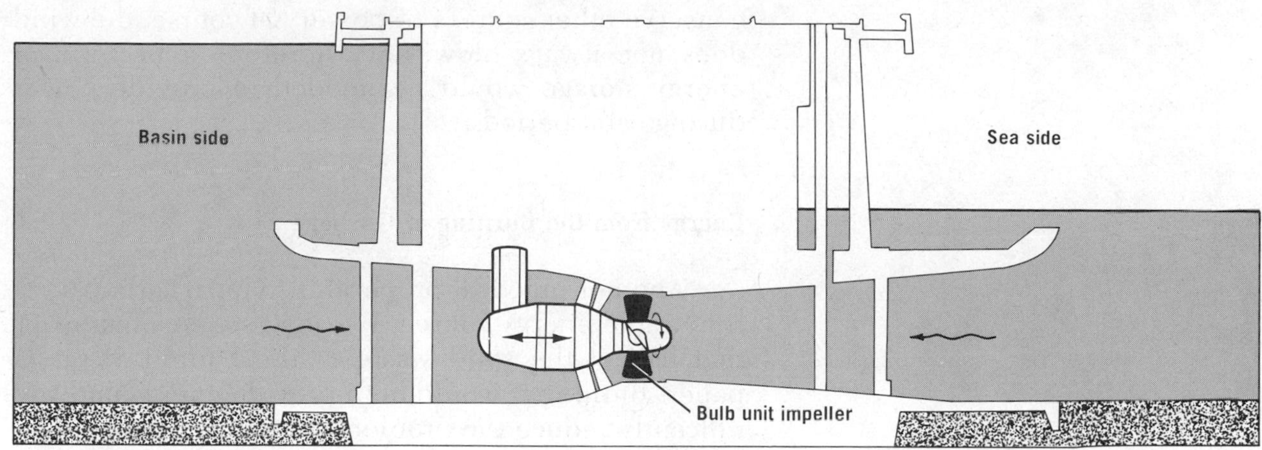

Tidal dam and turbine.

Energy from the Tides

The force of the tides as a source of power has long intrigued man. Imagine a turbine which is immersed in the sea near shore. Electricity could be generated as the tides turn the turbine blades, but the output for a single generator would be small. Moreover, lining the coast with millions of small generators would be uneconomical.

An alternative practical solution is to build a tidal dam across a bay or estuary where there is a large rise and fall of the tides. By damming and funneling the water, significant concentrations and economical power output can be realized. A tidal generating plant in France on the coast of Brittany generates 240,000 kw, about the output of a small fossil-fueled plant. If we should exploit this resource further, however, we shall have to reckon with the consequences on aquatic life that would inevitably result from the construction and maintenance of dams across the estuaries. Also, there are not enough properly oriented bays and estuaries to provide more than a small percentage of our energy requirements.

Energy from the Wind

The power of wind has been used since antiquity to pump water and grind grain. Construction of large-scale electric generating stations of this type would pose serious engineering problems because of their huge size, but small windmills as an auxiliary source of home electricity would be economical and would

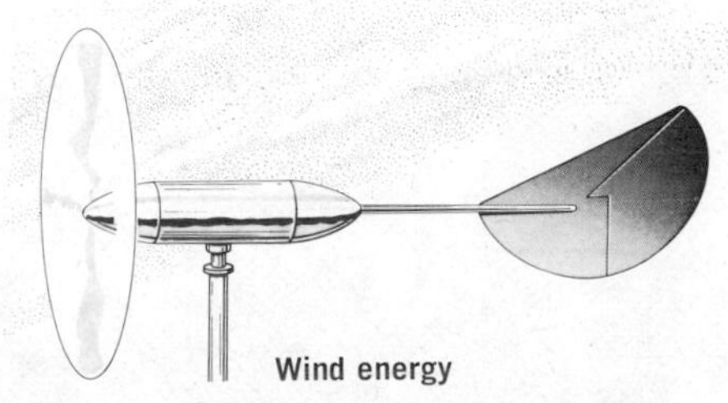

Wind energy

conserve other sources of power. Of course, the wind does not always blow, and therefore some form of energy storage would be needed to provide power during calm periods.

Energy from the Burning of Garbage

Another practical proposal is to burn garbage as a renewable energy source. We shall see in Chapter 11 that half of the solid waste of the United States is paper. Burning it would both provide energy and significantly reduce the problem of solid waste disposal. In fact, a Parisian company has been burning solid wastes to generate electricity for 50 years. At the present level of operation, a total of 1.5 million tons of garbage are consumed per year as an auxiliary fuel source. In the United States in 1971, there was an estimated 880 million tons of dry solid wastes, of which only 135 million tons are actually recoverable for use as a fuel, the remainder being difficult and uneconomical to collect and transport. But this 135 million tons could provide energy equivalent to 170 million barrels of oil (15 per cent of the 1971 consumption in the U.S.), or 1.35 trillion cubic feet of natural gas (38 per cent of the total). Saving 170 million barrels of oil in a year while simultaneously reducing solid waste pollution is a healthy approach to environmental problems. Why, then, aren't we burning garbage now?

One of the difficulties discussed in Chapter 11 is that of persuading large numbers of householders to classify their garbage. Another problem results from the use of large quantities of plastics in packaging. Plastics burn with more smoke than wood or paper; furthermore, chlorinated plastics, like polyvinyl chloride, generate hydrochloric acid and thus produce additional air pollution. Engineers could, at a price, remove noxious smokestack emissions, and housewives could, if they tried, (and perhaps if they were given economic incentives) keep plastics out of the wastebasket.

$$\left(\begin{array}{cc} H & H \\ -C- & C- \\ H & Cl \end{array} \right)_n + \text{oxygen} \rightarrow CO_2 + H_2O + HCl$$

Clearly, the burning of garbage is not a panacea. But if garbage, solar, geothermal, tidal, and wind energies were all exploited along with the implementation of various social remedies outlined later, the total contribution could be very substantial. In an age during which our fossil fuel reserves are rapidly being depleted and the transition to a fusion-based power grid may be very slow, we would be wise to buy a few more years' time by exploiting alternate energy sources.

5.11 THE ENERGY CRISIS

(Problems 15, 18)

If we have enough coal to last 350 to 400 years, and enough uranium to last well into the twenty-first century, and if the potential exists for harnessing large amounts of power from solar, geothermal, and wind sources, why are there energy shortages in large cities throughout the world? One reason is that our urban-suburban complexes are becoming so populated, their fuel requirements are so large, and the mining, transportation, generating, and transmission are so complex that the power industry cannot move quickly in response to changes or emergencies. A new nuclear power plant takes six years to complete and will probably become obsolete thirty years later. Fossil-fueled generators cannot be built much faster. A coal mine requires four years to open, and an oil field may take up to ten years to reach production. Therefore, if power consumption rises faster than had been predicted a decade earlier, the supplier's capability in an area will not be able to supply the demand. The unpredicted, rapid rise of home air conditioners in New York City has been partially responsible for the power shortages experienced there in recent summers. There is enough coal in the ground, and enough turbines being built in the factories to supply the required power, but they are not in use now.

The magnitude of our need poses other problems as well. The Commonwealth Edison Company of Chicago consumed twenty million tons of coal in 1970. That rate is approximately 55,000 tons per day. It takes such a tremendous industrial capacity to process all that coal and ship it to the power plants that it would require considerable time to provide enough *extra* capacity to take care of emergencies. Therefore, any breakdown, strike, or similar disruption can put Chicago out of power. Storage capabilities cannot maintain very many days' reserve and if, for example, the railroad goes on strike, no alternate source of transportation for such quantities can be mustered. Or suppose that a heat wave suddenly causes a 10 per cent increase in power requirements for air conditioning. Such a surge would call for 16,000

Alternatives

(From the Chicago Sun-Times.)

tons of additional fuel per day (assuming that power is delivered at 35 per cent efficiency; see page 192. If the fuel supply cannot be stepped up rapidly enough, a power shortage results.

In the preceding section we have shown how alternate methods of power production might, in the future, provide energy for our civilization. We should not underestimate the lag that will accompany any large-scale transition to new sources of energy. It is also important to reemphasize that although coal will be commercially available in the future, other fuels will soon be required in quantity. Quite possibly, the new sources of energy won't be available in time to replace fossil fuels, and the next 50 years may see a very serious fuel shortage. Because our civilization needs a continuous flow of energy to survive, we must encourage research now to avert future disaster.

But let us now look at the energy situation from a different viewpoint. Why are these huge quantities of fuel needed in the first place? In other words, what is the nature of our energy consumption, and can we use less energy instead of increasing the supply? These are rather new questions. For many years the emphasis was just the opposite—the utilities were urging us to use more and more energy. Ostensibly, the purpose of such exhortation was to show us the way to increased human comforts; coincidentally, it fostered the prosperous growth of the utilities. If we are to find answers to these questions, we must first consider the distribution of energy consumption. In 1971, 25 per cent of our total energy was applied toward transportation (about two-thirds of which was consumed by automobiles), 41 per cent was used industrially, and 34 per cent was used in household and commercial applications. We will examine these categories individually in the three following sections.

5.12 ENERGY CONSUMPTION FOR TRANSPORTATION

(Problems 19, 20)

The two *least* efficient modes of transportation, the automobile and the airplane, are the two fastest-growing industries in the transportation field. Consider six means of moving people: walking, bicycles, trains, automobiles, "jumbo" jet passenger planes, and supersonic transport (SST) planes. If the energy required to drive each of these is converted into passenger miles (pm) per gallon of petroleum fuel, we obtain the results shown in the margin of the facing page.

Consider also various means of moving freight. We will express these values in relative units that are dimensionally similar to those cited above, namely the number of miles that a given amount of freight can be transported by a gallon of fuel, using as a comparison standard an air-freight rate of one mile per gallon.* The data are shown in the margin.

The implications of these data are clear: reliance on inefficient methods of transportation contributes to the rapidly accelerating depletion of our fossil fuels. Even if the technology of the twenty-first century should provide us with vast supplies of energy, such wastage of fuel is harmful for several important reasons:

1. Every gallon of fuel consumed represents a multiple source of environmental disruption. Land is preempted and defaced in mining and in the establishment of transportation routes. The refining of petroleum and the consumption of fuel in internal combustion engines produce serious air pollution problems, as discussed in Chapter 9.

2. While new sources of power, such as nuclear reactors, are becoming increasingly available, they are not mobile sources. Fossil fuels are particularly valuable because they can be used in small engines. There are, of course, various possible ways to compensate for the dearth of fossil fuels by converting some of the abundant energy from stationary sources to more mobile forms of energy. The most obvious of these is the use of rechargeable batteries (see Chapter 9). Another appealing alternative is hydrogen, available in quantity from the dissociation of water: $2H_2O + \text{energy} \rightarrow 2H_2 + O_2$. Hydrogen is a convenient, mobile, and clean fuel that burns in air: $2H_2 + O_2 \rightarrow 2H_2O + \text{energy}$. Note that this equation is just the reverse of the preceding one. This chemical reciprocity means that hydrogen can serve as a medium for the transfer of energy from large stationary sources to small mobile engines. The problem is that we do not expect to be able to construct enough electric generating stations to produce hydrogen in quantity in this century, and that cheap petroleum reserves will be virtually depleted by 2020 or 2030. For a world that runs on oil, that's a slim margin.

Bicycle	**1000 pm/gal**
Walking	**660 pm/gal**
Intercity bus	**85 pm/gal**
Train	**50 to 200 pm/gal, ranging from poorly patronized intercity routes to commuter trains.**
Automobile	**10*n* to 40*n* pm/gal, depending on model and condition of car, where *n* = number of passengers.**
Jumbo jet	**20 to 25 pm/gal, for estimated average passenger loads.**
SST	**13.5 pm/gal, for estimated passenger loads.**

Pipeline	**93 mile/gal**
Railroad	**63 mile/gal**
Waterway	**62 mile/gal**
Truck	**11 mile/gal**
Airplane	**1 mile/gal**

*The actual value for air freight is 42,000 Btu per ton mile.

3. As natural fibers such as wood and cotton and metals such as iron and copper become scarcer, the economic pressure to shift to plastics continues to increase. Commerical plastics, by and large, are organic substances whose energy content is roughly at the level of fossil fuels. (Coal burns; so does polyethylene). The conversion of fossil fuels to plastics and to other valuable organic chemicals, therefore, does not demand unreasonable consumption of energy (although it does require ingenuity). If fossil fuels were exhausted, chemists could synthesize organic materials from energy-poor carbon sources such as atmospheric carbon dioxide and carbonate minerals, but the manufacture would require a replacement of the energy deficit at a correspondingly higher cost. Any shift to "abundant" metals will necessarily incur a similar deficit. The best example is aluminum, which is abundant in nature in the form of its oxide, Al_2O_3, the major constituent of bauxite ore. Aluminum oxide is an extremely stable (energy-poor) substance, and its conversion to metallic aluminum is comparable in energy demand to the conversion of carbonate rocks to plastics.

Although the petroleum crisis will become acute during our lifetimes, little attempt is being made to remedy the situation. Some of the fuel used for transportation could be conserved if one or more of the following measures were adopted:

On November 7, 1973, President Nixon proposed several approaches to conserve fuel used for transportation. He recommended a national speed limit of 50 mph, voluntary car pooling, and possible rationing of fuel. Such methods do indeed reduce energy consumption, but some of them are difficult to enforce. No references were made to mass transportation, or to the curtailing of road building.

1. Construct large-scale, efficient, mass transport systems, such as high-speed trains, in and between major cities.

2. Curtail road building. When public transportation becomes more convenient than driving private cars over crowded highways, public habits will change.

3. Place an environmental tax on fuel consumption, with the highest rates directed at the least efficient modes of transportation.

4. Accelerate the development of methods for obtaining oil by extracting it from shale or manufacturing it from coal.

Proposals of this sort will be dealt with in more detail in the last chapter.

5.13 ENERGY CONSUMPTION FOR INDUSTRY

Industrial power consumption involves a wide variety of products and processes. In Table 5.4 industrial consumers are listed in decreasing order of their energy demands. Manufacturing processes are using increasingly large amounts of energy because both the demand for goods and the energy needed per item are increasing. One reason why the energy needed per item is so high today is the trend toward increased automation to offset labor costs. Another factor, however, is more foreboding. As a natural resource like iron ore becomes depleted, lower-grade ores become commercially more attractive. The mining, purification, and metallurgy of low-grade ores requires more energy than the same operations performed on high-quality deposits.

(Problem 21)

TABLE 5.4 INDUSTRIAL ENERGY CONSUMERS IN THE UNITED STATES

TYPE OF INDUSTRY	*TOTAL ENERGY (TRILLIONS OF BTU PER YEAR)*
Primary metals	5298
Chemicals and chemical products	4937
Petroleum refining	2826
Food and kindred products	1328
Paper and allied products	1299
Stone, clay, glass and concrete	1222
All other industries	8050

One way to reduce industrial energy consumption is by increased recycling of material. In general, production of manufactured goods from recycled wastes consumes less energy than the production from raw materials. However, total manufacturing costs reflect the price of labor, transportation, and capital investment as well as fuel bills, and unfortunately, under our present economic structures, recycling operations are often more expensive than primary production. Unless industries are offered economic incentives or penalties, we can expect little immediate increase in industrial recycling.

Many aspects of the problem of industrial energy consumption are closely linked with the problems of solid waste disposal, for most manufactured goods are ultimately discarded. Thus reevaluation of planned obsolescence, reduction of paper products in packaging, and general reduced consumption would greatly help alleviate environmental stresses caused by manufacturing and waste disposal.

5.14 ENERGY CONSUMPTION FOR HOUSEHOLD AND COMMERCIAL USE

(Problems 22, 23, 24)

Most of the residential and commercial uses of power involve the burning of fossil fuels to heat air and water, and the consumption of electricity for other purposes.

Fire for heating was the most ancient use of fuel. Without space heating, people could not live in some areas they now inhabit. Today, space heating accounts for only 12 per cent of our total energy consumption (compared to 16 per cent for the automobile, for example) and remains one of the most efficient uses of fuel. Home furnaces can be built to operate at 65 to 85 per cent efficiency, a figure well above that for electric

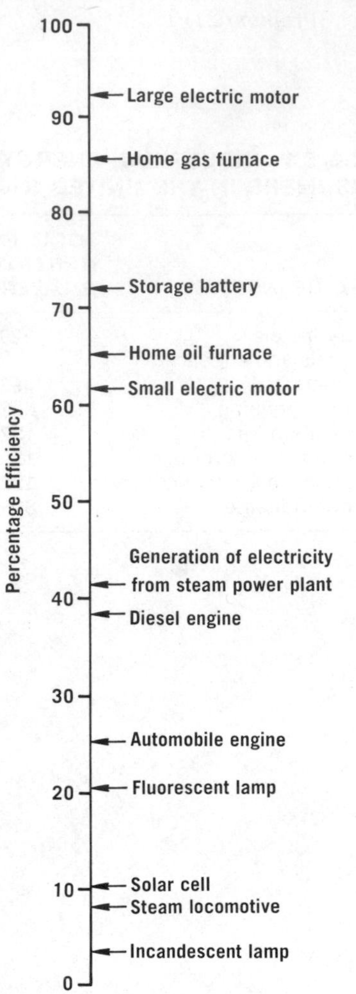

FIGURE 5.6 Efficiencies of some common machines, devices, and processes. The values for electrical devices do not include the energy losses at the generating station.

generating stations or automobiles (see Fig. 5.6). However, these efficiencies are seldom realized in the home because most furnaces are poorly adjusted. Moreover, most homes are poorly insulated. Building in most areas in the United States is regulated by strict legal codes. Yet the insulation requirements of the Federal Housing Authority are below the recommendations of the insulation manufacturers. Compliance with the manufacturers' suggestions would reduce fuel consumption and hence fuel bills, by about 33 per cent. The increased investment in insulation would be recovered in the first or second year.

Despite the high efficiency of home furnaces, gas and oil units do pollute the air, and the control of pollution on a house-by-house basis would be very costly. Therefore, there is an increasing trend toward the use of "clean, efficient, electric heat," as expressed by electric utility advertising slogans. If "electric heat" is taken to mean heat generated by electrical resistance, as in a giant electric oven, then the process in the home is, in fact, clean and 100 per cent efficient. Of course, the pollution is simply released at the site of the electric generating station instead of in the home. As we will see in Section 5.16, commercial electric generating stations are only 41 per cent efficient at best, and, if we include losses in transmission, we cannot expect more than about 33 per cent efficiency as delivered to the home.

The technology is available to make improvements toward the realization of clean, efficient electric heat, if we are willing to pay the price. Consider the household refrigerator, which we normally think of as being only a cooling machine, but which actually is a device that uses fuel to pump heat from the cold interior to the warmer exterior. Therefore, the refrigerator is just as much a warming machine as a cooling machine. The fuel is needed because, according to the Second Law of Thermodynamics, heat would normally flow the other way, from the hot outside room into the cold refrigerator, as in fact it does when the electricity fails. If one stands behind the refrigerator while is is operating, one can feel the warm air being pumped out. Heat pumps that can heat a room by cooling the outside are, in fact, commercially available, and can be designed to achieve efficiencies three times as great as electric resistance heaters. Thus, the overall efficiency of an electric heat pump, including calculation of the inefficiencies of the power plant, is competitive with home furnaces. However, high installation costs have discouraged the wide application of heat pumps.

Another device for heating the home is the solar collector, which must collect, store, and transmit the sun's energy. As shown in Figure 5.7, such a collector consists of coils of pipe welded to a flat surface. The pipe carries some working agent, usually water. Pipe and plate are painted black to absorb as much sunlight as possible. A few inches above the surface of the collector plate, one or more layers of glass or transparent plastic sheet are installed. Since the outer layer of glass or plastic is transparent to most of the sun's rays, they are absorbed by the blackened pipes, and the water is thereby heated. Normally, much of the heat absorbed by a surface is lost to the air by reradiation. However, the blackened surface radiates infrared, invisible light. Glass and plastics are opaque to infrared radiation. The result is that, as with a common greenhouse, the heat cannot escape and remains in the system. Flat plate collectors of this sort can

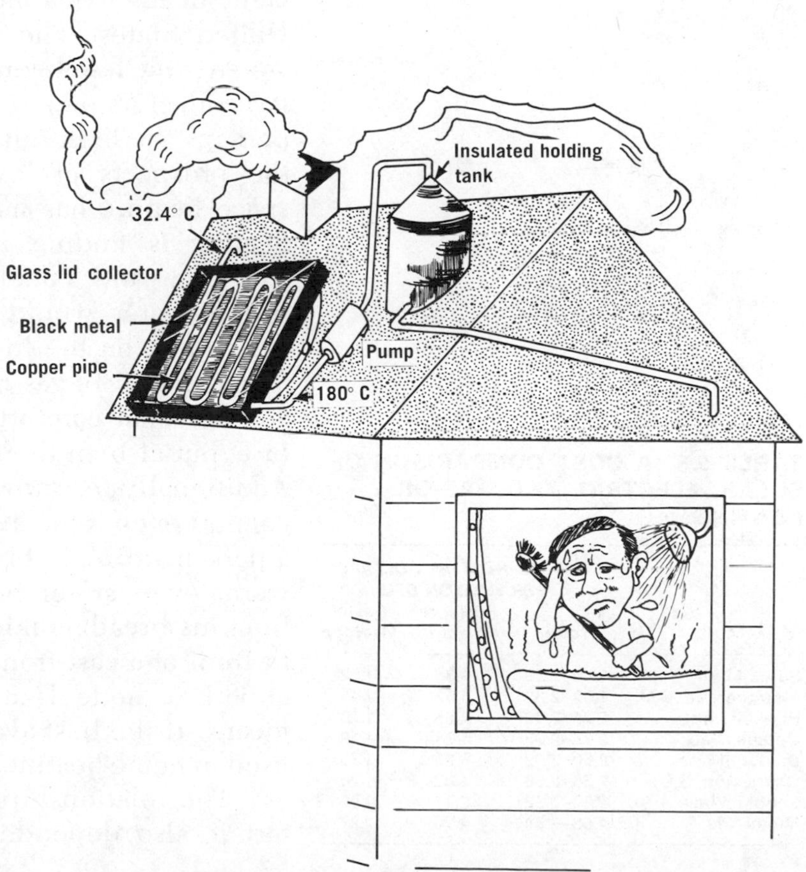

FIGURE 5.7 Solar water heating system.

heat water to boiling even in temperate zones. Under normal conditions, heating to 180°F can be attained routinely. Water at this temperature is ideal for use in home heating applications. Of course, a lot of hot water must be stored for use when there is no sunlight—at night or during cloudy weather. The required storage capacity to provide for prolonged spells of sunless weather would be prohibitively large. But, if our demands are more modest, and we construct solar units to store heat only through the night and one cloudy day at a time, and we use conventional furnaces during prolonged spells of bad weather, then solar heat becomes quite practical, despite the high initial cost. Table 5.5 compares the cost of solar with other forms of heating for several American cities, assuming that the initial outlay is depreciated over a period of 20 years and all operating costs, including interest, are taken into consideration. Of the cities studied, only in Santa Maria, California is solar heat the cheapest available form, but in all other cities studied it is cheaper than electric heat.

Solar units can certainly provide "clean, efficient heat," yet there is no present market in the United States. The reason is unclear; perhaps the advertising has been ineffective, perhaps the large, flat collectors on a roof would be considered ugly, or perhaps the high initial capital investment outweighs the prospects of low fuel bills. Worldwide, solar space heating has not been popular, but solar water heating is finding increasing popularity in Japan, Australia, and Israel. In the United States a solar water heater would cost between $100 and $500, depending on the location, compared to $50 to $125 for an electric or gas-fired unit.

TABLE 5.5 A COST COMPARISON OF SOLAR, ELECTRIC, AND GAS-OIL HEATING

	COST OF HEAT IN DOLLARS PER MILLION BTU		
Area	*Solar*	*Resistive Electric*	*Gas-oil Average*
Santa Maria, Calif.	1.10–1.59	4.36	1.52
Albuquerque, N.M.	1.60–2.32	4.62	1.48
Phoenix, Ariz.	2.05–3.09	4.25	1.20
Omaha, Neb.	2.45–2.98	3.24	1.18
Boston, Mass.	2.50–3.02	5.25	1.75
Charleston, S.C.	2.55–3.56	4.22	1.26
Seattle, Wash.	2.60–3.32	2.31	1.92
Miami, Fla.	4.05–4.64	4.90	2.27

Personal comfort depends not only on temperature, but also on the humidity and air flow in a room. Additionally, certain forms of radiation can produce comfort even if the ambient temperature is low. Thus, a person sitting in bright sun on a winter day can feel warm even at temperatures at which the moisture from his breath condenses. Likewise, a man standing in front of a cast-iron stove might feel as comfortable at 68°F as he feels at 75°F in a room heated by other means. If the best available human engineering were used in home heating, fuel bills could be lowered.

The relationship between temperature and comfort is also dependent on habits and customs. According to research carried out by The American Society of Ventilating Engineers in 1932, the preferred room temperature during the winter for a majority of subjects was 66°F. Similar research at later dates showed that the comfort range had risen to 67.5°F in 1941 and up to 68°F by 1945. In 1971 most homes, schools, and offices were heated to 70–72°F. Part of the change is due to the fact that 35 years ago people wore sweaters and long underwear indoors. If one lives through several winters of 66°F, then 70°F seems uncomfortably warm.

5.15 ELECTRICITY

(Problem 26)

Few issues in the field of environmental science have received so much attention as the problem of generating electric power. The production of electricity leads to water pollution, air pollution, solid wastes, radioactive wastes, land uglification, and thermal pollution. Yet our present society could not operate without electric power, because lighting, entertainment, home conveniences, many industrial processes, and thousands of cheap, relatively silent and trouble-free motors are dependent on it.

Electric consumption has risen faster than any other form of energy consumption in the past few years and is expected to continue to accelerate rapidly (see Fig. 5.8). One cannot reap the benefits of electricity without some environmental side effects,

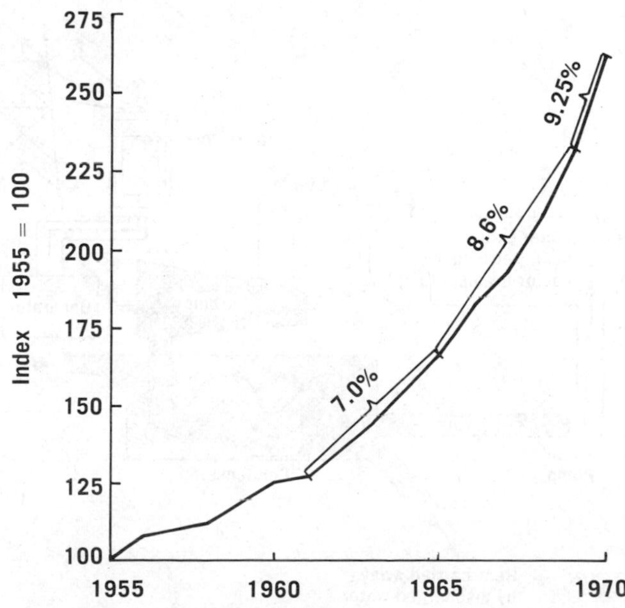

FIGURE 5.8 Consumption of electricity in the U.S. The percentages refer to growth rates over the periods indicated. (From *Scientific American*, September, 1971, p. 140. Copyright © by Scientific American, Inc. All rights reserved. From The Flow of Energy in an Industrial Society, by Earl Cook.)

but the use of innovative technology can certainly reduce the environmental insult.

Presently, most of our electricity is produced in steam turbines. The operating principle here is uncomplicated. Some power source, such as coal, gas, oil, or nuclear fuel, is used to heat water in a boiler and produce hot, high-pressure steam. This steam expands against the blades of a turbine and the spinning turbine operates a generator which produces electricity. The steam, its useful energy now spent, flows into a condenser where it is cooled and liquified, and the water is returned to the boiler to be reused. The cooling action of the condenser is vital to the whole generating process.

Let us return now for the last time to the caveman with his boulder, and the Second Law of Thermodynamics. The caveman learned that maximum efficiency was realized only when the temperature difference between the cold bar and the hot bar was greatest. Modern electric generators are much more complicated than the hot-bar engine, but the thermodynamic principles are the same. The steam, which is the working substance, expands against the turbine blades to do work. The steam cools as it expands, but it cannot cool below the temperature of its surroundings. Because the work output is proportional to the degree of expansion of the steam, and the expansion is proportional to the temperature difference between the hot and cold ends of the turbine, some provision

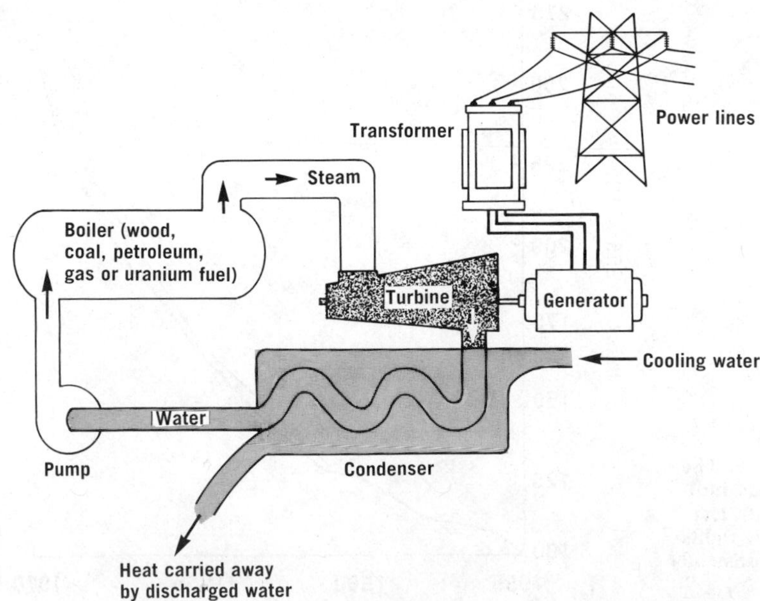

FIGURE 5.9 Power generator (schematic), showing discharge of waste heat.

must be made to maintain a cool environment at one end of the engine. These relationships are shown schematically in Figure 5.9. In practice, cooling is accomplished by circulating water around the condenser. It is obvious that maximum efficiency is reached with very hot steam and a very cool condenser. In practice, the nature of the metals in the turbine limits the temperature to a maximum of about 1000°F (538°C). The low temperature is limited by the cheapest coolant, generally river, lake, or ocean water. Within the constraints of these two limits, an efficiency of 60 per cent is theoretically possible, but uncontrollable variations in steam temperature, and miscellaneous heat losses, reduce the efficiency to about 40 per cent, even in the best installations. This level means that for every 100 units of potential energy in the form of fuel, 40 units of electrical energy are available as useful work and 60 units of energy are dissipated to the surroundings as heat. Moreover, because refrigeration of the condenser is too expensive, further increases in efficiency depend on the use of hotter steam. Since the maximum upper temperature is limited by the ability of metals to withstand the heat stress, efficiency cannot be expected to increase appreciably unless some new breakthrough in metallurgy takes place.

Nuclear-fueled power plants are less efficient than fossil-fueled units operating at the same cooling temperature. Since the metals which contain the fuel cells must survive radioactive stresses as well as heat stresses, lower operating temperatures are used, and as we have learned from the Second Law of Thermodynamics, the smaller the temperature difference between engine and coolant, the less efficient the engine.

5.16 TRANSMISSION OF ELECTRICITY

(Problem 27)

Electricity is carried by both low-voltage service lines and high-voltage trunk lines. Service or distribution lines are carried along familiar "telephone poles" and can be run underground for a minimal extra cost. High-voltage transmission lines, strung across large metal towers, carry current through a potential difference of 135,000 to 500,000 volts and require a right-of-way at least 100 feet wide. The transmission network of the United States is quite complex, for not only must power be brought from distant generating stations to local centers for distribution, but the major generating facilities must be interconnected into a unified grid to prevent major blackouts in the event a single plant breaks down or requires repairs. One can guess that the first high-voltage transmission lines to enter a rural county were a welcome harbinger of the comforts of electricity. But today, the crisscross of large electrical towers and cables is viewed as an eyesore, and the rights-of-way are seen as a waste of land. In fact, an area of land equal to that of the state of Connecticut is now removed from productive use because it is tied up in rights-of-way. Power companies find it expensive to buy 12 acres of land per mile of right-of-way in urban areas. The obvious solution, putting the lines underground, is fraught with technical problems.

Because the earth is a more efficient power sink than the air, if high-voltage transmission wires of the type now used overhead were buried underground, power losses would be prohibitive. Specially insulated wires must be used, but these cost roughly $840,000 per mile or 16 times as much for materials and installation as an overhead wire. However, underground lines require no land, for they can be located along roads or under farmlands.

Research is presently being conducted in the field of refrigerated underground lines. The principle here is twofold; at very cold temperatures power losses due to heat will be substantially reduced, and the conductivity of metals rises sharply at very cold temperatures. Such extremely cooled wires are believed to be able to operate at 200 to 300 million volts, and therefore to provide an extremely large capacity for transmission of power.

5.17 THERMAL POLLUTION

In the previous section we mentioned that an electrical generating plant uses a body of water as a coolant. A one-million-watt facility, running at 40 per cent efficiency, would heat 30 million gallons of water by 8.5 Celsius degrees (15 Fahrenheit degrees) every *hour*. A one-million-watt nuclear plant, operating at about 32 per cent efficiency, would require 50 million gallons of coolant water every hour to extract the same quantity of heat. It is not surprising that such large quantities of heat, added to aquatic systems, cause ecological disruptions. The term **thermal pollution** has been used to describe these heat effects.

The careful reader will note an apparent discrepancy here. If an operating efficiency of 40 per cent from a fossil-fuel plant yields 30 million gallons of warm water, then a nuclear plant at 32 per cent efficiency should warm 42.5 million gallons, not 50 million. The difference is accounted for by the fact that some of the heat from the fossil-fuel plant is not absorbed by water, but instead warms up a lot of *air*, by means of the hot gases discharged from the stacks. This is ecologically less disrupting than the warming of streams and lakes. Therefore, the difference in thermal pollution between the two types of plants is greater than that implied by thermodynamic considerations alone.

The processes of life involve chemical reactions,

and the rates of chemical reactions are very sensitive to changes in temperature. As a rough approximation, the rate of a chemical reaction doubles for every rise in temperature of 10 Celsius degrees (18 Fahrenheit degrees). We know that if our own body temperature rises by as much as 5°C (or 9°F, which would make a body temperature of 98.6 + 9 = 107.6°F) the fever may be fatal. What then happens to our system when the outside air temperature rises or falls by about 10 Celsius degrees? We adjust by internal regulatory mechanisms that maintain a constant body temperature. This ability is characteristic of warm-blooded animals, such as mammals and birds. Thus, the body temperature of a man or dog in a room cannot be determined by reading the wall thermometer. In contrast, non-mammalian aquatic organisms such as fish are cold-blooded: that is, they are unable to regulate their body temperatures as efficiently as warm-blooded animals.

(Problems 28, 29, 30, 31, 32, 33, 36)

How, then, does a fish respond to temperature increase? All its body processes (its metabolism) speed up, and its need for oxygen and its rate of respiration therefore rise. The increased need for oxygen is especially serious since hot water has a smaller capacity for holding dissolved oxygen than cold water. Above some maximum tolerable temperature, death occurs from failure of the nervous system, the respiratory system, or essential cell processes. According to the Federal Water Pollution Control Administration, almost no species of fish common to the United States can survive in waters warmer than 93°F. The brook trout, for example, swims more rapidly and becomes generally more active as the temperature rises from 40° to 48°F. In the range from 49° to 60°F, however, activity and swimming speed decrease, with a consequent decline in the trout's ability to catch the minnows on which it feeds. This inactivity is more critical because the trout *needs* more food to maintain its higher metabolic rate in the warmer water. Outright death occurs at about 77°F. In addition, spawning and other reproductive mechanisms of fish are triggered by such temperature changes as the warming of waters in the spring. Abnormal changes, to which the fish is not adapted, can upset the reproductive cycle.

In general, not only the fish, but entire aquatic ecosystems are rather sensitively affected by tempera-

For many years, the word pollute has meant "to impair the purity of," either morally* or physically. The terms air pollution and water pollution refer to the impairment of the normal compositions of air and water by the addition of foreign matter, such as sulfuric acid. Within the past few years two new expressions, thermal pollution and noise pollution, have become common. Neither of these refers to the impairment of purity by the addition of foreign matter. Thermal pollution is the impairment of the quality of environmental air or water by raising its temperature. The relative intensity of thermal pollution cannot be assessed with a thermometer, because what is pleasantly warm water for a man can be death to a trout. Thermal pollution must therefore be appraised by observing the effect on an ecosystem of a rise in temperature. Similarly, noise pollution has nothing to do with purity: foul air can be quiet, and pure air can be noisy. Noise pollution (to be discussed in Chapter 12) is the impairment of the environmental quality of air by noise.**

*(1857.) Buckle, *Civilization*, I., viii, p. 526: "The clergy . . . urging him to exterminate the heretics, whose presence they thought polluted France."

**(1585.) T. Washington, trans. *Nicholay's Voyage*, IV; ii; p. 115: "No drop of the bloud should fall into the water, least the same shuld thereby be polluted."

ture changes. Any disruption of the food chain, for example, may upset the entire system. If a change of temperature shifts the seasonal variations in the types and abundances of lower organisms, then the fish may lack the right food at the right time. For example, immature fish (the "fry" stage) can eat only small organisms, such as immature copepods. If the development of these organisms has been advanced or retarded by a temperature change, they may be absent just at the time that the fry are totally dependent on them. It may be very difficult to predict such effects by studying stream temperatures, because a very large portion of the total flow of a given stream may be by-passed through a power plant to carry away its waste heat. While fish are easily excluded by screens from such undesirable detours, it is not easy to keep out microscopic organisms that do make up an important part of the food chain. These organisms are subjected to temperatures that will exceed the maximum temperature of the stream. We do not yet know how serious the consequences may be—the indications of various studies range from a zero effect to a 95 per cent kill of the plankton.

Higher temperatures often prove to be more hospitable for pathogenic organisms, and thermal pollution may therefore convert a low incidence of fish disease to a massive fish kill as the pathogens become more virulent and the fish less resistant. Such situations have long been known in the confined environments of farm and hatchery ponds, which can warm up easily because the total amount of water involved is small. As thermal pollution in larger bodies of water increases, so will the potential for increased loss of fish by disease.

Aquatic ecosystems near power facilities are subject not only to the effects of an elevated average temperature, but also to the thermal shocks of unnaturally rapid temperature changes. Power generation and heat discharge vary considerably from a peak in the afternoon to a low in the hours between midnight and daybreak. Additionally, a complete shutdown of a day or longer occurs occasionally. Thus, the development of cold-water species is hindered by hot water, and the development of hot-water species is upset by the unpredictable flow of heat. In addition, both types of organisms are adversely affected by rapid temperature changes.

"It's not the humidity—it's the thermal pollution." (Amer. Scient. Sept–Oct. 1971.)

Additional disruptions can occur because hot water has a reduced oxygen content. Power plants are usually located near population centers, and many cities dump sewage into rivers. Since sewage decomposition is dependent on oxygen, hot rivers are less able to cleanse themselves than cold ones. The combination of thermal pollution with increased nutrients from undecomposed sewage can lead to rapid and excessive algal growth and eutrophication (see Chapter 10). Therefore, thermal pollution imposes the unhappy choice of dirtier rivers or more expensive sewage treatment plants.

Manmade poisons, too, become more dangerous to fish as the water temperature rises. First of all, toxic effects are accelerated at higher temperatures. Second, as we just mentioned, warm water favors increased growth of plant varieties such as algae. The algae tend to collect in the power-plant condensers and reduce water-flow efficiency. The electric company responds by periodically introducing chemical poisons into the cooling system to clean the pipes. These poisons are then mixed with the downstream effluent. Additionally, domestic and industrial water consumers are more apt to discharge treatment chemicals (such as copper sulfate) into water with high algae concentrations than into clean water. Thus, in warmer water, not only are fish less likely to resist poisons, but they are also likely to be exposed to them more.

The ecological balance in lakes can be particularly delicate, and the true biological effects are easily masked by misleading temperature data. Refer to Section 10.7 on lakes (Chap. 10) to understand how the temperature profiles and seasonal changes bring oxygen to the deeper waters (the **hypolimnion**). It is advantageous for power companies to draw cold water from these bottom layers and discharge heated water onto the surface. The natural temperature difference between the two layers, which is about 10 Celsius degrees, is roughly equal to the temperature rise imposed on the water as it circulates through the power plant. Therefore, the power plant operation can be effected without changing the *surface* temperature of the lake. This does not mean, however, that thermal pollution has somehow been miraculously avoided, for the relative volumes of the hot and cold layers, and therefore the *average* temperature of the lake, do change. The hot discharge increases the yearly plankton production by prolonging the growing season. However, benthic and deep-water species are adversely affected. The temperature of the hypolimnion is unchanged by the power plants, and therefore the *concentration* of oxygen will remain constant, but as the volume of the hypolimnion is decreased, the *total quantity* of dissolved oxygen is reduced. The lake becomes increasingly discolored and eutrophic as the increased surface production and the decreased underwater consumption cause organic matter to accumulate.

Studies of the actual effects of thermal pollution are contradictory. A study of the 10-mile stretch of the Connecticut River above and below the Connecticut Yankee atomic power plant has shown little environmental effect of thermal pollution. An average of 22 million gallons of river water enters this plant every hour and is warmed by about 11 Celsius degrees (20 Fahrenheit degrees) before it returns to the river. A comprehensive environmental study of the situation yielded the following conclusions:

1. Except for concentrations of algae near the hot discharge, there was little alteration of plankton populations in the river.

2. Adjacent to the intake pipes, the diversity and quantity of benthic life was markedly decreased, probably because of silt removal by the pumping operation.

3. Near the heated discharge, the quantity and diversity of benthic species increased relative to the undisturbed system, and there was a shift in the dominant species. Clams yielded to insect larvae and sea worms.

4. The population of the resident fishes—catfish, white and yellow perch, sunfish, shiners, bass, suckers, and killifish—did not seem to be seriously affected by the discharge.

5. Catfish caught near the hot discharge plume appeared to be in poorer physical condition than other river catfish.

6. A commercially important spawning fish, the shad, escapes harm by swimming under the hot water.

This report, showing only minor disruptions of river life at the Yankee atomic power plant, is certainly reassuring. However, large fish kills have been reported at or near other generating facilities. For example, 150,000 fish were killed in January, 1970 at a Consolidated Edison unit on the Hudson River; another 120,000 perch at the same location died three months later; "two truckloads" of fish were killed by thermal discharge near Port Jefferson, New York. Some of these fish were killed, not by heat, but by mechanical trauma at the intake pipes.

It is impossible from these data to predict with assurance the effects of hot water on any other river or the cumulative effects of two power plants on a river from data about one plant. For example, yellow perch, sunfish, and suckers can survive warmer water than trout, grayling, and salmon. The discharge of waste heat into natural waters carries the threat of ecological damage; therefore, the potential environmental effects should always be appraised before new thermal loads are imposed on any aquatic system.

As we have already explained, the electric power industry, and particularly nuclear power plants, necessarily produce waste heat, and such heat is most conveniently discharged into flowing waters. It is estimated that by the year 2000 the rate of cooling water needed by the power plants will be equivalent to one-third of the total rate of freshwater run-off in the United States. Since this heat could not easily be distributed uniformly among all the large and small bodies of water in the country, but would be concentrated either at shorelines or along rivers of sufficient capacity to accommodate large plants, the result will be that serious ecological damage will become more widespread.

TABLE 5.6 SOME HEAT SOURCES AND THEIR MAGNITUDE

DESCRIPTION OF HEAT SOURCE	*MAGNITUDE (watts/meter^2)*
Net solar radiation at earth's surface	approx. 100
Energy production in 1970 distributed evenly over all continents	0.054
Energy production in 1970 distributed evenly over whole globe	0.016
Heat production of man's activities averaged over all urban areas	12
Heat production of man's activities in:	
Cincinnati	26
Los Angeles	21
Moscow*	127
Manhattan	630

*Net solar radiation over Moscow is only 42 watts/meter^2.

5.18 WASTE HEAT AND CLIMATE

(Problem 37)

The burning of fossil fuels and the fission or fusion of atomic nuclei add heat to the environment above and beyond the energy received from the sun.

Will direct heat discharged into the atmosphere or waterways affect global or local climate? Table 5.6 shows that the sum of all human activities does not even begin to compete with the energy received from the sun, and even in the years to come a significant direct-heating effect is unlikely on a global scale. But, in many metropolitan areas, the heat output of man's activities already exceeds the heat received from the sun. The climate of some metropolitan areas is also influenced by the fact that the buildings and roadways do not absorb or reflect the sun's rays to the same degree as did the forests and prairies on which they were built. It is not surprising then that the climates of many large cities are measurably different from the climates of the surrounding countrysides (see Table 5.7).

The magnitude of these climatic changes will increase in the future, and the effect will undoubtedly be measurable over larger areas as cities merge into megalopolises. The question of possible changes in world climate is further discussed in the chapter on air pollution, Section 9.8.

5.19 SOLUTIONS TO THE PROBLEM OF THERMAL POLLUTION

(Problems 34, 35, 38, 42)

Recall that we cannot invent a process to avoid thermal pollution. The generation of power produces excess heat, and we must deal with this problem by non-miraculous means. Some suggestions are described in the following paragraphs.

How to Decrease the Thermal Load

As mentioned previously, mechanical losses account for a 15 to 20 per cent loss of efficiency in conventional power plants. Thus, thermal pollution could be reduced if power plants operated at higher efficiency. A promising new design that increases efficiency is based on the principle of magnetohydrodynamics (MHD). In this system, air is heated directly and is seeded with metals like potassium or sodium

$$\underset{\text{potassium atom}}{K} + \text{heat} \longrightarrow \underset{\text{potassium ion}}{K^+} + \underset{\text{negative electron}}{e^-}$$

which lose electrons at high temperatures. This hot, electrified air stream is allowed to travel through a large pipe that is ringed with magnets. It is the movement of these charged particles that constitutes the generated electric current. At the end of the passage of the hot air through the magnets, the expensive seeding materials must be recovered. Furthermore, the exhausted hot air can operate a conventional turbine with the production of additional electricity. The advantages of this system are twofold. First of all, the overall efficiency of the system is expected to reach 60 per cent, and secondly, most of the waste heat is dissipated directly into the air rather

TABLE 5.7 AVERAGE CHANGES IN CLIMATIC ELEMENTS CAUSED BY URBANIZATION

ELEMENT	*COMPARISON WITH RURAL ENVIRONMENT*
Cloudiness:	
cover	5 to 10% more
fog-winter	100% more
fog-summer	30% more
Precipitation, total	5 to 10% more
Relative humidity:	
winter	2% less
summer	8% less
Radiation:	
global	15 to 20% less
duration of sunshine	5 to 15% less
Temperature:	
annual mean	0.9°F to 1.8°F more
winter minimum (average)	1.8°F to 3.6°F more
Wind speed:	
annual mean	20 to 30% less
extreme gusts	10 to 20% less
calms	5 to 20% more

than into aquatic ecosystems. Lack of funds has led to the stagnation of the MHD program in the United States, but the feasibility of the technique has been shown by the Russians, who now operate a 250,000-watt MHD generator near Moscow.

Since solar energy does not impose any additional heat on the environment, increased reliance on sunlight will decrease thermal pollution. Sunlight can generate electricity directly from some materials by releasing electrons from their surfaces. Such materials are said to be photovoltaic, and devices that make use of this phenomenon are called **photovoltaic cells**. Such direct conversion of light to usable power is an ideal environmental goal—there are no moving mechanical parts, and there is no thermal pollution. Some photovoltaic materials are known and have been used successfully in spacecraft, where cost is not the controlling factor. However, they are as yet too expensive to be used in competition with the more familiar but "dirtier" energy sources. Considerable effort is needed to develop new, cheap materials that can make photovoltaic power generation competitive. However, little money is available for research. Even if large sums of money are allocated, developmental success of solar cells cannot be assured, but the benefits of success would be high. The decision to fund such research would be a measure of our willingness to pay for a clean environment.

Let us imagine that competitive solar cells were available. A one-million-watt plant would require about 16 square miles of collector area, depending of course on the latitude and cloudiness of the region. Put in other terms, 35,000 square miles of land would supply all the projected power needed for the continental United States in 1990. It has been suggested that "waste land" be used productively by building solar generating stations there. A marsh such as the great swamp in northern New Jersey is one possible site (see page 49 in Chapter 1). As we have already learned, it is impossible to generate electric power without some adverse environmental effects. In this case, destruction of that swamp, or other "wastelands" may lead to vast ecological disruptions.

Studies are under way to determine the feasibility of building a large photovoltaic collector in space and beaming the collected power to earth by microwave transmission.

How to Put Waste Heat to Good Use

Large quantities of hot water can be used for diverse purposes. Hot water can improve growth conditions in large greenhouses. The costs of the greenhouse and water handling systems are high, and the economic feasibility might ultimately depend on the accommodation that can be made between the farmer and the power companies. Unfortunately, greenhouse owners don't need hot water in the summer at a time when power demands for air conditioning are high and the deleterious effects of thermal pollution on aquatic life are apt to be most severe.

Success in utilizing hot water for irrigating open field crops has been reported in the State of Washington. Alternatively, hot water circulating in closed pipes can be used to heat the soil without irrigating it; the resulting benefit is a higher agricultural yield, as shown by the data in Table 5.8.

TABLE 5.8 EFFECTS OF SOIL HEATING WITHOUT IRRIGATION ON VEGETABLE PRODUCTION IN MUSCLE SHOALS, ALABAMA

CROP	*YIELD IN TONS PER ACRE*	
	Heat	*No Heat*
String beans	6.9	2.7
Sweet corn	6.2	3.2
Summer squash	20.6	17.6

The practice of adding chemical poisons to kill algae in the condensers is incompatible with the use of that water for irrigation, although not incompatible with closed-system soil heating. If a power plant were to shut down during a cold wave, the flow of hot water would suddenly halt and a whole crop could be lost. In order to mitigate the high average cost of raising food under these conditions, a substitute source of heat would have to be available on a standby basis.

The growth rate of certain fish can be enhanced markedly by raising them under carefully controlled temperatures. Proponents of aquaculture claim that high yields of food can be realized if cheap hot water is available. (Japanese aquaculturists support up to three million pounds of carp per acre!) Thus, the heated discharge that can be detrimental to a whole ecosystem can be advantageous to a single species. However, fish farmers have the same sorts of problems as greenhouse operators. In addition, they must somehow dispose of considerable quantities of warm water heavily polluted by fish wastes.

Pumping hot water into home radiators is attractive but, again, economically impractical in many instances. First of all, city residents prefer not to live close to power plants, and the costs and heat losses involved in piping hot water any significant distance are prohibitive. Second, the installation of an underground steam system would be an extremely complex task in a large established city. Such a proposal would be practical only for newly built areas. Third, the maximum power demands do not coincide with maximum heating demands. Peak electric utilization is in the afternoon, while late-night use is low. Some storage facility or auxiliary steam generator would be needed to supply heat in the evening and at night. Finally, home heating would not remove waste heat during warm seasons or in warm climates. Despite these difficulties the prospect of "free" heat is appealing, and the proposals are being studied in many cities.

Other proposals are to use the heat to speed the decomposition of sewage (without contact between the hot water and the sludge), or to desalinate seawater. Both these proposals require careful cost analysis.

An interesting example of the use of waste heat is provided by the relationship between the Bayway (New Jersey) refinery of the Humble Oil and Refining Company and the Linden (New Jersey) Generating Station. The Linden power plant is capable of producing electricity at 39 per cent efficiency. For the past 15 years, this efficiency has been lowered by a less than optimum cooling of the condenser, and some of the waste heat has been sold as steam to Humble. If we consider the two-plant operation as a single energy unit, the overall efficiency of power production has been raised to a level of 54 per cent. The process is beneficial to many: the companies save money, our fuel reserves are conserved, and thermal pollution of waterways is reduced 15 per cent. Similar operations are being planned or are in effect in other countries.

Though the technical problems of waste-heat utilization are not insurmountable in theory, the solutions are often economically discouraging. Typically, the waste water is hot enough to damage an aquatic ecosystem but not hot enough to be attractive for commercial use. Perhaps if the total environmental cost of thermal pollution were considered, waste-heat utilization would be more attractive.

How to Use the Atmosphere as a Sink

Another approach to the problem of thermal insult to our waterways is to dispose of the heat into the air. Air has much less capacity per unit volume for

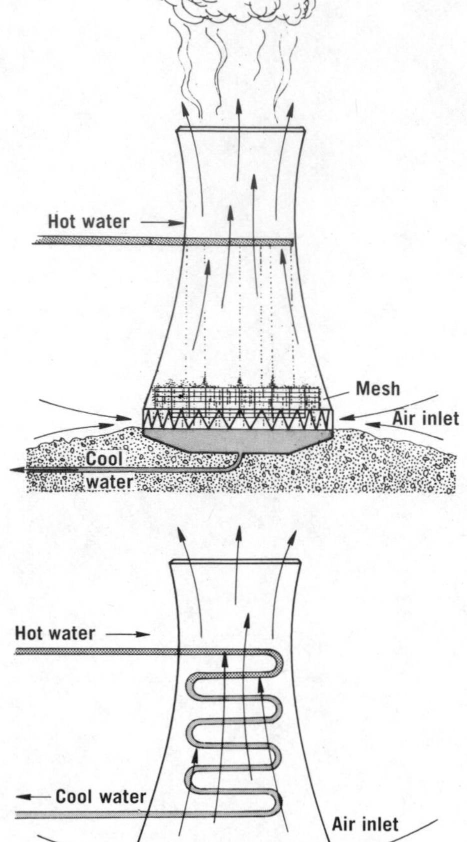

Dry cooling tower.

TABLE 5.9 COST OF THERMAL POLLUTION CONTROL FOR A FOSSIL-FUELED PLANT

TYPE OF CONTROL	*ESTIMATED AVERAGE COST mils per kwh**
Cooling Pond	0.08
Wet Tower, mechanical draft	0.10
natural draft	0.18
Dry Tower, mechanical draft	0.81
natural draft	0.99

*The cost of electricity is about one cent per kwh for consumers who use 1000 kwh per month, assuming that non-recycled water is used for cooling. The figures in this column therefore represent the *added* cost for control of thermal pollution. One mil = one-tenth of a cent.

absorbing heat than water does, so the direct action of air as the cooling medium in the condenser is not economically feasible. For this reason power plants must still be located near a source of water, the only other available coolant. However, the water can be made to lose some of its heat to the atmosphere and then can be recycled into the condenser. There are various devices available that can effect such a transfer.

The two cheapest techniques are based on the fact that evaporation of water is a cooling process. Many power plants simply maintain their own shallow lakes, called cooling ponds. Hot water is pumped into the pond where evaporation as well as direct contact with the air cools it, and the cool water is drawn into the condenser from some point distant from the discharge pipe. Water from outside sources must be added periodically to replenish evaporative losses. Cooling ponds are practical where land is cheap, but a one-million-watt plant needs one to two thousand acres of surface, and the land costs can be prohibitive.

A cooling tower, which can serve as a substitute for a cooling pond, is a large structure, about 600 feet in diameter at the base and 500 feet high. Hot water is pumped into the tower near the top and sprayed onto a wooden mesh. Air is pulled into the tower either by large fans or convection currents, and flows through the water mist. Evaporative cooling occurs and the cool water is collected at the bottom. No hot water is introduced into aquatic ecosystems, but a large cooling tower loses over one million gallons of water per day to evaporation. Thus, fogs and mists are common in the vicinity of these units, reducing the sunshine in nearby areas. Reaction of the water vapor with sulfur dioxide emissions from coal-fired power plants can cause the resultant air to carry sulfuric acid aerosols, as described in Chapter 9.

Environmental problems can be reduced if dry cooling towers are used instead of evaporative wet ones. A dry tower is nothing more than a huge version of an automobile radiator installed into a tower to promote a speedy flow of air past the cooling pipes. Dry towers are uneconomical because of the cost of the prodigious amount of piping required. The relative costs of the three cooling facilities are shown in Table 5.9.

5.20 ENERGY—AN OVERVIEW

(Problem 39)

From fuel mining to fuel transportation to fuel consumption to energy transmission, the environment always loses when electric power is generated. The problems are magnified by population density and increased demand on electricity. An electrical generating facility in Great Falls, Montana, in 1920 would have found all the problems inherent in a large modern facility, but space, air, and clean water were so plentiful, and the population was so sparse, that it would have been difficult to detect any environmental effects. No one would have believed that a serious problem could exist. A cowboy standing on a butte two days out of town could scan a circle a hundred miles in diameter and see almost no sign of man. This concept of inexhaustibility gave rise to what economist Kenneth Boulding calls "cowboy economics." Early Americans thought that since you can neither damage nor deplete the environment, you may as well take the cheapest route to your goal. Today we know that we can damage and deplete our planet. Old attitudes must therefore be re-evaluated.

The possibility of a serious crisis looms ahead. Three factors have combined to enhance the potential severity of this crisis: (1) Power companies have not always employed the best available technology to reduce pollution. Traditionally, they have been slow to employ environmentally satisfactory mining practices, to use the most effective air-cleaning devices, to construct cooling ponds or towers, and to engage in innovative research on environmental problems. (2) Population has risen. (3) Per capita consumption of electricity has also been steadily rising, as is shown graphically in Figure 5.10. Some of the components of this per capita increase are listed in Table 5.10. For example, in 1971, 51 per cent of the homes in the United States used electric blankets, rather than ordinary insulating ones; 44.5 per cent of the homes used room air conditioners, compared to 1.3 per cent in 1952; and 29 per cent of the homes had electric dishwashers, compared to three per cent in 1952. The severity of the energy crisis could be reduced by reversing any or all of these trends, but voluntary solutions to such problems are not notably successful. An alternative is some indirect or direct form of compulsion, such as a rise in the cost of energy, or legislation that regulates the excessive use of power. Among the many possible specific approaches, the

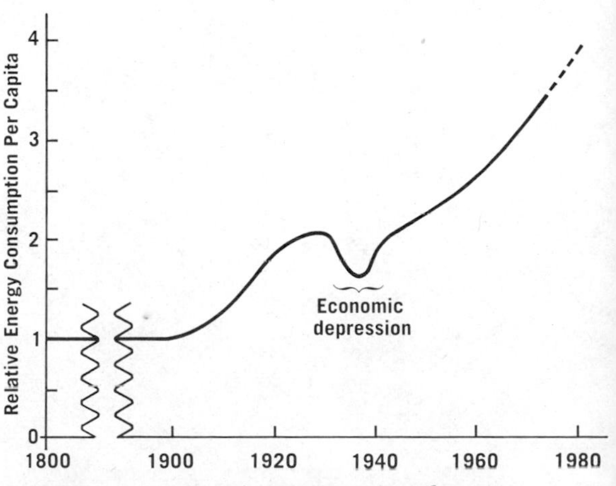

FIGURE 5.10 Per capita consumption of energy in the United States. (Vertical scale is based on: 1 unit = 10^8 B.T.U.'s per capita per year.)

TABLE 5.10 HOMES WITH SELECTED ELECTRICAL APPLIANCES: 1952 TO 1971*

ITEM	1952 NUMBER	1952 PER CENT	1960 NUMBER	1960 PER CENT	1965 NUMBER	1965 PER CENT	1970 NUMBER	1970 PER CENT	1971 NUMBER	1971 PER CENT
Total number of wired homes	42.3	100.0	51.7	100.0	57.6	100.0	64.0	100.0	65.6	100.0
Air-conditioners, room	0.6	1.3	7.8	15.1	13.9	24.2	26.0	40.6	29.2	44.5
Bed coverings	3.6	8.6	12.2	23.6	20.0	34.7	31.7	49.5	33.5	51.1
Blenders	1.5	3.5	4.2	8.0	7.5	13.0	23.4	36.5	26.2	40.0
Can openers	(NA)†	(NA)	2.5	4.8	14.2	24.7	29.1	45.5	31.5	48.1
Coffeemakers	21.6	51.0	30.2	58.3	41.3	71.7	56.7	88.6	59.7	91.0
Dishwashers	1.3	3.0	3.7	7.1	7.8	13.5	17.0	26.5	19.4	29.6
Disposers, food waste	1.4	3.3	5.4	10.5	7.9	13.6	16.3	25.5	18.6	28.4
Dryers, clothes (includes gas)	1.5	3.6	10.1	19.6	15.2	26.4	28.6	44.6	31.2	47.6
Freezers, home	4.9	11.5	12.1	23.4	15.7	27.2	20.0	31.2	21.4	32.7
Frypans	(NA)	(NA)	22.5	43.4	28.3	49.2	36.0	56.2	38.0	58.0
Hotplates and buffet ranges	9.0	21.2	12.5	24.2	13.1	22.7	15.7	24.5	16.3	24.8
Irons, total	37.9	89.6	45.7	88.4	57.1	99.1	63.8	99.7	65.4	99.8
Steam and steam/spray	8.3	19.5	30.6	59.2	44.6	77.5	56.5	88.2	59.1	90.1
Mixers	12.6	29.7	29.0	56.0	41.9	72.8	52.8	82.4	55.3	84.4
Radios	43.7	96.2	50.3	94.3	58.2	99.3	63.9	99.8	65.4	99.8
Ranges: Free-standing } Built-in	10.2	24.1	16.0	30.9	18.5	32.1	25.9	40.5	27.9	42.6
Built-in			3.3	6.4	5.9	10.3	9.6	15.0	10.3	15.7
Refrigerators	37.8	89.2	50.8	98.2	57.3	99.5	63.9	99.8	65.4	99.8
Television: Black and white	19.8	46.7	46.2	89.4	55.9	97.1	63.2	98.7	65.4	99.8
Color	(x)‡	(x)	(NA)	(NA)	5.5	9.5	27.2	42.5	33.5	51.1
Toasters	30.0	70.9	37.2	72.0	48.1	83.6	59.3	92.6	61.7	94.2
Vacuum cleaners	25.1	59.4	38.4	74.3	48.1	83.5	58.9	92.0	61.9	94.4
Washers, clothes	32.2	76.2	44.1	85.4	50.3	87.4	59.0	92.1	61.8	94.3
Water heaters	5.8	13.8	9.8	18.9	13.5	23.4	20.2	31.6	22.2	33.8

*Wired homes in millions, as of December 31, 1971. Percentages based on total number of homes wired for electricity. Prior to 1970, radio data based on total homes as follows: 53,300,000 in 1960, and 58,566,000 in 1965. (From *Merchandising Week,* annual statistical issues. New York, Billboard Publications, Inc., 1972.)

†Not available.

‡Not applicable.

following ones, though as yet unpopular, may soon be considered palatable:

1. More funds could be made available for research in various fields related to energy. This includes funding of "long shots" such as large-scale solar energy installations as well as more conservative programs such as the breeder reactors. Ultimately, it makes little difference in total cost whether the power companies support the research and the cost is reflected in higher electric bills or whether government funds are used and taxes are raised. Depending on the tax structure, however, the two methods of funding ultimately shift the burden of research cost onto different groups of people.

2. Low-cost loans could be made available for construction of solar heaters or installation of any device or improvement which conserves fuels.

3. The recent very rapid increase in the use of room air conditioners has accounted for much con-

sumption of power. There are various ways to reduce this consumption without sacrificing comfort. Buildings could be designed to promote convectional flow of air through them. This means more than just adding windows. The use of novel roof designs, the proper orientation of the house itself, and the installation of larger attic louvers can all increase natural cooling and reduce the need for power. Small air-conditioning units are half as efficient as central systems in apartment houses. All new apartments could be required to install central air conditioning, and those wishing to plug into it could pay to do so. There could be a ban on cooling below 75°F, and heating above 70°F. Finally, perhaps fashions could somehow be changed so men would no longer feel constrained to wear jackets, ties, and long pants in business offices during the summer!

4. Resistive electric heating could be made illegal or could be very heavily taxed in order to encourage more efficient heat pumps or home furnaces.

5. Fluorescent lights are three times as efficient as incandescent lighting, but most of our lighting fixtures accommodate only the conventional screw-in type incandescent bulbs. The substitution of screw-in fluorescent bulbs would save a lot of energy without

"The egg timer is pinging. The toaster is popping. The coffeepot is perking. Is this it, Alice? Is this the great American dream?" (Drawing by H. Martin; © 1973 The New Yorker Magazine, Inc.)

the necessity of converting most of the existing fixtures.

6. Excessive commercial power consumption for advertising and displays could be limited.

7. According to current pricing practice, electricity becomes cheaper per kilowatt hour as demand is increased. In a typical example, the first 50 kwh cost six cents per kilowatt hour, while any consumption above 900 kwh costs less than one cent per kilowatt hour. This rate structure naturally encourages additional use and should be re-evaluated. If the system were reversed, and prices per kilowatt hour rose with increased demand, people would be more reluctant to use electricity to perform tasks that were done manually 10 years ago. As a result, such a system would undoubtedly slow the rapid rise of per capita consumption. Naturally, different rate structures would be necessary for industrial consumers, or many manufactured goods would become too expensive to produce.

8. Of course, if population growth were stabilized, the energy requirements of the world would rise less sharply.

Even if these proposals to conserve power were adopted, our complex civilization would still consume large quantities of fuels, and many of the energy-related problems discussed in this chapter would remain with us. The most serious problems of power use—depletion, air pollution, radioactive waste disposal, thermal pollution, and destruction of wilderness—must be faced by innovative thinking in the physical sciences, economics, and social philosophy.

PROBLEMS

1. **Energy and power.** What is energy? Compare energy with heat, work, power.

2. **Energy and power.** Three farmers are faced with the problem of hauling a ton of hay up a hill. The first makes twenty trips, carrying the hay himself. The second loads a wagon and has his horse pull the hay up in four trips. The third farmer drives a truck up in one load. Which process—manpower, animal power, or machine power—has performed more external work? Which device is capable of exerting more power?

3. **Energy and power.** Your electric bill reflects the number of kilowatt hours that you have consumed. Are you being charged for power or energy? When we speak of the capability of a power plant we speak of its wattage. Are we speaking of power or energy? Explain.

4. **Energy and power.** Define foot-pounds per minute. Is this a unit of energy or power? What about pounds of ice melted? Pounds of ice melted per hour?

5. **First Law.** Write four statements of the First Law of Thermodynamics. (One statement appears in Section 10.6 of the chapter on water pollution.)

6. **Second Law.** Write four statements of the Second Law of Thermodynamics.

7. **First Law.** The drawing at left shows a perpetual motion machine based on osmosis. The two identical membranes are permeable to water but not to sugar. Water from the reservoir passes up through membrane 1 to a height that is determined by the osmotic pressure of the solution. Water also permeates membrane 2, falling back to the reservoir and doing work on the way down. Do you think this machine will work? If not, why not?

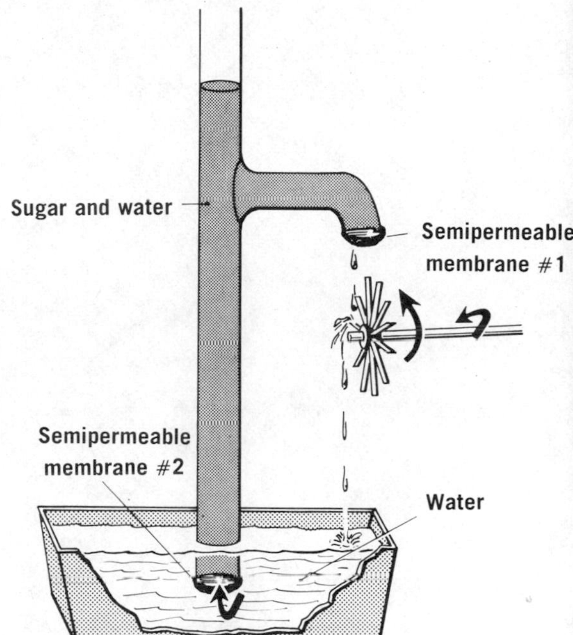

8. **Second Law.** Problem 8 in Chapter 1 reads, "We speak about nutrient cycles and energy flow. Explain why the words cycle and flow are used in their present context." Armed with the message of the Second Law, can you better explain the meaning of energy flow?

9. **Resources.** Predict the general shape of the curve showing world production of iron through the years. Explain your prediction.

10. **Oil shales.** Predict some environmental side effects of mining oil shales.

11. **Natural gas.** If the wholesale price of natural gas in Wyoming increases by a factor of three, how will that affect the retail price in New Jersey? Explain.

12. **Coal mining.** Discuss the relative advantages and disadvantages of strip and tunnel mining.

13. **Alaska pipeline.** Analyze critically the statements below.

    "[The Department of] Interior estimates that if performance of the oil tankers on the Valdez run were no better than the worldwide average, we could anticipate spills averaging 384 barrels a day." (From "The Living Wilderness," Summer 1970, page 8.)

    "The Copper River Basin has the highest density [of waterfowl]; however these high densities occur a mile or more from the route." (From "The Environment," published by the Alyska Pipeline Service Co.)

14. **Alaska pipeline.** Explain why a pipeline across the tundra is more difficult to construct and more disruptive to the local ecology of the area than a pipeline across Texas.

15. **Energy supply.** Outline the predicted problems of energy production and use in the year 2100. Compare with the factors outlined in the answer to Problem 14.

16. **Uranium mining.** Predict some environmental side effects of mining Sierra Nevada granites for their

uranium and thorium content. Compare your answer with the answer to Problem 10.

17. **Alternate energy sources.** Discuss briefly the prospects for solar, geothermal, tidal, wind, and garbage energies. What problems do these methods entail?

18. **Energy shortages.** In 1970, Consolidated Edison Company of New York was forced to cut the power delivered to consumers by five per cent on six different occasions. What do you think were some of the factors that caused these cutbacks?

19. **Transportation.** Presently, gasoline taxes are used for the construction of new roads. This practice has been considered fair because the roads are paid for by those who use them most. Increasingly, economists and social philosophers feel that many of our traditional concepts of fairness must be reevaluated in the light of environmental problems. Do you feel that there should be a reevaluation of road tax use? If so, how would you allocate funds? If not, explain.

20. **Hydrogen economy.** Explain how a hydrogen economy would function differently from a petroleum economy.

21. **Recycling.** Discuss some factors which might make recycling economically attractive.

22. **Space heating.** Why can a furnace be built to be more efficient than a steam engine?

23. **Heat pumps.** It has been suggested that small air conditioners be placed in the center of a room rather than in a window for more efficient central cooling. Do you think this is a good idea?

24. **Heat pumps.** What is the common name for a heat pump that heats the outdoors at the expense of the indoors?

25. **Solar energy.** Describe the construction and function of a flat plate collector.

26. **Electric power generation.** Explain the function of a condenser in a steam turbine. Is a condenser needed at a hydroelectric facility?

27. **Electrical transmission.** Explain the difference between transmission and distribution lines.

28. **Thermal pollution.** Define thermal pollution. How does it differ in principle from air or water pollution?

29. **Nuclear vs. fossil fuels.** Explain why nuclear-fueled power plants require more cooling water than fossil-fueled plants.

30. **Thermal pollution.** Dogs live on all parts of the Earth from the tropics to the Arctic. Trout, on the other hand, are confined to waters no warmer than about 60°F. Why are the permissible temperatures for trout so much more limited?

31. **Thermal pollution.** Since marine life is abundant in warm tropical waters, why should the warming of waters in temperate zones pose any threat to the environment?
32. **Thermal pollution.** Warm water carries less oxygen than cold water. This fact is responsible for a series of disturbances harmful to aquatic organisms. Discuss.
33. **Thermal pollution.** What would happen if cooling water were drawn from the surface and discharged into the hypolimnion? Would this change be more or less useful to the power companies than the current practice? How would it affect aquatic life?
34. **Thermal pollution.** On page 203 of this book we stated that the heat discharged into aquatic systems is likely to produce serious ecological damage. Yet on page 206 we discussed uses of hot water for increasing fish yields in aquaculture. Is this a contradiction? Explain.
35. **Hot water.** Discuss some difficulties with use of hot water for agriculture; for aquaculture. Discuss the potential benefits.
36. **Solar energy.** Explain how a solar-operated steam generator can thermally pollute a river without having an effect on the total heat flux of thc biosphere.
37. **Waste heat and climate.** What types of energy sources do not add any additional heat to the environment?
38. **MHD.** Explain the operation of an MHD generator. Why is it a desirable form of electrical generation?
39. **Economics.** Explain how "cowboy economics" in the United States is responsible for the fact that: (a) Recycling operations are, in general, not economical; (b) the automobile is the most popular form of transportation; (c) most power plants use once-through cooling; (d) merchant tankers discharge their ballast into the ocean. If the concept of "cowboy economics" were displaced by a more conservationist doctrine, the prices of many goods would rise, but the quality of our environment could improve. How would you go about making decisions in instances where the two systems are in opposition? Do you think it would be helpful to set a dollar value on the environmental improvement and then match that against the price rise? If so, how would you set such a value? If not, why not?

The following problems involve calculations.

40. **Work and energy.** A man drives a 4000 pound truck with a 2000 pound load 40 feet up a hill. The vehicle is 25 per cent efficient. How many foot-pounds of fuel energy are consumed? How many foot-pounds of useful work are done?
41. **Energy.** How many calories are needed to heat 1000 grams of water 30°C? How many kilocalories? Would you need the same amount of heat to warm 10,000 grams of water 3°C?

42. **Cost of thermal pollution control.** A typical home consumes 1200 kwh per month. Use Table 5.9 to determine the monthly cost to a family of (a) a cooling pond, (b) a mechanical wet tower, and (c) a mechanical dry tower.

(*Answers*: 40. 960,000 ft-lbs; 80,000 ft-lbs. 41. 30,000 cal or 30 kcal for both. 42. 9.6¢; 12¢; 96¢.)

BIBLIOGRAPHY

Four recent and comprehensive books on energy and the environment are:

Neil Fabricant and Robert M. Hallman: *Towards a Rational Power Policy: Energy, Politics, and Pollution.* New York, George Braziller, 1971. 292 pp.

John Holdren and Philip Herrera: *Energy.* San Francisco, The Sierra Club, 1971. 252 pp.

Hoyt C. Hottel and Jack B. Howard: *New Energy Technology.* Cambridge, Mass., M.I.T. Press, 1971. 363 pp.

Richard S. Lewis and Bernard I. Spinrad: *The Energy Crisis.* Chicago, Educational Foundation for Nuclear Science, 1972. 148 pp.

A fine book on some physical aspects of power production, thermal pollution and climate is:

Theodore L. Brown: *Energy and the Environment.* Columbus, C. E. Merrill, 1971. 141 pp.

The most comprehensive book devoted to climate modification is:

Inadvertent Climate Modification. Hosted by the Royal Swedish Academy of Sciences and the Royal Swedish Academy of Engineering Sciences. Cambridge, Mass., M.I.T. Press, 1971.

Two books on special topics are:

Farrington Daniels: *Direct Use of the Sun's Energy.* New Haven, Yale University Press, 1964. 374 pp.

Marvin M. Yarosh, ed.: *Waste Heat Utilization.* Springfield, Va., National Technical Information Service, 1971. 348 pp.

Two books with good articles on strip mining are:

Garry D. McKenzie and Russell O. Utgard, eds.: *Man and His Physical Environment.* Minneapolis, Burgess Publishing Co., 1972. 338 pp.

Fred C. Price, Steven Ross, and Robert L. Davidson, eds.: *McGraw-Hill's 1972 Report on Business and the Environment.* New York, McGraw-Hill, 1972. 505 pp.

Another book with a fine section on thermal pollution is:

H. Foreman: *Nuclear Power and the Public.* Minneapolis, University of Minnesota Press, 1970.

The subject of thermodynamics is covered in many standard texts at various levels, and no specific references need be given here.

The entire issue of *Scientific American* for September, 1971 was devoted to energy and power, and is an excellent reference.

Another general article on energy, written for a less technical audience, is found in the November, 1972 issue of *National Geographic.*

6

AGRICULTURAL SYSTEMS

6.1 INTRODUCTION

When man first evolved from his ape-like ancestors his diet depended on what he could manage to collect from day to day. He hunted and dragged his food back to his den or was hunted and dragged back to some other predator's lair. He competed with other herbivores for plant foods. Life was very hard during drought, flood, or pestilence. Because his technology was so limited, and his population so small, early man did not appreciably alter the earth's environment. His stone and wooden tools for digging and hunting were competitive with the tusk of the mammoth and the claw of the tiger, but certainly were not overwhelmingly superior.

Even when man first began to cultivate the land, his activities had little impact on global ecosystems. People lived close to their food supplies, and their wastes were returned to the farmlands directly (see Fig. 6.1). Thus, the consumption of nutrients was balanced by a return of nutrients. Of course, even a simple agricultural system is potentially disrupting, for it promotes the growth of a few species, where many species once existed, but the extent of the environmental alteration was small in early cultures. In fact, man was not alone in favoring those plants and animals which benefited him most; bees selectively pollinate the most succulent flowers, ants protect their herds of aphids, and sage grouse scratch the soil around the roots of sagebrush.

FIGURE 6.1 A primitive agricultural field in New Guinea. (From The Flow of Energy in an Agricultural Society, by Roy A. Rappaport. Scientific American, Sept., 1971, p. 116. Copyright by Scientific American, Inc. All rights reserved.)

When agricultural yields first enabled some individuals to pursue goals other than cultivation of the soil and to live far from sources of food, the cycle was broken, for nutrients were not all returned to the farmland from which they came. The result was an ecological imbalance. As food-growing technology became more efficient, the ecological disruptions became more severe.

Of course, man has understood for many years that his choices were either to refertilize the land or, eventually, to move elsewhere. In some areas, man has been quite successful in keeping farmlands fertile. For example, some regions of China, Japan, and Europe have been farmed successfully for thousands of years. During that period of time the soil has even been enriched by fertilization with human and animal manure and various other materials of biotic origin. However, in some places, as we will see, man still continues to destroy huge areas of previously fertile land.

The story of agriculture includes both successes and failures. While the total harvest of food for man has increased steadily for thousands of years, more than half of the world's people are victims of malnutrition, and famines due to crop failures continue to occur. Man's technological accomplishments, both in agriculture and in industry, have enabled him to rise to his present level of biological ascendancy, but many informed people feel that our present practices, despite their successes, are altering the earth's ecosystems so severely that the planet's future ability to produce high yields of food for man is being endangered.

6.2 AGRICULTURAL DISRUPTIONS

(Problems 1, 2, 3, 4, 5)

It is generally known that the agriculture of India cannot keep up with the needs of its people and that, consequently, famine is never more than a few dry seasons away. It is less well known that approximately two-thirds of the cropland has been completely or partially destroyed by erosion or soil depletion caused by man. The area of the province of Sind near the mouth of the Indus River is typical of a destroyed cropland. With the exception of areas under irrigation, Sind is now a barren, infertile, semi-desert region. Yet, archeological diggings have uncovered the remnants of highly civilized inhabitants who farmed there over 4000 years ago. Moreover, fossils of the native animals of the area show that elephants, water buffalo, tigers, bears, deer, parrots, wolves, and other similar forest dwellers lived there. The early settlers built temples with fire-baked bricks; undoubtedly, wood fed the fires, and the wood must have come from forests in the nearby areas that are now desert. The details of the transformation from forest to desert are largely conjectural because there are no recorded weather reports over these 4000 years. However, meteorology suggests a plausible mechanism. Remember that a forest maintains a cooler environment than a grassland. Part of this coolness is due to the water-holding ability of the forest ecosystem. Now, rain clouds passing over a hot tropical steppe will generally rise and precipitation will seldom occur, while rain clouds passing over cool jungle will be induced to drop their moisture. Thus, the change in climate which destroyed the land of Sind was probably caused by deforestation. This poses an interesting dilemma. Most farmers grow only those crops that produce an efficient quantity of food. Forests are poor producers of food for man. But the killing of the Sind forest ultimately destroyed the productivity of the land. Had the early inhabitants of Sind understood the importance of natural diversity, and had they been able to reach a workable apportionment of forest and cropland, they might have been able to grow enough food without destroying their ecosystem.

In the light of this hypothetical example it is interesting to observe what is happening in India today. As overgrazing destroys the ground cover, goat herders, using 20-foot-long sickles, cut the limbs of the trees for food for their goats. The trees die and the soil blows away. But try to tell a man who has a hungry baby at home to save the trees for future generations!

Another story of land destruction comes from the area of the fertile crescent, the "cradle of civilization." The Tigris-Euphrates valley gave birth to several great civilizations. We know that highly sophisticated systems of letters, mathematics, law, and astronomy originated in this area. Obviously, then, men had the time and energy to educate themselves and to philosophize. We can deduce that the food supply must have been adequate. Today much of this region is barren, semi-desert, badly eroded, and desolate. Archeologists dig up ancient irrigation canals, old hoes, and grinding stones in the middle of the desert. What must have happened?

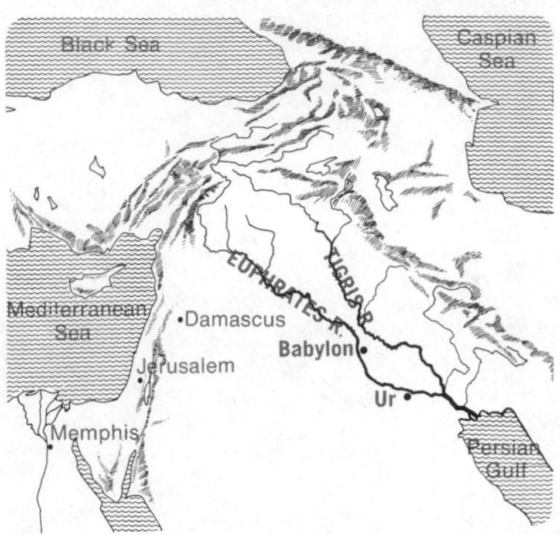

The fertile crescent.

Part of the story starts at the source of the great rivers in the Armenian highlands. The forests were cleared to make way for pastures, vineyards, and wheat fields. But croplands, especially if poorly managed, cannot hold the soil and the moisture year after year as well as natural forests or grasslands. As a result, large water run-offs such as those from the spring rains or melting mountain snows tended to flow down the hillsides rather than soak into the ground. These uncontrolled waters became spring floods. As a protective measure, canals were dug in the valley to drain the fields in the spring and to irrigate them in the summer and fall. Later, devastating wars resulted in the abandonment of the canals and croplands. The neglected canals turned to marsh, and the formation of marshes shrunk the rivers, and the diminished rivers could no longer be used to irrigate other areas of land. The water table sank, so that the yields per acre today are less than the yields 4000 years ago.

The story of land destruction can be repeated with monotonous regularity. Ancient Carthage was founded on the shores of the Mediterranean in North Africa amid dry but fertile grasslands. Grain was grown in abundance. Today much of this area has become part of the Sahara Desert. In fact, a large part of the Sahara is a by-product of over-plowing and overfarming fertile land, which led to a depletion of the amount of available moisture. To study this process in more detail we shall examine a great American fiasco—the Dust Bowl.

The early European settlers found millions of acres of virgin land in America. The eastern coast, where they first arrived, was so heavily forested that even by the mid-eighteenth century, a mariner approaching the shore could detect the fragrance of the pine trees about 180 nautical miles from land. The task of clearing land, pulling stumps, and planting crops was arduous. Especially in New England, long winters and rocky hillsides contributed to the difficulty of farming. It was natural that men should be lured by the West, for here, beyond the Mississippi, lay expanses of prairie as far as the eye could see. Deep, rich topsoil and rockless, treeless expanses promised easy plowing, sowing, and reaping. In 1889 the Oklahoma Territory was opened for homesteading. A few weeks later the population of white people there rose from almost nil to close to 60,000. By 1900 there were 390,000—a people living off the wealth of the soil. In 1924 a thick cloud of dust blew over the East Coast and into the Atlantic Ocean. This dust had been the topsoil of Oklahoma (see Fig. 6.2).

In each of the earlier examples one might contend that the destruction of the land was really caused by changes in climate rather than by man's mismanagement of the land. Indirect evidence, however, strongly implicates man. For instance, in the areas between the Tigris and Euphrates rivers where the canals were not destroyed, the land remains fertile. Similarly, some areas of North Africa near the Sahara still support trees believed planted by the Romans, and geological evidence indicates relatively constant weather patterns in these areas. However, in the case of the Oklahoma Dust Bowl we *know* that the land was destroyed by man and not by climate.

In Chapter 1 we learned that a natural prairie is a diversified ecosystem with homeostatic mechanisms to protect itself from spring floods and summer droughts. White man's contribution to the prairie was not particularly far-sighted. He planted large fields of single crops, thus destroying the naturally balanced system. He killed the bison to make room for his cattle, then killed the wolves and coyotes to prevent predation of his herds. Moreover, he often permitted his cattle to over-graze. In over-grazed land, the plants,

FIGURE 6.2 Dust storm. (© Arthur Rothstein, New Rochelle, N.Y.)

especially the annuals, become so sparse that they cannot reseed themselves. The land itself, therefore, becomes very susceptible to soil erosion during heavy rains. In addition, the water runs off the land instead of seeping in, resulting in a lower water table. Because the perennial plants depend upon the underground water levels, depletion of the water table means death for all prairie grasses. The whole process is further accelerated as the grazing cattle pack the earth down with their hooves and block the natural seepage of air and water through the soil.

Man's introduction of the plow to the prairie had an even more severe effect because the first step in turning a prairie into a farm is to plow the soil in preparation for seeding. At this point, of course, the soil is vulnerable, since the perennial grasses which normally hold the soil during drought have already been killed.

If the spring rains fail to arrive, then the new seeds won't grow and the soil will dry up and blow away. As an emergency measure during droughts, farmers often practice **dust-mulching.** By chopping a few inches of the surface soil into fine granules a thin layer of dust is formed. This dust layer aids the capillary action whereby underground water is brought to the surface. Thus the soil remains moist just below the dust, and the seeds sprout. However, before labeling dust-mulching a success, one must examine several other factors. Fertile soil is more than pulverized rock; it consists of decayed organisms and partially decayed organic matter. Moreover, prairie topsoils have evolved a balancing mechanism whereby the concentration of decayed organic matter remains approximately constant. Dust mulching brings to the surface much of the previously underground organic material. Once on the surface this organic matter reacts with the oxygen of the air (oxidizes) much more rapidly than it would have had it remained underground. The result is a decrease both in the concentration of organic matter in the soil and in the soil's fertility. As these losses continue, a time is finally reached when the soil is so barren that it cannot support even the growth of the hardy prairie grasses. Of course, the farmer can replace lost organic matter by spreading manure or other fertilizer, but the fact is that millions of acres of previously fertile farmland have been ruined because the farmers have not refertilized adequately.

FIGURE 6.3 Farmland in Mississippi ruined by soil erosion. (Photo courtesy of U.S. Forest Service.)

On the other hand, if, after plowing, the spring rains are too heavy, then the soil may easily wash away before the seeds have an opportunity to grow. Even after seeds have sprouted, the practice of pulling weeds between the rows leaves some soil susceptible to erosion by heavy rains.

The preceding discussion is relevant to the events in Oklahoma. Over a period of 20 to 35 years the soil fertility slowly decreased. Incomplete refertilization and loss of soil from wind and water erosion took their toll (see Fig. 6.3). Finally, when a prolonged drought struck, the seeds failed to sprout and a summer wind blew the topsoil over a thousand miles eastward into the Atlantic Ocean.

The droughts that killed the Oklahoma farms had no lasting effect on those prairies left untouched by man. In fact, these virgin lands are still fertile. In a few thousand years, perhaps, the wind-scarred Dust Bowl will regain its full fertility. We say "perhaps" because similar destruction of the North African prairie left the land so barren that nothing was left to hold the rain that did fall. Two thousand years after the farms failed one can stand in the center of the ruins of a wealthy country estate and watch the sands of the great Sahara blow by.

One might think that it would be impossible to destroy land by importing water (irrigating). Unfortunately, irrigation, too, can be destructive. When rain water falls on mountain sides, it collects in small streams above and below ground and it filters over, under, and through the rock formations. In the process of flowing into a large river, the water dissolves various mineral salts present in the mountain rock and soil. Usually these salts are concentrated in the oceans. However, if the river water is used for irrigation, man is bringing slightly salty water to his farm. When water evaporates the salt is left behind and, over the years, the salt content of the soil increases. Because most plants cannot grow in salty soil, the fertility of the land decreases. In Pakistan an increase in salinity decreased soil fertility alarmingly after a hundred years of irrigation. In parts of what is now the Syrian desert, archeologists have uncovered ruins of rich farming cultures. However, the land adjacent to the ancient irrigation canals is now too salty to support plant growth.

Irrigation is threatening agricultural production in parts of California, which produces about 40 per cent of the vegetables consumed in the United States. The richest vegetable-growing area in California is the San Joaquin valley where virtually all commercial operations rely heavily on irrigation. As the irrigation intensifies, so does the threat of destructive salination of the soil and the groundwater. Most of the valley is now serviced by a shallow underground drainage system to divert the brackish waters, but the old system now appears inadequate, and new drainage projects are being proposed.

In addition, extensive, poorly managed irrigation can raise the water table of the irrigated land. If the water table is too high, plant roots will become immersed in water and will die from lack of air. In the earlier 1960's waterlogging and salinity problems were causing the loss of 60,000 acres of crop land in Pakistan alone.

Occasionally, even an apparently well-planned upset of the balance of nature has produced disastrous results. For thousands of years the peasants of Egypt farmed the Nile Valley. Every spring the river flooded and brought with it water and fresh soil from the Abyssinian Mountains. When the water level subsided the farmers grew what they could during the long hot summer. To increase the agricultural yield for a growing population, engineers had long considered the feasibility of damming the Nile, thus storing the excess water during flood seasons for more efficient use during the summer. The Aswan High Dam, built over a period of 11 years and inaugurated in 1971, was to make this objective possible as well as to provide abundant hydroelectric power. Water storage started in 1964, when the Nile was diverted into a bypass channel during construction, and power has been produced since 1967. Other benefits obtained have included higher yields of cotton, grain, fruits, and vegetables by irrigation during the summer and by reclamation of previously barren land. But there have been problems. First, the sediment that the flood waters formerly washed out to sea, although useless to farmers, did nourish a rich variety of aquatic life. The absence of this nutrient has resulted in the annual loss of 18,000 tons of sardines. Second, the flood waters previously rinsed away soil salts that otherwise would have accumulated and made the soil less fertile. With the flooding under control, the salts are left in the soil. The consequent rise in soil salinity is a threat to the

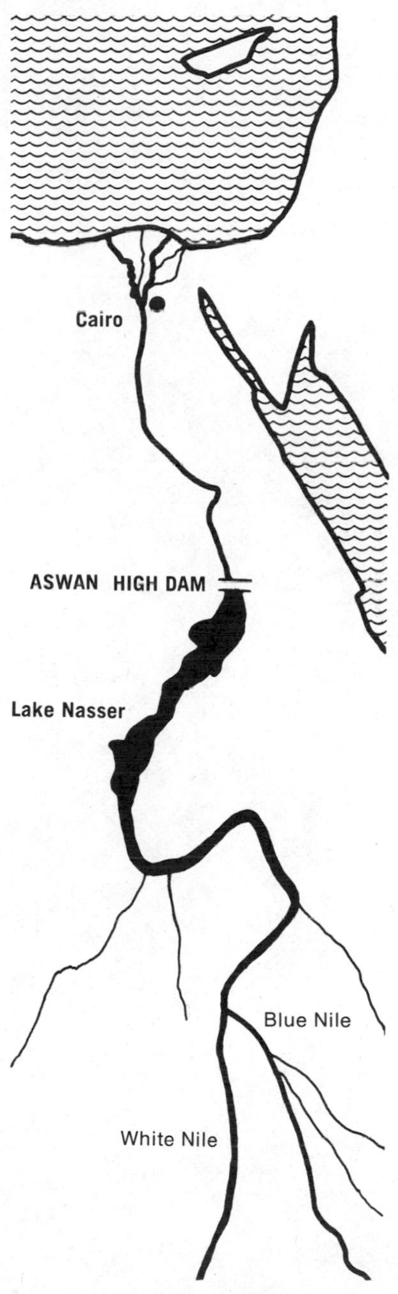

Aswan Dam region.

productivity of the land. Third, the sediment formerly protected the delta land in several ways: it served as an underground sealant to minimize seepage that would otherwise drain off various sweet-water lakes, and it helped strengthen natural sand dikes that protected the coastline against erosion by the powerful currents of the Mediterranean. These buffering actions, too, are gone, as the sediment from the Abyssinian Mountains instead sinks behind the High Dam, and the clear, silt-free waters rush downstream more rapidly than before, eroding the river banks and undermining the foundations of many of the bridges that span the Nile. Fourth, the loss of silt previously deposited on the downstream fields is a loss of nutrient, and must be compensated by chemical fertilizers. Fifth, the dry periods between floods, now eliminated by irrigation, used to limit the population of water snails which carry a worm that spreads easily to man, by depositing its larvae under the skin, whence they invade his intestinal and urinary tracts. The resulting debilitating disease, called **schistosomiasis,** follows along the paths of the irrigation canals, infecting about 80 per cent of the people who work in them. (See also Chapter 10, Sect. 10.) How can one balance lost manpower and increased human misery against higher crop yields?

Finally, the High Dam seems to be responsible for a serious loss of water, the one essential substance that was never to be in short supply. The lake behind the dam was to have been filled by 1970; it was not yet half full in 1971. Some limnologists (scientists who study the physical phenomena of lakes) predict that it will not be full for a century or more. Losses are suffered by seepage through porous rock and by evaporation in the hot winds of Upper Egypt. Of course, technology may come to the rescue. All the problems cited above are under intensive study; means may be found to control the erosion, the parasites, the seepage, and the salinity, and to make enough power to supply fertilizer plants to manufacture nutrients to replace those formerly carried down as silt from the mountains. There is no law of chemistry or physics that denies the possibility of such technological rescue of technology's disruptions. But thus far this goal is much more elusive than had been expected.

We have cited examples of some of man's past agricultural failures that have led to the loss of large areas of fertile cropland. At the present time, millions of acres are still being lost. In fact, farm soils are presently being destroyed faster than they are being formed by natural processes. In recent times, despite this overall loss of fertile areas, greatly improved farming techniques have increased agricultural yields across the world. However, these increases cannot be expected to continue indefinitely and some future leveling effect can be anticipated.

6.3 ENERGY FLOW IN INDUSTRIAL AGRICULTURE

(Problem 6)

Consider some differences between a wild oat plant growing in an unfarmed prairie and a domestic oat plant in a field. To survive, the wild oat must compete successfully for sunlight and moisture with its neighbors. A tall plant, one that sprouts early, or one with an effective root system, has a competitive advantage. The energy a plant needs to grow a tall stalk or a deep root must come from the sun. On the other hand, a farmer aids the survival of a cultivated oat. He waters it when necessary, removes plant competitors, and loosens the soil to stimulate the growth of root systems. Since all the seeds in the field are planted at the same time and are of the same variety, competition is minimal and the plant does not need a tall stalk or a unique and fast-growing root system to survive. In other words a wild plant must use some of its incident solar radiation for survival and some for the production of seeds or bulbs. A farmer is willing to add auxiliary sources of energy to help the plant survive if he can grow a variety that produces more food for man.

Now as we mentioned in Chapter 2, the genes of different individuals in a species are not identical; in particular, plants that have a high grain yield differ genetically from those which are low yielders. Thus, if a farmer replants seeds only from those plants that produce the most grain, regardless of the viability of that plant in an *untended* field, he should be able to breed high-yielding plants over the course of time. In today's laboratories, different varieties of plants are often cross-bred with the hope that a new combination of existing genes might produce progeny with desirable characteristics (see Fig. 6.4). Alternatively, the existing genes may themselves be changed by inducing mutations artificially. These genetic manipulations are often varied and quite sophisticated, but all of them lead to the same kinds of choices: The agrono-

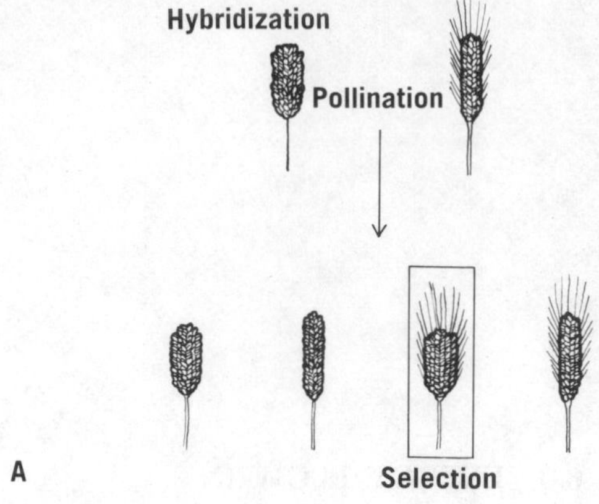

FIGURE 6.4 Scheme illustrating the general procedure for cross-breeding of plants.

mist must decide whether or not to select a plant for further breeding, and ultimately whether to adopt it for commercial planting. Traditionally, agronomists have chosen the plant varieties that yield the most desirable crop—usually with emphasis on quantity. As a result of such selections, however, crops have become increasingly dependent on cultivation for survival. Thus, if we compare a variety of Oriental rice developed in the late 1960's with wild Oriental rice, we find that the new variety produces more grain than the old, but is also shorter, less resistant to disease, and more dependent on irrigation than its predecessor. In addition, the new seeds have lost their biological clocks and can sprout any time they are planted, whereas native rices are geared to the seasons, and if left uncultivated, new seedlings will germinate only in the proper growing season. The new seeds require care, and care requires energy—energy to plant, to weed, to control pests and disease, and to irrigate.

Similarly, animal breeders have produced varieties of livestock that are more efficient than prairie grazers in converting grain to meat and to dairy products. But the farmer must invest energy to realize these high yields. A prairie chicken lays 10 to 20 eggs per year, all of them in the spring, then broods them and cares for the chicks. A domestic hen can produce over 200 eggs in a year, but most hens will not sit on a fertilized egg long enough for it to hatch. Therefore, most breeds of domestic chickens require artificial incubations for propagation.

Farmers in industrially developed countries have large supplies of fossil fuels available to them. They use these resources to supplement the energy that the plants get from solar radiation (see Fig. 6.5). Let us examine the overall energetics of a typical modern agricultural system—the mechanized production of beef. Remember that for every 12 calories of energy a plant receives from the sun, only about one-half calorie is available for the production of animal tissue by a herbivore that consumes the plant. Some of the 11.5 calories that are "lost" are required by the plant for its metabolism, some by the animal for its metabolism, and some of the energy is used by the animal in search of more food. In a feedlot, a steer is not required to move, but may stand in front of his feedbin, eating hay or grain that the farmer, with the help of his tractor, baled or threshed and brought in from the field. In addition, food additives and growth hormones, synthesized in factories powered by coal or oil, are used to increase growth. In a feedlot, about six calories of fossil fuel energy are used for every 12 calories that the plant matter receives from the sun. The total weight gain of the animal under these conditions corresponds to about five calories (see Fig. 6.6). The artificial system is more efficient than a natural system in converting both plant and total energy to meat. This enhancement of efficiency is the major benefit of the modern system of producing food. Production of beef in feedlots is based on a new cycle of manmade technology. High food production is needed for the urban workers who provide the technology required to maintain the high food production. The individual components of the cycle are interdependent. Large populations of human beings are dependent on high

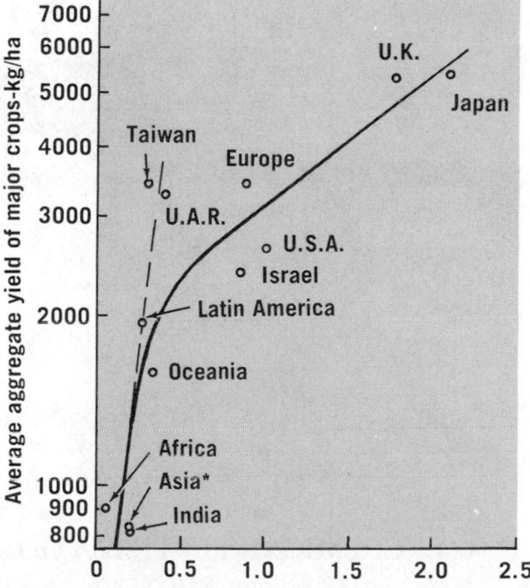

*Excluding Mainland China

FIGURE 6.5 The relationship between crop yields and energy.

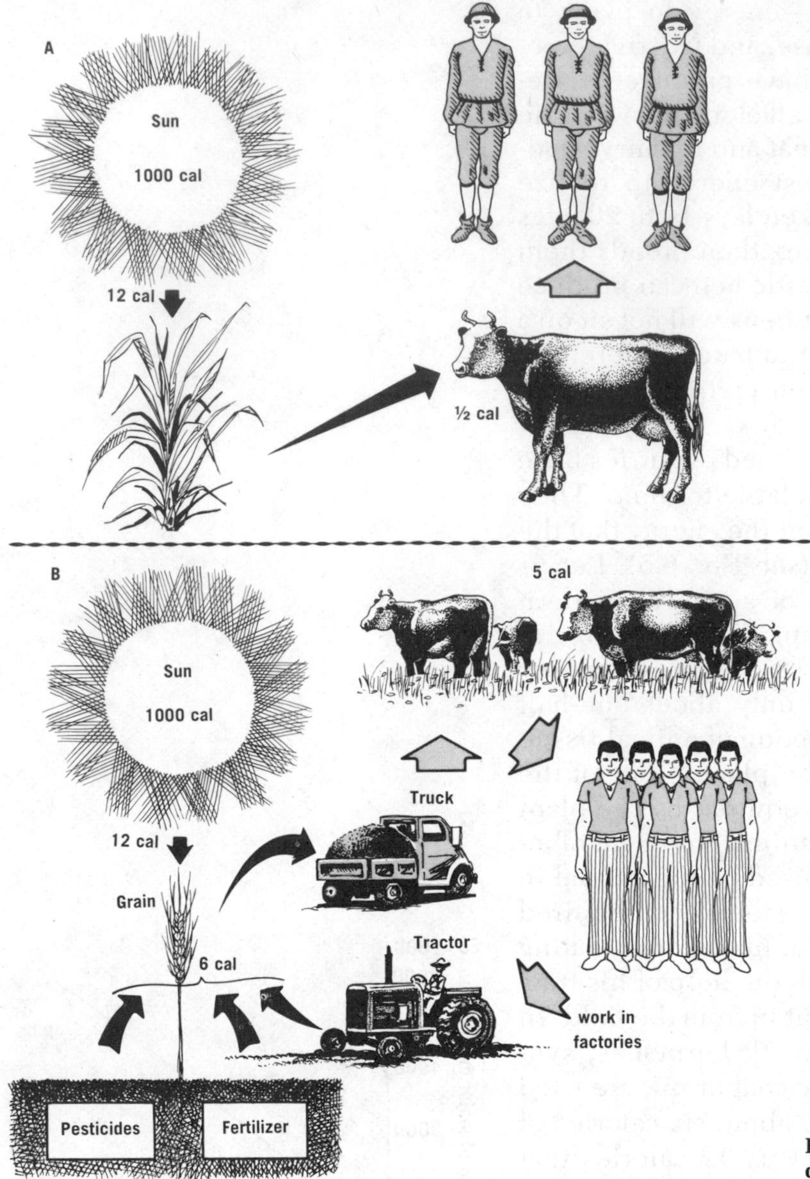

FIGURE 6.6 Comparison between beef production on the range and in a feedlot.

agricultural yields, while at the same time these high yields are dependent on high industrial outputs. If one link in this chain is cut, there cannot be reversion to the ancient range system without mass starvation.

6.4 CHEMICAL FERTILIZATION

(Problems 7, 8, 9, 10, 12, 13)

Industrial agriculture relies heavily on chemical control of disease, insects, and weeds, on chemical fertilization, on mechanization, and on irrigation. Insect, weed, and disease control will be treated separately in Chapter 7. Mechanization is discussed in Section 6.5, and irrigation in Section 6.6.

Man fertilized his crops with manure, straw, or

dead fish long before he understood the chemistry of fertilization. Today, mined and manufactured fertilizers are used so extensively (see Fig. 6.7) that the total energy requirement of the fertilizer industry constitutes a major portion of agriculture's large demand for fossil fuel. The chemistry of fertilization is essentially the same as the chemistry of the nutrient cycles discussed in Chapter 1. Oxygen and carbon are readily available to all plants that are exposed to air, and are of no concern to the fertilizer industry.

As mentioned earlier, atmospheric nitrogen is relatively inert and unavailable to most plants. Only a few organisms—for instance, the bacteria on the roots of legumes such as alfalfa, peas, and beans—are able to fix* atmospheric nitrogen, that is, to convert N_2 to ammonium ion, NH_4^+. Most grain crops lack this ability, and therefore must be supplied with fixed nitrogen from some outside source if the fields are to be harvested annually. Nitrogen fertilization can be effected by any of three different techniques. The simplest, and the method used most commonly by the ancients, is to recycle plant and animal matter. Another technique is by crop rotation, that is, planting legumes and grain in alternate years and thus maintaining soil nitrogen. Finally, man has recently learned to convert atmospheric nitrogen to plant fertilizer by producing ammonia ($N_2 + 3H_2 \rightarrow 2NH_3$), which is then readily converted to ammonium ion. Agriculturalists fear that an energy shortage might adversely affect the processes used to recover nitrogen from air.

Mineral fertilizing is qualitatively different from nitrogen fertilizing. If the minerals cannot be recycled by the reuse of dead plant matter, or by such material as bone meal, man must mine the needed minerals from some geological deposit. Thus, phosphorus, potassium, calcium, magnesium, and sulfur, all important constituents of a well-balanced fertilizer, are extracted from the earth. Sometimes these minerals are simply pulverized and dusted onto the soil, while sometimes they are chemically treated in order to enhance the speed or efficiency of uptake by plant roots. Of these five minerals, all but phosphorus are plentiful in the earth's crust. Known geological deposits of phosphate rock (an oxidized form of phos-

*To "fix" means to make firm or stable. In this sense, a gas (which is not firm) can be fixed by binding it in some form of solid or liquid.

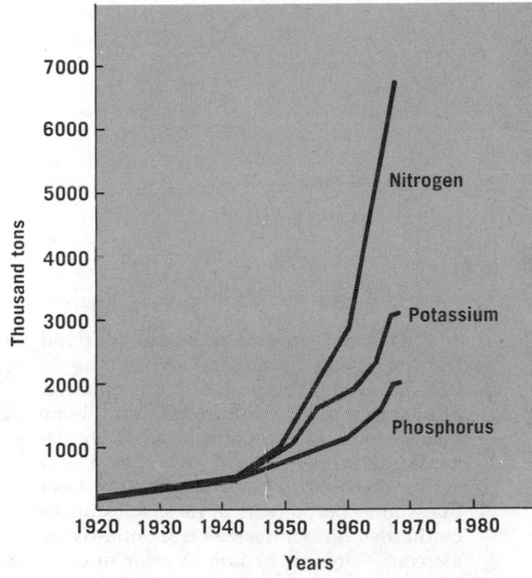

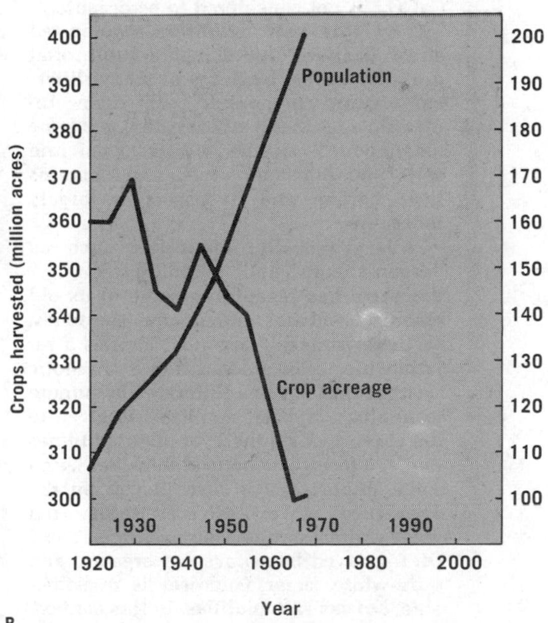

FIGURE 6.7 **Fertilizers in U.S. agriculture.** *A.* **Use of plant nutrients, 1920 to 1968.** *B.* **Acreage of 59 principal crops harvested plus acreages in fruits, tree nuts, and farm gardens. Total U.S. population, including persons in the armed forces.**

phorus) will be depleted early in the twenty-first century if current mining rates are continued. Without an adequate supply of phosphorus, highly productive agriculture will be impossible. New discoveries are not expected to increase known reserves appreciably.

When we speak of the depletion of a mineral deposit such as phosphate, we do not mean that all the elemental phosphorus on the earth will disappear, but rather that it will be dispersed. Our phosphate reserves will be considered to be depleted when the concentrated deposits have been mined, sold, and spread over agricultural systems. At this time, some fraction of the total phosphorus will have been absorbed by plants, consumed by humans, collected in sewage systems, and ultimately washed into rivers and from there to the oceans. Some will remain in the soil in chemical forms unusable by plants, and will slowly pass into groundwater and eventually return to the oceans. Mining phosphate dissolved in the ocean would be prohibitively expensive, for one would have to process several thousand pounds of sea water to extract one pound of phosphorus. In certain regions, various combinations of ocean currents and of gradients of chemical concentrations can bring about the deposition of phosphorus in mineral form. The long-term recycling of phosphorus has always been dependent on these geological processes, but the accumulation of phosphate deposits on the ocean floor and the uplifting of continental shelves to form dry land takes millions of years.

"Organic" originally meant "derived from a living organism." According to this definition, composted straw, manure, offal from a slaughterhouse, and bone meal would all be considered to be organic materials. After it was shown in a series of discoveries from 1828 to 1845 that some components of these substances could also be synthesized from non-living sources, "organic" came to refer to compounds containing C–C or C–H bonds, with or without other atoms. (Marble, $CaCO_3$, is not considered to be organic.)

By this new definition, composted straw, manure, and slaughterhouse offal are all complex mixtures of many different organic compounds (with a few inorganic ones in the mixture, too), but bone meal, which consists mostly of calcium salts and other minerals, and contains little carbon and hydrogen, is largely inorganic.

Very recently, in usages such as "organic foods" and "organic gardening," the word has recaptured some of its old meaning and has added new overtones. In these contexts "organic" implies a relationship to, or a derivation from, biotic systems that are unadulterated by human technology. Special emphasis is given to the absence of synthetic or other additives that are foreign to natural food webs, and some importance is also placed on the avoidance of excessive refining that greatly alters natural compositions. Thus, DDT-sprayed beans are not organic, and pure-white sugar (sucrose) is questionable, but sea salt qualifies. In this context, therefore, "organic" may be briefly defined as "naturally constituted." Thus, an organic gardener would include pulverized phosphate rock as well as bone meal in his list of organic materials.

In spite of the fact that manufactured fertilizers have raised agricultural yields across the world, and in spite of the general health and vitality of peoples raised on fertilized crops, the use of inorganic fertilizers is being questioned, and there are many outspoken advocates of "organic" gardening and farming. Exactly what is meant by "organic" gardening?

The major argument against non-"organic" farming is based on the chemistry and physical nature of the soil. Plants need soil with properly regulated nutrients, density, moisture, salinity, and acidity. Soil conditions have traditionally been regulated by the **humus**, a very complex mixture of compounds resulting from the decomposition of living tissue. A given piece of tissue, such as a leaf or a stalk of grass, is considered to be humus rather than debris when it has decomposed sufficiently in the soil system so that its origin becomes obscure. Compared to inorganic soil, humic soil is physically lighter, it holds moisture better, and it is more effectively buffered against rapid fluctuations of acidity. Additionally, certain chemicals present in humus aid the transfer and retention of nutrients. For example, calcium ions, Ca^{++}, can exist in water solutions from which they are usable by plants. However, atmospheric carbon dioxide reacts with water to form carbonate ion, $CO_3^{=}$, which reacts with calcium in water to form the sparingly soluble compound, calcium carbonate, $CaCO_3$. A striking consequence of its meager solubility has been the deposition of large masses of $CaCO_3$ on earth in the form of limestone and marble. But it is not *completely* insoluble; moreover, its solubility depends largely on the soil acidity, and under certain conditions, the calcium may be liberated. Under other conditions, it is very insoluble and the calcium may not be readily available to plants, even though it is present in the soil. Moreover, if dissolved calcium is not used immediately by plants, it may travel with water droplets down below the root zones where it becomes unavailable. This movement of free ions into the subsoil and underground reservoirs is called **leaching.**

$$CO_2 + H_2O \rightleftharpoons HCO_3^- + H^+$$

$$HCO_3^- \rightleftharpoons CO_3^{=} + H^+$$

$$CO_3^{=} \xrightarrow{Ca^{++}} CaCO_3$$

$$CaCO_3 + 2H^+ \rightarrow Ca^{++} + CO_2 + H_2O$$

Richly humic soils provide chemical reaction pathways for metal ions that are not available in simple inorganic solutions. Certain chemicals in the humus, which are known as chelating agents, react with inorganic ions such as Ca^{++} to form a special class of compounds known as chelation complexes. Ions bonded in chelation complexes are held tightly under some conditions, but are easily released under others. A calcium ion chelated by humus will not react readily with carbonate to form $CaCO_3$, nor will it leach easily. Rather it will tend to remain bonded to the chelating agent and thus be retained in the humic matter.

The word comes from the Greek *chele*, meaning "claw." Chelates bind atoms from two or more different directions, much as a lobster grabs on to its prey.

Plants and soil microorganisms have evolved mechanisms whereby they can release chelated ions readily and incorporate them into living tissue, as shown schematically in Figure 6.8. Naturally, the specific chelation chemistry is different for each soil nutrient, but in general, humus maintains a nutrient reservoir and enhances the ease with which nutrients are used by the soil organisms.

When soil is plowed, the humus which normally remains underground becomes directly exposed to air

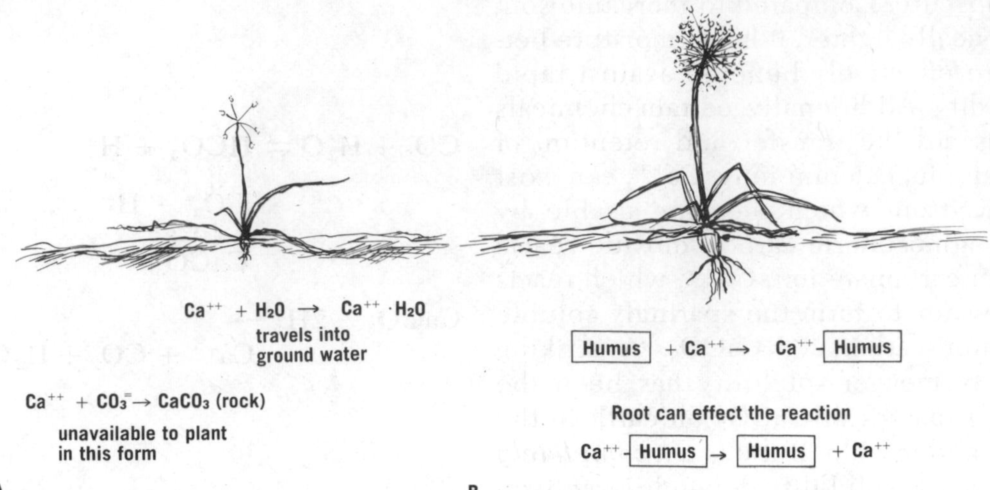

FIGURE 6.8 Schematic illustration of pathways for transfer of inorganic nutrients. (A) Without humus. (B) With humus.

where some of it oxidizes. Under these conditions the complex chemicals which contribute to the unique chemistry of humus decompose more rapidly than they do in undisturbed soil. Thus, when a virgin area is plowed and converted into cropland, the humus content of the soil decreases. In many areas of the world, organic fertilization has allowed land plowed for thousands of years to remain fertile.

In recent years, some potentially serious problems have arisen from the depletion of humus.

One hundred years ago the area that is now North and South Dakota, eastern Montana, and southern Saskatchewan was buffalo prairie; now it is cattle and wheat country. Currently, it is common to plow fields, leave them idle during the spring and early summer, then plant winter wheat in the late summer or early fall. The practice of leaving soil exposed to spring rains and summer heat has rapidly accelerated the decomposition of humus. Soil with little humus and no roots is vulnerable to leaching. The last ice age had deposited large quantities of salts in the subsoil of this region. Between the time of the last glacier and the twentieth century, the prairie roots and the humus had served as a barrier against the seepage of large quantities of rainwater down into the subsoil. At the present time, however, such filtering and leaching occurs more rapidly and the water table is rising each year. This rising water is carrying the glacial salts toward the surface, and if the salt water reaches the root zone, present agriculture will fail in this area.

When soil humus is not maintained, the efficiency of converting fertilizer to plant tissue is low and the ability of soils to store reserves of nutrients is poor. As a result, large quantities of nitrogen or mineral fertilizers are leached into surface and ground waters or tied into the soils in chemically unusable forms. This loss is a relatively minor problem to the farmer, for despite the inefficiency, inorganic fertilizers represent one of his best investments, returning about $1.20 worth of yield increase for every $1.00 spent. However, lost fertilizers collect in the waterways and significantly add to water pollution in agricultural areas. Compounds that are nutrients on land are also nutrients to aquatic life, and serve to feed algae and plankton, thereby upsetting the food webs in rivers and lakes. This subject will be discussed in more detail in Chapter 10 on water pollution. Nitrogen fertilizers are additionally harmful in that the nitrate ion, NO_3^-, may pose a health hazard.

NO_3^- (nitrate ion)
↓ biological reduction
NO_2^- (nitrite ion)
↓ R_2NH (amines in foods, etc.)
$R_2N—NO$
cancer-producing nitrosamines

Another argument against the practice of chemical farming is related to the nutritional value of the crops. Several studies have shown that the food value of many grains and vegetables decreases when high concentrations of fertilizers are used. In one study, alfalfa yields were increased dramatically with high applications of chemical fertilizers; however, cattle fed on that alfalfa became ill from cobalt and copper deficiencies. These trace minerals, necessary to the animal's health, were not present in the fertilizer, and were therefore lacking in the alfalfa. In a study examining the relationship between oats and phosphate fertilizer, both the total yield of grain and the phosphorus content of each oat plant increased with increasing quantities of phosphate fertilizer, but the ratio of weight of phosphorus per unit weight of oat grain decreased.

The lowering of nutritional values caused by chemical fertilization does not present a serious problem to most inhabitants of the developed nations. Indeed, many foods do not appear to be harmed by large fertilizer applications, and the decreased food value of many others is insignificant to people who can afford a rich and varied diet. Secondly, transportation and refrigeration enable shoppers to purchase foods from many different geographical areas. Because some soils in different regions are naturally endowed with an overabundance of certain trace minerals, a person eating foods raised in many different soils is likely to ingest an adequate supply of minerals. The problem of the nutritional value of individual foods is serious in areas where families eat a subsistence diet grown in one field.

Although commercial farming practices have had an overall record of past success, the problems associated with these practices are now demanding increased attention because the organic content of soils is continuing to drop. One solution would be to formulate more sophisticated artificial fertilizers including synthetic chelating agents. A second solution would be to initiate a massive program of naturally constituted fertilization through recycling. There is no doubt that the soils in the United States, for example, could be improved if the two billion tons of animal feces, slaughterhouse offal, and other organic residues produced each year were composted and returned to the land. Currently, some of this material is used as fertilizer. It is often cheaper, however, to buy and spread chemical fertilizers than to handle the natural materials. While the yield per dollar in a single growing season may often be maximized by discarding manure, the wastes wash into streams, thereby polluting them, and the cost of this pollution must be considered. Additionally, the agricultural yields for future generations must be assured, and many economists consider that we should pay our share of the costs for such provisions.

In many countries dried and pasteurized sewage sludge from municipal sources is marketed as fertilizer. However, short-sighted economic criteria have often favored disposing of wastes into nearby rivers.

6.5 MECHANIZATION IN AGRICULTURE

(Problem 11)

High agricultural yields depend on specific varieties of seeds, an active fertilizer program, and pest control. The use of tractors, on the other hand, does not improve yield per acre. For example, most Japanese farms vary in size from about one to five acres. Many farmers own small, two-wheeled rotary tillers, but very few own four-wheeled tractors, and the large combine is virtually unknown. Yet the yields per acre in Japanese agriculture are the highest in the world. However, these yields depend on great quantities of fertilizers and much labor. Therefore, the total auxiliary energy requirement of Japanese farming is high even though the use of heavy equipment is minimal.

The fact that a tractor can out-perform any animal

in pulling a plow is not the only advantage of mechanization. Another advantage is versatility. Farm equipment that can perform many intricate operations simultaneously is now available. A single device can plow, fertilize, and plant in one pass through a field, while another can pick the fruit, sort it according to size, bag it, weigh the bags, and release the labeled sacks for delivery to market. As a result of such extensive mechanization of agriculture, farms in the United States now use more petroleum than does any other single industry.

As mechanization becomes more versatile and more intricate, and as the individual devices get larger, agricultural yields per acre actually *decrease.* A machine that processes 10 rows of crops will necessarily be less precise than one that handles only two or three rows, because of unevenness of ground levels and other natural variations. The result will be that the larger machine causes more damage and hence more waste. A mechanical tomato picker requires a variety of tough-skinned tomatoes that all ripen at the same time, because the machine has no eyes to differentiate between red and green. Such varieties have in fact been developed, but their simultaneity of ripening is not perfect, and hence, again, the machines make waste. There are machines that shake apples from trees into collecting frames, but they bruise some apples. Furthermore, all of these problems are aggravated by the crowding of plants that is made possible by the greater productivity of fertilized soils.

6.6 IRRIGATION

(Problem 16)

Although irrigation has often destroyed land, it continues to be important in highly productive agriculture. Former success in irrigation has led many planners to look toward the deserts as future sources of food. The conversion of deserts into agricultural systems depends on huge quantities of water and, in many cases, on large amounts of energy to desalinate ocean water. Chapter 5 discussed the availability of energy sources in the future and the proliferation of nuclear reactors which could be used to desalinate seawater. Advocates of reactor-based irrigation systems point to the fact that if all the deserts that lie within 300 miles of the ocean were to be developed and planted to grain, the total grain-producing acreage of the world would rise from the present 1.6 billion acres to a phenomenal 6.8 billion acres. However, the required quantities of energy are not presently available, for now desalinated water is often too expensive to be practical, but, as energy technology improves, and as agricultural practices use water more conservatively, the economics of the system is likely to change.

In traditional irrigation systems water is pumped through ditches in open fields. About 98 to 99 per cent of the water is lost to evaporation either directly from the canals or indirectly through transpiration from the plant leaves. Most of this loss can be eliminated if crops are grown in plastic greenhouses under carefully controlled conditions. In the most advanced greenhouse technology, the atmosphere is maintained at close to 100 per cent humidity to minimize transpiration, and irrigation water is carried directly to the plant root system through plastic pipes to eliminate evaporative losses that occur in open systems. Additionally, carbon dioxide gas is pumped into the artificial environment to accelerate photosynthesis, and the soil is heavily fertilized. When the high incident solar radiation of desert areas is coupled with the enhanced water, CO_2, and nutrient inputs, some rather striking yields can be achieved. In a pilot project on the Arabian Peninsula, in one year experimenters have grown about a million pounds of vegetables on five acres (see Fig. 6.10). This harvest represents about a twenty-fold increase over the average produc-

FIGURE 6.9 Flood-water irrigation in the valley of Teotihuacán, Mexico, May 1955. (From Eric R. Wolf: *Sons of the Shaking Earth.* Phoenix Books, University of Chicago Press, 1959, p. 77. William T. Sanders, photographer.)

FIGURE 6.10 A desert greenhouse project. (Photo courtesy of Environmental Research Laboratory, University of Arizona.)

tion of the same crops in open fields. These yields must be balanced against the high capital investment for the desalination plant and the costs of the greenhouse operation, but the project appears to be economical for high-priced crops such as tomatoes and cucumbers.

The failures of some past irrigation projects (see Section 6.2) serve as a warning that future dependence on mammoth uses of imported water may not be reliable. As described in Section 6.2, heavily irrigated lands tend to become salty, and eventually unfit for agriculture.

6.7 AGRICULTURE IN THE LESS-DEVELOPED NATIONS AND THE GREEN REVOLUTION

(Problems 14, 15)

There has never been widespread famine in the United States. Barring a major war, a famine in the United States within this century is highly unlikely despite the potential problems discussed in the previous section which may make us worry about the durability of our agricultural systems. Agricultural difficulties could easily cause increases in the price of food, but there should be enough produce to supply the people of this nation in the near (and maybe far) future with adequate diets. Conversely, most school-age children in India have already experienced famine and might again in their lifetimes. In the less-developed countries of Asia, Africa, the Middle East, and Central and South America, the race between population growth and food production has been close; sometimes most of the people have enough food to survive, sometimes many don't. Even in "good" years, malnutrition and related illnesses are common among the living. It is a fallacy to believe that hunger exists in the less-developed nations solely because the population density is high. Looking at Figure 6.5, we see that India, Asia, Africa, and Latin America are four of the five least productive agricultural areas in the world, and it is this lack of productivity which is primarily responsible for famine. If rice farmers in India could achieve the success of their Japanese counterparts, there would be no hunger in India. Thus far, attempts to increase the efficiency of agriculture in these areas have been unsuccessful. Why so?

Oceania (Central and South Pacific island groups including Australia and New Zealand) is unproductive in terms of yields per acre, but the population density is so low that people are generally well fed.

The poorest people in the world are primarily concerned with survival, and often the decision to eat today jeopardizes the food supply for tomorrow. For example, Indian women follow cattle, collect the dung, dry it, and use it as fuel for cooking. The poor often have no alternative sources of heat, for there are virtually no trees in parts of India, and coal or oil is too expensive. But at the same time, few farmers can afford to purchase chemical fertilizers, and most of Indian agriculture is dependent on manure and other organic wastes. When the fertilizers are used as fuels to cook this year's crops, the yield of next year's harvest is being jeopardized, so a continuous downward spiral develops. Agriculture in these depressed areas must somehow reverse the current trends.

Many attempts by wealthy nations to aid the less-developed countries have met only limited success. Gifts or loans of machinery are seldom used efficiently for several reasons: (a) While almost all American or European farmers can service their own equipment, few farmers in the less-developed nations can fix a tractor. (b) The distribution and supply system for spare parts and for petroleum in many countries is practically non-existent. (c) Poor farmers often cannot afford petroleum or spare parts, even if they are available, and interest rates in some villages approach 50 per cent per year. Consequently, tractors often lie idle soon after they arrive.

Gifts or loans of fertilizers have been more useful, but problems have developed with these imports, too. The varieties of grains grown in most of the world 10 years ago were closely related to wild grasses, and are characterized by long, thin stalks and small grain clusters on top. If these varieties are fertilized heavily, the young plants will produce such abundant clusters of seed that the stalk cannot support the top, and the plant bends over (lodges) and falls to the ground before it matures.

In the mid-1960's American scientists working in Mexico and the Philippines developed new varieties of wheat and rice that were adaptable to tropical climates and were capable of producing higher yields than any native grains. These seeds were heralded as the means to make most countries self-sufficient, to free millions of farmers from their downward spiral and to buy time to solve the ultimate problem of population expansion. The era of the new varieties, called the **Green Revolution**, has opened with some truly spectacular successes. One such instance was the reversal of the crop failures and famine that occurred in Pakistan in 1965 and 1966, when the monsoon rains failed to fall on schedule. Massive shipments of new varieties of wheat seeds during the next two growing seasons, coupled with good weather, raised the total wheat harvest almost 60 per cent in two years, and Pakistan made strides toward self-sufficiency. Similarly, India's wheat crop increased by 50 per cent between 1965 and 1969. The incidence of famine in that country, too, has decreased. The Philippines had been importing rice for half a century and suddenly, with the cultivation of large acreages of new rice varieties, the trends have been reversed and Filipinos became rice exporters. The development of these high-yielding seeds has had more impact on the daily lives of millions living in poor countries than any other technological development of the past 15 years (see Fig. 6.11).

On the other hand, figures of national grain production do not always reflect the fate of groups of poor farmers within that nation. To understand the social impact of the Green Revolution, we must first understand that the new grain varieties, planted in an impoverished soil and dependent on variable rainfall for growth, produce equal or smaller yields than the native grains which have been cultivated in poor

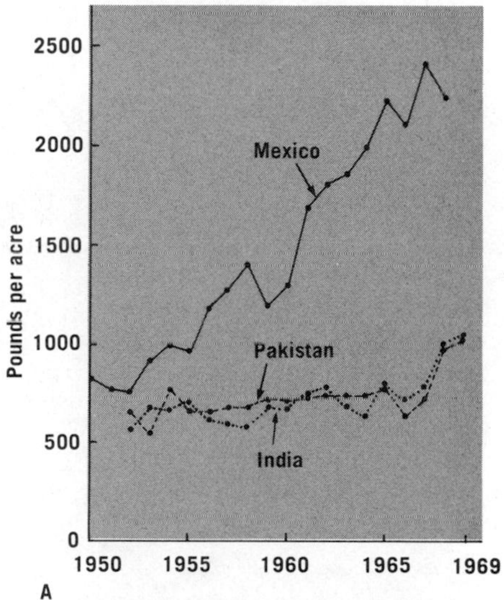

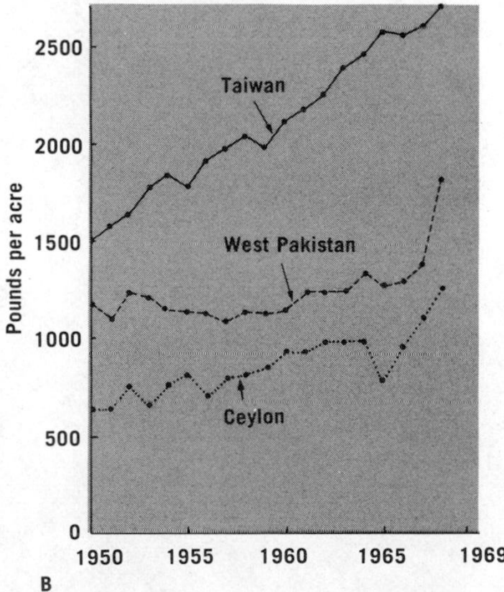

FIGURE 6.11 The yield takeoff of the Green Revolution. *A.* Wheat yields in Mexico, Pakistan, and India. *B.* Rice yields in West Pakistan, Ceylon, and Taiwan.

areas for centuries. The major advantage of the imported seeds is that they are more responsive to fertilizers than are the old seeds. The new varieties of wheat and rice have short, thick stalks and do not lodge when fertilized heavily. For example, the "miracle" rice variety IR-8 can absorb 120 pounds of fertilizer per acre economically, while native rices can only absorb 40 pounds of fertilizer per acre before the stalks bend or break. Moreover, IR-8 plants require fewer nutrients to build stalks, and hence the conversion of fertilizer to grain is twice as efficient as the conversion in tall-stemmed plants. The dependence of new seeds on high fertilizer applications means that a relatively poor farmer who can afford somehow to buy fertilizers or to obtain them on credit can gradually extract himself from poverty when his high yields are harvested. The truly poor, marginally subsistent farmer, however, cannot afford the investment that is required to rebuild his worn-out land and free himself from the threat of hunger.

Unfortunately, fertilizers alone do not ensure the success of the miracle seeds. These grain varieties are not so resistant to insect pests and fungal diseases as the traditional plants. As a result, farmers who invest in the seed and the fertilizer must also invest in pesticides. But even fertilizers and pesticides are not sufficient. Successful production of high yields with

Old way of harvesting wheat in India, by cutting off stalks at the base by hand. © Food and Agricultural organization of the United Nations. Via delle Terme di Caracalla 00100, Rome, Italy.

new grains is dependent on the availability of water and on a carefully controlled and regularly maintained irrigation system. The necessary equipment requires time, knowledge, and capital for investment in pipes, pumps, and fuel.

In reviewing the situation, one observer remarked that the tenant farmer in India has gone from secure poverty to insecure poverty, for he is now faced with a host of new problems. Many landlords, realizing that for the first time farming can be profitable, have simply evicted their tenants, invested in fertilizers, pesticides, and irrigation systems, and gone into business themselves. The evicted tenants suddenly have found themselves without homes or jobs. Current tenants live in fear of eviction, so that many are reluctant to invest in an irrigation system, even when low government loans are available.

Even the poor landowner is facing problems. The man who decides to continue his old secure system finds that because the total grain production of the country has increased, the market is flooded, the price of grain has dropped, and the value of the small quantity of wheat or rice that he has traditionally sold in good years has been deflated. Many of those farmers who are capable and willing to invest seed and the auxiliary inputs nevertheless also face insecurity. Grain prices in countries like India and Pakistan are notoriously unstable. Therefore, the price of grain at harvest time may not be sufficient to repay investments for fertilizers, sprays, and irrigation equipment. Moreover, few storage facilities are available in these countries, so a farmer cannot store his grain until its market value increases.

The use of the new seeds has led to many other problems, such as:

1. The taste and texture of hybrid rice IR-8 differ from those of traditional rice, and it commands a lower price. In Pakistan, many wholesalers refuse to handle this variety.

2. In many rice cultures, paddy farmers have raised fish in irrigation canals. Now fertilizer and pesticide pollution of the canals have killed fish. Thus, increased total caloric content of many people's diets has been associated with decreased protein content.

3. The problem of protein deficiencies has been augmented by the curtailment of production of soy

beans and other protein-rich legumes, which are now less profitable for some farmers than grains.

4. New grain varieties mature earlier than the old varieties. While this enables the prosperous farmer to grow two crops a year instead of one, many other farmers are finding that since harvest time now coincides with rainy weather, the grain can no longer be dried in the sun. Instead, mechanical driers are often needed.

5. National economic problems have also arisen because of frequently insufficient foreign exchange to purchase imported pesticides and chemical fertilizers. Even if the less-developed nations were to build their own chemical factories, many would have to buy fuels, phosphate rock, and other raw materials.

The consequences of the Green Revolution have been mixed. Food production in the less-developed nations has increased. Starvation has decreased, and millions are living with a new-found freedom from hunger and want. On the other hand, many rural people have sunk into deeper poverty, and homeless and untrained rural poor are migrating in large numbers to the city slums.

6.8 FOOD SUPPLIES FOR A GROWING POPULATION

(Problems 17, 18, 19, 20)

Worldwide food production and population growth have been in a nip-and-tuck race in recent decades. Widespread use of pesticides and fertilizers caused food production in the 1950's to increase faster than population. During most of the next decade, however, the situation was reversed; high birth rates and some poor crop years in many areas caused population growth to outstrip gains in agricultural yields. The latter part of the 1960's saw some tangible benefits of the Green Revolution and, once again, food production forged ahead of population growth. Gains were slight, however, and when a drought struck India during the 1972–73 growing season, famine returned to that country.

Demographers predict that population levels in the less-developed nations will grow at about 2.6 per cent per year through the decade 1970–1980. During the same time span grain production will be expected to grow at a rate of between 3.1 and 3.9 per cent. If both predictions hold, and the weather improves, the standard of living for many people will rise slightly in the near future. However, all such predictions are subject to large errors. Agricultural planners are now examining several different sources of future food supplies.

Continued Increases in Crop Yields

Japanese farms are six times as productive per acre as Indian farms. Yet, agronomists believe that with proper techniques many impoverished farms along the Ganges flood plain in India could be made at least as fertile as those of Japan. Even if yields in India rise at 3.5 to 4 per cent per year (an optimistic view), food production at the end of 25 years would still be well below the potential maximum. To reach this potential, the Indian farmer would have to multiply his energy and chemical inputs many times (see Fig. 6.5). At the moment, India lacks sufficient foreign exchange for this investment. Furthermore, the energy and raw material reserves of the planet may not be sufficient to support maximum production throughout India and the rest of the less-developed nations. If virtually unlimited supplies of energy become available from fusion reactors within a short time, and if agricultural techniques improve further, perhaps agricultural yields will rise significantly. The ultimate productivity of croplands is difficult to predict accurately.

Increases in Crop Acreage

Most of the prime, fertile land in the world is currently being farmed, so expansion of crop acreage will have to be in areas which were uneconomical to cultivate in the past, such as deserts, forests, and jungles. The farming of deserts will be feasible if massive quantities of inexpensive energy become available from nuclear sources and if salinity problems can be solved (see Section 6.4). Unfortunately, those nations which are technically and financially capable of entering the field of desert farming are not the countries that are facing immediate food shortages.

Many poor countries, especially those in South America and Africa, are looking hungrily at their jungles. However, jungles are not easily converted into fertile fields. They are ecologically peculiar, for decay and growth of plant tissue is so rapid that nutrients do not remain in the soil very long. Thus, the large biomass that exists is dependent on rapid recycling of tissue and not on large nutrient reservoirs in the soil. Moreover, simple fertilization of cleared land does not solve all the problems, for jungle soils harden when exposed to air, and they cannot be farmed with the techniques of temperate agriculture. When a farmer in these areas cuts down the trees and fights back the brush, he finds that the soil underneath is infertile and incapable of producing high-yield grain fields.

Most of the forests that lie between the tropics and the arctic have been cut back so that today only the mountain sides and the high plateaus are used for timber. If these areas are ever cleared for agriculture, alternative sources of fiber for construction and for writing material will be needed.

In summary, it is possible to increase crop acreage by exploiting marginal land, but the required energy and chemicals are too costly for the people who need new sources of food most urgently.

Food From the Sea

Man has often regarded the seas as a boundless source of food, especially protein. In truth, most of the central oceans are biological deserts (see p. 21), while the more productive continental shelf areas are already heavily fished. In 1967, approximately 60 million metric tons of fish were harvested from the seas. Marine ecologists believe that the oceans could support a sustained catch of about 100 to 150 million metric tons. Larger catches would dangerously deplete fish populations. Thus, if the seas were fished wisely and not exploited, the total food protein from the sea could be doubled.

Unfortunately, even the hope of a sustained yield of 100 to 150 million metric tons might not be realized, for the oceans are even now being threatened with overexploitation of many edible species. For example, since 1960, marine biologists have annually recommended to the International Whaling Commission to limit the catch of various species of whales. The biologists have warned that if whaling were intensified, the numbers of whales killed would exceed the reproductive capacity of the existing population, and the number of whales would decline. Since 1960, the International Whaling Commission has ignored the biologists' suggestions and set limits which greatly exceed the sustained catch levels. The Japanese and Russian whaling industries have practiced particularly intense whaling. Predictably, the total food yield from the whaling industry has decreased from 1960 to 1970. Recently, Russian whalers, finding that it is now unprofitable to hunt the big mammals, have started netting the Antarctic shrimp, called **krill,** which form the staple of the whale's diet. However, whales prove to be more efficient krill harvesters than man, and the protein harvest from the krill operation is less than that which would have been possible from sustained whaling a decade ago.

If the precedent set by the whaling industry is adopted by fishermen, the short-term catch may exceed the 100 to 150 million ton limit, but the future protein yield from the sea would then necessarily decrease. Unfortunately, efforts at international marine cooperation have not been particularly successful, and stocks of sardines, salmon, tuna, flatfish, and other sources of marine protein are presently being overexploited.

Some nations, notably Japan and the United States, have practiced **aquaculture**, or fish farming. Proponents of this method point to already successful ventures with oysters and yellowtail fish and claim that aquaculture will revolutionize ocean-based food supplies just as animal husbandry revolutionized food gathering on land. In aquaculture, the animals are housed in underwater cages, fed, and slaughtered at maturity. Food production is maximized because every mature fish can be "caught," and slaughtering can be done at a time which maximizes the ratio of weight gain to food consumed by the fish. If certain problems, such as the control of disease epidemics in artificially concentrated and caged schools can be solved, aquaculture will probably expand rapidly for a short time. Ultimately, however, these ventures will be limited by the food requirements of the animals. If the fish are fed seaweed and plankton, then man must grow or collect these plants, and these operations are difficult (see p. 249). If the feed is manufactured from terrestrial vegetation, then once again man is dependent on his croplands for food.

The Japanese have demonstrated the feasibility of seaweed culture in shallow waters, and undoubtedly an extension of coastal farming could increase man's worldwide food supply if care were taken not to disrupt estuary nurseries. One serious drawback to seaweed farming is that fertilizing the fields efficiently is difficult, for ocean currents often carry the nutrients away. The coastal areas suitable for efficient farming are small, and seaweed culture, though helpful, is not expected to yield prodigious quantities of food. The prospect of collecting or cultivating plankton is discouraging. First, it is difficult, expensive, and energy-demanding to collect billions of small organisms, concentrate them, and somehow render them palatable. Second, plankton form the base of salt-water food webs, and plankton harvests would necessarily reduce wild fish populations. If man first over-exploits fish populations, and then turns to the small organisms for food, it is likely that plankton harvests would yield less total food protein than present fishing harvests do, just as krill collection is now less efficient than whaling once was. Finally, any attempt to fertilize the central oceans to increase plankton yields is an undertaking well beyond our present technology.

Judicious regulation of the fishing industry, coupled with some expansion of seaweed and fish farming, could increase to some extent the productivity of the oceans. However, it is also possible that man will pollute the ocean beyond its ability to support even the present life systems. This subject is treated in more detail in Section 10.9.

Another serious threat to the oceans is the urbanization of coastlines and the destruction of estuaries. In the United States, especially on the eastern seaboard, it is often cheaper to purchase a salt-water marsh and convert it to solid land than to purchase prime industrial sites. Ecologically, the destruction of an estuarine marsh is disastrous, for, as was mentioned in Chapter 1, estuaries are nursery grounds for many deeper water fish. It has been estimated that one acre of estuary produces enough fry per day to grow into 240 pounds of marketable fish. Legally, the problem of conversion of marsh to factory site is quite complex. One such dispute involves the Ryker Industrial Corporation of Bridgeport, Connecticut, which purchased 700 acres of marsh in 1948, paid taxes on them for 25 years, and now wants to develop the coastline. Opposition lawyers argue that because coastal waters are under the jurisdiction of the Federal Government, and a salt marsh is an extension of coastal waters, Ryker never had a legal claim to the estuary, does not own it now, and cannot develop it.

The Cultivation of Algae

In Figure 1.7, we saw that terrestrial plants, which operate at about one per cent efficiency, are poor converters of sunlight to leaf tissue. Underwater

PLATE 5

With the topsoil gone, only the heartiest of plants can survive in the eroded wasteland. (Courtesy of James S. Packer.)

(A) Natural forested areas protect and increase the topsoil on mountain slopes. (H. Armstrong Roberts)

PLATE 6

(B) Uncontrolled and repeated burning exposes the topsoil to erosion by wind and rain. (H. Armstrong Roberts)

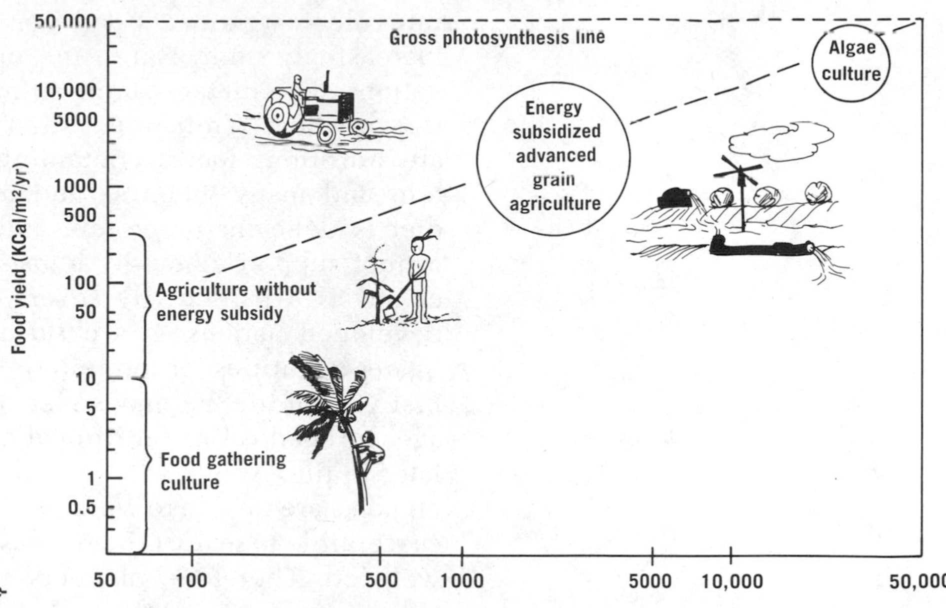

FIGURE 6.12 The relation between food yields and auxiliary energy.

algae are much more efficient converters of solar energy to organic matter. If the underwater efficiencies could be maintained in concentrated cultures at full sunlight, a farm half a square meter in size could feed a man on a sustained basis. While this calculation is accurate, it does not follow that algae farming will solve man's food problems in the future, for algae do not grow as well in concentrated surface cultures as they do under water. As light intensity is increased, the conversion rate decreases, so that at full sunlight, algae are only about four times as efficient as wheat or rice. Furthermore, the advantage of this efficiency is countered by the fact that algae are not well adapted to concentrated growing conditions, and tremendous work inputs are required to maintain high yields. Figure 6.12 compares the overall efficiencies of algae cultures and other means of producing food.

Processed and Manufactured Foods

Food chemists have pursued three different concepts in developing new food products: (1) alteration of good natural foods so as to make them more acceptable or more marketable; (2) conversion of agricultural wastes to edible food; and (3) the chemical synthesis of food. Let us discuss each in turn.

Alteration of Natural Foods

Perhaps the most familiar example to us is the conversion of vegetable oils into the butter substitute oleomargarine. Food conversion is becoming increasingly important in many less-developed nations. The dietary staple in most poor countries is rice or wheat. While unpolished rice and whole wheat are nutritious foods, containing carbohydrates, protein and many vitamins and minerals, an all-grain diet is deficient in protein and some vitamins, and cannot support a healthy human being. Protein deficiency is a particularly severe problem in the less-developed nations. If a child does not receive adequate quantities of protein in his diet during the first year of life, he may suffer permanent brain damage; no medical or nutritional aid can repair the loss later in life. Adults who have developed protein deficiencies are unable to work productively and are more susceptible to many other diseases than those who are well fed. Therefore, planners are attempting to find ways to increase protein intake for the poor.

In some areas such high-quality vegetable protein sources as soybeans, lentils, and other legumes, though readily available, are unpopular. On the other hand, soft drinks are quite popular, even among the very poor. One solution has been to use soy protein as an ingredient in soft drinks, and various sweetened, carbonated, soy beverages are now sold in Asia, South America, and Africa. In the developed nations, soy is used as a base for imitation milk and bacon.

Conversion of Agricultural Wastes

When cooking oils are extracted from peanuts, soybeans, coconuts, cottonseeds, corn, or any other vegetable seed, the residue is a waste mash high in vegetable protein. Presently, most of these residues are fed to cattle or discarded, but it is relatively easy to incorporate the proteins into synthetic foods. Cottonseed mash contains a natural insect repellent which is toxic to humans. Research is under way to find an economical process to remove this poison.

Sawdust, straw, and other inedible plant parts also represent a major source of waste materials. The major chemical component of sawdust and straw is cellulose. Cellulose consists of very large molecules which, when heated with dilute acid and water, will decompose into a mixture of various sugars. Yeasts can thrive on this sugar mixture if other nutrients such as urea (a source of nitrogen) and mineral salts are added to the culture medium. In turn, yeasts are sources of

```
 /      CH2OH                \
|  H    C———————O       O     |
|   \  /H        \     /      |
|    C            C           |
|   /  \OH     H/  \          |
|  O    C———————C     H       |
 \      H       OH           /n
```

cellulose

Simulated ground meat made from soybean flour. (Courtesy of Archer Daniels–Midland Company.)

high-quality protein and can be eaten directly. Alternatively, yeast protein extract can be used as an additive in manufactured or processed foods. For example, whole wheat is deficient in the amino acid lysine which is necessary for protein synthesis in the human body. Lysine can be extracted from yeast cultures and sold for about $1.00 per pound, and in India it is added to dough in government-owned bakeries for the production of enriched bread. Lysine fortification requires a much larger quantity of yeast than is needed to make bread rise.

Yeast culture, like algae culture, requires large quantities of auxiliary energy to synthesize the fertilizers and to maintain the growth chambers. Yeasts are heterotrophs, while algae are autotrophs and can fix energy from the sun. Thus, yeasts do not represent a primary source of food, but rather are organisms which can convert low-quality organic compounds into high-quality foods if sufficient additional energy is supplied. The acids and microorganisms present in the digestive system of cattle are also capable of converting cellulose and urea into protein, and future livestock feed may contain only sawdust, urea, and inorganic chemicals.

Chemical Synthesis

The high costs of maintaining yeast organisms or feedlot cattle are prompting chemists to completely bypass the heterotrophic organisms and synthesize amino acids directly from coal and petroleum. Presently, lysine synthesized chemically costs about the same as lysine extracted from yeast cultures, and requires about as much energy for production. Other synthetic foods include vitamins used for food additives and artificial cellulose derived from petrochemicals used as a pie filler.

It is encouraging to know that bread in New Delhi is more nutritious than ever before, but we must recognize that such dietary advances are concomitant with the trend from the range cow to the feedlot cow to synthetic amino acids, and that man's existence is thus becoming increasingly dependent on his technology.

PROBLEMS

1. **Agricultural ecosystems.** Explain why agriculture is more disruptive to natural ecosystems than are hunting and gathering of food.

2. **Destruction of ecosystems.** Chemical defoliation was introduced as a tactic of war in Southeast Asia in the 1960's, and a study group of the American Association for the Advancement of Science concluded that "it is to be expected that in any future wars of this nature, more extensive use will be made of it (defoliation)." Mangrove trees that grow in the river delta areas can be killed by a single application of defoliant,

The destruction of Indochina. (From *Bulletin of the Atomic Scientists,* **May, 1971.)**

and many sections of mangrove forest have been rendered barren by this tactic. Outline the ecological factors that will determine whether or not such forests will eventually reestablish themselves. Do you think such reestablishment is certain, doubtful, or hopeless? Defend your answer.

3. **Destruction of ecosystems.** A common logging practice in the United States today is **clearcutting**, the removal of all trees and bushes to make room for logging roads across hillsides (see below). Conservation groups attack clearcutting because it destroys wilderness recreation areas. Will clearcutting affect people other than those who enjoy wilderness? Explain.
4. **Longevity of agriculture.** If a given land area is farmed for many years, does its productivity necessarily decrease? Justify your answer.
5. **Natural vs. agricultural ecosystems.** Briefly discuss the relative importance of each of the following characteristics to a plant species existing (a) in a natural prairie; (b) in a primitive agricultural system; (c) in an industrial agricultural system: (i) resistance to insects, (ii) a tall stalk, (iii) frost resistance, (iv) winged seeds, (v) thorns, (vi) biological clocks to regulate seed sprouting, (vii) succulent flowers, (viii) large, heavy clusters of fruit at maturity, (ix) ability to withstand droughts.
6. **Energy.** List the energy inputs that are added to a wheat field in industrial agriculture.

"Happy?"

(Drawing by Koren; © 1973 The New Yorker Magazine, Inc.)

7. **Phosphorus reserves.** The depletion of both natural gas and of phosphorus reserves is likely to be a serious problem in the early part of the twenty-first century. Explain how the meaning of the word depletion varies when discussing these two different resources.

8. **Definition of "organic."** (a) Write all the definitions of the word "organic" that you can find in a good dictionary and that were not given in this text. (b) Obtain definitions of the word "organic" from organic gardening publications or from the proprietors of health food stores, and compare them critically with the definitions given in this text.

9. **"Organic" foods.** "Organic" food is grown and processed without using any unnatural product. Reliable commercial sources contend that 10 times as much "organic" food is sold as is produced. (a) "Organic" food can be sold to the consumer at a relatively high price. How is that relevant to the above contention? (b) Suppose a farmer grows a crop of tomatoes "organically." He can show his field to several potential buyers and promise each he will deliver tomatoes from that field. What if the farmer promises more than the field delivers? How can the buyer be certain he has received the organically grown tomatoes and not produce from some other field? (How would you test whether a tomato has been grown organically?) (c) How can a consumer be certain that his purchased "organic" tomatoes are genuine? (d) If you wanted to buy a jar of organic honey, what would convince you that the product was unadulterated? Its appearance? taste? the reputability of the store you bought from? the label? the price? (e) "Organic" granola is a cereal made from many ingredients; the more elaborate formulations include oatmeal, wheat germ, other grains, honey and other natural sweeteners, almonds, sesame seeds, and dried fruit. At what points in the growing of the raw materials and in the processing of the granola itself might unnatural foods or contaminants enter the product? Do you feel contamination would be likely to be accidental or deliberate? Do you think the FDA or some other governmental body should be charged with the duty of insuring that the label "organic" refers to an uncontaminated product? Consider the cost of policing and the relevance of natural foods to nutritional health in the United States.

10. **Humus.** Briefly discuss some of the chemical and physical properties of humus.

11. **Industrial agriculture.** What are the major techniques of industrial agriculture? Under what circumstances of use or misuse may each of these methods engender a loss of fertility of the land?

12. **Fertilizers.** Assess the advantages and disadvantages of using manure as a fertilizer. Do you feel manure

should be used to fertilize crops? (Remember that a field fertilized with manure stinks; however, such odors are not harmful to health.)

13. **Manure.** (a) A father and two sons walk along a plain in India, driving several cows before them. The mother follows, carrying a basket on her head. When a cow defecates, the mother scoops up the droppings into her basket. Later, she plasters these droppings onto a wall to dry. Most of the dried dung is then traded for rice; some is kept by the family for use as a fuel to cook the rice. Outline a possible sequence that might have led to the use of cow dung as fuel. Does the situation described above represent the lowest condition of poverty of the land, or does it lead to still further disruptions? Explain. (b) Many of the buffalo-hunting Plains Indians of North America used buffalo chips (dried buffalo dung) as fuel. Would you describe the situation of those Indian tribes as being comparable to the situation of the Indian family described in part (a)? Defend your answer.

14. **Agriculture in India.** Which of the following inventions would you believe has had significant effects on agriculture in India? Which has had a moderate effect? No effect? (a) pesticides, (b) the tractor, (c) inorganic fertilizers, (d) the electric motor, (e) the high-pressure irrigation pipe.

15. **The Green Revolution.** Examine the curve for Pakistani wheat yields in Figure 6.12. If no significant new advances occur in agricultural sciences, how would you expect the curve to grow in the near future? (Hint: Recall some of the discussion of graphical extrapolation in Chapter 4.)

16. **Agriculture in marginal lands.** Discuss some problems inherent in farming deserts; jungles; temperate forest hillsides; the Arctic.

17. **Food from the sea.** Jacques Cousteau has said, "The shores of the rivers are the roots of the oceans." What do you think he meant?

18. **Aquaculture.** Aquaculture, or seaweed farming, is likely to be practiced along shorelines and in coastal bays or marshes. Outline a series of events which might lead to a situation where successful aquaculture would *decrease* the total food harvest from the sea. Do you think that this set of events is likely to occur?

19. **Synthetic foods.** It is reasonable to believe that food chemists will be able to synthesize economically a complete and balanced diet from coal in the near future. Outline some advantages and disadvantages of the chemical synthesis of foods.

20. **Synthetic foods.** How does the probable energy requirement necessary to make a soy drink compare to the direct use of soybean soups?

21. **Sources of protein.** Mushrooms are high in protein and some varieties can be grown to maturity in 10 days to two weeks. Discuss the advantages and disadvantages of mushrooms as a source of protein. (*Hint*: Mushrooms and yeasts are both varieties of fungi.)

22. **Synthetic food.** Discuss the advantages and disadvantages of the food production sequence outlined below.

$$CO_2 + H_2O \xrightarrow[\text{using nuclear energy}]{\text{factory process}} C_6H_{12}O_6 \text{ (sugar)}$$

$$\text{sugar} + \text{urea} + \text{yeasts} \longrightarrow \text{protein}$$

$$\text{protein} + \text{sugar} \xrightarrow{\text{industrial processes}} \text{imitation bread, meat, string beans, etc.}$$

BIBLIOGRAPHY

An excellent book which discusses the power base of industrial agriculture is:

Howard T. Odum: *Environment, Power, and Society.* New York, Wiley-Interscience, 1971. 331 pp.

The reader who wishes to investigate the Green Revolution and food in the future should refer to:

Lester R. Brown: *Seeds of Change: The Green Revolution and Development in the 1970's.* New York, Encyclopaedia Brittanica, Praeger Publishers, 1970. 205 pp.

Willard W. Cochrane: *The World Food Problem: A Guardedly Optimistic View.* New York, Thomas Y. Crowell Co., 1969. 331 pp.

Rene Dumont and Bernard Rosier: *The Hungry Future.* New York, Praeger Publishers, 1969. 271 pp.

Francine R. Frankel: *India's Green Revolution.* Princeton, N.J., Princeton University Press, 1971. 232 pp.

Two recent excellent books on agriculture and food resources are:

Kusum Nair: *The Lonely Furrow: Farming in the United States, Japan, and India.* Ann Arbor, University of Michigan Press, 1970. 336 pp.

N. W. Pirie: *Food Resources, Conventional and Novel.* New York, Penguin Books, 1970. 208 pp.

A delightful book about human nutrition from prehistoric to modern times is:

Lloyd B. Jensen: *Man's Foods.* Champaign, Ill., Garrard Publishing Co., 1953. 278 pp.

Some of the pollution problems associated with fertilizer applications are discussed in:

Barry Commoner: *The Closing Circle.* New York, Alfred A. Knopf, 1971. 326 pp.

Ted Willrich and George E. Smith, eds.: *Agricultural Practices and Water Quality.* Ames, The Iowa State University Press, 1970. 415 pp.

7

CONTROL OF PESTS AND WEEDS

7.1 COMPETITION FOR FOOD BETWEEN SMALL HERBIVORES AND MAN

(Problem 1)

Perhaps the most persistent and difficult problem in the agricultural production of large quantities of food has been competition from small herbivores. If a cow wanders into a field, the farmer's son can chase her away. An insect, a field mouse, the spore of a fungus, or a tiny root-eating worm (a nematode) is more difficult to deal with. Since these small organisms reproduce rapidly, their total eating capacity is very great. In addition to their voracity, these pests may be carriers of disease. The bubonic plague, carried by a flea which lives on rats, swept through medieval Europe, killing as much as one-third of the total population in a single epidemic. Malaria and yellow fever, spread by mosquitos, have killed more people than have all of man's wars.

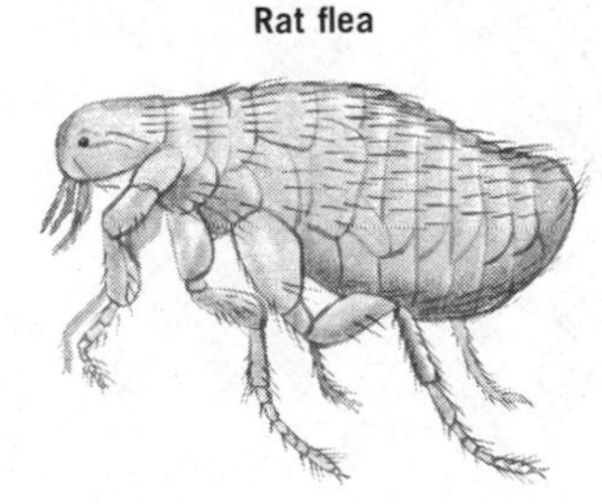
Rat flea

However, not all insects, rodents, fungi, and nematodes are pests. Most do not interfere with man and many are directly helpful. Millions of nematodes live in a single square yard of healthy soil. Most are necessary to the process of decay and hence to the recycling of nutrients. Fungi, too, are essential to the process of decay in all the world's ecosystems. Certain species of fungi are used by man for the production of antibiotics, bread, beer, and cheese. Some rodents are part of natural ecosystems; thus, squirrels help spread pine seeds, and lemmings provide a staple food for almost all carnivores in parts of the Arctic. The role of insects in the maintenance of the biosphere is extensive and versatile. For instance, bees are essential to the life cycle of most flowering plants. In their search for food, bees inadvertently transfer pollen from flower to flower, and thereby ensure fertilization. In fact, it is a common agricultural practice to move hives into crop areas at time of bloom to assist or ensure pollination. Many insects, such as species of springtails, are part of the process of decay.

Insects are the prime food source of many animals that are vital, in turn, to the maintenance of natural balance. For example, the diet of many species of birds includes both insects and fruit. Fruit seeds transferred intact through the bird's digestive system and deposited at distant locations have been an important contributing factor in the continuing existence of certain plants. Thus, insects are necessary to the survival of many species of birds that are necessary in the life cycle of many wild fruit trees, which, in turn, help to support wildlife. In addition, many carnivorous or parasitic insects feed on insects that eat man's crops. The periodic invasion of some African villages by driver ants is a fascinating illustration of insect ecology. Many disease-carrying rodents and insects live in the village houses, posing a constant threat to the human population. At periodic intervals, however, millions of large driver ants invade the villages, chase away the inhabitants and eat everything that remains. When the people return, they find that their stored food supply is gone, but so are all the cockroaches, rats, and other pests—everything has been eaten.

An insect is an animal; therefore, an insect-eater is a carnivore.

Species of pests have lived side by side with man for millions of years. Several factors have made this accommodation possible, if not pleasant. First, during most of man's existence, the total human population, and hence the total human food requirement, has been much smaller than it is now. Also, many species of food plants produce insecticides for their own protection. For instance, the roots of some East Indian legumes contain the insecticide **rotenone**; and many wild cereal plants such as wheat, corn, rye, and oats are naturally resistant to fungal attack. Such natural accommodations are now insufficient for man. One reason is that the natural controls do not always work too well, especially in the short run; thus, throughout history natural periodic cycles of pest populations have occurred, resulting in blooms of these species

Body louse

and famine for man. Perhaps even more critical is the rapid growth of the human population and hence its increasing food requirements.

7.2 INSECTICIDES—INTRODUCTION AND GENERAL SURVEY

(Problem 2)

Chemical control of insects is not new. Marco Polo introduced pyrethrum to Europe after learning of its use by farmers in the Far East. However, systematic spraying of crops was not carried out until the early twentieth century. By the early 1900's, rotenone, pyrethrum, nicotine, kerosene, fish oil, and compounds of sulfur, lead, arsenic, and mercury were in common use. These formulations did not change appreciably until the 1940's. The most significant breakthrough occurred around the start of World War II, when the insecticidal properties of a chemical called DDT were discovered. DDT was far cheaper and more effective against almost all insects than the previously known control methods. The use of DDT led to dramatic early successes: it squelched a threatened typhus epidemic among the Allied army in Italy; antimosquito programs saved millions from death from malaria and yellow fever; and pest control, leading to increased crop yields all over the world, saved millions more from death by starvation.

Enthusiastic supporters of DDT predicted the complete destruction of all pest insects within the foreseeable future; the chemist who first discovered its insecticidal properties received a Nobel Prize. But within 30 years the promise of insect-free abundance had been broken, and the "miracle" chemical that was to have achieved it had fallen from grace. On January 1, 1973, all interstate sale and transport of DDT in the United States was banned except for use in emergency situations in which life is immediately threatened. In this Chapter we will review the story of this transition, a classic example of the interplay between the activities of technological man and the dynamic forces of the environment which he tries to control.

To focus attention only on DDT, however, would be incomplete, for many other chemical insecticides have been in use during the past 20 years. Some of the most common of these are listed in Table 7.1. Note the first three classes of compounds—organochlorides, organophosphates, and carbamates. These words designate categories of structurally related chemicals. Chemical compounds usually act on living systems or

TABLE 7.1 COMMON CHEMICAL INSECTICIDES

COMPOUND CLASS	DESIGNATION
Organochlorides (also called chlorinated hydrocarbons)	Aldrin
	Chlordane
	DDD
	DDT
	Dieldrin
	Endosulfan
	Endrin
	Heptachlor
	Lindane
	Toxaphene
Organophosphates	Diazinon
	Malathion
	Parathion
Carbamates	Sevin (carbaryl)
Naturally occurring or organic pesticides	Nicotine
	Pyrethrum
	Rotenone

Dichlorodiphenyltrichloroethane (DDT)

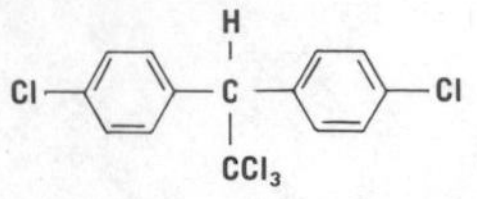

Dichlorodiphenyldichloroethane (DDD)

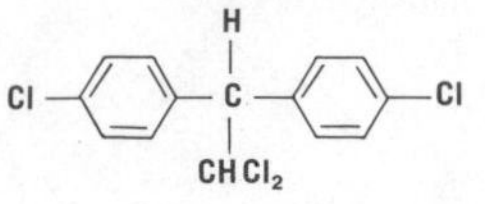

Aldrin

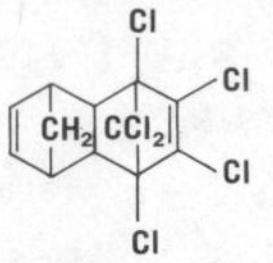

Lindane (containing 99% γ-isomer)

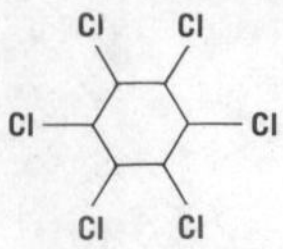

Parathion

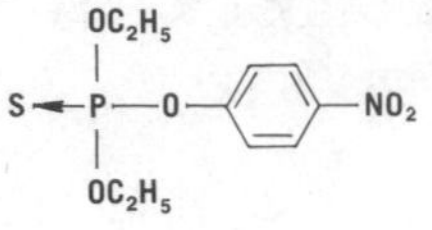

Malathion

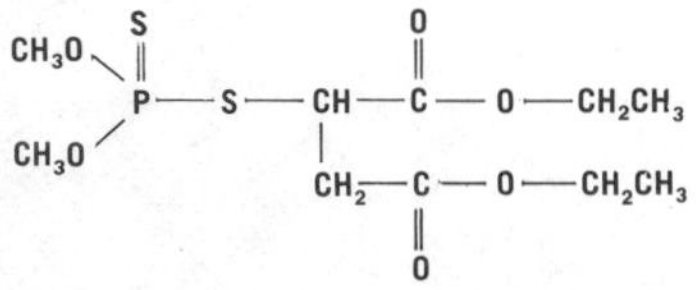

Sevin® (1-naphthyl-*N*-methylcarbamate)

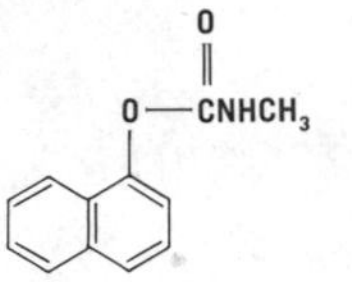

on other compounds in a manner that is roughly predictable from the composition and the structure of their molecules. It is therefore convenient to classify compounds according to molecular composition and structure. Roughly speaking, all organochloride insecticides have similar mechanisms of action in living tissues, and they are all about equally soluble (or insoluble) in various solvents such as water, alcohol, or fats. Organophosphates, as a group, also demonstrate a recognizably coherent set of properties that are quite different from those of the organochloride compounds.

The insecticides in these three groups are not produced by living organisms anywhere on the earth, but are *synthetic* products of laboratories and chemical factories. The last grouping in Table 7.1, naturally occurring or "organic" (see page 232) pesticides, is not a set of chemically related compounds. Nicotine, pyrethrum, and rotenone are similar to each other in that they are all of biological origin, that is, they are extracted from the living tissue of certain naturally occurring plants.

The names of the individual compounds in Table 7.1, with a few exceptions, are trade names. Thus the word "Aldrin" gives no more chemical meaning to the nature of the insecticide than the word "Detroit" gives a geographical meaning to the size and location of that city.

7.3 THE ACTION OF CHEMICAL PESTICIDES—BROAD-SPECTRUM POISONING

(Problems 3, 4, 18)

Chemical insecticides were initially received with great enthusiasm because they are inexpensive, easy to use, fast-acting, and effective against a wide range of pests. Their promise engendered an uncritical optimism. It was imagined, for example, that a tomato farmer, faced with a mid-season invasion of some insects that started to eat the green fruit, would not even have to identify the pest. He would simply call in an aerial spray company and would expect 90 per cent destruction of his pests by the next day or two. A simple problem, a simple solution—or so it seemed.

Yet if we examine the problem more closely, complexities emerge. A tomato field under attack by some insect is not merely a two-species system. The tomatoes and pests are but members of a large agricultural ecosystem of thousands of species that include predator insects, bacteria, parasites, and many types of soil dwellers, as well as carnivorous, herbivorous, and omnivorous birds and other migratory animals. Therefore, despite the undeniable fact that innumerable successes of spray programs have been recorded during the past 25 years, it is important to look more closely into the intricacies of the problem.

The use of non-selective sprays has often led to the destruction of the natural controls on relative population sizes. As an example, when DDT and two other chlorinated hydrocarbons were used extensively in pest control in a valley in Peru, the initial success gave way to a delayed disaster. In only four years, cotton production rose from 440 to 650 pounds per acre. However, one year later, the yield dropped precipitously to 350 pounds per acre, almost 100 pounds per acre less than before the insecticides were introduced. Studies indicated that the insecticide had destroyed predator insects and birds as well as insect pests, so with natural controls eliminated, the pest population thrived better than it had ever done before.

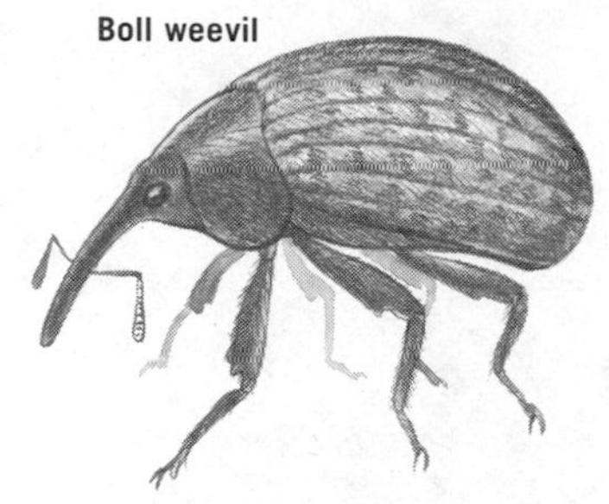

But, you may ask, how could the pests stage a comeback? Why didn't the predators stage an equal comeback? Why couldn't the farmers combat the pest resurgence with more spraying?

One of the major problems with chemical insecticides is that many insects become resistant to the poisons. In other words, a given insecticide at a given concentration often becomes less effective after some years of use. It appears as though the chemical has diminished in potency, although the composition has, of course, remained unchanged. To understand the reason for this phenomenon, recall from Chapter 2 that the chemistry of plants and animals changes from time to time as a result of random mutations of their reproductive cells. A mutant has a good chance of survival if its particular mutation protects it from a hostile environment. This mechanism of random mutation has allowed insects to adapt to their environment for millions of years, and it is this process that protects insects from pesticides. In areas where spraying is heavy, strains of insects have evolved which are genetically resistant to a particular chemical. Genetic resistance to insecticides is an extremely serious problem. By 1945 at least a dozen species had developed some resistance to DDT; by 1967 the number had increased to 165 species. About 30 of these resistant species carry disease and about 75 others are serious agricultural pests. Since resistant parents tend to pass this character on to succeeding generations, the old pesticides are rendered ineffective. This effect was directly demonstrated in an experiment in which DDT-resistant bedbugs were placed on cloth impregnated with DDT. They thrived, mated, and the females layed eggs normally. The young, born on a coating of DDT, grew up and were healthy. Attempts to change pesticides have in many cases simply produced strains of insects that are resistant to more than one chemical.

Insect resistance to poisons is a serious problem in itself, but it can be compounded if the pest becomes resistant and various predators do not. If this happens, the pests have a new biological advantage and can thrive in greatly increased numbers. This is favored by three factors:

1. *Insect pests are frequently smaller and reproduce more frequently than their predators.* More frequent reproduction engenders more frequent mutations, and thus improves the chances that resistant strains will arise. Furthermore, once a resistant population of insect pests appears, it can repopulate its ecological niche much faster than can a larger, more slowly reproducing species. In effect, the pest species develops resistance faster than the predator.

2. *Predators generally eat a diet richer in insecticides than that of the original pests.* Because chemical poisons are not immediately excreted by herbivorous insects (or any other organism for that matter), the concentration of the chemical in their bodies becomes greater as more contact is made with the poison either directly or through the ingestion of sprayed leaf tissue. Since death may be delayed for some time after poisoning, many poisoned but living insects will be eaten by their natural enemies. In this way predators eat a diet that is more concentrated in poison than that of the herbivores, the original pests.

3. *There are always fewer predators than herbivores (including pests) in an ecosystem.* A species

with a small population runs the risk of extinction, because it is easier for a group containing only a few individuals to be wiped out by some disaster than it is for a group containing many. Therefore, the species of herbivores (the pests, which are more numerous) have a greater chance of survival than the species of predators.

We can now reconstruct what happened in Peru. Although the pesticides caused a rapid decrease in pest population size, resistant mutants soon displaced the susceptible individuals. By this time the situation had become worse than it was originally, because the natural predators did not achieve immunity so well as the pests, and therefore the pests were under less control than ever.

Another effect of broad-range pesticides is that they sometimes lead to new pest populations. Consider the story of the spider mite in the forests of the western United States. The spider mite feeds on the chlorophyll of leaves and evergreen needles. Because in a normal forest ecosystem predators and competition have kept the number of mites low, they have never been a serious problem. However, in 1956 the U.S. Forest Service sprayed with DDT in a campaign to kill another pest, the spruce budworm. The result was that the budworms died, but so did such natural enemies of spider mites as ladybugs, gall midges, and various predator mites. The next year the forests were plagued with a spider mite invasion. Although the spruce budworm had been temporarily controlled, the new spider mite problem proved to be more disastrous.

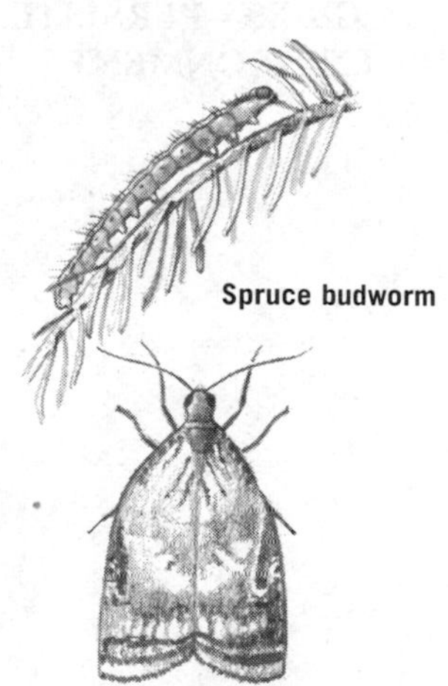
Spruce budworm

These examples are two of the many cases in which broad-range insecticides have engendered more serious problems than they have solved, and where the goal of increased agricultural production has ultimately been frustrated. The response to such failures has often been to increase the number of sprayings or the amount of insecticide per spraying.

There have been thousands of suicides, homicides, and accidental deaths attributed directly to insecticide poisoning. (See Section 7.7 for a discussion of health effects.) While these deaths are certainly tragic, it must be remembered that insecticides have saved millions of lives by preventing starvation and illness. Moreover, the rules of safe practice in handling pesticides have been well developed, and "acceptable" levels of pesticide residues on foods have been established. People are advised to stay clear of spray zones, and government agencies condemn foods that retain excessive insecticide levels. Unfortunately, wild animals, and sometimes even livestock, are not nearly so well protected. As a consequence, many natural populations of vertebrates have been decimated by DDT. For example, in 1954, several communities in eastern Illinois were sprayed aerially in an effort to stop the westward movement of the Japanese beetle. As a result, many species of birds were almost completely annihilated in the sprayed area, ground squirrels were almost eradicated, 90 per cent of all farm cats died, some sheep were killed, and muskrats, rabbits, and pheasants were poisoned. These unwanted side effects might have been considered a necessary price to pay for pesticidal success, but the cost did not yield the desired benefit; the Japanese beetle population continued its westward advance.

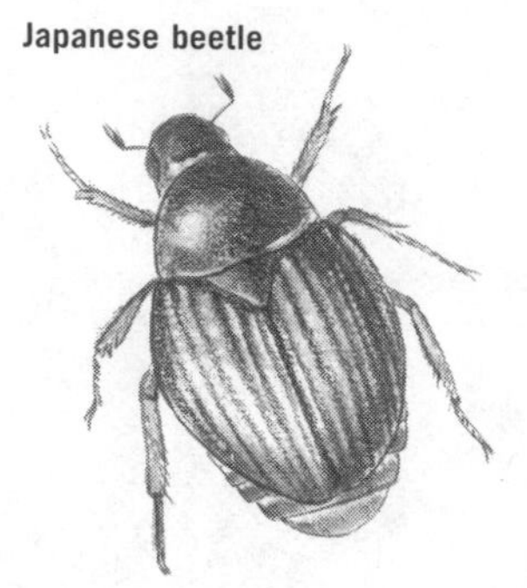
Japanese beetle

Even when such a program is successful in eradicating pests, it is extremely difficult to assess whether the increased crop yields are economically and socially worth the destruction of wildlife.

7.4 THE ACTION OF CHEMICAL PESTICIDES—PERSISTENCE IN THE ENVIRONMENT

(Problem 6)

Most naturally occurring chemical compounds are **biodegradable,** that is, capable of being broken down by some organisms. Biodegradability is a phenomenon that has developed with the evolution of species; organisms evolve to consume compounds with energy or raw material. However, most of the chemical insecticides are synthetic and have only recently appeared in the environment. Some of these are similar enough to naturally occurring compounds to be easily biodegradable. The others decompose more or less rapidly through the chemical action of water, environmental acids and bases, atmospheric oxygen (especially in the presence of sunlight), and perhaps other agents. Generally speaking, organophosphates, carbamates, and "organic" insecticides are non-persistent, that is, half of the initial chemical applied to an agricultural system is decomposed within a few weeks, and all but a small amount of the remainder decomposes within a few months. Of course, the decomposition rate varies considerably with temperature, soil type, moisture, humus content,

Decomposition rates are characterized by "half-lives"; see Section 8.3.

and a host of other chemical factors. By contrast, the persistence of organochlorides is measured in years; in some soils DDT or Aldrin has been detected in appreciable quantities 15 years after a single spray application. Such persistence is a serious ecological problem, for not only does a single spray application subject non-target organisms to continuous exposure to poisons for many years, but with continued annual spraying, insecticide accumulates in the environment.

7.5 PHYSICAL DISPERSION OF ORGANOCHLORIDE INSECTICIDES

(Problems 5, 9, 10)

Organochlorides not only persist in the immediate agricultural environment, but their stability enables them to endure while they are being mechanically or biologically transported throughout the biosphere. To understand the mobility of insecticides, we must first review some of their physical characteristics. Organochlorides are relatively insoluble in water, evaporate slowly, have a strong tendency to adhere to tiny particles of dust, dirt, or salts, and are soluble in the fatty tissues of living organisms.

Let us now consider the fate of an application of DDT or one of its sister compounds sprayed from an airplane onto a crop. Typically, the insecticide will be mixed with some inert ingredient so that at a calculated airplane speed and spray rate, a dose of between one-half and five pounds of insecticide will be applied per acre, depending on the type of insecticide, pest, and the nature of the crop. Not all the spray will hit the exact target. The accuracy is dependent on such factors as the skill of the pilot, the number and position of electrical wires and trees which he must dodge, and the wind direction, velocity, and turbulence. Some of the insecticide that misses the desired target lands on nearby houses, roads, streams, lakes, and woodlots, while some is carried into the air either as a gas or as a "hitchhiker" on particles of dust or on water droplets. The quantity of insecticide in the air tends toward an equilibrium with the concentration of the chemicals on the ground, but true equilibrium is never really attained; dust storms and evaporation increase the atmospheric concentration, and settling decreases it. The net result of these processes is that a great deal of insecticide travels long distances in the air. Indeed, all organisms on earth are subject daily to some exposure to these chemicals. Almost all rainwater contains measurable concentration of organo-

Crop duster dusting sulphur to retard mildew on grapevines 20 miles south of Fresno, California. (Courtesy of EPA-DOCUMERICA, photographer, Gene Daniels.)

"There I was, coming in low at a hundred and seventy-five m.p.h. with forty-two acres of broccoli to the left of me, eighteen acres of asparagus to the right of me, eighty acres of carrots straight ahead! Power lines all around! My target: seven acres of badly infested garlic smack in the center. . . ." (Drawing by Dedini; © 1972, *The New Yorker Magazine, Inc.*)

chloride insecticides. Additionally, some of these poisons enter the trade-wind currents and circulate around the globe.

Patterns of dispersal of organochlorides and some average DDT concentrations are shown in Figure 7.1. The effects on non-target species (including humans) are taken up in the following two sections.

7.6 BIOLOGICAL CONCENTRATION OF ORGANOCHLORIDE INSECTICIDES—EFFECTS ON NON-TARGET SPECIES

(Problems 8, 11, 12, 13, 14, 26)

The danger of high concentrations of poisons in the soil arises from the fact that fertile soil contains much living matter. One pound of rich farm earth contains up to one trillion bacteria, 200 million fungi, 25 million algae, and 15 million protozoa, as well as worms, insects, and mites. These organisms are vital for continued fertility of the soil. They fix nitrogen, they break down rock and thus make minerals available to the plants, they retain moisture, they aerate the soil, and they bring about the essential process of decay. Without these organisms, the plants above ground usually die. The effect on these organisms of an in-

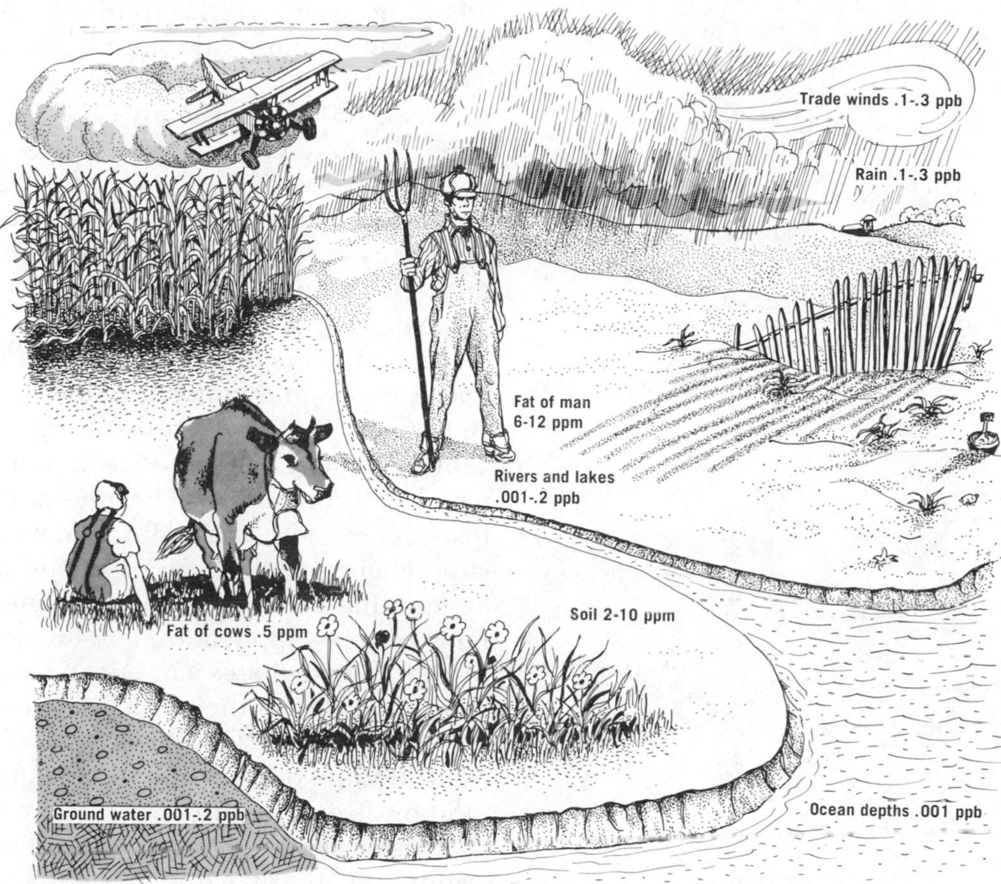

FIGURE 7.1 Physical dispersion of insecticides with some average values for DDT levels.

creasing concentration of poison in the soil is largely unknown. In many heavily sprayed areas of the world, man is harvesting more food per acre than ever before. Yet some facts are coming to light which may presage future disaster. Studies in Florida have shown that some chlorinated pesticides seriously inhibit nitrification by soil bacteria. Termites have not been able to survive in soils that were sprayed with Toxaphene 10 years previously. Similarly, Endrin present in as low a concentration as one part per million caused significant changes in the population of soil organisms, and, consequently, in the relative concentrations and availability of important soil minerals. As a result, beans and corn grown in soils treated with Endrin contained different nutrient values than the same crops grown in untreated soils – some minerals were increased, some decreased. The net result, however, was a decrease in the total plant growth. In another instance, Aldrin routinely sprayed on a golf course measurably depleted the earthworm population.

As is the case for many types of ecological disruptions, the long-term effects of insecticides in soils are not known. Perhaps many soil organisms are or will become resistant to the pesticide accumulations. But the stakes in the gamble are large, for if life in the soil dies, plants will soon succumb.

Not all the pesticides in the soil remain fixed there. Some fraction of the total seeps into groundwater reservoirs and hence into drinking water supplies. In fact, a government study of private wells in Illinois showed that all surveyed contained water contaminated with insecticide. A large fraction of soil-based insecticides gets carried into the world's surface water supply along with the tons of sediment eroded from agricultural systems. As a consequence of the current rates of spray application and erosion, all major rivers in the United States contain measurable insecticide concentrations in the parts per billion range. In river water at 72°C (45°F), which is normal temperature for trout, 1.4 ppb of Endrin will kill half of a population of rainbow trout in three days. Similarly, many other species of fish cannot survive insecticides in concentrations greater than about one to ten parts per billion. Trout in the United States today live in the more mountainous regions which are not so polluted, while those that were once native to the Great Lake water systems or the lower Missouri River can no longer be found. Under severe conditions, such as occurred during 1950 in parts of the South when heavy rains followed heavy spray applications, agricultural runoff was so high that the residual pesticide concentrations were raised a thousand fold. Nearly all of the fish in many watersheds were killed.

Inevitably, if insecticides are present in major rivers, they must also be present in the ocean, and the concentrations must be highest in estuary systems, that is, at river mouths and in coastal bays. One of the most serious problems in estuary systems is that insecticides reduce photosynthesis carried out by plankton. In one study it was shown that DDT at a concentration of 1 ppb reduced phytoplankton activity by 10 per cent and at 100 ppb by 40 per cent, as compared with that of plankton grown in unpolluted waters.

We have already seen how the effect of a given spray application is more severe on carnivorous insects than on the original pest. In light of the fact that persistent pesticides are present throughout the biosphere—in air, soil, water, and on plant tissue—it is not surprising that biological magnification of DDT and its sister compounds occurs throughout numerous food webs, with carnivores of all kinds receiving the highest concentration in their tissues. In fact, the chemistry of organochloride compounds is particularly amenable to this biological transport. Their persistence makes them survive long enough to travel. Moreover, they are fat-soluble, and, therefore, when they enter an organism, they dissolve and are stored in fatty tissue.

The normal cleansing mechanisms of the body depend on the transport of unwanted substances through aqueous media such as blood and urine. These mechanisms therefore operate very poorly on organochloride compounds, with serious detrimental consequences. Consider the widely publicized case of the spraying of Clear Lake in California in the 1950's with DDD for the control of biting pests (Fig. 7.2). After the project was completed, the water contained 0.02 ppm of DDD. However, plankton living on the surface of the water contained 5 ppm of the insecticide in their tissues. Plant-eating fish had magnified this concentration to 40–300 ppm (depending on the species); some carnivorous fish and birds at the top of the food chain had as much as 2000 ppm of DDD in their tissues. Thus, it is not surprising that fish and birds can be poisoned by an insecticide that originates in waters in which its concentration seems innocently low. In fact, in Clear Lake as well as in other aquatic systems, it has been observed that a year or so after introduction of a pesticide, the chemical cannot be detected at all in the water, but is still present in the food web.

The effects of the high levels of insecticide on carnivorous fish and birds have been particularly severe. In many cases the animals die soon after application of the insecticide. In others, sublethal doses impair the normal activity or reproduction of the animals, resulting in either a delayed death of part of the population or a decreased population growth. For example, in many sprayed areas, insect-eating fish, such as salmon fry, are subjected not only to the poi-

FIGURE 7.2 Biological magnification of DDD (Clear Lake, California).

son itself, but also to starvation when their food supply is destroyed. Trout and salmon with insecticide in their systems avoid cold waters and tend to lay their eggs in warmer areas where the young will have a smaller chance of survival. Pesticides can be fatal to animals that store food energy in fat for use during winter months. Trout build up a layer of fat during the summer months when food is plentiful. In areas where the land has been sprayed, this fat contains high concentrations of DDT. During winter, the fat is used as a source of energy. The DDT released into the bloodstream upon fat breakdown has been known to kill the animal. The eggs of fish also contain a considerable amount of fat which is used as food by the unborn fish. In one case, 700,000 hatching salmon were poisoned by the DDT in their own eggs. Finally, the evidence seems undeniable that DDT poisoning is responsible for the sharp decline in populations of many birds, especially those that are secondary or tertiary consumers (see Fig. 7.3). One manifestation of the early stages of DDT poisoning is the inability to metabolize calcium properly. In birds, this has led to the production of thin-shelled eggs. Often these weakened eggs crack or break in the nest, with resulting prenatal death. Laboratory birds that are fed low concentrations of DDT in their diet produce eggs with thin and weakened shells. The populations of several species of birds—among them, the peregrine falcon, the pelican, and some eagles—are declining so rapidly that many conservationists fear that they will become extinct in the near future.

FIGURE 7.3 Biological magnification of DDT after spray application for Dutch elm disease.

7.7 EFFECTS OF INSECTICIDES ON HUMAN HEALTH

(Problems 15, 16)

Since the mid-1940's, the incidence of acute fatal poisonings due to pesticides has been about constant at one per million people per year. At present, more than half these cases are children who are exposed to these toxic chemicals through carelessness in packing or storage. The balance of fatal poisonings has occurred largely to those who work with the chemicals either in production or in agriculture. Unless greater care is exercised in the handling of these chemicals, the number of acute fatal poisonings is certain to increase over the next few years as the very toxic organophosphates are substituted for the less toxic DDT.

The other effects of pesticides are the chronic effects, those of long duration. We confine ourselves

Rachel Carson (1907–1964) was an American biologist and science writer. Her last book, *Silent Spring*, published in 1962, awakened the public to the environmental effects of pesticides.

here to the chronic effects of DDT, for little is known about chronic effects of other insecticides. The recent debate about the use of DDT has centered around possible long-term damage to man. We shall examine the controversy over DDT in some detail, for this whole question is a perfect illustration of the difficulty in setting public policy on the basis of fragmentary scientific evidence.

The possible long-term effects of the chlorinated hydrocarbons have been a matter of concern ever since Rachel Carson suggested in *Silent Spring* that exposure to DDT might increase the occurrence of cancer in human beings. It can safely be said that after a decade of study, proof for or against the carcinogenicity (cancer-producing qualities) of DDT in man is still lacking. The following gives some flavor of the types of studies that have been carried out to date:

1. Direct feeding of DDT to human subjects. When DDT was ingested by healthy volunteers at doses as high as 35 mg per day for nearly 22 months and the subjects were then observed, some for about five years following the end of the treatment period, no adverse clinical or chemical effects were seen.

2. Examination of men who have had long-term occupational exposures to DDT. Several studies on a population of workers, who were estimated to have daily intakes of DDT nearly 400 times the "normal," have shown absolutely no evidence of any toxic effect. These studies have not, of course, gone unchallenged. The major (and completely justified) criticisms are that the numbers of people observed were too small to draw any sort of valid conclusion about anything but the grossest kind of effect. In addition, the length of time that most of the subjects have been observed is too short to warrant confident conclusions about an effect like carcinogenesis, for cancer can take many years to develop after exposure to a cancer-causing agent. Other studies have suggested that exposure to insecticides may be related to increased abnormalities of pulmonary function, but here, too, it is difficult to establish whether or not there are long-term effects.

3. Experiments with laboratory animals. Within the last decade, several groups have shown that laboratory rodents fed DDT develop tumors more frequently than those not fed DDT. These results sound simple enough, but proper interpretation is difficult for several reasons. (A) Questions have been raised about faulty experimental design in the most important of these experiments. (B) The doses of DDT fed to the animals were huge in comparison to those to which human beings are exposed. (C) The strains of laboratory mice used were highly inbred and have a very high *spontaneous* incidence of tumors; hence they may not react to chemicals as would "normal" outbred strains. (D) Most of the induced tumors were a certain type of liver tumor. Expert pathologists disagree about whether these tumors are typical cancers. In at least a few cases, however, the tumors did seem to behave similarly to malignant tumors in human beings. (E) There is always great difficulty in extrapolating any kind of results across species lines, for things that are true of mice are not necessarily true of man.

This is the strongest available evidence concerning the chronic health effects of DDT. From a purely scientific standpoint, we simply do not know whether DDT has any chronic effects in man. During the late 1960's, however, the pressure by environmental groups promoting the banning of DDT began steadily mounting. After a long and frequently acrimonious series of hearings on the subject, on June 30, 1972, William Ruckelshaus, then head of the Environmental Protection Agency (EPA), issued an order, effective January 1, 1973, banning all domestic uses and transport of DDT, except in cases of health emergencies.

This decision was made not only because of the suspicion of carcinogenicity, but also because of all the other effects on various aspects of the biosphere that we have mentioned previously, and because many experts felt that the organophosphates were satisfactory alternatives to the use of DDT. This decision has been both praised as farsighted, and damned as nearsighted. Critics have argued that much the same objectives could have been achieved from restricting the use of DDT, rather than imposing a total ban. Some have suggested that an absolute ban by the United States will inevitably result in similar prohibitions by poorer countries that can ill afford such bans, since the alternative insecticides are so much more expensive than DDT. Still others have emphasized the paucity of positive data and have criticized the EPA for knuckling under to the "environmentalist lobby." Perhaps the most vociferous critic has been

Norman E. Borlaug (1914–) is an American plant scientist best known for his work in plant pathology, pesticides, and wheat breeding. His results led to methods for increasing agricultural yields. He was awarded the Nobel Peace Prize in 1970.

Dr. Norman E. Borlaug, the winner of the 1970 Nobel Peace Prize for his work in developing high-yield wheat strains. He has emphatically denied that such chemicals as DDT are significant contributors to the deterioration of the environment and fears that all other pesticides will soon be banned, leading to failures in agriculture and possible mass starvation. These predictions seem overly gloomy, but they do illustrate the levels of passion that this whole question has aroused.

On the other hand, those who support the ban on DDT as well as restrictions on the use of other pesticides take the viewpoint that any substances as biologically active, persistent, and foreign to natural food webs as the organochloride insecticides can be *presumed* to be harmful to human health. They cite examples such as cigarette smoke and asbestos dust, whose injurious effects often emerge only several decades after exposure, to illustrate how insidious these dangers can be. Furthermore, they add, it is difficult to predict whether the switch to organophosphates, which provide a less persistent but more toxic substitute, will make matters better or worse in the long run. Finally, they point out that a permissive attitude to chemical pesticides encourages their proliferation, whereas a more restrictive stance serves as a drive to develop other, more environmentally sound methods of pest control, such as those that are described in the following section.

When experts disagree, it is often very discouraging to the layman, who develops a feeling of hopelessness about ever being able to determine which decision is the "right" one. Probably even more important than having a definite opinion, however, is having a reasonable grasp of the foundations on which decisions are based.

7.8 OTHER METHODS OF PEST CONTROL

Use of Natural Enemies: Predators, Parasites, and Pathogens

(Problems 17, 19, 20, 21, 22, 25)

We have shown how destruction of predator populations by indiscriminate spraying may cause upsurges in pest populations. It is reasonable to assume that the opposite treatment—importation of predators—may be an effective control measure.

Excellent success with such techniques has been achieved in several cases. We mentioned earlier that DDT spraying against the Japanese beetle in Illinois caused widespread havoc among other species. The

Japanese beetle, a native of the Orient, was inadvertently imported with a shipment of some Asiatic plants. In the absence of any effective natural control, the beetles thrived on the eastern seaboard of the United States and gradually became a major pest. Scientists then searched for natural predators and imported several likely species. One of these, an Oriental wasp, provides food for its young by paralyzing the Japanese beetle grub and attaching an egg to it. When the young wasp hatches, it eats the grub as its first food. The life cycle of the wasp is dependent upon the grub of the Japanese beetle; it does not naturally breed on the grubs of other insects. Therefore, this type of control is species-specific and does not seriously affect the rest of the ecosystem. Importation of a species of bacteria which infects Japanese beetles in their native environment has also helped control the pest. Presently, the spores of the bacteria, known as milky disease, are commercially available.

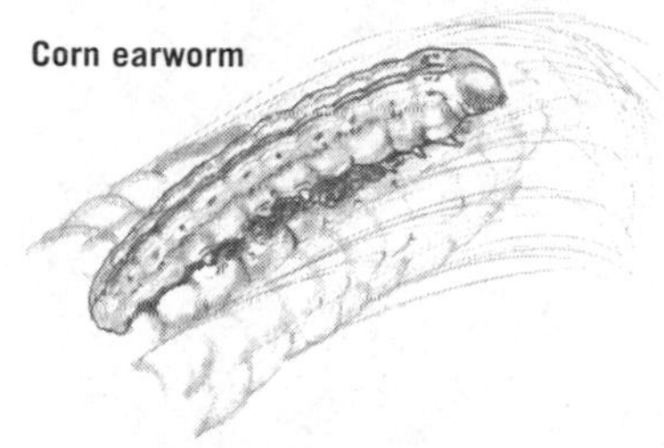
Corn earworm

Many similar successes have been registered against other major pests. Recently, a virus effective against the cotton bollworm and the corn earworm has been approved by federal regulatory agencies. In many cases, simply augmenting existing populations of natural predators such as the ladybug or the praying mantis has been rewarding.

There are many advantages to importing pest enemies. Because these agents are living organisms, they reproduce naturally and one application can last for many years. Most insect parasites and pathogens are very specific and do not interfere with the health of vertebrates. No harmful or questionable chemicals are introduced into the environment.

With these encouraging advantages, one may wonder why insect enemies are not used more frequently. The major objections to the use of predators, parasites, and pathogens are often based on social and economic considerations (see Section 7.10).

Another difficulty with the use of natural enemies to control insect pests is that the technique is neither fast-acting nor simple to apply. The farmer must know something about the life cycles of the pest and of the control organism to choose the proper time to release the predators, pathogens, or parasites. Although ultimately a high level of control can be reached, results cannot be expected immediately, for it takes time for organisms to act. In the interim between the onslaught of a pest epidemic and the growth of the enemy population, worms might be found in crops, and the crop size would be smaller. Certainly it is a common human trait to choose the chance of avoiding an immediate, visible loss, even if it involves the risk of later harm by some invisible poison. But the prevalence of such behavior does not necessarily make it the best strategy.

Sterilization Techniques

The screwworm (the parasitic larvae of the screwworm fly) is a serious cattle pest that has been responsible for large financial losses to many ranches. Some years ago, the United States Department of Agriculture initiated a program in the southeastern states to raise male screwworm flies, sterilize them by irradiation, and release them in their natural breeding grounds (see Fig. 7.4). The female mates of the irradiated males could not lay fertile eggs. Within two years of the start of the project, the screwworm menace had been brought under control in the target area. This, then, is another technique that is specific to the target pest and therefore not disruptive to the entire ecosystem.

Screwworm fly

Unfortunately, many technical problems must be solved before control by sterile males can be applied to other pests. First, a way must be found to sterilize the male insect without impairing his chance of finding a mate. In practice this has been hard to do. Luckily, the sperm of the screwworm fly can be killed without harming the adult; more frequently, however, the sterilization process debilitates the male so that he cannot compete successfully with untreated, naturally occurring insects.

Even if the sterilization problem can be solved,

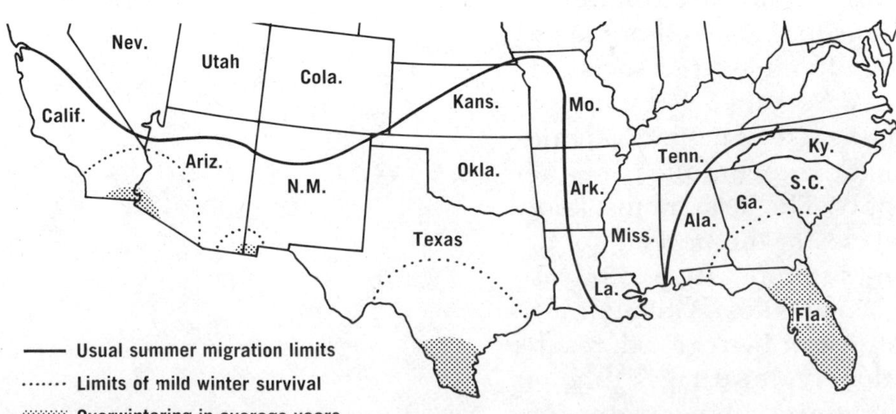

FIGURE 7.4 Screwworm distribution in the United States.

TABLE 7.2 MATHEMATICAL POPULATION DECLINE WHEN A CONSTANT NUMBER OF STERILE MALES ARE RELEASED AMONG AN INDIGENOUS POPULATION OF ONE MILLION FEMALES AND ONE MILLION MALES

GENERATION	*NUMBER OF VIRGIN FEMALES*	*NUMBER OF STERILE MALES RELEASED*	*RATIO OF STERILE TO FERTILE MALES*	*NUMBER OF FERTILE FEMALES IN THE NEXT GENERATION*
1	1,000,000	2,000,000	2:1	333,333
2	333,333	2,000,000	6:1	47,619
3	47,619	2,000,000	42:1	1,107
4	1,107	2,000,000	1,807:1	less than 1

the situation must be adjusted so that those fertile males that were part of the naturally occurring population have a small probability of finding and impregnating a female. In the case of female insects who copulate many times with different partners, it is virtually impossible to reduce the chance of fertile matings to levels that are low enough to effect control. The case of the screwworm is particularly favorable because a virgin female fly copulates only once in her lifetime, and if the ratio of sterilized to nonsterilized males is sufficiently high, then control is possible (see Table 7.2). In practice, the numbers of sterilized males required to achieve this flooding is often prohibitively great. The screwworm program was successful because the flies migrate into small, well-defined southern areas during the cold months, and the program was initiated during a severe winter which had already decimated the native population. When sterilized males were released against the codling moth, a serious pest of apple and pear orchards, the initial control was successful, but migration of fertile males from nearby areas spoiled the early gains. Future sterilization techniques will probably either be coordinated with spray programs or accomplished by field application of chemical sterilizing agents (**chemosterilants**). In the former case, an initial heavy spray would reduce the indigenous population, and a regularly applied sterilized-male approach would maintain control. Such a program would almost necessarily be government-sponsored, for to be successful against the probability of migration, the target area would have to be extensive. Chemosterilants offer the promise that a spray application which could sterilize 90 per cent of either the males or females would, in the long run, be more effective than a spray which killed 90 per cent of the population. Such programs must be initiated with very great care, however, to be sure

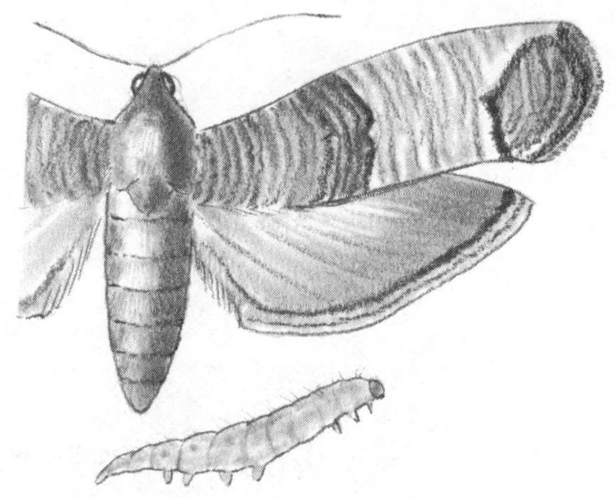

Codling moth

that the biologically powerful agents used do not cause chromosome damage to non-target organisms.

The Use of Insect Hormones

At some point in their development, insects metamorphose from a larval to an adult stage (see Fig. 7.5). While an insect is in the larval state, it continuously produces a so-called "juvenile hormone." It is only when the flow of that biochemical agent stops that the animal metamorphoses. Scientists have been able to inhibit the metamorphosis of insects in the laboratory by spraying them with the juvenile hormone of their species. Because insects can neither mate nor survive long as larvae, such a spray application is eventually lethal.

Preliminary field tests have revealed some technical problems with hormone control. One difficulty is that although natural juvenile hormone is stable in the body of a caterpillar, it is not stable in the environment and often breaks down chemically before it can act. This problem has been circumvented by the discovery of organic chemicals which are structurally similar to natural juvenile hormone, biologically active, but more stable. Another difficulty is in choosing the best time for the spray application. In a test study of the effectiveness of this technique in Colorado against a potato beetle pest, the hormone prevented metamorphosis effectively, but the caterpillars didn't die soon enough; they grew to gigantic size and ate the entire crop.

The technical problems are being solved and commercially available products might be on the market by the time this book is published. It is likely that widespread use of juvenile hormone or its analogues would produce minimum environmental insult because these chemicals are biodegradable and active only against specific insects. Therefore, with the aid of careful timing, pest populations could be destroyed without decimating their predators.

Sex Attractants

In many species of insect, when a virgin female is ready to mate, she signals by emitting a small amount of a species-specific chemical sex attractant, called a **pheromone**. The males detect (smell) very minute quantities of pheromone and follow the odor to its source. Attempts have been made to bait traps either with the chemical or, more simply, with live virgin

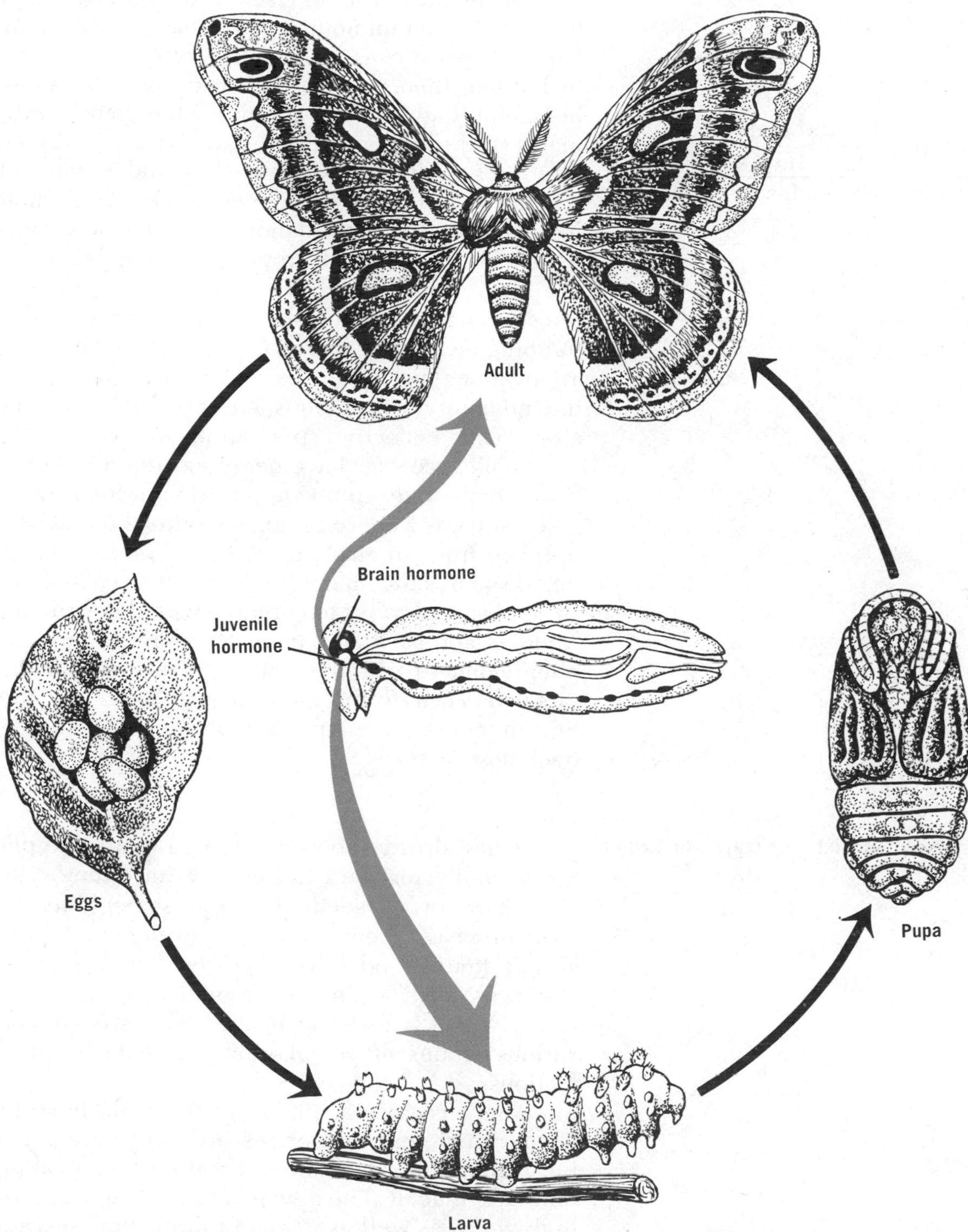

FIGURE 7.5 The role of hormones in insect metamorphosis. The juvenile hormone, secreted mainly during the larva stage, keeps the caterpillar in this immature state until it is ready to metamorphose into a pupa and adult. The hormones must be secreted in the right amounts at the right time. (Redrawn from Fortune, July, 1968.)

$$\frac{1000 \text{ traps} \times \dfrac{100 \text{ males}}{\text{trap}}}{1{,}000{,}000 \text{ males}} \times 100\% = 10\%$$

females. In this program, only minute quantities of natural chemicals are released into the air by evaporation, and environmental perturbations are nonexistent. However, as in the case of the sterile male technique, there is a statistical problem in areas of heavy infestation. Very simply, if there are a million males in the area, and 1000 traps are set, and if each trap catches 100 males before they find a female, the control project is only 10 per cent effective. Nonetheless, the traps and the labor of setting them up are expensive. In parts of Europe, the cost of this program has been reduced by having school children set the traps as a part of an educational program. Still, the trapping technique is probably best suited for special applications such as for the control of migration where the migratory populations are small, or for control after a single effective spray application.

Some successes have been registered by the confusion technique. In one application millions of cardboard squares were impregnated with pheromone and dropped from an airplane. With a high ratio of cardboard squares to females, males become confused and were observed to try to copulate with the cardboards. Another approach has been to spray the area with pheromone so that the entire atmosphere smells of virgin females (to the male insect, that is, not to the human nose). Then the males would be unable to track down a mate.

Use of Resistant Strains of Crops

It has already been mentioned that some plants are naturally resistant to pests because they synthesize their own insecticides. For some time, plant breeders have been actively developing more resistant strains, and have succeeded in a number of instances, such as in the production of a variety of alfalfa which is resistant to the alfalfa weevil, and of various strains of cereal crops resistant to rust infections.

The successes achieved in this field have been encouraging, and further research deserves support, but it should be remembered that the technical problems are difficult. The new plant variety must produce high yields as well as maintain the natural resistance to other diseases. Moreover, genetic adaptation is not stagnant, for throughout biological time genetic defense in one species has been traditionally met by genetic changes in the attacking organism to neutralize the defense. Thus, many resistant crops lose their immunity after several years and new resistant varieties must be developed.

Control by Cultivation

Uniformity is not typical of virgin land masses. The systems created by modern agriculture differ from natural systems in that farms tend to specialize in a very few species of plants, whereas areas untouched by man do not. For instance, in Kansas and Nebraska, areas covering thousands of acres of land are covered almost exclusively with wheat fields. The result of such specialization is that a mold, a fungus, or an insect that consumes wheat has an almost unlimited food supply and an extremely hospitable environment. No barriers to spreading exist, and the pest can grow quickly in uncontrolled proportions. The advantage of diversity in nature was discussed in relation to the ability of an ecosystem to survive perturbations (see Chapter 1). One such perturbation is attack by specific consumers. A fungus that attacks wheat will spread more slowly if half the plants in a field are something other than wheat because the spores have less of a chance of landing on a wheat plant. Moreover, if the disease spreads slowly, there is more time for the development of natural enemies of the fungus or of strains of wheat that are naturally resistant to it. Therefore, one solution to the problem of pests is to grow plants in small fields, with different species grown in adjacent fields.

Unfortunately, the mechanization that is essential to modern agriculture is best suited to large fields of single crops.

Simply planting small fields of different crops is effective in itself, but by judiciously choosing the companion plants, additional success can be realized. For example, the grape leafhopper, a pest of vineyards, can be controlled by a species of egg parasite that winters in blackberry bushes. Knowledgeable grape-farmers therefore maintain blackberry thickets. Conversely, one variety of stem rust that attacks North American cereal crops must live part of its life cycle on barberry bushes, and selective destruction of these plants will reduce rust infestations.

Other methods of cultivation which have been successful in controlling pests include (a) crop rotation so that a given pest species cannot establish a permanent home in one field, (b) specific plowing and planting schedules to favor predators over pests, and (c) planting certain weeds which some omnivorous insects need for food during part of their life cycle.

Miscellaneous Control Techniques

The Egyptian Nile sparrow is a grain-eating bird that appears at harvest time in huge and destructive flocks. Modern control practices have been frustrated by the fact that one cannot poison the grain just prior to harvesting. However, the Chinese have successfully used an interesting technique against their own sparrows since the beginning of recorded history. They have discovered that the birds take to the air when bothered by noise, and furthermore, that they die of exhaustion if forced to fly for more than 15 minutes at a time. When a sparrow infestation threatens, farmers over a wide area race through fields yelling, beating gongs, or generally making noise in a coordinated effort. The technique is often successful in destroying entire flocks.

Some more modern, but no more successful, pest control practices are: (1) use of sound or light to attract insects to traps; (2) use of sound as a repellent; (3) use of proper sanitation such as draining of swamps for mosquito control; (4) introduction of innocuous insects which are effective competitors of pest species. For example, some populations of biting insects such as gnats can be controlled by introducing species which don't bite man but which otherwise occupy niches similar to those of the pests.

Integrated Control

If man has learned any lesson during the DDT era, it is that insects are not passive to control measures, and that any approach to pest management is subject to resistant reactions from the insect world. Therefore, many scientists now believe that we should never again rely on unilateral attack, but rather we should use as many available control measures as possible in an integrated and well-planned manner. Integrated control will never be so conceptually or technically simple as straightforward chemical control can be. It will be slower-acting and more expensive initially, although we hope it will be more effective and cheaper in the long run. Despite its disadvantages, integrated control has been successful in many areas.

In Section 7.3 we discussed the reduced yields of the cotton growers in a valley in Peru after several years of insecticide use. Following the crop failure, a new program was initiated. First of all, predacious and parasitic insects were imported from nearby valleys. Second, all cultivation of marginal soils was banned in an effort to cull out weak plants which are often breeding centers for disease. Third, the planting and irrigation cycles were adjusted so as to interrupt most effectively the life cycles of the pests. Last, all spraying with synthetic organic insecticides was banned except under the permission and supervision of a panel of scientists. The result was that, despite the fact that some land was purposely not planted and very little insecticide was used, cotton yields were higher than had ever been recorded in the area.

7.9 HERBICIDES

(Problems 23, 24)

During the past 10 years, the use of herbicides for the chemical control of unwanted plants of all types has been so increasingly popular that it has resulted in a multimillion-dollar industry. In many respects, the use of herbicides has closely paralleled insecticide uses. The discovery of potent chemicals for application against noxious weeds has led to volume sales, and a great deal of indiscriminate use, resulting in chemical pollution of the environment despite only partial success of the control programs.

There are about 75 different herbicidal chemicals in common use, but the wide variety of structural formulas makes classification into a few neat chemical groupings impossible. The two most popular herbicides, 2,4-D and 2,4,5-T are closely related chemically. Both are toxic to mammals. They are not nearly so persistent in the environment as organochloride insecticides, for they decompose in a few months. Although herbicides, like organochloride insecticides, become concentrated in the food web, the severity of the problem is reduced by the degradability of herbicides.

Reports of the teratogenicity (the tendency to cause birth defects) of 2,4,5-T have caused considerable concern. Although the relationship between commercial preparations of 2,4,5-T and birth defects is undisputed, there is considerable argument as to whether the teratogen is actually 2,4,5-T or some impurity produced in the manufacturing process. Meanwhile, as the argument continues, the 2,4,5-T, impurity and all, can be purchased at your local hardware store.

Uses of Herbicides

Agriculture

Weed control, like pest control, is necessary for efficient agriculture. In an age in which farmers can no longer afford to hire small armies of men to hoe fields, weed irrigation ditches, and cut brush along fence lines, it is inevitable that chemical control should be in common use. For example, herbicides have been used extensively to kill sagebrush in semiarid range lands of the western United States in an effort to encourage larger yields of grasses. But the sagebrush ecosystems have evolved uniquely and effectively to this harsh climate of dry summers and cold winters, and it is debatable which system supplies more year-round food for cattle, the sagebrush–natural grass system or the farm-hay system. Perhaps in such a delicate situation the probability of chemical pollution should swing the decision in favor of the undisturbed range.

Right-of-Way Clearing

Herbicides are used extensively by industry and by all levels of government to restrain the natural advance of brush and forest from encroaching on various rights-of-way. Transmission lines, railroads, highways, county roads, fire control roads in national forests, and hikers' paths in wilderness areas have all been subjected to chemical weed control. Despite obvious agreement that brush and trees must be kept back from roadways to allow drivers to see traffic patterns around turns, and despite the fact that other rights-of-way must be maintained, the present system of maintaining these openings by repeated mowing and spraying might not be the cheapest and most effective technique available. Shrub and tree seedlings sprout easily in cut grass so the objective of preserving lawns perpetuates the need for further maintenance. On the other hand, if the grasses and shrubs are allowed to grow and once a year, or once every several years, the tall shrubs and trees are selectively killed, the rights-of-way can be maintained for a fraction of the cost and a fraction of the pollution. The question of esthetics is subjective; some feel that a field of natural grasses and shrubs, green in the spring and yellowed by late August is beautiful, while others feel it looks unkempt and ugly.

Home Gardening

Prodigious quantities of herbicides are sold yearly to homeowners for control of crabgrass and other lawn and garden weeds. Consider two neighbors, one who believes that it is his right to sit in his backyard on a Sundy afternoon and breathe clean air and the other who believes that it is his right to spray against crabgrass. Whose right takes precedence? If the spray drifts downwind from one person's garden to another's hammock, does the latter have any recourse to action? The question is not simple either legally or morally, but the problem must be faced.

Warfare

During the course of the war in Southeast Asia, extensive areas of forests and croplands were chemically defoliated. A large number of birth defects have most probably been caused by high concentrations of 2,4,5-T in the air, water, and food supplies of that area. Destruction of natural ecosystems, and consequent erosion and revegetation with early successional plants such as bamboo, is likely to change the character of the land for centuries to come.

7.10 ECONOMIC FACTORS IN PEST CONTROL

In Section 7.8 we outlined various methods of pest control that offer environmentally attractive alternatives to the use of persistent chemical pesticides. The reader may wonder, then, why such advantageous methods are used so infrequently. One possible answer is that the more specific, and hence environmentally more desirable, control methods are often the least profitable to market. For example, a natural predator, parasite, or pathogen, once deployed against a target insect pest, tends to maintain its own population by reproduction. Therefore, a private company that provides them to a farmer will not enjoy the same rate of reorders as will a manufacturer of a chemical pesticide, which does not reproduce itself. Furthermore, a new chemical can be protected by a patent, whereas a microbe or a ladybug cannot.

Even among chemical pesticides, a compound that is effective against a single species, and, therefore, less generally disrupting to the environment, will enjoy a smaller sales market than will a broad-spectrum insecticide that can be used for a variety of crops. The most highly specific agents, such as pheromones, will be used in the smallest quantities.

The commercial development of any new product is expensive. In the 1970's, the cost of research, testing, and obtaining approval for use either of a new chemical insecticide or a new biological control measure is about 5 to 10 million dollars, and requires several years' time. Understandably, the economic incentive favors the product that has the greater sales potential as well as the protection of a patent. This incompatibility of sound economic and ecological practice presents a serious problem of environmental management. Some of the general aspects of such contradictions are discussed in Chapter 13. What is called for in the area of pest control, at the very least, is a greatly increased level of government support of research and development that private companies lack the incentive to support.

PROBLEMS

1. **Insects and man.** What are the harmful effects and the benefits that insects bring to man? By what mechanisms did man accommodate himself to insects before the production of modern pesticides?

2. **Chemistry of insecticides.** Consider the following table:

Insecticide	Compound Class	Chemical Formula	Melting Point, °C
DDT	organochloride	$C_{14}H_9Cl_5$	108.5
Parathion	organophosphate	$C_{10}H_{14}NO_3PS$	6.1

(a) A compound has the formula $C_{12}H_8Cl_6$ and the melting point 104°C. What compound class would you guess that it belongs to?

(b) Another compound is a common insecticide and has the melting point 2.8°C. With no further information, could you make a guess as to whether it is an organochloride or an organophosphate chemical? Would you be sure of your answer?

3. **Resistance to pesticides.** Explain why it often becomes necessary as time goes on to use larger quantities of a given pesticide to achieve the same results.

4. **Pesticides and predators.** Pesticides have been known to be more harmful to predators than to the pests they are designed to control. What factors could account for this selectivity?

5. **Chemical control.** Explain why it is ecologically unsound to completely replace complex natural insect controls with chemical ones. Explain why it has worked nevertheless in many instances.
6. **Biodegradability.** The first sentence in Section 7.4 (page 264) states that most natural compounds are biodegradable. Is this statement correct or should it have said *all*? Does the existence of coal and oil deposits on Earth have any bearing on this question? Defend your answer.
7. **Biodegradability.** Compounds synthesized by man are less likely to be biodegradable than naturally-occurring compounds. Account for this difference.
8. **Wrong targets.** Explain how soil runoff from agricultural systems can be deadly to fish in nearby streams.
9. **Pesticide solubility.** What advantages and disadvantages would result from the use of a pesticide that was soluble in water? Do you think it likely that it would be practical to use a water-soluble pesticide? Defend your answer.
10. **DDT solubility.** Imagine that you are going to eat a fish that is contaminated with DDT. Which of the following methods of preparation would be most effective in reducing the insecticide concentration: (a) broiling on a charcoal fire; (b) deep fat frying; (c) boiling in water; (d) steaming; (e) baking in an oven; (f) none of the above, that is, just eating it raw. Defend your choice.
11. **Pesticides in aquatic systems.** Explain why carnivorous fish are generally more susceptible to low levels of pesticides in the water than are herbivorous fish.
12. **Pesticides in aquatic systems.** If no measurable quantities of DDT were found in the water of a pond, would that necessarily mean that the aquatic ecosystem was unpolluted by DDT? Explain.
13. **Sublethal doses.** An ecologist studying the wildlife populations before and after a heavy spray application determined that since no animals were directly killed by the spray, no harm had resulted. Would you agree? Explain.
14. **Birds.** Do you think that peregrine falcons might become resistant to DDT? Would you think that resistance might save the birds from extinction? Defend your answers.
15. **Health effects.** Explain why it is difficult to determine whether there is a relationship between DDT and cancer.
16. **DDT ban.** Outline the arguments for and against the DDT ban.

17. **Alternatives.** What methods of pest control are available as alternatives to the use of chemical sprays?

18. **Resistance.** Would you think that the development of resistance to insect parasites, predators, and pathogens would be a serious problem if widespread use of these natural enemies were initiated?

19. **Sterile males.** Give three reasons why the sterilized male approach was successful for the control of screwworms. Explain how these unique conditions might not be met with other pests.

20. **Sterile females.** What would you think of a program to release sterilized females instead of sterilized males?

21. **Integrated control.** Explain why one indiscriminate spraying with DDT could destroy the effectiveness of an integrated control program.

22. **Integrated control.** Birds often become major pests in vineyards because they eat the grapes. (a) Which of the following control programs would you recommend for bird control: (i) spreading poison; (ii) broadcasting noise from a loudspeaker system; (iii) shooting; (iv) covering the vineyard with some fencing material? Discuss. (b) Do you think that it might be wise to initiate research directed toward: (i) developing a sterilization program against the birds, or (ii) developing new strains of grape which would be unpalatable or poisonous to birds? Discuss.

23. **Herbicides.** Ragweed is considered to be a major plant pest because its pollen causes misery to hayfever victims. In natural systems ragweed is characterized as an early successional plant. A few years ago, the State of New Jersey initiated a program to eliminate ragweed. Thousands of acres of roadways and old fields were sprayed with herbicides. Why do you think this program failed?

24. **Herbicides.** Outline briefly some of the parallel problems that exist between herbicide use and insecticide use.

25. **Nematodicides.** Another category of broad-spectrum pesticides, known as **nematodicides,** are used against soil pests such as cutworms. What types of ecological problems would you guess would develop from use of these chemicals?

The following question requires arithmetic reasoning.

26. **Pesticides in food chains.** Data are given in Section 7.6 on the increasing concentration of the pesticide DDD as it moves up the food chain from plankton to fish to birds. Are these increases constant, accelerating, or decelerating? Can you suggest a reason for this progression of concentration?

BIBLIOGRAPHY

The best general, but technical, reference on pesticide use is:

U.S. Department of Health, Education, and Welfare: *Report of the Secretary's Commission on Pesticides and Their Relationship to Environmental Health.* Washington, D.C., 1969. 677 pp.

Another useful government publication which is a collection of data with an absolute minimum of text (very dull to read but an excellent reference) is:

David Pimentel: *Ecological Effects of Pesticides on Non-target Species.* Washington, D.C., U.S. Government Printing Office, 1971. 219 pp.

An authoritative book on the control of pests with minimum use of chemicals is:

W. W. Kilgore and R. L. Doutt, eds.: *Pest Control.* New York, Academic Press, 1967. 477 pp.

The book that started much of our current concern about pesticides, and a more recent sequel to it, are the following:

Rachel Carson: *Silent Spring.* Boston, Houghton Mifflin Co., 1962. 368 pp.

Frank Graham, Jr.: *Since Silent Spring.* Boston, Houghton Mifflin Co., 1970. 333 pp.

Two topical books are:

R. C. Muirhead-Thomson: *Pesticides and Freshwater Fauna.* New York, Academic Press, 1971. 248 pp.

T. L. Willrich and G. E. Smith, eds.: *Agricultural Practices and Water Quality.* Ames, Iowa State University Press, 1970. 407 pp.

The use of herbicides by the United States, with particular emphasis on herbicides in Vietnam, is covered in:

T. Whiteside: *The Withering Rain.* New York, E. P. Dutton, 1971. 223 pp.

8

RADIOACTIVE POLLUTANTS

8.1 INTRODUCTION

"The doctrine of atoms is still alive although it came into being about a hundred years ago. It has been proved to be illogical just as the ether that fills all space has been shown to be incapable of existence. Properties must be ascribed to the atom that it cannot possess and the same is true of the ether. What are we to do? Throw over the atom and the ether? Although both have been convicted of being illogical, I do not think it would be logical to give them up, for they are helpful in spite of their shortcomings, and in some way they suggest great truths. They are symbolic."

The atomic theory was first proposed on a philosophical basis in the fifth century B.C. by the Greeks Leucippus and Democritus. In 1803 the Englishman John Dalton reaffirmed the theory and put it on a quantitative basis by assigning relative masses to atoms, thereby accounting for the weight relationships in chemical reactions. The great advances in chemistry during the nineteenth century, such as the structural theory of molecules from which organic chemistry is derived, or the concept of the periodic table which systematizes the properties of chemical elements, were based on atomic theory. All this might suggest to you that by the start of the twentieth century everyone believed in atoms. Not so. The struggle over the acceptance of the existence of atoms is apparent from the words of Ira Remsen in 1902, taken from his presidential address to the American Chemical Society.

In 1896, six years before Remsen's speech, the French physicist Henri Becquerel discovered natural radioactivity, thus initiating a new chain of discoveries that led to a great refinement of our ideas about the atom, and that eventually convinced practically everyone that atoms do exist. These discoveries, in turn, facilitated the invention of artificial radioactivity, and led still others to the discovery of atomic fission, which made it possible to design and build the atomic bombs and nuclear reactors that generate radioactive waste products.

No one questions the fact that the radioactive wastes produced by man's activities constitute a potentially serious disturbance to the processes of life on earth. There are very sharp differences of opinion, however, as to whether the precautions now being taken by the nuclear power industry to prevent such disturbances are adequate. Another way of expressing the same doubt is to ask the question, "Will the benefits to man from his involvement with nuclear reactors outweigh the costs resulting from the disturbances to the global ecosystem?" The opponents of nuclear power say no, the benefits are not worth the costs—we disrupt the environment to produce energy that we should not have, such as energy to support too many people, or to overheat our homes when we could keep them cooler and wear sweaters indoors, or to power electric pencil sharpeners. Furthermore, some opponents add, we do not know how serious the costs may become; the deleterious effects may be subtle but long-lasting, and therefore more serious than we can now realize. In this view, no tightening of precautions can be considered adequate. Instead, we should minimize or abandon nuclear energy in favor of truly nonpolluting sources such as solar energy.

The proponents of nuclear energy take a diametrically opposite set of views. They point out that no other phase of man's technology was introduced with such sophisticated safety precautions as have been used by the nuclear power industry. The bulk of our radioactive wastes are isolated in portions of the earth that will remain undisturbed for periods that are measured on geological time scales, and, therefore, cannot be said to constitute a potential disturbance to terrestrial processes. The small proportion of artificial radioactive matter that does enter the biosphere is less than that which results from cosmic rays received by Earth from outer space, or from natural sources such as radioactive elements in rocks, to which life has already become adjusted. Furthermore, the nuclear proponents add, we need nuclear energy for many purposes that are not at all trivial—to power various anti-pollution devices, such as equipment that purifies air and water, and to substitute for other energy sources, such as coal or oil, whose burning produces more harmful immediate environmental effects than do nuclear reactors.

The scientific qualifications of those holding opposing views are impressive, and their arguments are by no means trivial. It is therefore important for all of us to understand the technical and social basis for these opposing positions. To reach this conceptual level, we must review the nature of atomic (nuclear) processes, learn something about the operation of a nuclear power plant, with special attention to safety and waste-disposal procedures, and we must evaluate the potential hazards that might result from malfunctions or accidents.

The required lessons must include an outline of atomic theory, and of the nature of natural and artificial radioactivity. These topics will be taken up in the three following sections: 8.2, 8.3, and 8.4. If you are already familiar with this material, you may proceed directly to Section 8.5 on nuclear reactors.

8.2 ATOMIC STRUCTURE

(Problems 1, 20)

Evidence from various electrochemical experiments supports the conclusion that all matter of the kind we are familiar with on earth is a collection of fundamental units that we call **atoms,** and that these atoms consist of electrically negative, positive, and neutral particles. These particles, respectively, are electrons, protons, and neutrons.

The Atomic Nucleus

Imagine that you had a piece of solid gold that was exactly the size and shape of this book (if the book were closed). This sample of gold would, of course, have a certain mass (weight). If you tried to stick a pin through the thickness of this gold book, you would find it difficult. The difficulty would be about the same wherever you tried to stick the pin. The conclusion of this crude experiment is that, so far as you can tell, the matter in a piece of solid gold is distributed equally in all the space occupied by the gold. Now imagine that you repeat the experiment in a much more delicate way. You hammer the gold into a very thin sheet (much thinner than a page of this book). You now try to puncture this thin gold sheet, not with a pin, but with a beam of very small particles and you observe whether the particles go through the gold sheet or bounce off. The result of this experiment is that most of the particles go right through the gold sheet, but that a small proportion (one in several thousand) bounce off. The conclusion of this experiment is just

The particles originally used in this experiment were alpha particles (nuclei of helium atoms). The work was done by Lord Rutherford in 1911. It is not necessary to know the details of this experiment to understand this section.

the opposite of that of the crude test with the pin described earlier. The results of this new experiment tell us that matter is *not* distributed uniformly in a solid piece of gold, but is concentrated here and there in dense masses. The particles that bounce off the gold sheet must be those that happen to hit the dense masses.

This and similar experiments provide the fundamental evidence from which we conclude that the mass in any sample of matter is concentrated in what we call atomic nuclei. We observe the same kind of result with other types of thin sheets, such as those made of copper or of silver. We believe that the mass of all matter as we know it on earth is predominantly concentrated in atomic nuclei.

The Atom

The nucleus of the atom contains the protons and the neutrons, and hence all of the positive charge and practically all of the mass. (The electrons do account for a small portion of the mass of the atom; the actual mass of an electron is about 1/2000 of that of the proton). The nucleus occupies comparatively little volume, however. If the nucleus of a gold atom were as large as the dot on the letter *i*, then the distance between two gold nuclei in a piece of solid gold would be about one foot. The electrons of the atom contain all the negative charge and are dispersed over most of the space, but they contain very little of the mass. An atom contains an equal number of protons and electrons; hence it is electrically neutral.

Atomic nuclei can be described by two numbers, which are defined in the margin at right.

Atomic number = the nuclear charge = the number of protons in the nucleus.

Mass number = the number of protons in the nucleus plus the number of neutrons in the nucleus.

PARTICLE	ELECTRICAL CHARGE	MASS NUMBER
Electron	−1	0
Proton	+1	1
Neutron	0	1

A chemical element is a substance that consists of atoms of the same atomic number. Elements are considered to be kinds of ultimate chemical raw materials; all substances are composed of elements. Each element has a common name and is represented by one or two letters. Eighty-eight elements occur naturally on Earth. These are elements 1 to 92 (with numbers 43, 61, 85, and 87 missing). The four missing elements, together with twelve more above 92, have been made by man, so that the total number now known is 104. Here are some examples:

Name	Nuclear composition		Atomic number (protons)	Number of electrons	Mass number (protons + neutrons)	Symbol
	Protons	Neutrons				
Carbon-12	6	6	6	6	12	^{12}C
Carbon-14	6	8	6	6	14	^{14}C
Calcium-40	20	20	20	20	40	^{40}Ca
Calcium-44	20	24	20	20	44	^{44}Ca
Uranium-235	92	143	92	92	235	^{235}U
Uranium-238	92	146	92	92	238	^{238}U
Uranium-239	92	147	92	92	239	^{239}U

Mass is a fundamental property of matter that is somewhat difficult to define. It is perhaps best to think of the mass of a sample as the quantity of matter in the sample. Thus, the mass of a cat is (approximately) 100 times the mass of a mouse. This means that there is 100 times as much matter in a cat as there is in a mouse. On Earth the weight of a body is determined by its mass; therefore, we can say that a cat weighs 100 times as much as a mouse. If we place a cat on one side of a balance, and 100 mice on the other, the balance will indicate that the weight (mass) of the cat equals the weight (mass) of 100 mice. However, this experiment must be done on Earth. In a space ship far removed from any attracting body, there is no gravity and therefore no weight. But there is mass, and in a space ship the mass of a cat is still 100 times the mass of a mouse.

Atoms, too, have mass, although we cannot weigh atoms individually on a balance. Atomic masses, more commonly called atomic weights, are based on the assignment of an atomic weight of 12 to carbon-12. Therefore, for carbon-12, the mass number and the atomic weight are exactly equal. For other isotopes, the two numbers are very nearly equal. For example, the atomic weight of oxygen-16 is 15.99491.

Note that each element is symbolically represented by its letter(s) and its mass number.

Note also that a given element may have more than one mass number. Atoms of the same element (that is, atoms with the same atomic number) that have different mass numbers are called **isotopes.** Thus, ^{12}C and ^{14}C are carbon isotopes. Both ^{12}C and ^{14}C represent carbon atoms (or carbon nuclei), because they both have six nuclear protons. They are isotopes because they have different mass numbers. This difference results from the fact that there are different numbers of neutrons in the two nuclei. Isotopes of an element are chemically equivalent (or very nearly so): they have the same ability to combine with other atoms. Thus, ^{12}C and ^{14}C are both chemically the same substance, carbon. One of the properties of carbon is its ability to burn in air. Therefore, we may write the following equations for burning the carbon isotopes:

$$^{12}C + O_2 \rightarrow {}^{12}CO_2$$
$$^{14}C + O_2 \rightarrow {}^{14}CO_2$$

Note that the two carbon nuclei, ^{12}C and ^{14}C, are not altered by the chemical reaction, and therefore persist in the carbon dioxide molecules produced by combination with oxygen.

8.3 NATURAL RADIOACTIVITY

(Problems 2, 21, 22)

Some atomic nuclei are unstable, that is, they tend to decompose spontaneously. Such nuclei are called **radioisotopes,** and the substances in which they exist are said to be **radioactive.** Some of these unstable nuclei occur naturally on earth; others have been made by man. An example of a naturally-occurring radioisotope is radium-226, or ^{226}Ra. As we stated earlier, natural radioactivity was discovered accidentally in 1896, when the Frenchman Henri Becquerel found that uranium minerals emit a radiation somewhat similar to X-rays. Later, it was learned that the energy of this emission was stored somehow in the atomic nuclei themselves. The emissions were found to consist of electrically positive particles, electrically negative particles, and electrically neutral rays (similar to X-rays) that we call gamma-rays. When a radioisotope decomposes, a new atom is left behind. In some cases this leftover atom is stable (nonradioactive); in others it is radioactive. For example, the radium-226 nucleus decomposes to produce a positive particle (an alpha-particle; see note on page 292) and a new element, radon-222, which is a radioactive gas:

$$^{226}Ra \rightarrow \text{alpha-particle} + {}^{222}Rn$$

Radon-222 also decomposes to produce an alpha-particle and still another radioisotope, polonium-218. The decompositions continue for seven more steps until a stable isotope, lead-206, or ^{206}Pb, is formed.

$$^{222}Rn \downarrow {}^{218}Po \downarrow {}^{214}Pb \downarrow {}^{214}Bi \downarrow {}^{214}Po \downarrow {}^{210}Pb \downarrow {}^{210}Bi \downarrow {}^{210}Po \downarrow {}^{206}Pb \text{ (stable)}$$

Now we ask two related questions: First, if ^{226}Ra nuclei are unstable, why are there any left on earth? Second, if we had just one atom of ^{226}Ra, containing 1 nucleus, when would it decompose? The second question cannot be answered at all. Think of the ^{226}Ra nucleus as an energetic bundle of electrically charged matter; it may or may not break apart at any time. But we do know what the *chances* are that the ^{226}Ra nucleus will decompose in any given minute, or day, or year, or century. To understand this idea better, let us leave the Ra nucleus for a moment and consider a more familiar example. A bird lays an egg and leaves it behind, unattended, in the nest. At any time a predator, for example, a lynx, may come along and destroy the egg. How long will the egg last? Obviously we cannot answer the question; the particular egg we are considering may be lucky or unlucky. But if we understand the system completely we can tell what the chances of egg destruction are for any given period of time. Let us say that in any given day one egg has a 50–50 chance of surviving; thus the chance of survival equals the chance of destruction. Now imagine that we see 64 such individual eggs, each with a 50–50 chance of surviving any given day. If we return a day later, we expect to find half of them destroyed, and half intact. Therefore, we expect to find 32 eggs. After another day the number will be halved again, and only 16 eggs will be expected to remain; the third day we would expect to see eight, then four, and so on. Because half of the eggs are expected to be destroyed in any given day, this time interval (in this case, a day) is called the **half-life.**

The half-life concept applies to radioisotopes. One nucleus of a ^{226}Ra atom has a 50-50 chance of surviving in any given interval of 1600 years. This means that the half-life of ^{226}Ra is 1600 years. Therefore, if one gram of ^{226}Ra were placed in a container in 1974, there would be only one-half gram left after 1600 years (in the year 3574), and only one-quarter gram after another 1600 years (in the year 5174), and so on. This process is called **radioactive decay.**

The concept of half-life does not imply that after 1600 quiet years half of the ^{226}Ra will suddenly decompose. Since there is a 50–50 chance that any one ^{226}Ra atom will decompose in 1600 years, we would expect one ^{226}Ra atom in 1600 atoms to decompose in one year—or one out of 1600×365, which is one out of 584,000, to decompose in one day, or about one out of 50,000,000,000 to decompose in one second. But a gram of ^{226}Ra contains about 2,650,000,000,000,000,000,000 atoms. This number is so large compared with the proportion of atoms expected to decompose every second that the decay process appears to be continuous, and any nearby Geiger counter will respond by clicking all the time.

The rate at which alpha particles are emitted from a sample of ^{226}Ra depends on the quantity of ^{226}Ra. Since this quantity is constantly decreasing, the rate of emission, or radioactivity, of the sample of ^{226}Ra is also decreasing. Remember, however, that ^{226}Ra produces other radioisotopes when it decomposes. Any sample that has been decomposing for some finite time will contain some of the original ^{226}Ra plus some of each of its radioactive "daughters," as well as some of the stable final product, lead-206. These radioisotopes have different half-lives, ranging from fractions of a second to about 20 years. Therefore, the total radioactivity produced by a sample of radium together with its radioactive waste products is more than that produced by the ^{226}Ra sample alone.

The radioactivity present on earth before the twentieth century was derived from radioisotopes that have survived over the history of the earth. Therefore, these radioisotopes must have very long half-lives. The half-life of natural uranium-238, ^{238}U, for example, is 4,500,000,000 years. The radiations from such materials plus the effect of radiation that comes to the earth from outer space is called the **background radiation.** It has been proved that radiation accelerates the process of genetic mutation. Life on earth has always existed in the presence of a background radiation and, in fact, radiation-induced mutations must have been a factor in the development of species.

In recent years man has vastly increased the quantity of radioactive materials in various parts of the earth. As we mentioned at the beginning of this chapter, we cannot invent anything to stop this radioactivity. It slows down by radioactive decay at a rate determined by the half-lives of the radioisotopes involved.

8.4 HOW MAN HAS PRODUCED MORE RADIOACTIVE MATTER

(Problem 3)

Lord Rutherford and Frederick Soddy first proposed in 1902 that radioactive decay results in the change of atoms of one element to atoms of another element. Seventeen years after that, in 1919, Rutherford produced a transmutation by exposing nitrogen to alpha-particles to produce oxygen-17: The ^{17}O, however, is not radioactive.

$$^{14}N + \text{alpha-particle} \rightarrow {}^{17}O + \text{proton}$$

Fifteen years later, in 1934, Irène and Frédéric Joliot-Curie, Mme. Curie's daughter and son-in-law, bombarded boron with alpha-particles and produced nitrogen-13, which is radioactive: ^{13}N was the first artificially produced radioisotope. This process, therefore, resulted in a man-made increase in radioactivity on earth. It was the first production of atomic waste. However, the quantity of radioactivity produced by an experiment such as the one cited above has an inconsequential effect on the earth, because only very small quantities of radioactive matter are involved. The alpha-particles used to bring about the transformation come from naturally radioactive sources such as radium, and, consequently, their availability is limited.

$$^{10}B + \text{alpha-particle} \rightarrow {}^{13}N + \text{neutron}$$

The discovery of the **nuclear chain reaction**, which occurs in nuclear fission, led to the production of vast quantities of radioactive matter. For our purposes, it is important to understand first what a chain reaction is. Then we shall see why the chain reaction makes the production of radioactive wastes a real problem for life on earth.

"Soddy . . . turned to his colleague and blurted: 'Rutherford, this is transmutation!' Rutherford rejoined: 'For Mike's sake, Soddy, don't call it transmutation. They'll have our heads off as alchemists.' Rutherford and Soddy were careful to use the term 'transformation' rather than 'transmutation.'" (*Scientific American*, **August, 1966, p. 91.)**

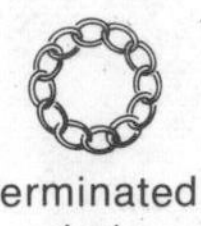

A chain is a series of links. Think of the process of making a chain; it involves the successive addition of links. The process of adding links to a chain is called **chain lengthening.** If the end of a chain links onto the beginning, it forms a cycle, and the chain ends. This is one form of **chain termination.** If more than one link is added to a given link, various arms of the chain develop; this is called **chain branching.**

A series of steps in a process that occur one after the other, in sequence, each step being added to the preceding step like the links in a chain, is called a chain process or **chain reaction** (see Fig. 8.1). Chemical chain reactions can also undergo branching. An example of a branching chemical chain reaction is a forest fire. The heat from one tree may initiate the reaction (burning) of two or three trees, each of which, in turn, may ignite several others. If the lengthening of one chain proceeds at a given rate, the production of 10 branches means that 10 reactions are going on at the same time, so that the rate has increased tenfold. A chemical chain reaction that continues to branch can produce an explosion. The condition under which a chain reaction just continues at a steady rate, neither accelerating nor slowing down, is called the **critical condition.**

The production of the atomic (fission) bomb and of nuclear reactors depends on branching nuclear chain reactions. The process is initiated when a neutron strikes a ^{235}U nucleus and can proceed in any of several different ways. Two examples are shown here:

$$^{235}\text{uranium} + 1 \text{ neutron} \rightarrow {}^{142}\text{barium} + {}^{91}\text{krypton} + 3 \text{ neutrons}$$

$$^{235}\text{uranium} + 1 \text{ neutron} \rightarrow {}^{137}\text{iodine} + {}^{97}\text{yttrium} + 2 \text{ neutrons}$$

Note the following important points about these equations:

1. The reaction is started by one neutron, but produces two or three neutrons. These neutrons can initiate two or three new reactions, which in turn produce more neutrons, and so forth. This is, therefore, a branching chain reaction. Thus, repetition of the first reaction could be written as follows:

$$3\ {}^{235}\text{U} + 3 \text{ neutrons} \rightarrow 3\ {}^{142}\text{Ba} + 3\ {}^{91}\text{Kr} + 9 \text{ neutrons}$$

$$9\ {}^{235}\text{U} + 9 \text{ neutrons} \rightarrow 9\ {}^{142}\text{Ba} + 9\ {}^{91}\text{Kr} + 27 \text{ neutrons}$$

2. The ^{235}U nucleus is split in half (roughly) by these reactions. This is called **atomic** or **nuclear fission.** Fission releases energy because the uranium nuclei are less stable than their breakdown products. The amounts of energy involved are very large compared with those of chemical reactions. If the branching chain reaction continues very rapidly, we have an atomic explosion. If the chain branching is carefully controlled, energy can be released slowly, and we have a nuclear reactor which can be used for power production.

3. The equations are balanced in mass numbers:

$$235 + 1 = 142 + 91 + 3$$

and

$$235 + 1 = 137 + 97 + 2$$

4. Fission reactions produce radioactive wastes. ^{142}Ba, ^{91}Kr, ^{137}I, and ^{97}Y, the products shown in the equations above, are all radioactive. Furthermore, the reactions represented by these equations are only two out of many that occur in atomic fission. Many different radioisotopes are produced by atomic fission. Also, the fission products, as a group, are much more radioactive than the uranium that is the raw material. As we mentioned previously, the half-life of the naturally abundant uranium isotope, ^{238}U, is 4½ billion years.* Many American homes contain old (pre-World War II) orange-colored kitchen pottery prepared from a uranium oxide pigment. The radiation level from such materials is low. (However, it would be well to get them out of your kitchen. Donate them to the nearest university.) But the half-lives of the fission products are much shorter; some are measured in centuries, some in years, days, minutes, seconds, or fractions of a second. Therefore, their radiation levels are much higher. These materials, produced in variety and abundance by nuclear chain reactions, are the atomic wastes that concern us.

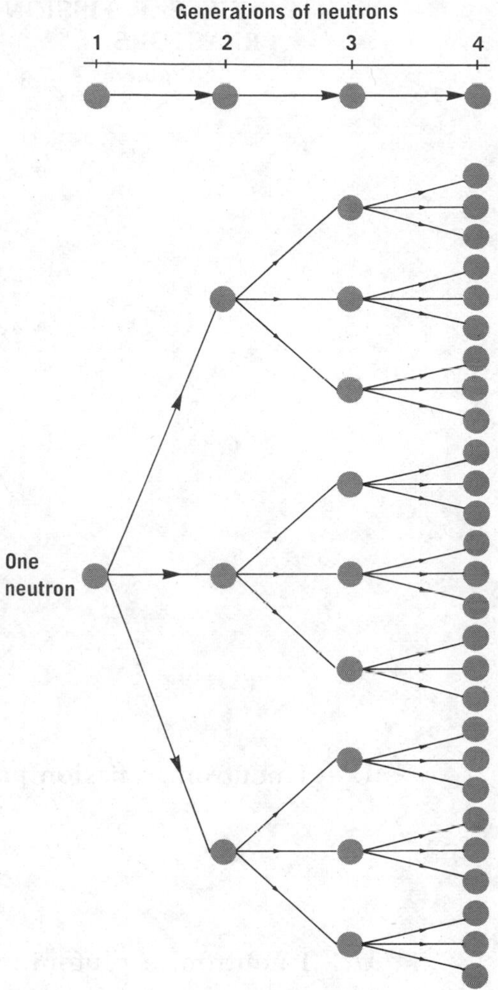

FIGURE 8.1 Neutron chain reactions. *Top:* **unbranched chain.** *Bottom:* **three-for-one branching.**

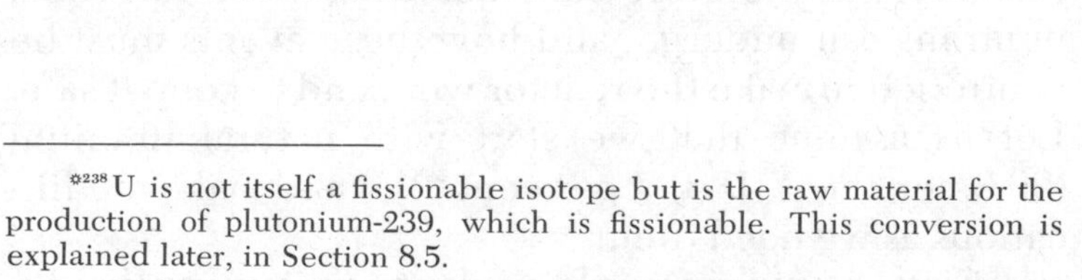

*^{238}U is not itself a fissionable isotope but is the raw material for the production of plutonium-239, which is fissionable. This conversion is explained later, in Section 8.5.

8.5 NUCLEAR FISSION REACTORS

(Problems 4, 5, 6)

In this section we will describe the fundamental features of nuclear reactors. Our purpose is to provide a conceptual framework that will make it easier for you to confront the environmental problems that are associated with nuclear energy, and to make evaluations concerning the social and economic issues that arise. Therefore, we will place special emphasis on such features as safety systems and the discharges of wastes to the environment.

Nuclear fission reactors require fuel, and the fuel must be a substance whose nuclei can undergo fission. There are two significant nuclear fission fuels, uranium-235 (^{235}U) and plutonium-239 (^{239}Pu). These are not the only known fissionable isotopes, but they are the ones that are available in sufficient quantity to be used in nuclear power plants.

^{235}U occurs in nature; it constitutes 0.7 per cent of natural uranium. The remaining 99.3 per cent is the heavier isotope, ^{238}U, only very little of which undergoes fission in the reactor. The predominant fission reaction may be written:

$$^{235}U + 1 \text{ neutron} \rightarrow \text{fission products} + 2 \text{ to } 3 \text{ neutrons} + \text{energy} \quad (1)$$

The second fuel, ^{239}Pu, does not occur in nature, but it is produced by the reaction of ^{238}U with neutrons. The reactions may be written as follows:

$$^{238}U + 1 \text{ neutron} \rightarrow \text{plutonium-239, or } ^{239}Pu \quad (2)$$

$$^{239}Pu + 1 \text{ neutron} \rightarrow \text{fission products} + 2 \text{ to } 3 \text{ neutrons} + \text{energy} \quad (3)$$

Thus, the two important naturally occurring sources of fission energy are ^{235}U (fissionable but not abundant) and ^{238}U (abundant but not fissionable until it is converted to ^{239}Pu).

The chain reaction is initiated by an artificial neutron source.

It is apparent from the fission equations that nuclear reactors require another essential ingredient besides fuel, namely, neutrons. In fact, it is the behavior of neutrons, more than any other single factor, that determines the design of a nuclear power plant. Let us therefore consider what possible events neutrons can undergo, and how these events must be controlled to make the reactor work and to keep it safe. Let us assume that we start with natural uranium (99.3 per cent ^{238}U, 0.7 per cent ^{235}U) and make modifications as we need them.

First, a neutron could undergo *fission capture* by

^{235}U (Equation 1). This would be fine; it gives us more neutrons to branch the chain, and the reaction yields energy. But there are some problems. The ^{235}U nuclei capture only *slow* neutrons to undergo fission, but release *fast* neutrons. Some medium is therefore needed to slow down the emitted neutrons to make them more amenable to capture. Such a medium is called a neutron **moderator.** The first material used was graphite; water is also suitable. Another problem arises from the meager abundance of ^{235}U in natural uranium. The rate of fissions, and hence the production of energy, can be speeded up by increasing the proportion of ^{235}U in the fuel. Such enrichment must be limited, however, because the chain reaction must not be allowed to branch out of control.

Second, a neutron can undergo *non-fission capture by ^{238}U,* thereby producing ^{239}Pu (Equation 2), which can in turn undergo fission (Equation 3). This capture occurs only with fast neutrons, not slow ones, and will therefore be impeded by moderators used to slow down neutrons for ^{235}U fission capture. On the other hand, the production of ^{239}Pu is, in effect, a "breeding" of new fuel, and, therefore, is very attractive as a source of energy for man's needs. These circumstances imply that the design of the reactor will depend to a great extent on the process we wish to favor.

Third, a neutron can undergo *non-fission capture by impurities.* This causes loss of neutrons and dampening of the chain, but it also offers us a convenient means of controlling the reaction. Fission products accumulate as impurities (Equations 1 and 3), and because they soak up neutrons, the fuel elements that they contaminate must eventually be removed and purified. But any time we wish to reduce the neutron flux, we may insert a stick of neutron-absorbing impurity. The more impurity we push in, the more we slow down the reaction. Devices which do this are called **control rods**; they usually contain cobalt or boron and other metals, and they can be inserted into or withdrawn from the reactor core so as to regulate the neutron flux with great precision.

Dampen means to inhibit by the application of moisture. The usage is appropriate because a spreading fire is a branching chain reaction, and water slows down a fire by terminating many of the reaction branches.

Finally, a neutron, traveling as it does in a straight line, may simply miss the other nuclei in the reactor and *escape.* Obviously, the larger the reactor, the more atoms will stand in the way, the greater will be the chance of neutron capture and the less the chance of escape. This circumstance imposes lower limits on reactor size—we shall never have pocket-sized fission generators, nor even fission engines for motorcycles. This escaping tendency also demands adequate shielding to prevent neutron leakage.

Now, how do we construct the reactor so as to satisfy all these requirements simultaneously? We must decide first whether we wish to produce energy from ^{235}U or whether we also wish to "breed" ^{239}Pu. Let us start with the non-breeder.

Non-Breeders

Refer to Figure 8.2 as you read the following description. We need a reacting core of fuel, moderator, and control rods. Since considerable heat is being produced, we also need to circulate some medium to carry the heat away. The fuel is typically a ceramic form of uranium dioxide. This compound is much better able than the pure metal to retain most fission products, even when overheated. The uranium fuel itself is prepared by conventional (non-nuclear) chemical treatment and consists of the natural non-fissionable isotope, ^{238}U, enriched with the fissionable ^{235}U by a factor only three or four times above its naturally occurring level. This low level of enrichment provides an automatic protection against an increased fission rate if the temperature should rise accidentally, such as by failure of the coolant system. The fuel is inserted into the reactor core in the form of long, thin cartridges (see Fig. 8.3). These fuel cartridges are coated (or "clad," as they say in the nuclear power plants) with stainless steel or other alloys.

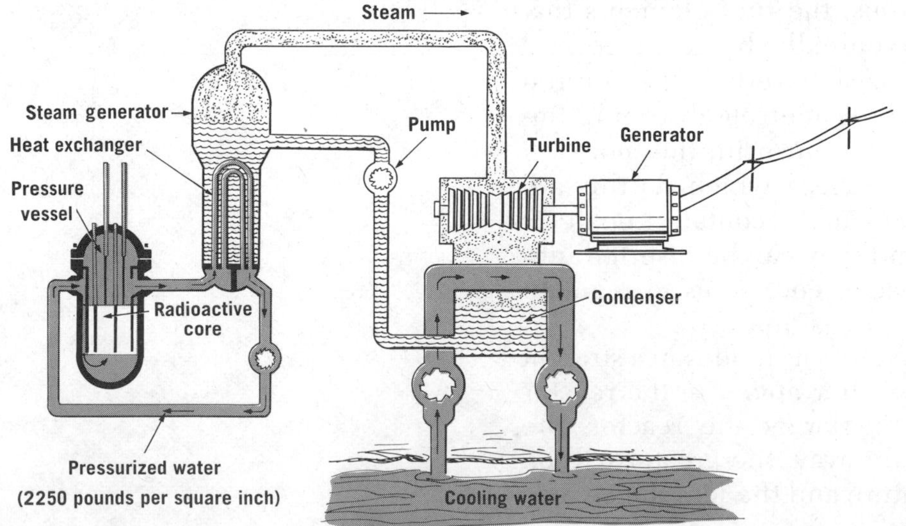

FIGURE 8.2 Schematic illustration of a nuclear power plant.

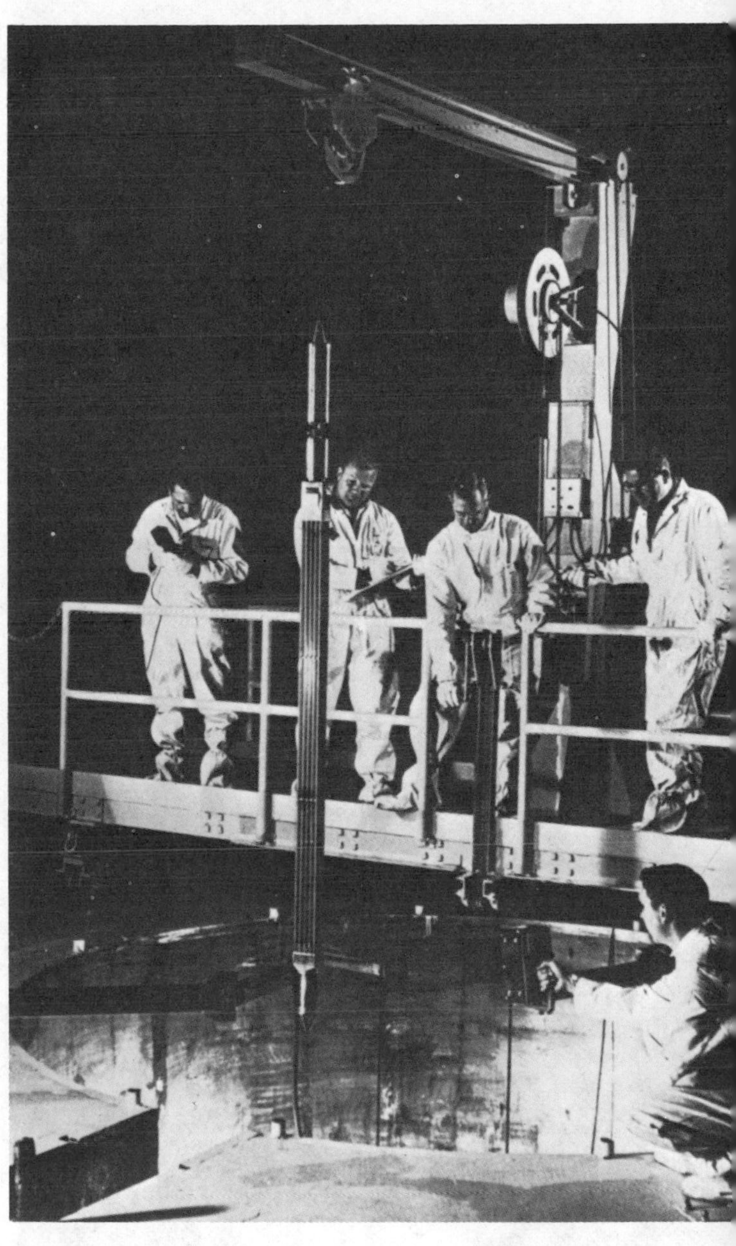

FIGURE 8.3 A typical fuel assembly, consisting of many fuel rods, being lowered into place in a reactor. (From D. R. Inglis: *Nuclear Energy: Its Physics and Its Social Challenge.* **Reading, Mass., Addison-Wesley Publishing Company, 1973, p. 91.)**

The fuel elements are surrounded by the moderator, whose function, remember, is to slow down the neutrons so that they will undergo fission-capture. Water is advantageous because it is both a moderator and a coolant. In the boiling water reactor, the fission heat converts the water to steam. In the pressurized-water reactor, shown in Fig. 8.2, the water is maintained in the liquid state under high pressure. It then releases its energy to another body of water in a heat exchanger. The mode of generation of electricity is the same as in fossil fuel plants. The steam drives a turbine that powers an electric generator; the waste steam is then cooled and returned to the heat exchanger.

Interspersed into the matrix of fuel cartridges, moderator, and coolant are the control rods. They serve not only to regulate neutron flux but also as an emergency shut-off system (see Fig. 8.4). To speed up the chain reaction, the rods are partially withdrawn; to slow it down, they are inserted more deeply. In the event of malfunction, the rods are pushed rapidly all the way into the core to capture as many neutrons as possible and quench the chain reaction.

Now, what about safety precautions? The nuclear

FIGURE 8.4 Control-rod driving mechanism of a pressurized water reactor (PWR), partially disassembled. (From Inglis, *ibid.*, p. 93.)

FIGURE 8.5 The Yankee nuclear power station at Rowe, Massachusetts. (From Inglis, *ibid.*, p. 100.)

power industry considers that its procedures embody a triple layer of defense that makes any serious accident almost unthinkable.

The first line of defense is the incorporation of safety factors into the basic design, construction, and operation of the reactor. These are the design features we have just described, including the modest enrichment of the uranium fuel, and the control rods. A modern nuclear reactor (cf. Fig. 8.5) is therefore nothing like an atomic bomb, not even potentially so, even if all safety systems should fail. The use of ordinary water as both the coolant and neutron moderator also provides an automatic safeguard: an unexpected rise in power level would boil away some of the water, and this loss is used as a signal to initiate a reactor shutdown sequence. This description of the "first line of defense" is far from exhaustive; it is intended only to illustrate that safety must initially be inherent in the process itself.

The second line of defense is based on the assumption that the first line, contrary to all expectations, may somehow fail. It is rather like having a well-designed electrical system in a hospital to provide power for essential services such as operating

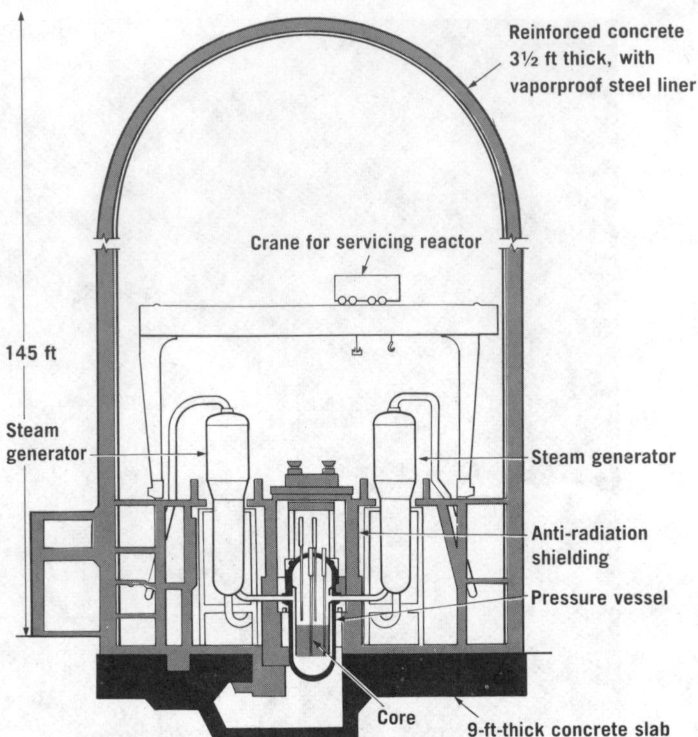

FIGURE 8.6 Containment structure for a nuclear power plant. (© 1971 by The New York Times Company. Reprinted by permission.)

rooms, and, in addition, having a standby gasoline-powered generator in case the primary system fails. Thus, if the mechanisms that cool the reactor core should fail, at least *two* other independent cooling systems are available. If the power system on which the emergency measures depend should fail, an off-site source of power can be used. If *that* fails, on-site diesel generators or gas turbines can take over. Secondary systems of this type are quite complex and are interrelated in such a way that their responses are specifically appropriate to the nature of the emergency. Furthermore, these responses are fully automatic; they do not have to be initiated by a human operator.

The third line of defense assumes that both the first and second lines may fail. This final barrier to radioactive contamination of the environment is a massive containment structure (Fig. 8.6) which is a thick, vapor-proof, reinforced concrete housing that shields the reactor and steam generators. This barrier is designed to withstand earthquakes and hurricanes and to contain all matter that might be released inside, even if the biggest primary piping system in the reactor were to shatter instantaneously.

Breeders

Return now to page 300 for another look at Equations 2 and 3. This process produces fissionable ^{239}Pu from abundant ^{238}U, and, therefore, makes it possible for one reactor to provide fuel for another. (Hence the reactors "breed.") The ^{238}U is called a "fertile" material, to keep up the biological metaphor. The time required for a breeder to double its quantity of fissionable material depends on its design and operation; the values obtained by theoretical calculation range from 6 to 20 years. A doubling time about midway between these extremes is considered by many to be a reasonable expectation.

The $^{238}U \rightarrow {}^{239}Pu$ reaction requires fast neutrons. This means that a moderator, which slows down neutrons, must be excluded. The reactor core, then, consists of ^{238}U that is highly enriched with ^{235}U or ^{239}Pu (the sources of the fast neutrons), and no moderator. The space vacated by the moderator can accommodate additional enriched fuel; thus, the overall concentrations of *both* the fertile and fissionable materials in the breeder core are much greater than they are in a non-breeder. This situation is inherently more dangerous because energy is released in a more concentrated form. Thus, in case of a malfunction, there is greater danger of overheating and melting of the core, which would concentrate the fuel still further and release radioactive products still more rapidly.

The compactness of the breeder core demands a very rapid removal of heat; water is disadvantageous because it boils at relatively low temperatures even under high pressures, and steam is a poor heat conductor. The coolant of choice is liquid sodium. So-

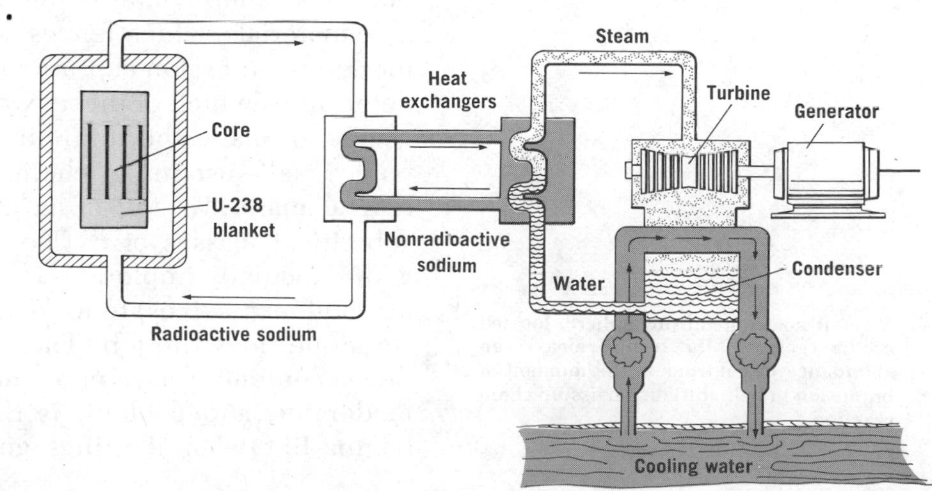

FIGURE 8.7 Schematic diagram of a fast breeder reactor.

dium is a silvery, soft, chemically active metal. It reacts with water to produce hydrogen gas; if air is present, the heat of the reaction can spark the explosion of the hydrogen. The sodium becomes highly radioactive when exposed to the reactor core. But its saving virtue is its ability to carry heat away from the reactor core rapidly since it is an excellent heat conductor and it remains in the liquid state over a very wide temperature range, from 98°C to 890°C at normal atmospheric pressure.

The heat exchanger in which steam is produced to drive the turbine must be shielded from the radioactive sodium. This is accomplished by an intermediate loop of non-radioactive sodium. The entire arrangement is shown schematically in Figure 8.7.

8.6 ENVIRONMENTAL HAZARDS FROM FISSION REACTORS

(Problems 7, 8, 9, 10, 11, 23)

Four sets of questions must be discussed in assessing the environmental hazards associated with nuclear fission plants: (1) What are the chances of a serious accident? (2) What are the extent and the environmental effect of the routine emissions of radioactive materials under normal conditions of operations? (3) What are the problems associated with the disposal of radioactive wastes? (4) What are the environmental effects of the waste heat released from the nuclear plants? The last of these effects, which is thermal pollution, was dealt with in Chapter 5. Here we consider the first three.

Chances of a Serious Accident

To begin with, we should understand that a nuclear fission plant is not designed like an atomic bomb. A bomb contains highly concentrated fissionable materials, which leaves only two significant fates for neutrons: fission capture or escape. The factor that determines which of these two fates will predominate is size, or mass; the minimum mass required to support a self-sustaining chain reaction is called the **critical mass.** To set off an atomic bomb, therefore, subcritical masses of ^{235}U or ^{239}Pu are slammed together (actually imploded by shaped charges of chemical high explosives) to make a supercritical mass; this act alone does the job. The nuclear reactor, with its dilute content of fissionable material, its control rods, moderator, and coolant, is therefore nothing like a bomb. But what if things go wrong? For example,

A small neutron-emitting sphere, located in the center of the bomb, releases an abundance of neutrons at the moment of implosion to help initiate the fission chain reaction.

what if an operator in the plant went mad and attacked all the pushbuttons on the control panels, punching them at random? Or worse yet, what if he deliberately tried to sabotage the operation? Or what if there were a severe earthquake on the premises, or if the facility were hit by a bomb or a meteorite? Any one of these highly improbable events would certainly make a mess, and in the worst instances there could be a severe release of radioactive matter to the atmosphere or to the cooling water effluent. But none of these events would have the effect of *concentrating* the fissionable matter in the fuel, and since that is the only way to make a bomb, there could be no true nuclear explosion.

The situation in a "breeder," however, is potentially more dangerous. The breeder core has no moderator to temper the neutrons' energies. The fuel is more compactly spaced, and it is richer in fissionable isotopes and, therefore, is closer to the critical mass than it is in the non-breeder. Furthermore, the short neutron lifetimes make everything happen faster in the breeder, and this means that there is less time for control mechanisms to take effect. To add to these concerns, the liquid sodium imposes another set of hazards, because if it escaped and came into contact with water, the resulting hydrogen gas could explode in air, and the force of the explosion might breach the containment vessel.

There is no limit to the levels of improbabilities we can imagine: for example, what if all the events we cited happened *at the same time*? It is unproductive to pursue these speculations indefinitely. What we must do instead is to assess the risks as well as we can, to estimate the benefits, and to weigh one against the other according to our best judgments.

Accidents have actually occurred. One notable instance in 1966 involved the breeder reactor of the Enrico Fermi Power Plant near Monroe, Michigan, not far from Detroit. Two pieces of sheet metal in the liquid sodium cooling system tore loose and blocked the flow of coolant. This made the fuel elements overheat and melt, and the rising radiation level inside the reactor building set off the alarms. The operators (*not* the automatic safety devices) inserted the control rods and thus shut down the reactor. It took four years to start it up again. This incident has been cited on both sides of the nuclear safety controversy. The opponents

Enrico Fermi (1901–1954) was an Italian physicist who was awarded the Nobel Prize in 1938 for his work on neutron bombardment of uranium. He emigrated to the United States in the same year. Fermi was in charge of the construction of the first chain-reacting structure, or "nuclear pile," which went into operation on Dec. 2, 1942 in a squash court of the University of Chicago.

of nuclear power have emphasized the inherent uncertainty of even the most advanced safety procedures, and have pointed out the threat of major catastrophe to the city of Detroit. On the other side, the proponents of nuclear power have emphasized that the final lines of defense *did* work, because no radioactive cloud was in fact released.

In all, four nuclear reactor accidents have occurred, none catastrophic, and none that involved the nonbreeding water-moderated reactors now in commercial use. There had been a total of about 100 reactor-years of operation up to 1971. No one can convincingly predict the probability of serious nuclear reactor accidents in the future—estimates range from as high as about one in 1000 reactor-years to as low as one in 100,000. And such an accident, some experts say, would kill thousands of people; others project "only" 100 or so.

"Routine" Radioactive Emissions

Even with the best design and with accident-free operation, some radioactivity is routinely released to the air and water outside the plant. In the boiling-water reactor, for example, the water passes directly through the reactor core and thus circulates around the fuel elements. Some of the fuel claddings, which are very thin (about 0.02 inch), inevitably develop little leaks, which permit direct transfer of radioactive fission products to the water. It is not practical to stop operations for a little leak here and there among many fuel rods; in fact, the proportion of faulty rods is usually permitted to reach one per cent before the plant is shut down. Even in the absence of leaks, some neutrons do get through to the water and make some of its impurities radioactive, and this effect, too, is a source of "routine" emissions to the watercourses that serve as the ultimate coolants for the power plant.

Radioactive material can also be gaseous. Krypton-85, for example, is a radioactive fission product (half-life, nine years) that is insoluble in water and escapes to the atmosphere through a tall stack. These emissions, taken together, are so small that their effect may be compared with the background radiation from cosmic rays and naturally radioactive materials in the earth's crust to which all life is subject. Are we willing to accept any radiation in addition to the natural dose? This problem must be approached in terms of weighing health hazards against the benefits of having abundant electrical energy. The health aspects of the question are explored in Section 8.8.

The Disposal of Radioactive Wastes

When enough impurities have accumulated in the fuel cartridges (usually after about a year of operation), the absorption of neutrons by the fission products slows down the chain reaction and the fuel cartridges must be removed. It is here that the wastes are most concentrated, most radioactive, and therefore most dangerous. The spent fuel must be reprocessed, the uranium and plutonium recovered, and the wastes properly disposed. The disposal method involves a complex series of operations. The ultimate objectives are to concentrate the wastes as much as possible in order to save space and to store them where they will do no harm. Of course, the scheduling of the entire procedure must be responsive to the time sequence of radioactive decay. The wastes contain many radioactive isotopes with widely differing half-lives, and decays often occur in a series of steps before a stable isotope is produced, but for simplicity let us imagine a single waste substance with a half-life of one month. Figure 8.8 shows the shape of the decay curve. (Note that the radioactivity is decreased by half each month but that the curve is smooth; it does not go down in steps.) For the first two months or so, the material is highly radioactive. As time goes on, the radioactivity decreases to low levels, but the *rate* of decrease be-

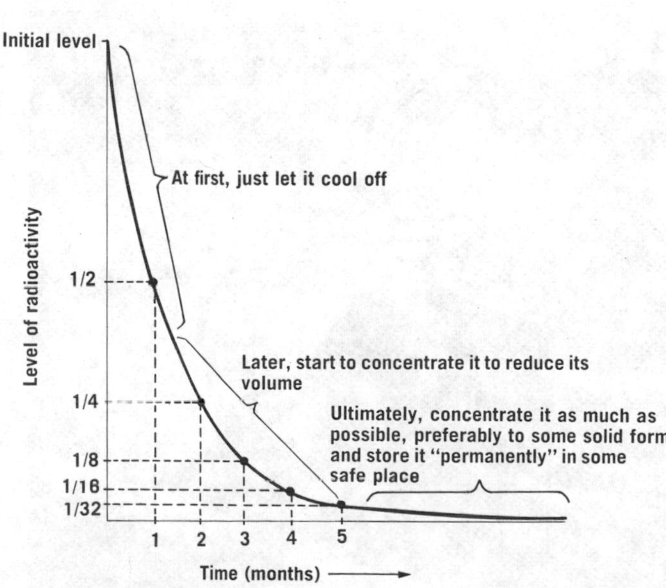

FIGURE 8.8 Disposal of a hypothetical radioactive waste product with a one-month half-life.

comes slower and slower, and the curve never reaches zero. This decay pattern implies that the kind of treatment that is best applied to the waste will change as time goes on.

Such, indeed, is the case. The spent fuel is first stored under water at a site on the plant premises for a few months to allow the most highly radioactive components to decay. The partially decayed fuel is then shipped to a waste-processing plant. It is here that the valuable uranium and plutonium must be recovered and separated from the impurities. The first step is to dissolve the solid fuel in a concentrated solution of strong acids.

Some of the radioactive wastes in these solutions have half-lives measured in tens of years; these are the most troublesome because they continue to be highly dangerous for a long time (see Section 8.8). If the fission power industry continues to operate, these wastes will continue to be produced, and will therefore always be with us. There are several things we can do with them, but the choices are not appealing. First, we can put them in special tanks (Fig. 8.9) that are as well protected from leakage as we can make them, and preferably are stored underground. But these tanks are handling no ordinary solution; their contents are corrosive, hot, and radioactive. Twenty years is about the best they can do; after that, the tanks begin to succumb to mechanical and chemical stresses

FIGURE 8.9 A "tank farm" under construction, showing large tanks to hold high-level liquid wastes. (From Inglis, *ibid.*, p. 142.)

FIGURE 8.10 "Fission's graveyard."

and they start to leak. Of course, we can then pump the contents into new tanks, and continue the game of "musical tanks" indefinitely. It is obvious that this is very far from an ideal procedure.

A second possibility is to concentrate the wastes until they are solid (Fig. 8.10) and then to put them somewhere. Where? Disposal at sea was formerly used extensively, until it was banned by international convention. The most appealing location has been considered to be an abandoned salt mine or salt cavern (Fig. 8.11). These geologic formations are attractive because some of them have been stable for millions of years and have not been in contact with subsurface waters. The U.S. Atomic Energy Commission seriously considered a salt mine under the town of Lyons, Kansas (population about 5000). Lyons sits over two salt mines, one abandoned, the other in daily operation. Their main tunnels come to within 1800 feet of meeting each other. Some of the previously mined cavities were left filled with water. These circumstances raised such serious doubts about the ability of the abandoned mine to maintain its dry integrity for "millions of years" that, in 1972, the Atomic

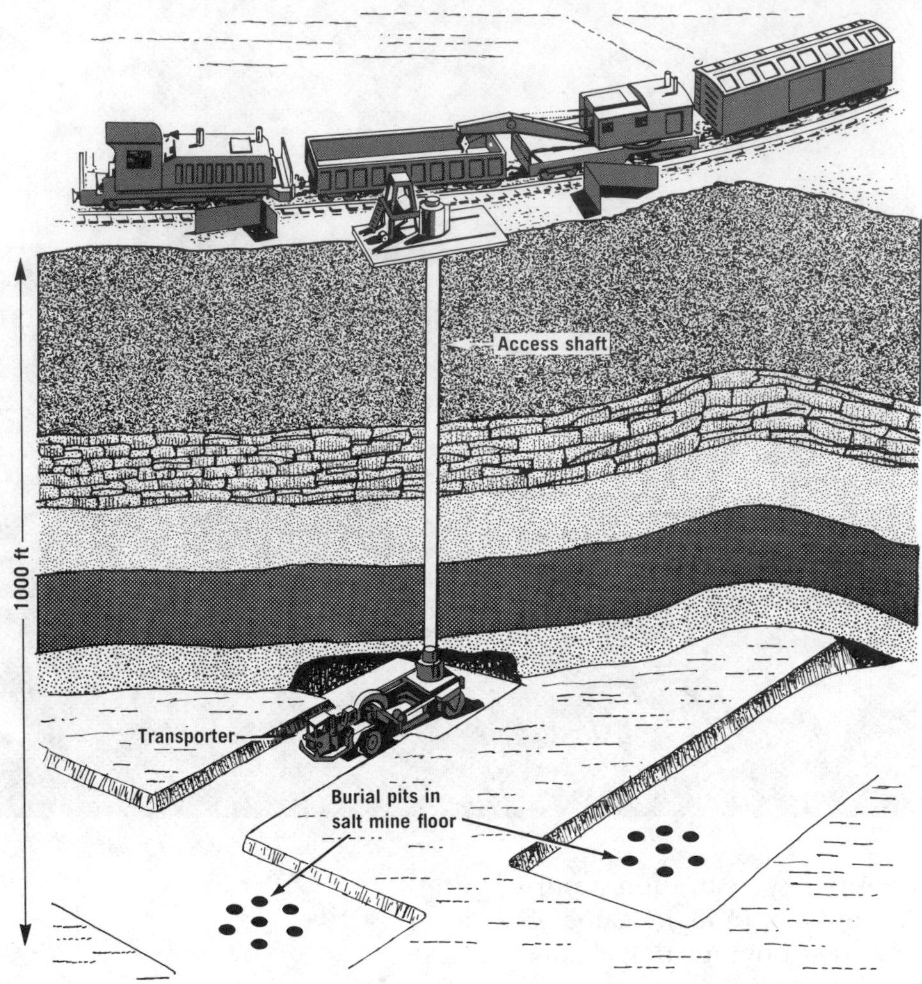

FIGURE 8.11 Final disposal of radioactive wastes in a salt mine. (© 1971 by The New York Times Company. Reprinted by permission.)

Energy Commission abandoned the Lyons site. The general approach, however, is still considered reasonable, and the search for suitable sites continues.

One recent suggestion has been to bury the wastes under the Antarctic ice cap. The method is appealing. If hard radioactive masses were placed on the surface of the ice, they would melt their own way down to bedrock. The hole would then rapidly reseal itself by freezing. It is estimated that the average Antarctic temperature has remained below freezing for over one million years, so that the radioactive wastes may be expected to be securely isolated from the biosphere for a similar period into the future. Critics point out, however, that once the canisters reach bedrock, they will continue to give off heat and will melt the ice around them. The resulting water, contaminated with radioactive matter, may flow out from beneath the ice sheet into the open ocean. Moreover, the entire ice sheet may eventually be able to slide over its warm, wet bed (pulverizing the canisters along the way) and eventually raise the sea level by five meters or so. Proponents of the method think these dangers are remote, but all agree that a very careful environmental impact study is required.

A possible alternate procedure is to make up a kind of cement that incorporates the radioactive waste matter. This "hot cement" is then injected into underground seams of geologically stable rock, where it hardens and becomes part of the solid structure of the earth's crust.

But our roster of anxieties is not complete. There is the problem of transportation of wastes from reactor to processing plant, and from processing plant to disposal site. Trains can be derailed and trucks can crash and overturn. In an attempt to counteract the effect of such possible disasters, the wastes are shipped in strong, heavy, radiation-shielded caskets (Fig. 8.12). How good has the record been? Very good indeed; since World War II there has never been a recorded major accident in which the radioactive contents of a waste shipment has killed or injured anyone. But there have been some hair-raisers (see Fig. 8.13). For example, on March 31, 1971 at 9:35 P.M., a trainmaster for the Burlington Northern Railroad in Missoula, Montana smelled smoke from a freight train that had just pulled in from the nuclear facility in Hanford, Washington. Investigating, he found a smoking box car labeled "Danger: Radioactive Material." The car contained crude uranium scraps. The fire was ultimately extinguished, and no one was harmed, but it is this sort of event that makes many experts believe that a shipping accident *could* happen. Of course, as with

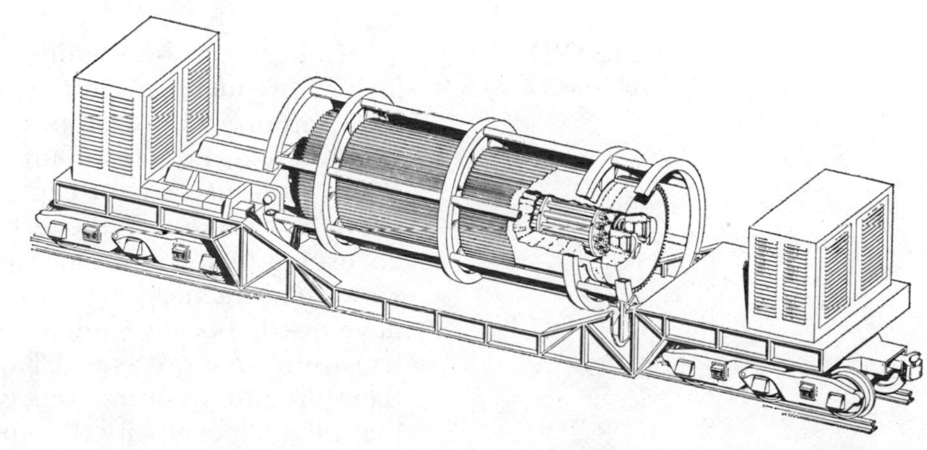

FIGURE 8.12 Liquid metal fast breeder reactor spent-fuel shipping cask with 18 assemblies (From Science 177:31, July 7, 1972. Copyright 1972 by the American Association for the Advancement of Science.)

FIGURE 8.13 Railroad accident showing a cylinder of uranium hexafluoride that fell off a flatcar in Wingate, North Carolina, on April 29, 1971. The cylinder was not damaged. (From Science 172:1319, June 25, 1971. Copyright 1971 by the American Association for the Advancement of Science.)

the possibility of accidents in the power plants themselves, we can chase the spiral of smaller and smaller chances of greater and greater catastrophes to the never-never land of statistical indeterminacy.

Finally, there is the matter of loss or theft of radioactive materials. Uranium and plutonium inventories are not always fully accounted for, although the discrepancies are more likely to be in the books than in the atoms. Shipments are delivered to the wrong city, sometimes even the wrong country. These circumstances are annoying, perhaps even disconcerting, but so far no atomic criminal has set up a mechanism for nuclear blackmail.

8.7 NUCLEAR FUSION

(Problems 12, 13, 14)

Energy can be obtained by the combination, or **fusion**, of nuclei of certain light elements, particularly certain of the isotopes of hydrogen. The fusion reaction does not create any environmental problems at this time (except for explosions of the "hydrogen" bomb in warfare or in the testing of weapons), because no useful fusion reactors have yet been developed. However, practically everyone hopes that they will be developed, because nuclear fusion promises "inexhaustible" energy (see Chapter 5). It is important, therefore, to examine the environmental problems that might accompany this source of energy.

In our consideration of fission we learned that the nuclei of some heavy elements break down because they are unstable; that is, they are less stable than elements of medium atomic weight. There is instability at the other end of the scale, too, for the very light elements are also less stable than elements of medium atomic weight. The very light elements, therefore, can release energy when their nuclei coalesce, or fuse, to produce heavier nuclei. By far the most promising of these reactions—the only ones, in fact, that merit serious considerations—are the fusion of hydrogen isotopes to produce isotopes of helium. There are three isotopes of hydrogen:

Name of isotope	Name of nucleus	Symbol*	Atomic Number	Mass Number	Natural Abundance
Protium, also called "light hydrogen," or "ordinary hydrogen," or, usually, just "hydrogen."	proton	${}^1_1H$	1	1	99.985%
Deuterium, sometimes called "heavy hydrogen."	deuteron	${}^2_1D$, or ${}^2_1H$	1	2	0.015%
Tritium (radioactive; 12-year half-life)	triton	${}^3_1T$, or ${}^3_1H$	1	3	none

*The subscripts designate the atomic number.

The nuclear equations (at right) represent different fusion reactions of hydrogen isotopes. All of these reactions release energy, and the total potential supply of the raw materials is so great that if the earth's population rose to 30 billion, and if everyone used as much energy as the average American does today, then the ${}^2_1D$ in the oceans could provide energy for about a billion years. This statement, by itself, does not describe the kind of world that such an abundance would create, but it does imply the likelihood of drastic social and environmental change.

$$
\begin{aligned}
{}^1_1H + {}^2_1D &\longrightarrow {}^3_2He \\
{}^2_1D + {}^2_1D &\longrightarrow {}^3_1T + {}^1_1H \\
{}^2_1D + {}^2_1D &\longrightarrow {}^3_2He + {}^1_0n \\
{}^2_1D + {}^3_1T &\longrightarrow {}^4_2He + {}^1_0n
\end{aligned}
$$

The symbol ${}^1_0n$ refers to a neutron. The subscript 0 means it has no charge, and the superscript 1 means its mass number is 1.

Unlike the fission reaction, however, fusion cannot be triggered by neutrons; what is required instead is that the fusing nuclei be fired at each other at very high speeds so as to overcome their initial repulsions. In other words, the temperature must be very high; the resulting fusion is therefore called a **thermonuclear reaction.** If a large mass of hydrogen isotopes fuse in a very short time, the reaction cannot be contained and it goes out of control; this is the explosion of the "hydrogen bomb." If, instead, we wish to ex-

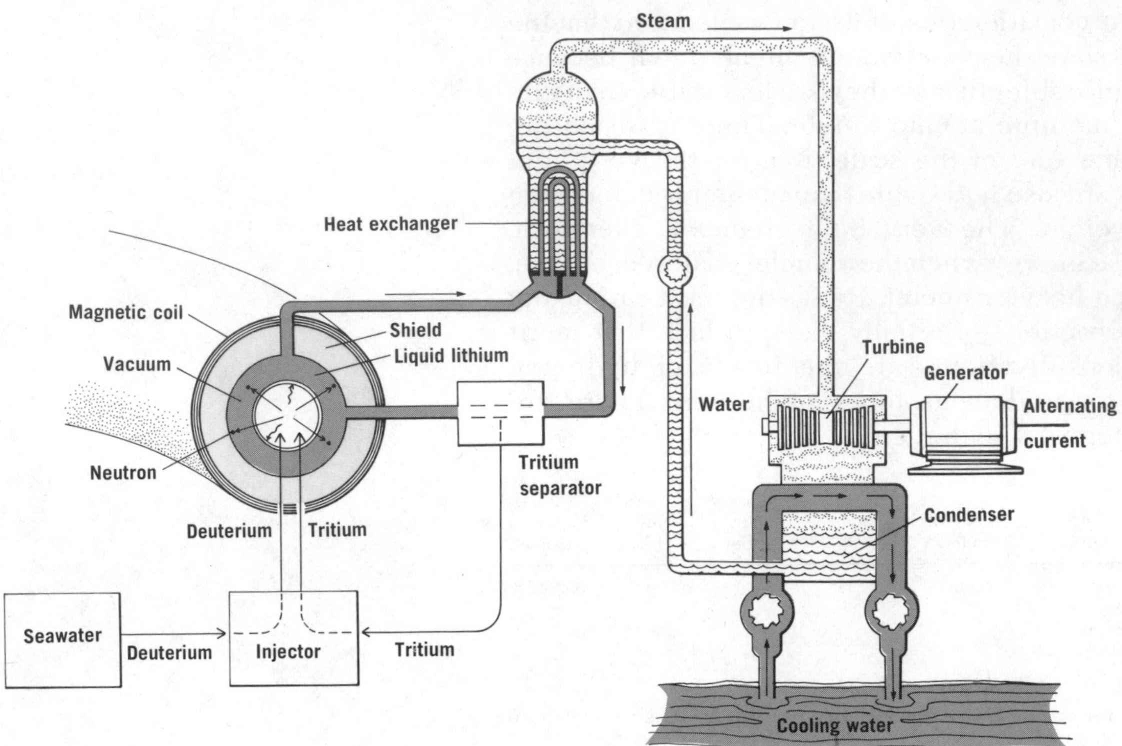

FIGURE 8.14 Schematic drawing of a thermonuclear power plant.

tract useful energy from fusion, we must devise what is called "controlled thermonuclear reaction," and it is this problem that we are concerned with here (see Fig. 8.14). The required temperatures are about 40 million degrees Celsius for the D-T reaction and 400 million degrees for the D-D reaction. Major problems are imposed by these requirements. Of course, the problems are much more severe at the higher temperature needed for the D-D reaction—so much more severe, in fact, that our present efforts to develop fusion reactors are directed toward the "cooler" D-T reaction. But then we need a source of tritium, which is not naturally available on earth, and this requirement, too, poses problems.

Thermonuclear reactions are extremely difficult to control because no materials can withstand the required temperatures. Instead, all substances decompose into their free atoms, and the atoms themselves decompose into a mixture of positive nuclei and free electrons that is called a **plasma.** No rigid container exists that can survive long enough to confine a plasma for the useful production of thermonuclear energy. Instead, what is envisaged is a sort of "magnetic bottle," which does not consist of a physical substance at all, but rather is a magnetic field so designed that it will confine the charged particles of the plasma in which the thermonuclear reaction is going on. The useful energy will have to be extracted from the process in the form of the kinetic energy of the evolved neutrons. Since the neutrons carry no charge, they will penetrate the magnetic field and escape from the plasma. The energy of the speeding neutrons can then be extracted by a moderator, just as in a fission reactor. If the moderator is water, the energy will create steam which can drive a turbine.

We come now to the tritium problem. Remember that the D-T reaction is cooler than the D-D reaction, and hence preferable to it, but tritium is not naturally available. However, tritium is produced by the reaction between neutrons and the light isotope of the metal lithium, ${}^{6}_{3}Li$, which constitutes 7.6 per cent of natural lithium. To exploit this process, the entire fusion reactor would be encased in a sheath or blanket in which molten lithium is continuously circulated. The lithium would absorb the neutrons, supply the tritium and then release its heat to water in a heat exchanger.

$${}^{6}_{3}Li + {}^{1}_{0}n \longrightarrow {}^{3}_{1}T + {}^{4}_{2}He$$

It should not be surprising that all these requirements impose engineering problems so difficult that they have not yet been solved (see Fig. 8.14). But if they were solved, what kind of environmental problems would be created?

Explosion

First, could the fusion reactor get out of control and go off like a hydrogen bomb? We are entirely confident that the answer is no, an explosion could not occur. The reason is that the hydrogen isotopes are continuously fed into the reactor and are continuously consumed; they do not accumulate. The total quantity of fuel in the plasma at any one time would be very small—about two grams or so—very far below the critical mass required for a runaway reaction. If the temperature were to drop, or the plasma somehow dispersed itself, the reaction would stop; in effect, the fusion would turn itself off. The situation is rather analogous to that of a burning candle; if something goes wrong, the flame goes out, the candle does not explode.

Radioactive Wastes

The primary potential problems would be the release of tritium or of neutrons. Tritium could be a problem because it is radioactive (12-year half-life) and is chemically reactive just as ordinary hydrogen is. Therefore, it can combine with oxygen to form water. (See equations in margin.)

$2H_2 + O_2 \rightarrow 2H_2O$ (ordinary water)
$2T_2 + O_2 \rightarrow 2T_2O$ (radioactive water)

The tritiated water could conceivably enter food webs and harm living organisms. However, the energy released by tritium (in the form of beta particles or electrons) is so weak that it would be virtually harmless if its source were outside the body. Furthermore, sound engineering practice should be able to restrict any tritium emissions to inconsequential levels.

Neutron release is another potential hazard, but we must remember that the neutrons stop when the reaction stops. However, neutrons are absorbed by atomic nuclei, and the new atoms that are thereby produced are sometimes radioactive. Nonetheless, proper shielding should be able to prevent any significant releases to the environment.

Therefore, as far as the environmental risks of radioactive wastes are concerned, fusion would be a tremendous advance over fission. There would be no accumulation of long-lived residues to bury, and the biological hazard of tritium is about one-millionth that of the radioactive iodine from a fission reactor. However, this optimistic outlook does not mean that the environment would be unperturbed by the proliferation of fusion reactors. The most serious problem could well be that of thermal pollution, which was treated in Chapter 5.

8.8 BIOLOGICAL EFFECTS OF RADIATION

How Radiation is Delivered: External and Internal Irradiation

(Problems 15, 16, 17, 18, 19)

There are two ways in which an animal can be irradiated. The first is from an external source of radiation, for example an x-ray tube or a material like cobalt-60, which emits gamma rays. The second is via the ingestion or inhalation of radioactive materials; this is called internal irradiation since the sources of radiation are inside the body. It should be clear that this internal type poses the more important problem in connection with the health hazards of radioactive waste products. Whether or not a radioactive material poses a significant biological hazard depends on two quite independent properties, discussed below.

The Chemical Nature of the Substance

As explained earlier, this will determine whether the substance can enter the food chain and be taken up by animals, plants, and ultimately man. For example, ^{90}Sr, a common radioactive by-product of atmospheric nuclear weapons testing, is chemically similar to calcium. It is thus taken up by plants, where it is ingested by herbivores such as cows. Like calcium, it is then concentrated in the milk of the animals and from there it becomes a part of the human diet. In man, it becomes a structural part of bone along with ingested calcium. Therefore, as radioactive decay continues, the cells in the bone and the bone marrow become the prime targets of the radiation from this source.

Another significant fallout isotope is cesium-137 (half-life = 33 years), which is chemically similar to potassium. ^{137}Cs accumulates in muscle tissue, while ^{90}Sr accumulates in bone and milk. ^{137}Cs is therefore the more significant source of radioactivity in the diet of Eskimos, who eat caribou muscle.

Remember also that the food web has the ability not only to transport, but also to concentrate various materials. Concentration of radioisotopes in the food chain in the Arctic regions is particularly efficient. The effect of this concentration was noted when it was found that Eskimos absorbed more fallout radioactivity than did people who lived in other zones of the earth where more fallout actually occurred. The first step in this highly efficient concentration is provided by the Arctic lichen, a plant that gets its mineral nourishment directly from dust particles that settle on it. For this reason, the lichen collects fallout dust particularly efficiently. In summer, caribou migrate north to the tundra where they wander over large areas in search of lichen, which becomes an important part of their diet. The effect is as if someone sent out the caribou to collect and bring back the ^{90}Sr, and they accomplish this task very well. Then, of course, the Eskimos eat the caribou, sometimes as their only food, and so, at the top of the food chain, they get the most concentrated radiation.

The Half-Life of the Radionuclide

It should be clear that, even if the substance does get into the food chain in the manner described above, if the half-life is very short (for example, seconds or minutes), it will pose no hazard at all, since most of the material will have decayed away before entry into the food chain. Similarly, if the half-life is measured in millions of years, no significant amount of radiation will be generated during the lifetime of the animal. In our previous example, the half-life of

^{90}Sr is about 25 years. Thus, this isotope has the ability to be incorporated into living tissue and, provided enough is absorbed, may generate significant radiation.

Effects of Radiation on Living Cells

We have learned that nuclear reactions can release energy. In our emphasis on chain reactions, we have paid most attention to energy in the form of speeding neutrons. However, nuclear reactions also release energy in the form of x-rays and gamma rays, both of which are highly energetic varieties of electromagnetic radiation.

When x-rays or gamma rays are directed against a target material, an important result is the production of ions in the material. That is, the energy of the radiation separates electrons from the atoms with which they are associated, and these atoms are then left with a net electrical charge. The electrons may themselves escape from the material, or be recaptured by other ions, or collide with other atoms causing further ionizations. It is also possible, under certain circumstances, for x-rays to be produced when an electron interacts with an atom. To summarize, when radiation strikes a target material, ions, electrons, and x-rays can be produced.

These potent carriers of energy are dangerous to living cells. The effects are produced in any of the following ways: (1) The ionized atoms may be component parts of molecules critical to normal cellular functioning. If the molecule thus loses the ability to function normally, the cell may die. (2) The radiation may interact with the water in the cell, giving rise to highly reactive species of molecules that are called **radicals.** These may in turn interact with and destroy the function of critical intracellular molecules. (3) Interaction of the radiation with water may produce hydrogen peroxide (H_2O_2), which is a powerful oxidizing agent and may poison the system which generates energy for cellular metabolism.

$$H_2O + \text{oxygen} \xrightarrow[\text{energy}]{\text{radiant}} H_2O_2$$

$$H_2O_2 + \begin{matrix}\text{living}\\ \text{matter}\end{matrix} \rightarrow \begin{matrix}\text{oxidation}\\ \text{products}\end{matrix} + H_2O$$

In all these cases you will note that the ultimate effect on cells is due to a prior destruction of the function of molecules that are necessary to the cell. One such molecule is DNA, which contains all the genetic information that is required for the growth and maintenance of the cell (see Chapter 2). DNA is a sensitive target for radiation; when a cell is irradiated, the DNA strands tend to break into fragments. If the rate of delivery of the radiation is low, then the cell's repair mechanisms can seal the breaks in the strands, but above a certain dose rate, the repair process cannot keep up, and the DNA fragmentation becomes irreversible.

Cell types differ greatly in their sensitivity to radiation. In general, those cell types which divide most rapidly are the ones most easily killed by radiation. These include cells of the bone marrow, which make the red cells, white cells, and platelets of the blood, those that line the gastrointestinal tract and the hair follicles, and the sperm-producing cells of the testes. Conversely, muscle and nerve cells, which do not divide in the adult, are quite resistant even to large doses of radiation. The rule, however, is not invariable. Neither lymphocytes (the cells that control the immune response) nor the egg cells of the ovary divide under ordinary circumstances but both are exquisitely sensitive to killing by radiation.

What about non-lethal effects? You might expect, quite correctly, that anything as potent as radiation would exert some effect on cells, even if delivered in sub-lethal doses. It has been known for many years that radiation is a powerful inducer of mutations. Mutations are produced when the DNA is altered in some way. The alterations may sometimes be so gross that the chromosomes may look abnormal when viewed under the microscope. Other changes may be far more subtle and occur over only a minute length of the DNA (point mutations). The importance of mutations, whichever type occurs, is that (1) they may cause changes in the function of the genes in which they occur; and (2) they are passed on to all daughter cells. A simple example of a mutation might be the loss of the ability of a cell to transport a sugar such as glucose from the outside of the cell to the inside. A more complex example (but perhaps only because we do not yet understand the mechanism) is the change by which a normal cell becomes a cancer cell.

Effects on the Whole Animal

We now turn our attention to the effects of radiation not on single cells but on whole animals. It is convenient to divide these into *somatic* effects, or those which are confined to the population exposed to the radiation, and *genetic* effects, or those which are inherited by subsequent generations.

Early Somatic Effects: Radiation Sickness

On several occasions during the last 75 years, groups of people have been exposed to large doses of ionizing radiation over periods of time ranging from a few seconds to a few minutes. The holocausts at Hiroshima and Nagasaki, together with accidents at civilian nuclear installations, have provided much information about what radiation can do when a lot of it is administered to the entire body over a short period of time. Let us consider first the simplest and most drastic measure of radiation effect—death. Figure 8.15 shows the relation between the dose administered to a population of animals and the percent of the population remaining alive three weeks or more after the exposure. Up to a dose of about 250 rads (see Table 8.1 for definition of a rad), virtually everyone survives. When the dose is increased above this point, survival begins to drop sharply, and above a dose of 700 rads, everyone dies.

Does this mean that below doses of 250 rads there is no observable effect? Not at all. For even if the exposed individuals do not die, they may become quite ill. At doses between 100 and 250 rads most of the people will develop fatigue, nausea, vomiting, diarrhea, and some loss of hair within a few days of the exposure; the vast majority, however, will recover completely from the acute illness. For doses around 400 to 500 rads, however, the outlook is not so rosy. During the first few days, the illness is similar to that of the previous group. The symptoms may then go away almost completely for a time, but beginning about three weeks after the exposure, they will return

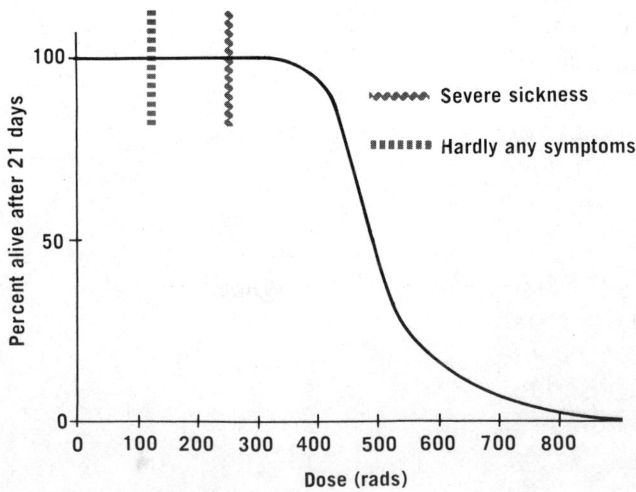

FIGURE 8.15 Curve showing the approximate relationship between the dose of radiation administered to a whole animal like a mouse or a man and the percent of the treated population which survives three weeks afterward. Mice have, of course, been intensively studied in the laboratory; accidents in industry and the nuclear explosions in Japan have provided the approximate data for man. (From American Scientist 57:206, 1969. Copyright 1969 by Sigma Xi National Science Honorary.)

TABLE 8.1 UNITS RELATED TO RADIOACTIVITY

UNIT	ABBREVIATION	DEFINITION AND APPLICATION
Disintegration per second	dps	A rate of radioactivity in which one nucleus disintegrates every second. The natural background radiation for a human body is about 2 to 3 dps. This does not include "fallout" from man-made sources such as atomic bombs.
Curie	Ci	Another measure of radioactivity. One Ci = 37 billion dps.
Microcurie	μCi	A millionth of a curie, or 37,000 dps.
Roentgen	R	A measure of the intensity of X-rays or gamma-rays, in terms of the energy of such radiation absorbed by a body. (One R delivers 84 ergs* of energy to 1 gram of air.) The roentgen may be considered a measure of the radioactive dose received by a body. The dose from natural radioactivity for a human being is 5 R during the first 30 years of life. A single dental X-ray gives about 1 R, a full mouth X-ray series, about 15 R.
Rad		Another measure of radiation dosage, equivalent to the absorption of 100 ergs per gram of biological tissue.
Rem		A measure of the effect on man of exposure to radiation. It takes into account both the radiation dosage and the potential for biological damage of the radiation. The damage potential is based on the following scale of factors: x-rays, gamma rays, electrons : 1 neutrons, protons, alpha particles: 10 high-speed heavy nuclei : 20 The rem is then defined by the relationship: Rems = Rads × Biological damage factor. Therefore, 100 ergs per gram (x-rays) = 1 rad × 1 = 1 rem, but 100 ergs per gram (neutrons) = 1 rad × 10 = 10 rems.

*See Appendix for a discussion of units of energy.

again. In addition, because the radiation has impaired bone marrow function, the number of white cells and platelets in the blood will fall. This is of great significance, since without white cells the body cannot fight infection and without platelets the blood will not clot. Looking at Figure 8.15 again, we can see that about 50 per cent of those exposed in this dose range will die, and of these, most will die of either infection or bleeding.

If, instead, the dose administered is about 2000 rads, the first weeks of the illness will again be the same as in the previous groups, but rather than waiting three weeks for symptoms to return, these people become very ill in the second week, with severe diarrhea, dehydration, and infection leading to death. At these dose levels the cells of the gastrointestinal tract are affected before the bone marrow toxicity has a chance to become severe, and these patients may die even before their blood counts have dropped to life-threatening levels.

At doses above 10,000 rads, animal experiments

have shown that death, which may occur within hours of administration of the dose, is due to injury of the brain and heart.

Delayed Somatic Effects

Of the late somatic effects of radiation (that is, those occurring months or years following the exposure), none is better studied or of more concern than the increased incidence of cancer in those with a history of prior exposure to radiation. Although the molecular mechanisms at work here are still largely obscure, the evidence that radiation does increase the incidence of cancer in exposed populations is overwhelming. Before the dangers of radiation were appreciated, early workers were careless in their handling of radioactive materials and suffered a greatly increased incidence of skin cancers. The famous case of the radium dial workers in the 1920's also deserves mention. These women were responsible for painting the dials of watches with the phosphorescent radium paint in use at that time and routinely tipped the end of the brush in their mouths before applying the paint to the dial face. In later years this group experienced a very high incidence of bone tumors. (Ingested radium, like strontium, is preferentially incorporated into bone.) The survivors of the atomic attacks at Hiroshima and Nagasaki, within a decade of the attacks, had far more leukemia than one would have expected from a group of this size, and subsequently the incidence of a variety of malignancies other than leukemia seems to be rising as well.

"I hope you weren't planning on a big family, Miss Whipple."

From: Industrial Research, September 1973.

Medical therapeutics itself also provides lessons. A painful disabling type of arthritis called ankylosing spondylitis affects the vertebrae of the neck and in the past was treated with irradiation, which was effective in relieving the symptoms. In the years following treatment, however, the incidence of leukemia in irradiated patients has significantly exceeded that in non-irradiated patients with this disease. As a final example in what could easily be an even longer list, evidence is accumulating that the children born to women who have had x-rays of their pelvis during pregnancy have a higher risk of developing leukemia than those whose mothers have not had such exposures.

Induction of malignancies is not the only late somatic effect. Irradiated animals also have a propensity for cataract formation in the lenses of the eye.

Also, for some reason, all irradiated species studied show a shortening of the lifespan. We shall discuss the implications of all these findings in a later paragraph.

Genetic Effects

We now consider those effects of radiation which do not manifest themselves in the irradiated individual himself but result in mutations in the genetic material of the **germ cells** (the sperm cells of the testis and the egg cells of the ovary) that are passed on to succeeding generations. In every experimental system studied in the laboratory, in organisms as diverse as viruses, bacteria, fruit flies, and mice, radiation has been shown to be a potent inducer of mutations. Though both ethical and practical considerations militate against genetic experiments with man, scientists have attempted to find evidence of an increased mutation rate in irradiated populations such as the atomic bomb survivors. These investigations have yielded only equivocal results. Nonetheless, there is no reason to believe that man should behave differently in this respect from every other well-studied species and, thus, scientists and policy-makers must assume that radiation is also mutagenic in man.

Linus C. Pauling (1901–) is an American chemist who was awarded the Nobel Prize in chemistry in 1954 for his work on molecular structure. Since World War II, Pauling has fought against the nuclear danger to humanity. He was awarded the 1963 Nobel Peace Prize for the effect that his efforts had in ending the testing of nuclear weapons by the major nuclear powers.

You may be wondering why we care so much about the mutation rate; after all, aren't mutations as likely to be beneficial as deleterious? The answer is certainly not. To understand why, we must recall that as species evolve in the course of time, those characteristics are preserved that best enable the individuals carrying them to compete successfully and survive. Such characteristics are usually produced by the complex interaction of many different genes. A random mutation would be far more likely to impair the system than to improve it.

Some Solutions

If, then, we wish to minimize the somatic and genetic effects of radiation, the task is clear—we must minimize unnecessary radiation exposure. To do this intelligently, we must first have some idea of the contribution of the various sources of radiation to man. By far the greatest source is natural background radiation; this amounts to about 0.125 rads to the gonads per person per year, and comes from sources in outer space, the earth's crust, and building materials. Now that atmospheric testing of nuclear weapons has been largely curtailed, radioactive fallout accounts for very little increase over the background. In the Western world, by far the largest addition to background comes from the diagnostic uses of medical x-rays; the best estimates are that, on the average, diagnostic x-ray studies increase the genetically significant radiation load to the population by about 50 per cent of background. ("Genetically significant radiation" is that which reaches the gonads of people who are still in the reproductive age group.)

The ICRP has no legal status, but its recommendations are generally adopted by countries within which radiation is used.

Since 1928 the International Commission on Radiological Protection, a group composed of scientists from many countries, has been promulgating radiation standards, usually in the form of maximum permissible doses for total body irradiation of members of a population. The general way in which this is done is that the results of animal experiments with high-dose radiation are extrapolated back to low doses to get an estimate of the likely effect of low-dose irradiation on whole animals or people. In doing so, the assumption is always made that there is no safe threshold; that is, there is no low-dose level of radiation below which radiation is completely harmless. Most, but not all, scientists agree that this "no threshold" assumption is valid; in any case, it is the safest assumption in the present state of our ignorance. Currently, for members of the general population (that is, people who do not work with radiation on a daily basis), the ICRP recommends a maximum dose of five rads per generation, which amounts to about 170 millirads a year. This is to include all radiation except natural background and medical sources. The ICRP has estimated that if the entire population of the United States were to be exposed to this maximum level, there would be about 2500 additional cases of cancer each year. Linus Pauling has calculated that the number is closer to 96,000 new cases. Drs. Arthur Tamplin and John Gofman, two radiation scientists who have been particularly vocal critics of the 170 millirad/year guideline, put the number at 30,000. The United States Atomic Energy Commission has been quick to point out that there is no obvious way that all the people in the country could be exposed to this level; rather, only those who live or work on the borders of a nuclear reactor site might approach this limit, and if they did, those further removed from the site would receive much less radiation.

We must emphasize that all these calculations

are based on assumptions which seem reasonable but are unproved, and they all involve large approximations in arriving at the final estimates. But the lesson is clear—we will not have nuclear reactors without some price, whether the number of additional cancer cases each year in the United States is 100 or 100,000.

We have seen that radiation poses many hazards to human health. Some are easily assayed (for example, acute radiation sickness) and some only with the greatest difficulty (genetic effects in future generations). The responsibility on those who use radiation is thus enormous, for the implications of what we do now extend far into the future to those yet unborn. If radiation were *only* dangerous, of course, there would be little difficulty in making policy decisions, but things are not so simple. The benefits and potential benefits of radiation are very great. Modern medical diagnosis is unthinkable without x-rays. We have said that radiation can increase the incidence of cancer in exposed populations, but we should also mention that in many patients who already have cancer, therapy with radiation can sometimes cure the disease and can often provide substantial palliation of suffering. The potential benefits of atomic power have already been discussed.

There is no way to balance known benefits in the present with unknowable dangers in the future. Nonetheless, there are reasonable steps to take. An x-ray examination should be undertaken only when truly necessary, particularly for children and for adults of child-bearing age. Whenever possible, gonadal shields should be provided to the patient undergoing the examination. Except in emergencies, women of child-bearing age should have x-ray examinations only during the first two weeks of their menstrual cycle, since pregnancy is most unlikely during this interval.

Technical improvements in x-ray equipment and greater sophistication in those using the equipment may serve to minimize the dose of x-rays needed for each examination. Increasing international pressure on France and the People's Republic of China, the two nations still testing nuclear weapons in the atmosphere, would work toward elimination of this last source of atmospheric contamination with radioactive materials. Finally, the licensing, installation, and operation of all nuclear power plants must be carefully monitored.

8.9 PROBLEMS AND ISSUES RELATED TO RADIOACTIVE POLLUTANTS

The questions involved in the use of radioactive materials require decisions of crucial importance for all mankind. First, we will not consider the development and testing of nuclear weapons as a valid subject for cost-benefit analysis. If "mutual assured destruction" existing between two countries is to be considered a "benefit" because peace is guaranteed by fear of reprisal, then the cheapest way to accomplish this objective would be for each country to mine the other's cities and to maintain control over the "destruct button." This would cost far less than fleets of jets, missiles, or submarines. The fact that no such agreement would ever be made illustrates its absurdity, and if it is absurd to do this cheaply, it cannot be wise to do it expensively.

The development of nuclear power, however, is quite another matter. We can no more "renounce" power than we can agriculture. Therefore, we must balance the nuclear risks against the air pollution that might be produced by the burning of gas, oil, or coal (fossil fuels) to yield an equivalent amount of energy. We must also consider how the Earth's resources are to be exploited. Gas, oil, and coal are not only fuels, but also essential raw materials for production of the many organic chemicals, such as medicinals, plastics, and dyes, that are useful to man. To burn them for production of energy is to waste them for chemical synthesis, as we pointed out in Chapter 5.

Given these benchmarks, there are various kinds of judgments we have to make.

First, the reliability of long-term storage of radioactive wastes must be considered in terms beyond the possible leakage of tanks or the wetting of salt caverns. It must also be understood that we are setting up procedures that we are depending on generations of our successors to follow. We are therefore relying on a continuity of political systems that will be adequate to the task of following or improving on the methods that we establish.

Second, we should reckon with the fact that successful development of fusion reactors is possible but not assured. That is, we are not attempting to violate any "laws of nature" such as the conservation laws, and, indeed, technology has come to our rescue often before, but that is no guarantee for the future. Some experts in the field are fairly optimistic, but others see such difficulties ahead that their mood is almost hopelessness.

Third, we must face the question of determining where the responsibility shall lie for protecting the public against radiation hazards. For many years the U.S. Atomic Energy Commission has established its own safety standards. This situation has been likened to the fox watching the henhouse. Is surveillance best left to a separate governmental body, such as the Environmental Protection Agency, to industry groups, to organizations of private citizens, or to some consortium of these interests?

Fourth, what criteria shall we use to judge health effects? Shall we consider only acute illness, or shall we include more subtle and less easily demonstrable effects, such as estimated excess cases of cancer?

Finally, how do we rank our needs for abundant energy in comparison with other human wants? Dr. Glenn Seaborg, former president of the Atomic Energy Commission, has made the point that we need nuclear energy to power the various anti-pollution devices occasioned by the use of other technologies. But then, how important is the proliferation of other technologies if they produce pollutants that must be controlled by mechanisms that use energy that must be produced by nuclear processes which in turn produce environmental stresses?

Glenn T. Seaborg (1912–) is an American chemist who was awarded the Nobel Prize in 1951 for his work on the trans-uranium elements. In 1961, he became chairman of the U.S. Atomic Energy Commission, where he served until 1970. Seaborg has been an advocate of the development of nuclear power to serve man's energy needs.

PROBLEMS

1. **Mass.** Explain the difference between the mass and the mass number of an atom.
2. **Definitions.** Define radioactivity; radioisotope; isotope.
3. **Chain reaction.** What is a chain reaction? Explain chain propagation (lengthening); chain branching; chain termination; critical condition. Can the spread of a rumor among a large group of people function as a chain reaction? If so, illustrate how the chain could branch or terminate. Define the critical condition of such a system.
4. **Reactors.** What are essential features of a nuclear fission reactor?
5. **Neutrons.** List the possible fates of neutrons in a fission reactor. Which of these events should be favored, and which should be inhibited, in order to (a) shut down a reactor, (b) breed new fissionable fuel, and (c) produce more energy?
6. **Natural nuclear reactor.** It has been found that the ^{235}U content of certain natural uranium deposits is abnormally low, and it has been speculated that such

depletion might have been the result of a naturally occurring chain reaction. (a) How could a natural rock formation approximate a nuclear reactor? Be specific in your answer: What factors might have conserved neutrons? What substance could have served as the neutron moderator? (b) What do you think is the chance of a naturally occurring nuclear explosion? Explain.

7. **Nuclear safety.** Outline the general concept and approach to safety used in nuclear power plants. Can you think of any specific series of events that would cause all of the safety features to fail and a radioactive cloud to be released to the atmosphere? If so, describe them.

8. **Nuclear safety.** Do you think it would be reasonable to set safety limits in nuclear power plants that would prohibit *any* release of radioactive matter? Defend your answer. If your answer is no, what criteria would you use to set the limits?

9. **Nuclear safety.** Compare or contrast the control systems in a nuclear fission plant with the homeostatic mechanisms of a natural ecosystem, giving special attention to the following points: (a) Are the two sets of mechanisms comparable in that they both respond in a manner that tends to alleviate stress, or are they conceptually different and therefore not to be compared? (b) Is the effectiveness of typical natural homeostases a reasonable goal for the nuclear industry, or must we do better? Give some example of perturbations in natural systems and state whether variations from the norm of similar magnitude would be acceptable in a nuclear power plant.

10. **Radioactive wastes.** Sand-like radioactive leftovers from uranium ore processing mills, called "mill tailings," have been used to make cement for the construction of houses in Colorado, Arizona, New Mexico, Utah, Wyoming, Texas, South Dakota, and Washington. These tailings contain radium (half-life 1620 years) and its daughter radon (a gas, half-life 3.8 days), as well as radioactive forms of polonium, bismuth, and lead. Radon gas seeps through concrete, but is chemically inert. Are the following statements true or false? Defend your answer in each case:

    (a) Since radon has such a short half-life, the hazard will disappear quickly; old tailings, therefore, do not pose any health problems.

    (b) Even if the radon gas is present, it cannot be a health problem because it is inert and does not enter into any chemical reactions in the body.

    (c) Continuous ventilation that would blow the radon gas outdoors would decrease the health hazard inside such a house.

11. **Nuclear explosion.** Explain why a critical mass of pure fissionable material must be exceeded if an ex-

plosion is to occur. Why is it thought that a nuclear reactor could not explode in the manner of a bomb?

12. **Hydrogen isotopes.** If we use the symbols shown on page 320, it is possible to write six formulas for water. Write them. Identify the three that represent radioactive forms of water.

13. **Fusion.** Suppose that someone claims to have found a material that can serve as a rigid container for a thermonuclear reactor. Would such a claim merit examination, or should it be ignored as a "crackpot" idea not worth the time to investigate? Defend your answer.

14. **Fusion.** Outline the reasons why fusion reactors are expected to be far less serious sources of radioactive pollutants than fission reactors.

15. **Health.** Outline the types of damage to the body that can result from exposure to high-energy radiation. Can there ever be any benefits? Explain.

16. **Health.** Give two reasons why a 70-year-old woman does not face so serious a problem regarding the health effects of radiation as a 17-year-old.

17. **Health.** The formula for thyroxin, an essential chemical growth regulator, is $C_{15}H_{11}I_4NO_4$. Which of the following radioactive waste products of nuclear reactors pose a particular threat to the thyroid gland: ^{90}Sr; CH_3I containing I-131; or radon? Justify your choice.

18. **Health.** Let us suppose that the establishment of a nuclear reactor system in the United States were to result in the deaths of a certain number of people each year from cancers that they would not have contracted if the radiation levels were lower. What effects of nuclear power plants can you imagine that might *increase* the general level of health?

19. **Health.** Relate the type of cells most easily affected by radiation to the symptoms of early somatic radiation sickness.

The following questions involve some calculations.

20. **Atomic structure.** Complete the table to the right by substituting the correct numerical value where a question mark appears. (Note that atomic numbers, but not mass numbers, appear in the table of atomic weights in the appendix.)

21. **Half-life.** A Geiger counter registers 512 counts per second near a sample of radioactive substance; 48 hours later the rate is 256 cps (counts per second). What is the half-life of the radioactive substance? What rate will the counter register after an additional 144 hours?

22. **Half-life.** Cesium-137 is a radioactive waste product whose half-life is 30 years. It is chemically similar to

Isotope	Atomic Number	Mass Number	Number of Neutrons in Nucleus
Oxygen-18	?	?	?
Strontium-90	?	?	?
Uranium-?	?	?	141
Iodine-?	?	131	?

Answers

20. 8, 18, 10 neutrons; 38, 90, 52 neutrons; Uranium-233, 92, 233; Iodine-131, 53, 78 neutrons

21. 48 hours; 32 cps

potassium, which is an essential element in plants and animals. It compounds are readily soluble in water.

(a) How long will it take for 1000 mg of cesium-137 to decay to 125 mg?

(b) Since cesium compounds are soluble, would it be wise to dump this isotope into an open holding pond and let it dissolve and decay until only negligible quantities remain? Defend your answer.

22. 90 years

23. Radioactive wastes. Assume that a radioactive waste product with a one-year half-life is produced at the rate of 16 tons per year. (a) Will the *total* quantity of waste product continue to rise indefinitely, level off, or gradually diminish? If your answer is that the quantity of waste levels off, what value will it approach? Complete the table to the left to help you answer the question.

(b) What does your answer imply with regard to the ultimate requirements for radioactive storage facilities if production levels off?

Year	*Tons of waste*
1974	16 = 16
1975	8 + 16 = 24
1976	4 + 8 + 16 = 28
1977	2 + 4 + ? = ?

23. 32 tons

BIBLIOGRAPHY

There are many books on atomic energy and nuclear engineering; many of them assume previous training in physics and chemistry. The following two texts, however, present somewhat more elementary introductions:

Alvin Glassner: *Introduction to Nuclear Science.* New York, Litton Educational Publisher, Van Nostrand-Reinhold Books, 1961.

Samuel Glasstone: *Sourcebook on Atomic Energy.* New York, Litton Educational Publisher, Van Nostrand-Reinhold Books, 1958.

(Glassner's book is based on a short course given at Argonne National Laboratory since 1957, and presupposes only one year of college physics. Glasstone is more comprehensive and offers more introductory matter.)

For more advanced books, refer to the following:

George I. Bell: *Nuclear Reactor Theory.* New York, Van Nostrand-Reinhold Books, 1970. 619 pp.

Peter John Grant: *Elementary Reactor Physics.* New York, Pergamon Press, 1966. 190 pp.

Two recent excellent books that integrate social and technical aspects of the problems of nuclear energy are:

David Rittenhouse Inglis: *Nuclear Energy: Its Physics and Its Social Challenge.* Reading, Mass., Addison-Wesley Publishing Co., 1973. 395 pp.

Henry Foreman, ed. *Nuclear Power and the Public.* Minneapolis: Univ. of Minnesota Press, 1970, 272 pp.

Specific discussion of nuclear hazards may be found in:

G. C. Collins: *Radioactive Wastes: Their Treatment and Disposal.* New York, John Wiley & Sons, 1961. 239 pp.

Thomas Craig Sinclair: *Control of Hazards in Nuclear Reactors.* London, Temple Press Books, 1963. 84 pp.

Finally, the excellent periodical *Science and Public Affairs (Bulletin of the Atomic Scientists)* carries many articles on the problem.

9

AIR POLLUTION

9.1 INTRODUCTION

The predominantly gaseous envelope that surrounds the Earth is its **atmosphere**, and the stuff of which it consists is **air**. We say "predominantly" gaseous because the atmosphere also contains solid particles and liquid droplets, to which we shall refer later.

Air is essential for life on Earth, and the preservation of its quality is therefore a matter of utmost concern. If we think of the quality of air as its fitness to support life, then the air at the top of Mount Everest is quite poor because there is scarcely enough to breathe. The "air" that is exhausted from a diesel engine is also too poor to breathe, although it contains more oxygen per cubic foot than is available at the summit of Everest. Clearly, the two samples of air are poor in different ways—the one is diminished in concentration, the other contains harmful foreign matter, and is therefore said to be **polluted**. In this chapter we are concerned primarily with air pollutants, but we shall also consider environmental stresses that threaten to diminish the quality of the air by altering the concentrations of its normal components rather than by introducing impurities.

9.2 THE PHYSICAL NATURE OF AIR POLLUTANTS

(Problems 2, 4, 32)

Imagine an enclosed space at constant temperature, such as a perfectly insulated empty room. We now seal the room completely—let us say we line the walls, floor, and ceiling with gold foil, leaving no openings anywhere. The inside seems still, lifeless. What happens to the air in the room? The dust particles will settle out, some more rapidly than others, and gradually will accumulate as a thin deposit on the floor. But nothing else will happen. If a person should step into the quiet darkness after a year, or after a century, the air will still be fit to breathe; it will not have settled to the floor to join the dust; in fact, its components will not have separated from each other to any discernible degree. This is the way of a true gas. The gas consists of molecules; each air molecule is an aggregate of a very few atoms—typically two or three; in some cases it is only one. The molecules are in constant motion; they move in straight lines between collisions with each other or with the walls, floor, or ceiling. The molecules are so small, and so numerous, and their collisions so frequent, that nothing in our experience (such as, say, a swarm of gnats) can be called upon to help us visualize them in a quantitative way. (The number of molecules in a liter, for example, is about 3×10^{22}.) But there are two kinds of numbers that are comfortable for us to think about, and that later will help us understand some of the technical aspects of air pollution and its control. One of these numbers is the average speed of the molecules: a typical speed is about 900 miles per hour for nitrogen molecules at 0°C (although the average distance between collisions is very short). Another number is the proportion of "empty" space in the room—that is, space not occupied by the molecules themselves. The concept of the actual volumes of the molecules may be confusing to think about because the molecules are not "solid" in a macroscopic sense. But we can easily visualize the following imaginary experiment. Let us strip off the insulation and put our sealed room into a very cold giant-sized refrigerator. When the air inside the room becomes cold enough (below about −195°C), it will liquefy, and the liquid air will collect on the floor with the dust. If the volume of the room were 1000 cubic feet, the liquid air would occupy only slightly more than one cubic foot. It is therefore conservative to say that the molecules of air under ordinary terrestrial conditions (normal atmospheric pressure) occupy only about 0.1 per cent of the total space; the other 99.9 per cent is empty.

Of course, when we speak of "one cubic foot of gas," we include both the molecules and the space in which they dart around. In fact, at any given temperature and pressure, the volume of any gas is directly proportional to the *number* of molecules it contains, and is independent of the *kinds* of molecules present.

This statement is known as Avogadro's Law, after Amedeo Avogadro, who suggested it in 1811.

Now let us return our room to its original state, back up an automobile against it, and drill a hole in the wall just large enough to accommodate the tailpipe. We run the engine for a minute, during which time 33 cubic feet of exhaust gas are introduced into what has now become our polluted room. Let us say that the auto exhaust contains three per cent carbon monoxide, CO (a typical value for an engine that is not provided with control devices). This number means that three per cent of the molecules in the exhaust are CO molecules, or that the exhaust contains three per cent CO by volume. Let our room bulge by this added volume.

We would get the same result if the room were rigid, but the pressure would be higher. In dealing with air pollutants, we assume the pressure is constant because we are concerned with unconfined atmospheres.

The percent CO in the room will be:

$$\frac{33\ \text{ft}^3 \times 0.03}{(1000 + 33)\text{ft}^3} = 0.001 = 0.1\%$$

The concentrations of gaseous pollutants are more frequently expressed as "parts per million" (ppm) and sometimes "parts per billion" (ppb), by volume. The meanings of all these expressions are:

per cent by volume	= volume of pollutant per 100 volumes of air	= number of molecules of pollutant per 100 molecules of air
ppm by volume	= volume of pollutant per 1,000,000 volumes of air	= number of molecules of pollutant per 1,000,000 molecules of air
ppb by volume	= volume of pollutant per 1,000,000,000 volumes of air	= number of molecules of pollutant per 1,000,000,000 molecules of air

Auto exhaust into polluted room.

> *Conversion:*
> **To change per cent to ppm, multiply by 10,000.**
> **To change ppm to ppb, multiply by 1,000.**

Let us try to get some feeling for the sizes of these numbers. Is 1 part per billion a large concentration or a small one? One billion pennies laid out rim to rim in a straight line would extend about 12,000 miles, or a distance almost equal to halfway around the Earth's equator. If one of these pennies were a bad one (a "pollutant") it would indeed be hard to find. This example makes 1 ppb sound small. But if air were contaminated with 1 ppb of SO_2 (much less than in city atmospheres), there would be 27,000,000,000 (27 billion) molecules of SO_2 per cubic centimeter. This example makes 1 ppb sound large. Don't let either of these examples fool you. Concentration values by themselves do not provide information on the effects of pollutants. For some substances, 1 ppb is inconsequential; for others it is highly significant. We must therefore concern ourselves both with concentrations of pollutants and their effects on living organisms and on materials.

Not all air pollutants are gases. Some are airborne solid particles or liquid droplets, which are much larger bodies than individual molecules. For example, the diameter of a particle of dust may be 100,000 times that of a gas molecule. Concentrations of particles are usually expressed in terms of weight of pollutant per unit volume of air. Figure 9.1 illustrates the relative sizes involved.

We sometimes measure the rate at which particles settle out from the air. You may have read reports of air pollution expressed in units such as "tons of dustfall per square mile per month." However, such figures tell very little about total concentrations in air.

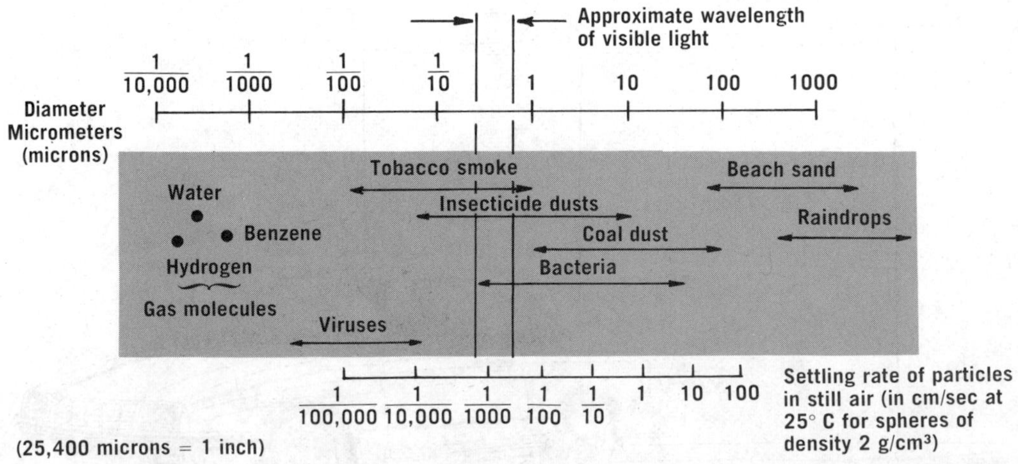

FIGURE 9.1 Small particles in air.

Note from Figure 9.1, for example, that a 10-micrometer coal dust particle settles at about 1 centimeter per second, whereas a 1/10-micrometer smoke particle settles at a rate several thousand times slower, and gas molecules do not settle at all.

9.3 PURE AIR AND POLLUTED AIR

With the arithmetic in hand, let us look at the Earth's atmosphere and its pollutants. The most variable component of air is water vapor, or moisture, whose concentration may range from a negligibly small value in a desert to about five per cent in a steaming jungle. If we neglect the moisture, the composition by volume of dry air is roughly 78 per cent nitrogen, 21 per cent oxygen, and one per cent of other gases. A more detailed breakdown is given in Table 9.1. Table 9.1 does not include non-gaseous, or "particulate," components. The "natural" concentrations of particulate matter in the air vary much more than those of gaseous matter. Thus, if we analyzed air in various parts of the Earth away from man's activities, the composition of the gases would be very close to the values in Table 9.1. But the particulate matter would vary widely from place to place. It would include nonviable (not capable of living) particles such as airborne soil granules, volcanic dust, and salts from evaporation of sea spray. It would also include viable particles such as plant and insect matter.

TABLE 9.1 GASEOUS COMPOSITION OF NATURAL DRY AIR

	GAS	CONCENTRATION (BY VOLUME) ppm	per cent
"Pure" Air	Nitrogen, N_2	780,900	78.09
	Oxygen, O_2	209,400	20.94
	Inert gases, mostly argon, (9300 ppm) with much smaller concentrations of neon (18 ppm), helium (5 ppm), krypton and xenon (1 ppm each)	9,325	0.93
	Carbon dioxide, CO_2	315	0.03
	Methane, CH_4, a natural part of the carbon cycle of the biosphere; therefore, not a pollutant although sometimes confused with other hydrocarbons in estimating total pollution	1	
	Hydrogen, H_2	0.5	
Natural Pollutants	Oxides of nitrogen, mostly N_2O (0.5 ppm) and NO_2 (0.02 ppm), both produced by solar radiation and by lightning	0.52	
	Carbon monoxide, CO, from oxidation of methane and other natural sources	0.3	
	Ozone, O_3, produced by solar radiation and by lightning	0.02	

Some people define air pollutants to be substances not considered "natural" components of air. However, a hay fever sufferer may consider ragweed pollen an air pollutant even though it is a natural component of air in various parts of the Earth, and SO_2 from a volcano is indistinguishable from SO_2 from a smokestack. Other people would consider air pollutants to be only those impurities which are present in sufficient concentration to produce some measurable adverse effects on either living organisms or on materials. But our knowledge of adverse effects changes as we study them, and we prefer to reject such inconstancy. Instead, we shall arbitrarily define a substance called "pure air," and we will classify any other component as a pollutant. Let us say, then, that pure air is a gaseous mixture of the first six components of Table 9.1: nitrogen, oxygen, inert gases, carbon dioxide, methane, and hydrogen in the concentrations shown in the table, or close to them, plus any additional moisture that may be present. Of course, any significant variation in these compositions could be harmful; for example, air containing 10 per cent CO_2 would be poisonous, and air containing 10 per cent H_2 or 10 per cent CH_4 would be explosive. Thus, CO_2 in high concentrations is a pollutant. We will consider all other gases, regardless of concentration, whether of human or non-human origin, as well as all particulate matter, to be pollutants. Note also that NO_2, CO, and O_3, though they occur naturally, are listed as pollutants.

9.4 GASEOUS AIR POLLUTANTS

(Problems 3, 6, 8)

In this section we will describe the major classes of gaseous pollutants and some important individual pollutant compounds.

Carbon Oxides

Carbon dioxide, CO_2, is a normal component of air (see Table 9.1) and a part of the carbon cycle of the biosphere; therefore, it is not ordinarily considered to be a pollutant. However, the burning of coal, oil, and natural gas as fuels produces large quantities of CO_2. The chemical equations are shown at left.

$C + O_2 \rightarrow CO_2$ (burning of coal)

$CH_4 + 2O_2 \rightarrow CO_2 + 2H_2O$ (burning of natural gas)

The present rate of increase of the worldwide CO_2 concentration is about 0.7 ppm per year. In Section 9.8, we will consider the possible effects of a continued increase in CO_2 concentration on the Earth's atmosphere.

Carbon monoxide, CO, is a product of the incomplete combustion of carbon or of carbon compounds. CO is notorious as a gas that is colorless, odorless, and non-irritating, yet very toxic. We read of accidental deaths from CO escaping into a room from a faulty gas heater, or of suicides carried out by remaining in a closed garage with an idling automobile. Most of the carbon monoxide in the Earth's atmosphere, however, comes from natural sources, specifically from the oxidation of methane that is emitted by decaying organic matter, and to a lesser extent directly from the decay and growth of chlorophyll. Taken together, natural sources of CO account for more than 90 per cent of the worldwide average level of between 0.1 and 0.5 ppm. Since the worldwide concentration of CO is relatively stable, there must be some mechanism for its removal; in other words, it must cycle. The prime reactive agent seems to be microorganisms in ordinary soil; sterilized soil is ineffective.

$$2C + O_2 \rightarrow 2CO$$

$$2CH_4 + 3O_2 \rightarrow 2CO + 4H_2O$$

These facts do not imply that atmospheric CO is always benign. The global averages do not apply, for example, to cities, where 95 to 98 per cent of the CO comes from combustion of fuel by man and where the prevailing levels can be 50 to 100 times the worldwide concentration. For example, the level inside an automobile moving in a heavy stream of traffic on a multilane highway will be in the neighborhood of 25 to 50 ppm. The CO concentration in the exhaust itself, as we mentioned earlier, can reach three per cent, or 30,000 ppm. The maximum allowable concentration for healthy workers in industry, for an eight-hour working day, is 50 ppm. A concentration of 1000 ppm can produce unconsciousness in one hour and death in four hours.

Compounds That Contain Sulfur

The important sulfur oxides are **sulfur dioxide**, SO_2, and **sulfur trioxide**, SO_3. From the viewpoint of its harmful effect on man and the difficulties involved in preventing its discharge into the atmosphere, SO_2 is probably the most significant single air pollutant. High SO_2 concentrations have been associated with major air pollution disasters of the type that have occurred in large cities, such as London, and that were responsible for numerous deaths. SO_2 is produced when sulfur or fuel containing sulfur is burned,

$$S + O_2 \rightarrow SO_2$$

or when sulfide ores, such as copper sulfide, are "roasted" in air.

$$2CuS + 3O_2 \rightarrow 2CuO + 2SO_2$$

Since sulfur is present in coal and oil, the burning of these materials for heat and power produces SO_2.

The other important sulfur oxide, SO_3, is produced in the atmosphere by the oxidation of SO_2 under the influence of sunlight.

$$2SO_2 + O_2 \rightarrow 2SO_3$$

In addition, some SO_3 is introduced directly from combustion processes along with SO_2. The moisture in the air reacts rapidly with SO_3 to form a mist of sulfuric acid.

$$SO_3 + H_2O \rightarrow H_2SO_4$$

When such conversions occur, the material originally introduced to the atmosphere is called a **primary air pollutant**. The new materials produced by chemical reaction in the air are called **secondary air pollutants**.

SO_2 concentrations in air vary over wide ranges in different locations, during different seasons of the year, and on different days and hours. For example, the average levels in rural atmospheres or "clean" cities will be less than 0.01 ppm, while in heavily industrial urban districts a typical one-day average may be about 0.1 ppm and, on a very bad day, the level may rise to about 0.5 ppm. Such high concentrations pose significant health hazards (see Section 9.9).

The conversion of SO_2 to H_2SO_4 produces a strongly acidic atmospheric mist that corrodes metals, destroys living tissue and nylon stockings, and deteriorates various building materials. Marble is particularly vulnerable to such attack; as a result, many build-

FIGURE 9.2 Paint damage from H_2S emitted from polluted San Francisco Bay waters. (Photo by J. E. Yocom. From Stern: *Air Pollution*. 2nd Ed. New York, Academic Press, 1968.)

ings and monuments that have endured for centuries are now being rapidly eroded. Sulfuric acid mist in air consists of droplets that are usually about one to four micrometers in diameter; this particular size range favors the deep penetration of the acid into the lungs, with consequent damaging effects.

Another important sulfur-containing compound is hydrogen sulfide, H_2S, which has the odor of rotten eggs. It blackens lead paints (Fig. 9.2), and is even more poisonous than carbon monoxide. The occurrence of H_2S in damaging concentrations is usually associated with some specific source, such as decomposing organic matter, sewage, or some industrial operation. The fact that H_2S is a product of biochemical action (see Chapter 10) implies that the total production on Earth is high. Figure 9.3 shows that man-made sources account for only about a third of the Earth's total atmospheric sulfur.

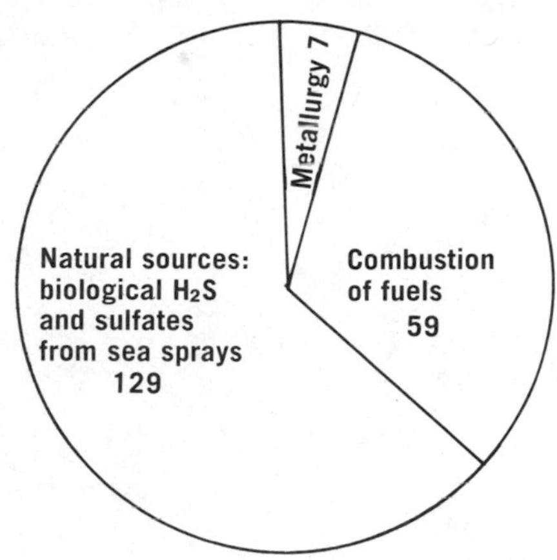

FIGURE 9.3 Gaseous sulfur pollutants from all sources, in millions of metric tons of sulfur per year.

Compounds That Contain Nitrogen

The important oxides of nitrogen that occur in the air as pollutants are nitrogen oxide, NO, and nitrogen dioxide, NO_2. Both are produced by any combustion process that occurs in air, because some oxidation of atmospheric nitrogen occurs at flame temperatures:

$$N_2 + O_2 \rightarrow 2NO$$

$$2NO + O_2 \rightarrow 2NO_2$$

For purposes of convenience in reporting analytical results, chemists frequently group NO and NO_2 together under the general formula NO_x.

Biological processes produce much more NO_x (mostly NO and N_2O) than comes from combustion processes but, again, it is the locally high concentrations that are responsible for adverse effects on living organisms and on materials. Typical rural levels are only a few parts per billion; urban concentrations are usually about 10 to 30 parts per billion, but under severely adverse conditions they may rise to peaks approaching 1 ppm. NO_2 is a reddish-brown gas whose pungent odor can be detected at concentrations above about 0.1 ppm, and it therefore contributes to the "browning" and the smell of some polluted urban atmospheres.

In considering the toxicity of oxides of nitrogen, it is usually sufficient to focus attention on NO_2, because NO is converted to NO_2 in air. The effects of NO_2 on man range from unpleasant odor and mild irritation to serious lung congestion to death, depending on the concentration of the NO_2 and the duration of exposure. NO_2 concentrations in polluted outdoor air are not usually high enough to produce acute toxic effects, but they may produce or contribute to chronic respiratory ailments (see Section 9.9). NO_2 is also significant as an air pollutant because it is one of the key substances that enters into a chain of chemical reactions to produce "smog," discussed in the following section.

Some organic nitrogen compounds, called **amines**, are strongly malodorous, smelling like rotten fish.

Hydrocarbons

These are compounds of carbon and hydrogen; the number of carbon atoms per molecule ranges from one in methane, CH_4, to hundreds or thousands in tars and asphalts. As we have seen, methane is a normal component of air; it is not toxic in its prevailing atmospheric concentrations. The "natural gas" that is used as a fuel is mostly methane.

Octanes are representative hydrocarbon components of gasoline; there are 18 of them, and all have the same molecular formula, C_8H_{18}. Their boiling points range from 99°C to 126°C, which implies that they can evaporate readily and persist as vapors in air. Generally, the larger the molecules, the less volatile is the substance, and, therefore, the atmospheric concentrations of gaseous hydrocarbons much heavier than octanes are negligibly small. The gaseous hydrocarbons, for the most part, are harmless themselves even at the concentration levels prevalent in urban atmospheres, but as we shall see later, they can be precursors of more damaging pollutants.

Some other gaseous hydrocarbons:

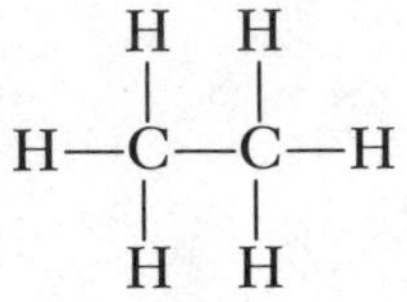

```
   H  H          H           H
   |  |           \         /
H—C—C—H            C=C
   |  |           /         \
   H  H          H           H

  ethane           ethylene

   H  H  H
   |  |  |
H—C—C—C—H     H—C≡C—H
   |  |  |
   H  H  H

  propane         acetylene
```

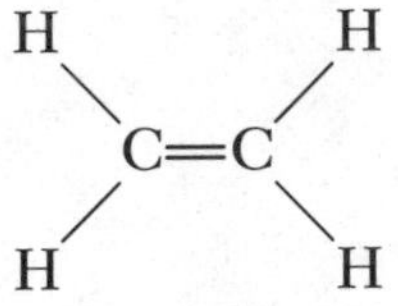

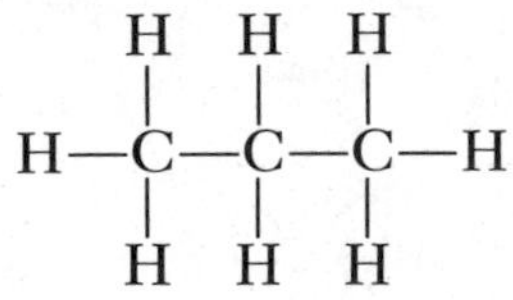

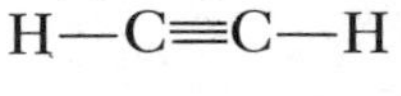

Some liquid hydrocarbons:

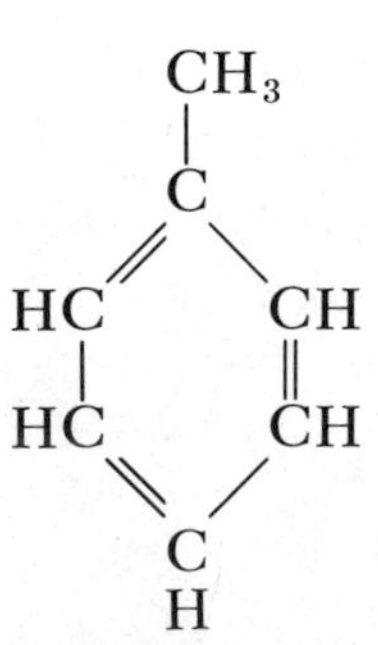

```
                              H
                             /
     CH3            H—C=C
      |               |     \
      C               C      H
    //  \           //  \
  HC     CH       HC     CH
  |      ||       |      ||
  HC     CH       HC     CH
    \\  /           \\  /
      C               C
      H               H
   toluene         styrene
```

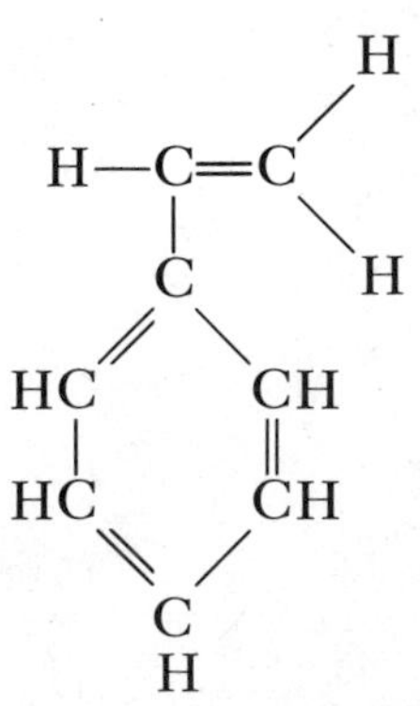

"Oxygenates"

These are compounds that contain carbon, hydrogen, and oxygen; they include such well-known classes as alcohols and organic acids. Such substances are produced by the incomplete combustion of hydrocarbons such as gasoline and diesel fuel. Another source is the evaporation of solvents from various industrial operations.

Ozone and Oxidants

Ozone, O_3, occurs to some extent in "normal" air (see Table 9.1), but in higher concentrations it is a

(A) White Cascade petunia leaves show injury from ozone. White-flowered varieties tend to be more susceptible than those with colored flowers. (Courtesy of the United States Department of Agriculture.)

(B) Tomato leaflet grown in laboratory in unfiltered air containing oxidants. (Courtesy of the United States Department of Agriculture.)

PLATE 7

(C) Peony displaying symptoms of fluoride damage. (Courtesy the United States Department of Agriculture.)

(D) Typical sulfur dioxide injury to white birch leaf at rig following two hours of exposure to two parts per million. (Cou tesy of the United States Department of Agriculture.)

PLATE 8

Patterns of smoke plumes. (Courtesy of R. S. Scorer.)

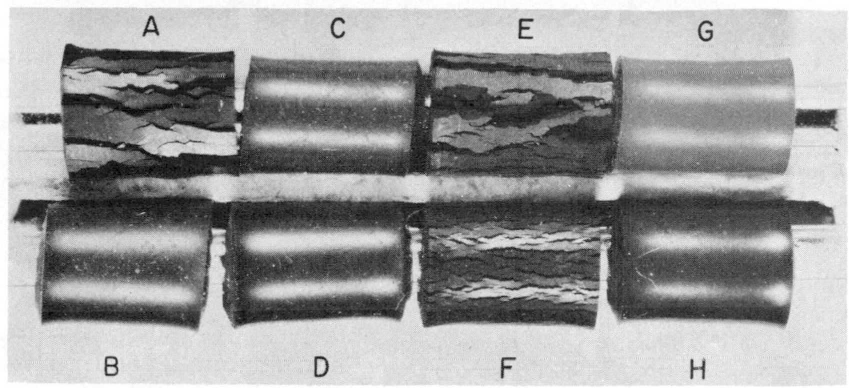

FIGURE 9.4 Effect of ozone exposure on samples of various rubber components. *A*, **GR-S;** *B*, **Butyl;** *C* **and** *D*, **Neoprene;** *E*, **"Buna-N";** *F*, **Natural rubber;** *G*, **Silicone;** *H*, **"Hypalon." (Photo courtesy of F. H. Winslow, Bell Telephone Laboratories. From Stern:** *Air Pollution.* **2nd Ed. New York, Academic Press, 1968.)**

toxic substance. The maximum allowable concentration for a healthy industrial worker over an eight-hour working day is 0.1 ppm.

It is somewhat curious that ozone has come to be popularly associated with pure air and that devices which produce ozone have been regarded as "air purifiers." Ozone is, in fact, produced naturally in outdoor air by lightning, and its characteristic pungent odor under such conditions has probably been associated with the outdoors and with the cleansing action of rainstorms. These circumstances have led real estate developers to call their properties such fanciful but chemically inappropriate names as "Ozone Park," and the authors of travel folders to write such nonsense as ". . . A warm, a blue, an ever-changing sea, sparkling in the warm summer sunshine, and filling the air with ozone." (If it were true, the tourists would all be dead.) More serious and less laughable are the home and hospital appliances that purport to purify the air by producing ozone. Again, the confusion is rooted in some well-known but irrelevant chemical phenomena. Ozone is a chemically reactive substance that is sometimes used to deodorize exhaust gases (such as those emanating from sewage treatment) by oxidizing them to less objectionably odorous products before they are released to the atmosphere. The ozone concentrations needed for this treatment range between 10 and 20 ppm. Such concentrations would be rapidly fatal to human beings. The ozone produced by home appliance devices, on the other hand, is too dilute (around 0.1 ppm) to affect ordinary household odors to any measurable degree. Ozone has also been known for many years as a germicidal agent, and ozone generators are therefore sometimes used in the hope that they will suppress the spread of infectious organisms. The practical

Three views of downtown Los Angeles. *Top:* **a clear day.** *Middle:* **pollution trapped beneath an inversion layer at 250 feet.** *Bottom:* **pollution distribution under an inversion layer at 1500 feet. (Photos from Los Angeles Air Pollution Control District.)**

value of this scheme was questioned as early as 1913, when it was shown that ozone concentrations capable of killing bacterial pathogens killed guinea pigs even more rapidly. A more recent U.S. government study has concluded that "ozone, in low concentrations which do not otherwise cause irritation of the human respiratory tract, cannot be expected to provide any effective protection against airborne bacterial infection through direct inactivation of the infectious carrier particulates." Thus, home appliances that produce ozone do not purify the air, but pollute it.

U.S. Department of Health, Education, and Welfare, National Air Pollution Control Administration: *Air Quality Criteria for Photochemical Oxidants*. Washington, D.C., 1970. pp. 6–18.

There are various other pollutant gases chemically related to ozone that are collectively called **oxidants**. The properties they have in common include certain toxic and irritating effects on people, various patterns of damage to vegetation, and the ability to produce cracks in natural rubber (see Figure 9.4). These materials are generally produced by the reactions of hydrocarbons and other organic vapors with oxides of nitrogen in sunlight, and hence are typical pollutants of the atmospheres of sunny urban areas with considerable automobile traffic, such as Los Angeles. The visible component of such pollution is commonly called "smog."

Chemical formulas of ozone and a typical chemical oxidant

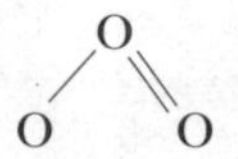

Ozone, O_3

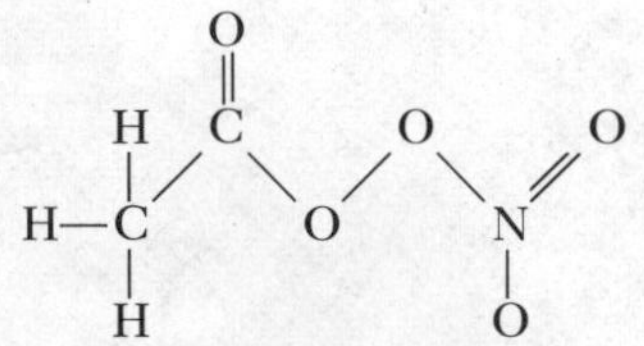

Peroxyacetyl nitrate, $C_2H_3O_5N$, an oxidant component of "smog"

Hydrogen Fluoride (HF)

This gas is an important pollutant because it causes serious and widespread damage to vegetation. However, it is not generally a component of polluted atmospheres, but rather originates from various specific industrial operations, such as the production of aluminum.

9.5 PARTICULATE AIR POLLUTANTS

(Problems 1, 5, 7, 31)

The names used to describe airborne particulate matter are somewhat confused and inconsistent, referring sometimes to size, sometimes to source, and

*DIAMETER LESS THAN 1 MICROMETER**		*DIAMETER GREATER THAN 1 MICROMETER**
Aerosols Smokes Fumes	May be solid or liquid, depending on their origin	Dusts (solid particles) Mists (liquid droplets)

*25,400 micrometers = 1 inch. A micrometer is also called a micron (see Appendix).

sometimes to a solid or liquid state. The preceding classification is a rough but useful description of this matter.

The word **aerosol** is used very commonly and should be remembered; it refers generally to any small particle in air. The 1-micrometer distinction between aerosol and dust or mist is not at all precise; many workers in air pollution refer, for example, to 10-micron (micrometer) aerosol particles.

Particle size is positively related to the speed of settling (see Fig. 9.1). Large dust particles are therefore troublesome only at relatively short distances from their source. Very small particles settle so slowly that they persist in the air for long periods of time and may be carried over long distances, often many miles.

There is great diversity among the types of particles in air. It is convenient for purposes of discussion to classify them into three categories: viable (capable of living), nonviable, and radioactive. We shall discuss the first two; radioactivity has been taken up in Chapter 8.

"Gee! The seventy-sixth floor! Right up where the photochemical particulants meet the clouds!"

(From *Wall Street Journal*, April, 1972.)

Viable Particles

These include pollen grains, microorganisms such as bacteria, fungi, molds, or spores, and insects or parts of insects such as hairs, wings, and legs. Viable particles are responsible for many effects that are detrimental to man, including hay fever, some kinds of bronchial asthma, various fungal infections, and airborne bacterial diseases.

Nonviable Particles

This group comprises a vast array of materials, some from natural sources and others from man's activities. The natural materials include sand and soil particles, salty droplets near the seashore, volcanic dust, and even particles of extraterrestrial origin. Man-made particulate pollutants include both organic and inorganic matter. Much of the organic particulate matter is in smoke from the incomplete combustion of coal, oil, wood, and garbage.

Consider for a moment what happens when a hydrocarbon, such as methane, gasoline, or paraffin wax burns freely in air. A luminous yellow flame appears. The luminosity originates from incandescent carbon particles; a cold porcelain dish that is held in the flame will rapidly blacken with the deposited soot. This experiment demonstrates that the combustion is incomplete. The following equations show examples of complete and of incomplete combustion:

$$CH_4 + 2O_2 \rightarrow CO_2 + 2H_2O \quad \text{(complete combustion)}$$
$$CH_4 + O_2 \rightarrow \underset{\text{soot}}{C} + 2H_2O \quad \text{(incomplete combustion)}$$

The soot particles are very small and become airborne pollutants. But the situation is much more complex than is represented by the two equations above. The products of incomplete combustion of hydrocarbons are many and varied; some are gaseous, others particulate. Among the latter are hydrocarbons related to benzene (C_6H_6), such as anthracene ($C_{14}H_{10}$) or phenanthrene ($C_{14}H_{10}$). Note that the conversion of methane to, say, naphthalene is also an incomplete oxidation. (See equation at left.)

$$10CH_4 + 8O_2 \rightarrow \underset{\text{naphthalene}}{C_{10}H_8} + 16H_2O$$

These transformations are important because some of these benzene-related hydrocarbons are carcinogenic (cancer-producing), and *all* sooty smoke contains some of these substances. Therefore, airborne particulate hydrocarbons must be regarded as potentially hazardous to health.

Benzene

Anthracene

Phenanthrene

Other airborne organic particles are insecticide dusts and some products released from food processing and chemical manufacturing. Inorganic particulate matter originates largely from metallurgical operations, non-metallic mineral production industries, inorganic chemical manufacturing, and the lead used in gasoline.

From the viewpoint of air pollution, the most significant metallurgical operations are those involved in the production of iron and steel, copper, lead, zinc, and aluminum. The particulate matter discharged to the atmosphere in any given metallurgical process is not the pure metal itself but one or more of its compounds, some of which may be poisonous to living organisms. Non-metallic mineral products include

Chrysene

3, 4 Benzopyrene

1, 2 Benzanthracene

cement, glass, ceramics, and asbestos. Operations in the manufacture of these products that are prone to produce airborne particles are blasting, drilling, crushing, grinding, mixing, and drying. Other inorganic particulate pollutants are specific to various chemical manufacturing operations, such as acid mists from production of nitric or sulfuric acids and phosphate rock dust from the manufacture of phosphate fertilizers.

Much of the lead in the air comes from the compounds used as anti-knock agents in gasoline, tetraethyl lead and tetramethyl lead. The total quantities produced in the United States in 1962 were 494 million and 18 million pounds, respectively, of tetraethyl lead and tetramethyl lead. These lead compounds are mixed with some simple chlorinated and brominated hydrocarbons before being added to the gasoline. What happens to all this lead? About 70 to 80 per cent of it is exhausted into the atmosphere in the form of small particles (ranging from a few hundredths up to several micrometers in size) of lead compounds, usually lead combined with chlorine and bromine, such as PbClBr. Of the remaining 20 to 30 per cent, about half is scavenged into the lubricating oil and half is retained in the engine and exhaust system. We know that lead compounds are poisonous, but we do not know precisely how damaging to the environment are the lead particles from automobile exhaust.

```
                 |
               —C—
                 |
               —C—
     |    |      |     |    |
   —C — C — Pb — C — C—
     |    |      |     |    |
               —C—
                 |
               —C—
                 |
```

Tetraethyl lead

9.6 THE METEOROLOGY OF AIR POLLUTION

(Problems 9, 10, 11)

To assess the extent to which your health is affected by, say, sulfur dioxide, you must be concerned in part with the quantity or concentration that you inhale. The total quantity of SO_2 in the earth's atmosphere, or the concentration in the exhaust gas of some particular copper smelter, does not affect you directly, because you do not breathe all the world's air, nor do you stick your head into the smokestack. Therefore, you must consider how pollutants are transported in the atmosphere, and how atmospheric conditions affect their concentrations. To understand these processes, you must study some special aspects of meteorology, the science concerned with atmospheric phenomena.

Let us start, then, with another imaginary experiment. Take some air and wrap it up in a package. Let this package be something like an extremely thin but perfectly insulated bag; this means that our parcel of air will be at the same pressure as the outside atmosphere but that no heat will be transferred into or out of the parcel. We now move the bag up from ground level higher and higher toward the upper atmosphere. As the altitude gets higher, the atmospheric pressure becomes lower, and our parcel of air expands. Air cools as it expands, even if no heat is removed during the process, and, therefore, our parcel of air also cools as it ascends. This rate of cooling, which is very close to one Celsius degree for every rise of 1000 meters, is called the **adiabatic lapse rate** (Fig. 9.5A).

Adiabatic means "without loss or gain of heat."

Now let us do another experiment: we will retrace the ascent with only a thermometer, and measure the temperature of the real atmosphere through which we pass. The energy gains and losses of the atmosphere are not necessarily in balance, and therefore the lapse rate is not necessarily adiabatic. What if the atmospheric lapse rate were superadiabatic, that is, the temperature dropped faster than for our insulated parcel of air? Such a situation is shown in Figure 9.5*B*. At any given elevation, our insulated (adiabatic) parcel of air will be warmer than the surroundings; being warmer it will be lighter; being lighter it will

FIGURE 9.5 Atmospheric lapse rates.

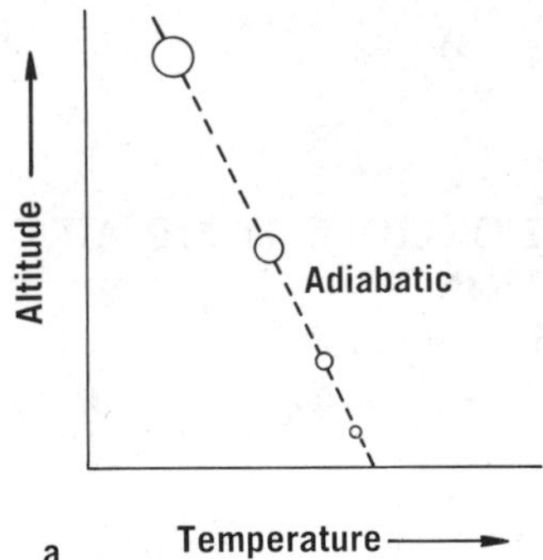

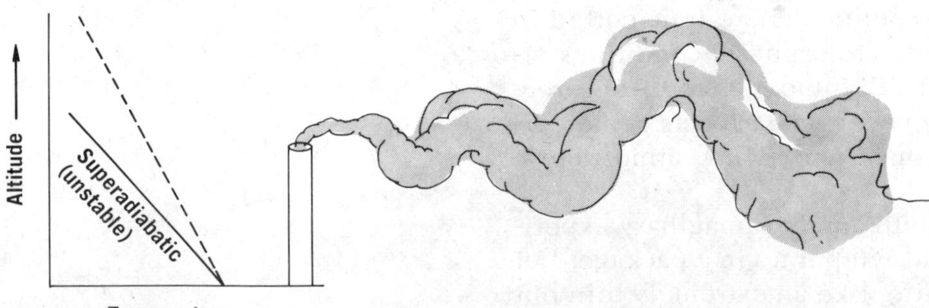

Figure 9.5 continued on opposite page.

tend to rise. If our parcel of air is large enough, its ascent will stir up the atmosphere and produce an unstable condition which favors turbulent mixing. Now let us say that our "parcel of air," instead of being imaginary, is something real—for example, an emission of pollutant gases from a smokestack. The instability associated with the superadiabatic condition of Figure 9.5*B* will be very favorable for the dilution of the pollutant, and the stack effluent will be effectively dispersed in a looping pattern.

Let us now consider what would happen if the atmospheric lapse rate were subadiabatic, that is, if the temperature dropped *less* sharply than our insulated parcel of air, as represented in Figure 9.5*C*. Our parcel of air will then be cooler than the surroundings and it will not be able to rise and disperse so well. The atmospheric condition is then more stable and pollutants disperse less readily. The shape of a smoke plume being discharged from a stack will be approximately conical.

When the temperature *increases* with increasing altitude, the condition is called an **atmospheric inversion.** Such a situation could start to develop an hour or two before sunset after a sunny day, when the ground

FIGURE 9.5 *Continued.*

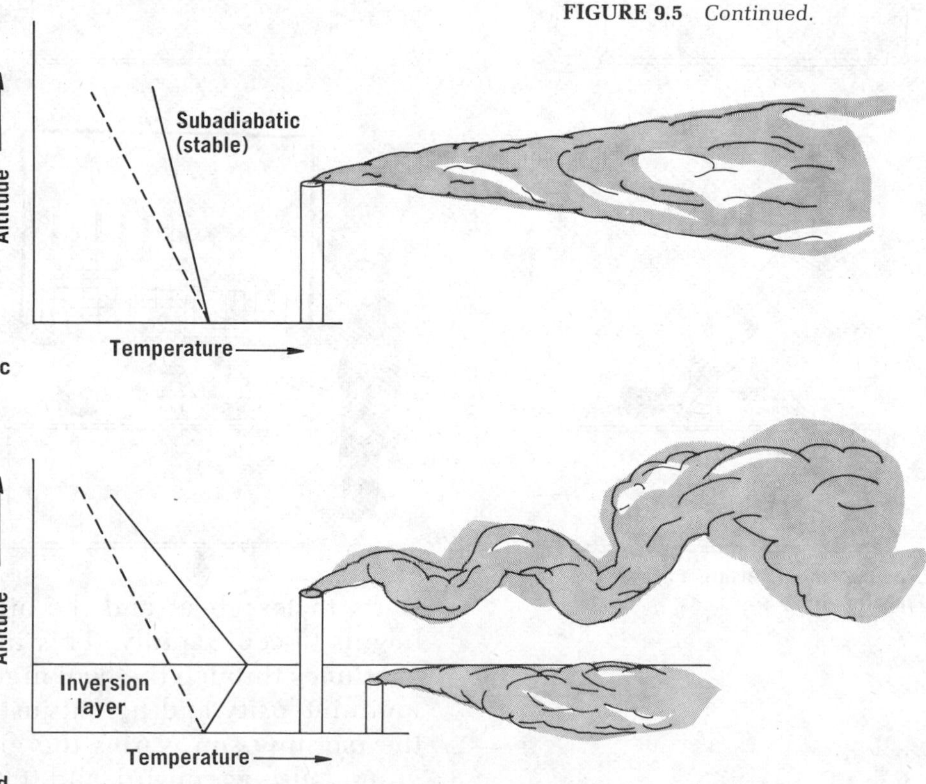

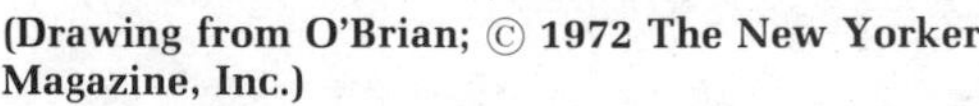

(Drawing from O'Brian; © 1972 The New Yorker Magazine, Inc.)

starts to lose heat and the air near the ground also begins to cool rapidly. This inverted condition often continues through the cool night and reaches its maximum intensity and height just before sunrise. When the morning sun warms the ground, the air near the ground also warms up, and it rises. Within an hour or two the inversion may be broken up by turbulent mixing. Sometimes, however, atmospheric stagnation allows the inversion to persist all day or even for several days. Consider now what would happen in an atmosphere in which such an inversion layer existed between the ground and 600 feet, but not at higher altitudes. Any "parcels" of polluted air discharged into the atmosphere at elevations below 600 feet would not be able to rise; the 600-foot level would act as a ceiling, or lid, that traps the pollutants below it. In fact, even if the lower air were turbulently stirred by warm currents, the total volume available for mixing would be limited by the 600-foot ceiling, and any pollutants discharged at lower levels would concentrate beneath the ceiling. The longer the inversion lasted, the greater the build-up. However, a chimney tall enough to penetrate the 600-foot barrier would discharge its effluents into the superadiabatic atmospheric layers where they would be diluted in a much larger volume (Fig. 9.5*D*).

The world's tallest chimney (as of 1972), built for the Copper Cliff smelter in the Sudbury District of Ontario, Canada, is as tall as the Empire State Building. It is illustrated in Figure 9.6 together with three

FIGURE 9.6 World's tallest chimney. Built in Ontario, Canada, at a cost of $5.5 million, this chimney stands 1,250 feet high. (Photo courtesy of M. W. Kellogg Company, division of Pullman Inc.)

smaller ones. Such heroic structures can be quite effective in reducing ground-level concentrations of pollutants. For example, during a 10-year period studied by the Central Electric Generating Board in Great Britain, the SO_2 emissions from power stations increased by 35 per cent, but, as a result of the construction of tall stacks, the ground level concentrations decreased by as much as 30 per cent. However, tall chimneys do not collect or destroy anything, and, therefore, they do not reduce the total quantity of pollutants in the Earth's atmosphere.

9.7 THE EFFECTS OF AIR POLLUTION – INTRODUCTION

(Problem 17)

The idea that polluted air can be harmful to man dates back at least to the Middle Ages and to the concept of poisonous airs, or "miasmas." The Italian expression for bad air is *mala aria,* from which is derived the word "malaria," a disease once erroneously associated with the odors of swamps rather than with the germs carried by the mosquitoes that breed there. More direct evidence of bad effects from polluted air began to accumulate after the first use of coal, around the beginning of the fourteenth century. The dark smoke, the unpleasant odors, the blackening of buildings and monuments (see Fig. 9.14), all clearly resulted from the addition of unnatural and unwholesome substances to the atmosphere.

The first noticeable effect of air pollution is that visibility is reduced, as illustrated by Figures 9.7 and 9.8. At times in London and in some American cities

FIGURE 9.7 Air pollution in the form of visibility-restricting particulate materials engulfs San Francisco, California, on June 2, 1960. (San Francisco News-Call Bulletin photo. From Stern: ***Air Pollution.*** **2nd Ed. New York, Academic Press, 1968.)**

'. . . And from the top there'll be a spectacular view.'

FIGURE 9.8 The Gateway To The West arch, in St. Louis, Missouri. Left: before its completion, a cartoonist speculates. Right: the finished product. (Cartoon by Engelhardt, in the *St. Louis Post-Dispatch,* **from U.S. Public Health Service Publication No. 1561:** *No Laughing Matter.* **Photo from O'Sullivan: Air Pollution.** *Chemical and Engineering News,* **June 8, 1970.)**

the effect has been severe enough to curtail the flow of traffic. The reduction of visibility is caused by the scattering of light by small particles in the air; the most pronounced reduction is produced by particles between 0.5 and one micrometer in diameter.

Only in the past few decades have we begun to realize the extent and complexity of air pollution effects and the imprecise nature of our knowledge about them. We may classify these effects into six categories: (a) the effect on climate, (b) the effect on human health, (c) damage to vegetation, (d) injury to animals, (e) deterioration of materials, and (f) aesthetic effects.

9.8 EFFECTS OF AIR POLLUTION ON THE ATMOSPHERE AND ON CLIMATE

(Problems 12, 13, 14, 15, 16)

The possibility that the Earth's climate, particularly its temperature, may be drifting away from its previous range of conditions is alarming. A warming portends a melting of glaciers and a flooding of the populous coastal plains. A cooling presages a new ice age. Changes in precipitation may not be quite so cataclysmic, but, nonetheless, a greater or lesser rainfall, or its redistribution, can have profound effects on agriculture, and so on people.

Can the pollution of the Earth's atmosphere, then, affect its climate? To approach this question, we must first consider what happens to the solar radiation that the Earth receives. As we pointed out in Chapter 1, all of this energy is eventually radiated back into space; the fact that the Earth, in spite of various temperature fluctuations, is not suffering an unchecked heating or cooling attests to this complete exchange of radiant energy. Figure 9.9 shows how this balance is distributed. Only 21 per cent of the incident solar radiation strikes the solid earth directly. The other 79 per cent is intercepted by the atmosphere—the clouds, gases, and aerosol particles. Some of this radiation is

FIGURE 9.9 Energy balance of the Earth. The sets of numbers in the dashed areas total 100 per cent.

reflected back to space, some absorbed, and some scattered down to the surface of the Earth. The surface of the Earth, in turn, loses energy by evaporation of water, by conduction, and by radiation. Much of this loss is absorbed by the atmosphere, from which it is radiated either out to space or back to the Earth. The *temperature* of the Earth's surface or atmosphere depends on the amount of solar radiation that it *absorbs;* that which is reflected is, in effect, "lost." This reflected fraction is called the **albedo**; the value shown in Figure 9.9 is 31 per cent. Obviously, therefore, more reflection portends a colder Earth; more absorption a warmer one. Let us consider both effects in turn.

Reflection of Radiation

Energy is scattered into space by reflection. Dust reflects light. Pollution makes dust. Therefore, one would expect that the effect of dust pollution is to cool the earth. What evidence do we have to bear on the accuracy of this conclusion? So far, major volcanic eruptions far outstrip man in producing dust (but man is catching up). The most spectacular eruption in modern times was that of Krakatoa (near Java) in 1883; its dust particles stayed in the atmosphere for five years. Summers seemed to be cooler in the Northern Hemisphere during this period, although the extent of temperature fluctuations makes even this observation somewhat doubtful. Volcanic eruptions occur continually here and there on the earth, and they do inject large quantities of small particles into the upper atmosphere. The best evidence seems to be that such particles do account for some back-scattering of solar radiation. What, then, about man-made dust? Of course, much of the larger particulate matter falls to the ground (see Fig. 9.1) or is washed down by rain. There is some persistent introduction of small particles into the upper atmosphere where they join the volcanic dust in increasing the earth's albedo. However, this effect is as yet puny compared with that of volcanic dusts. The effect of pollutant particles in the lower atmosphere, where most of them are concentrated, is more complicated. Particles can absorb as well as reflect radiation, and the net heat effect of man-made dusts near the ground is not at all easy to interpret. In any event, temperatures in the lower atmosphere seem to be rising slowly rather than falling. There is therefore no convincing evidence that pol-

"Cerro Negro Volcano (Nicaragua) blanketed the countryside and the city of Leon (17 miles away) with ash from October 23 to December 7, 1968. Reprinted by permission of *Science & Public Affairs,* the Bulletin of the Atomic Scientists. Copyright © by the Educational Foundation for Nuclear Science.

lutant dusts in the lower atmosphere have any important effect on the Earth's temperature. However, the dust that settles to the ground is another matter. Most of us have seen how snow in the city can become dirty after a few days. Dust that falls on snow and ice in mountainous and polar regions depresses their albedos. Snow and ice that retain heat may melt, and if this occurs extensively, it may produce a rise in sea level with significant global consequences.

Absorption of Radiation: The "Greenhouse Effect"

If there were no atmosphere, the view from the Earth would be much like that which the astronauts see from the moon—a terrain where starkly bright surfaces contrast with deep shadows, and a black sky from which the sun glares and the stars shine but do not twinkle. The atmosphere protects us by serving as a light-scattering and heat-mediating blanket. As shown in Figure 9.9, about half of the incident radiation from the sun passes through the atmosphere to the Earth; the rest is reflected or absorbed in the atmosphere. Of the heat emitted from the Earth, a large portion (represented by the arrow that curves down) is reabsorbed by the atmosphere and is, in effect, conserved, with the result that the surface of the Earth is warmer than it would otherwise be. This warming is called the "**greenhouse effect**," by analogy with the ability of a greenhouse to keep its inside warmer than the outside during the daytime. The energy emitted from the Earth is, of course, invisible (the Earth does not glow); it is largely infrared radiation, sometimes known as heat rays. The atmospheric gases that absorb infrared are ozone, water, and carbon dioxide. Water plays the major role in absorbing infrared because it is so abundant. Ozone is the least important because there is so little of it. Carbon dioxide is also important, particularly because our combustion of fossil fuels produces more and more of it, and its concentration in the atmosphere is increasing. This trend presages a warming of the Earth's surface temperature. We want to learn what actually has been happening, and what we can expect to happen if we continue to burn fossil fuels.

The term "greenhouse effect" is partially a misnomer. It was once thought that a greenhouse keeps its inside warm because its glass walls are transparent to sunlight but opaque to the infrared energy that is reradiated from the interior. However, it has been shown that most of the heat conservation is accounted for by the fact that the closed space prevents the exchange of warm air with the colder air outside. The atmosphere, of course, is open and therefore imposes no such restriction.

The CO_2 concentration in any one area fluctuates between night and day by about 70 ppm during the growing season. For this reason, and because the older analytical methods were not sufficiently precise, we do not know the pre-Industrial Revolution levels as accurately as we would like. In recent years we have been measuring CO_2 concentrations in polar regions and on high mountains, locations where they do not vary so much. Our best estimate is that the worldwide CO_2 concentration has increased by 10 to 15 per cent during the past century, that is, from about 280 ppm to its present level of 315 ppm, and that this rate of increase is rising. We don't know just how to extrapolate this trend because we are not sure how the world ecosystem will respond to it—that is, we don't know enough about the operations of the earth's CO_2 sinks. If it is a matter of dissolving CO_2 in the oceans, then the prospect is poor, because the solubility of CO_2 decreases as the temperature goes up, and warmer waters will therefore hold less. This implies that the worse it gets, the worse it will get. If, on the other hand, the major sink is the total biomass of plant matter, or the solid carbonates of the earth's crust, the upward trend could be counteracted.

A simplified equation for such conversion is:

$$CO_2 + Ca(OH)_2 \rightarrow CaCO_3 + H_2O$$

Scientists estimate that, if the rate of removal remains constant, a doubling of the present CO_2 concentration, which is an increase of about 300 ppm, would require 400 years and would warm the surface of the earth by an average of about two Celsius degrees (3.6 Fahrenheit degrees). Fluctuations of this magnitude have occurred in recorded history, and the glaciers didn't melt, so don't sell your beach property on this account.

Precipitation

The aerosol particles in cities whose atmospheres are polluted serve as nuclei for the condensation of moisture. Fogginess, cloudiness, and perhaps rainfall are usually increased considerably in contrast with the less polluted countrysides. However, the rise of warm air from cities, and the circulation of cooler surrounding airs that is thereby induced, can sometimes reverse this situation.

The worldwide effects of pollutants on precipitation are even more difficult to assess. The concentration of small particles in the lower atmosphere is increasing over wide areas. The lead compounds from automobile exhaust are of particular concern because some of them, particularly lead chloride and bromide (see page 349) are chemically similar to cloud-seeding compounds. That is to say, they serve particularly well as nuclei for the condensation of water droplets or ice

Ice fog from emissions from industrial processes, Edmonton, Alberta. (Photo by Dr. James W. Smith, Toronto, Canada.)

crystals. This action could lead to more precipitation, if the droplets coalesce, or less precipitation, if the nuclei are so small that the droplets stabilize and coalescence is inhibited.

Jet Aircraft

There has been concern that jet aircraft, and particularly the supersonic transport (SST), whose flight patterns are in the 60,000- to 70,000-foot range, could engender stratospheric air pollution with consequent changes in climate. Jet exhaust contains water, CO_2, oxides of nitrogen, and particulate matter. The question is, can the effects of these pollutants be harmful? The answer, at this time, is speculative. For example, it is estimated that a fleet of 500 SST's over a period of years could increase the water content of the stratosphere by 50 to 100 per cent, which could result in a rise in average temperature of the surface of the Earth of perhaps 0.2 Celsius degrees and could cause destruction of some of the stratospheric ozone that protects the Earth from ultraviolet radiation. The particulate matter would nucleate clouds of ice crystals and would increase the stratospheric albedo to some extent. This could lead to a cooling of the Earth's temperature, probably of only a small magnitude. One visible effect of such continued high-altitude pollution

Jet exhaust. (Photo by David F. Hall; courtesy of the Connecticut Lung Association.)

would be that the skies would gradually become hazier and lose some of their blueness.

Much of the discussion in this section has been conjectural, because we know so little. Therefore, we cannot predict with confidence whether the effects of man-made pollution on the Earth's climate will be trivial or serious, soon or late, rapid or slow. Environmentalists take the position that we should exert all the restraint we can against trifling with the Earth's atmosphere, and that, for example, we should not trade such risks to save a few hours of long-distance travel time for a tiny fraction of the population. The recent defeat of the SST appropriation bill in the U.S. Congress is consistent with such a view. Others take the position that technical developments make life more comfortable, that no serious worldwide climatic effects have yet been demonstrated, and that technological remedies can always be developed to get us out of trouble. The general aspects of such questions are discussed in the last chapter.

9.9 EFFECTS OF AIR POLLUTION ON HUMAN HEALTH

(Problems 18, 19)

In the previous sections of this chapter we have discussed the sources of air pollution and the types of pollutants that occur commonly in industrial societies

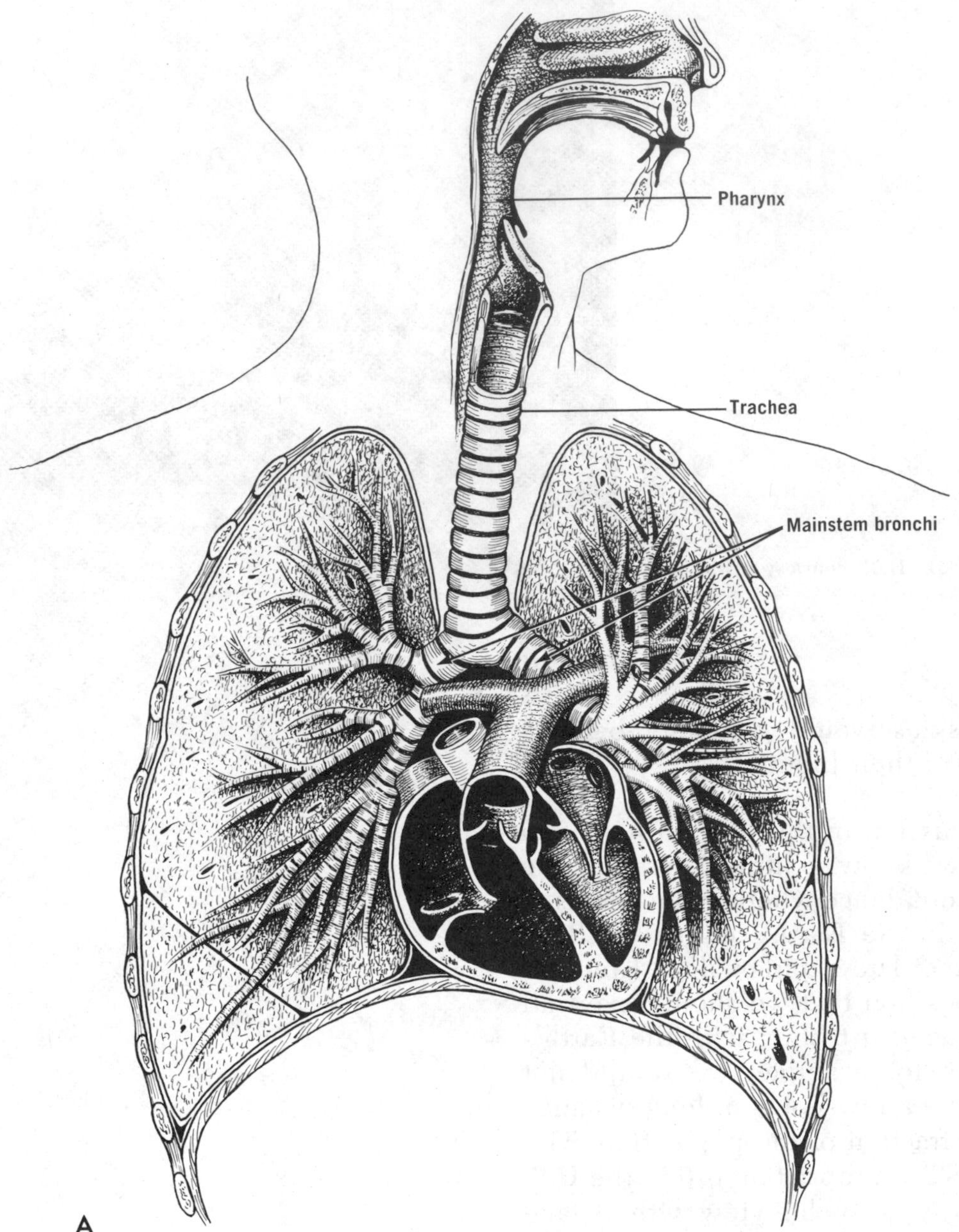

FIGURE 9.10 **A diagram of the human respiratory tract. Air is breathed into the pharynx and from there travels through the *trachea*, or windpipe, which divides into the two *mainstem bronchi*. Each of these subdivides into airways which in turn subdivide into conduits of increasingly smaller caliber until they terminate in the *alveoli*, or air sacs. It is in the air sacs that gases in the inspired air exchange with the gases dissolved in the blood. This gas exchange is the function of the lung. Lining the walls of the airways are cells, some of which form glands that secrete mucus. The walls are also covered with hairlike projections called *cilia*, which wave in synchrony with one another and serve to propel the mucus and any impurities from the air that may be suspended or dissolved in it up toward the *pharynx*, where it is usually swallowed.**

Figure 9.10 continued on opposite page.

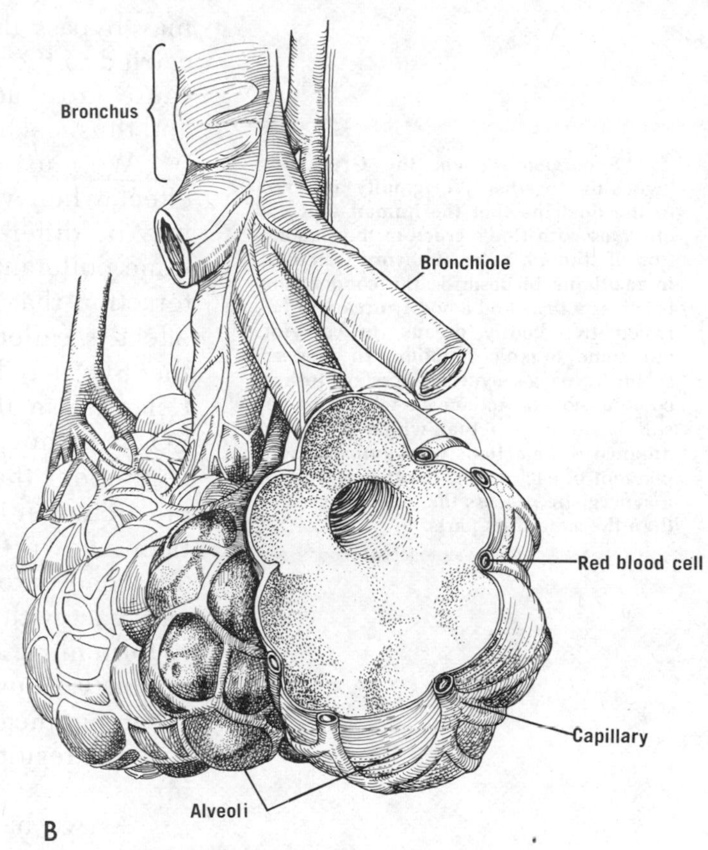

FIGURE 9.10 *Continued.*

whenever large numbers of people live and work in relatively small areas. As mentioned previously, air pollutants may be either particulate or gaseous. The fate of a particle once it is inhaled depends largely on its diameter. If it is greater than about two micrometers (microns), it is usually trapped in the nasal passage or in the mucus of the bronchi; in the bronchi the action of the cilia drives the mucus up toward the trachea whence it is either coughed up or swallowed. If the diameter is less than two micrometers, the particle may be carried all the way through the air passages into the alveoli, or air sacs, of the lung where it may be trapped by specialized cells lining the alveoli or absorbed into the bloodstream (see Fig. 9.10).

For gases the determining factor is largely the solubility of the gas in water. Since biological tissues are rich in water, a water-soluble gas like SO_2 will rapidly dissolve in the soft tissues of the mouth, nose, throat, bronchi, and eyes where it produces the characteristic dry mouth, scratchy throat, and smarting eyes that most city dwellers have experienced at intervals. By contrast, NO_2, which is relatively insoluble, may bypass this part of the respiratory tract and be carried to the alveoli where in very high doses it may cause gross accumulation of fluid in the air spaces, and thus make effective lung function impossible.

"Synergism" (from the Greek for "working together") originally referred to the doctrine that the human will co-operates with God's grace in the redemption of sinners. Later, the word was used in medicine to describe the cooperation between a drug and a bodily organ, or between two bodily organs, in carrying out some organic function. In modern scientific usage, synergism refers to any combination of actions in which the result is more than that which would be attained if the actions were entirely independent of each other. In other words, in a synergistic process the whole is greater than the sum of its parts.

We must emphasize, however, that the net toxic effect when various pollutants are inhaled together may be different from the sum of the effects of these same pollutants if they are inhaled separately. An interaction that produces *more* than a merely additive effect is called **synergism.** It is known, for example, that SO_2 may be adsorbed onto particles; if these particles are smaller than two micrometers, molecules of SO_2 may thus gain access to the alveoli in greater concentrations than they would otherwise. The retention of carcinogenic hydrocarbons in the human body has been shown to be greatly enhanced if they are first adsorbed onto soot particles. Also, oxygen and water can react with SO_2 to form sulfuric acid and with NO_2 to form nitric acid. It is clear, then, that the interaction of the various pollutants with each other may have great significance in determining the types of toxicities that result when a human being breathes polluted air.

As we have noted, most city residents have frequent personal experiences with what we might call the nuisance effects of dirty air. Indeed, if we define ill health simply as the absence of a feeling of physical well-being, then city residents are almost universally the victims of air pollution. Whether the dry mouth, smarting eyes, and offended nose constitute grounds in themselves for limiting pollution of the air is not a trivial question. As we shall see, however, many instances have occurred in which increased air pollution levels have had very serious consequences on the level of health of the exposed population.

Acute Effects—Donora and London

In 1948 Donora, Pennsylvania, was a small industrial city of 14,000 people located in the Monongahela River Valley. Operating in the city at that time were a large steel mill, a plant manufacturing sulfuric acid, and a zinc production plant. In the fall of that year a temperature inversion occurred, thus preventing the normal dissipation of the effluents from the various industrial establishments for several days. As described by Roueché in his book *Eleven Blue Men,* "The weather was raw, cloudy, and dead calm, and

B. Roueché: *Eleven Blue Men.* Boston, Little, Brown, 1953.

it stayed that way as the fog piled up all that day and the next. By Thursday, it had stiffened adhesively into a motionless clot of smoke [see Fig. 9.11]. That afternoon it was just possible to see across the street, and except for the stacks, the mills had vanished. The air began to have a sickening smell, almost a taste. It was the bittersweet reek of sulfur dioxide. Everyone who was out that day remarked on it, but no one was much concerned. The smell of sulfur dioxide, a scratchy gas given off by burning coal and melting ore, is a normal concomitant of any durable fog in Donora. This time it merely seemed more penetrating than usual." The population soon became concerned, however. Over the three-day period of maximal pollution, nearly half the population became ill, and 20 died. Most of the deaths were people with pre-existing pulmonary or cardiac disease.

In 1952 the population of London experienced for four days a similar build-up of the usual air contaminants that affected the British Isles. During that period, and continuing for the next two to three

FIGURE 9.11 Region of Donora, Pennsylvania.

weeks, there were nearly 4000 more deaths in London than would have been expected for that time of year. As in Donora, those most affected were in the older age groups and generally had disease of the heart or lungs prior to the pollution episode. A much larger number of individuals became ill, chiefly with exacerbations of chronic respiratory and cardiac ailments.

Such disasters are fortunately infrequent. They all have certain common patterns. First, those who become severely ill usually are elderly or have chronic heart or lung disease. Secondly, as the London disaster showed, the increase in illness and mortality following the acute episode actually lasted for weeks. What probably happened was that the pollutants interfered with the normal mechanisms of the respiratory tract for dealing with invading pathogenic organisms, and that this lowered resistance continued for a time after subsidence of the intensified air pollution.

Personal Air Pollution—Cigarettes

In 1964 the Office of the Surgeon General of the United States Public Health Service published a volume entitled *Smoking and Health.* This represented the attempt of a distinguished panel of medical scientists, chemists, and statisticians to evaluate all the existing evidence concerning the effects of smoking on health. The aim of the panel was not simply to define whether there was an *association* between smoking and various diseases, for such associations have been known for years. The true purpose was to discover whether there were strong grounds for believing that a cause-and-effect relationship existed between smoking and disease. Ideally, in science we prove cause-and-effect relationships by doing carefully controlled laboratory experiments in which only one factor is varied at a time; in this way, causes can be isolated. Since in most cases both technical and ethical considerations preclude these sorts of experiments on causation of chronic diseases in man, we must rely instead on a variety of indirect approaches. In our case, these include (a) evidence of a strong statistical association between smoking and disease; (b) evidence that within a population whose members smoke, the probability of developing a disease is related to the amount of smoking; and (c) supporting chemical and biological evidence in animal experiments. Let us consider these categories one at a time.

The association of two events does not by itself constitute evidence for a cause-and-effect relationship. For example, it turns out that the annual birthrate in Great Britain from 1875 to 1920 is very strongly correlated with the annual production of pig iron in the United States during the same period! We reject the notion that the rising pig iron production in the U.S. caused the rise in the English birthrate for reasons that have nothing to do with the strength of the correlation, but rather because an analysis of the problem of birthrates in England does not suggest that American pig iron production is a reasonable determinant.

Association of Smoking and Disease

The best current figures show that average smokers have an approximately ten-fold risk of developing and dying of lung cancer, a six-fold risk of death from pulmonary disease, and a nearly two-fold risk of dying from coronary heart disease, as compared with a non-smoking population. There are also a host of other illnesses which are significantly, though less strikingly, associated with smoking.

Dose Effect

Within the group of people who smoke, rates of incidence of lung cancer can be shown to increase with all of the following factors: number of years that the individual has smoked, number of cigarettes a day, earliness of the age when the individual started smoking, and amount inhaled. Conversely, the incidence of cancer drops roughly in proportion to the length of time since discontinuance of smoking. In other words, the more you smoke, the greater your chances of getting lung cancer.

Supporting Laboratory Evidence

Chemical analysis of tobacco and cigarette smoke has revealed the presence of at least seven distinct polycyclic hydrocarbons which have been shown to produce cancer in animals under certain conditions. Cigarette smoke also contains polonium-210, a radioactive substance that may be carcinogenic. It should be mentioned that intensive efforts have thus far failed to induce lung tumors in experimental animals with cigarette smoke. This failure does not shift the weight of evidence very heavily since there are certainly significant species differences in susceptibility to carcinogens, and failure to induce tumors in mice may say nothing about the ability of the same smoke to induce cancers in a man or woman who has been inhaling it for decades.

To summarize simply, the weight of evidence is that the smoking of cigarettes is the prime cause of this century's lung cancer "epidemic" and very important in the genesis of heart and pulmonary diseases as well.

The adjective epidemic **(from the Greek** epi, **upon, and** dēmos, **people), referring to a disease, means "common or prevalent among a people or community at a special time." Lung cancer, pulmonary disease, and coronary heart disease are therefore epidemic because they are extremely** common **illnesses. For example, if the 10-fold increase in lung cancer in smokers is in fact caused by cigarette smoking, as most scientists believe for the reasons enumerated above, then in 1961 about 36,000 people died of lung cancer that would not have developed the disease had they not been smokers.**

Chronic Effects of Community Air Pollution

Another question of concern to medical scientists for many years has been the issue of whether chronic exposure to polluted air results in a higher rate of illness and death than is experienced by those who breathe relatively clean air. Many illnesses have been examined, but the ones that have prompted more research than any others have been lung cancer, chronic bronchitis, and emphysema. By bronchitis, we mean an inflammatory condition of the bronchi. When the bronchi become inflamed, the cell layers that line them become thickened and the mucus content of the air passages increases markedly. The small airways usually become constricted, and segments of the lung may become infected with bacteria or viruses. Emphysema, by contrast, refers to a breakdown in the walls of the alveoli themselves. Since emphysema and chronic bronchitis usually coexist in the same patient, we generally refer to the combination as chronic obstructive pulmonary disease (COPD).

Several populations have an especially high risk of developing these conditions. One such population is that of industrialized cities. Some studies show that the prevalence of COPD in cities is up to four times that in the country. Before incriminating air pollution alone, however, we must first consider in what other respects cities differ from rural areas. Clearly, there are many differences. Population density, occupations of the inhabitants, socioeconomic levels, ethnic composition, and even possibly smoking habits, age distribution of the population, and other factors may vary from city to country. If they do differ, we cannot assume that differences in the incidence of any disease from city to country may not be due to one of these, or other, unidentified factors. Many of the better studies of this problem take one or more of these variables into account. The preponderance of the evidence indicates that, after most of these factors have been taken into consideration, there still remains an excess incidence of both lung cancer and COPD in urban environments. The most likely element of the "urban factor" seems to be air pollution, though conclusive data is lacking. Compared with cigarette smoking, however, the effect is small.

Moreover, cigarettes and the urban environment seem to have a more than additive effect on the incidence of lung cancer. That is, if one compares urban non-smokers with rural non-smokers, the difference in incidence of lung cancer or COPD is small; but urban smokers have a much greater incidence of the disease than rural smokers. These studies indicate that the combination of smoking and living in large cities is particularly pernicious.

Recent advances in the analysis of particulate matter and oxides of nitrogen and sulfur have made possible better quantitation of air pollution. It should now be possible to approach the important question of whether the development of respiratory illness is quantitatively related to levels of air pollution, that is, whether the same kind of dose-response relationship prevails with air pollution as with cigarette smoking.

Thus far, the only effects of air pollution we have considered are those on mortality from lung cancer and COPD. It is quite clear, though, that much of the illness brought about by breathing polluted air may not cause death, but rather just increased suffering and debility. Moreover, studies from Japan and the United States both indicate that effects on the respiratory system may in fact begin in childhood. Children living in highly polluted areas in both countries have been shown to have a much increased incidence of acute respiratory infections, and the Japanese studies have also documented increased airway constriction during periods of high pollution.

Those with pre-existing cardiac disease, particularly the type caused by narrowing of the blood vessels to the heart (so-called coronary heart disease) are, as we have mentioned, at high risk of serious illness when subjected to very high levels of air pollution. Recently, scientists studied chronic heart patients as they were driven in automobiles along a freeway in the Los Angeles area during heavy morning traffic. After a 90-minute drive, the subjects showed significantly less ability to do a controlled series of exercises without experiencing pain in the chest (this is a standard test to assess the severity of coronary heart disease); their pulmonary function also showed deterioration, and their blood levels of carbon monoxide had risen significantly. In order to show that these effects were related to the breathing of polluted air rather than, say, the harrowing experience of simply driving along a freeway in southern California during rush hour, the experiment was repeated the next day, but this time the subjects breathed purified compressed air during the trip. After this part of the experiment, their exercise testing and lung function test showed no deterioration at all. These studies show clearly that air polluted by ordinary automobile exhaust at a level that is probably attained most days during the year can have deleterious effects on the cardiovascular function of people whose heart is already compromised by other disease. One wonders how many people have died (or killed others) on crowded highways from heart attacks triggered by dirty air.

9.10 DAMAGE TO VEGETATION

Air pollution has caused widespread damage to trees, fruits, vegetables, and ornamental flowers (see Color Plate 9). In fact, the total annual cost of plant damage in the United States has been estimated at close to one billion dollars. The most dramatic early instances of such effects were seen in the total destruction of vegetation by sulfur dioxide in the areas surrounding smelters (Fig. 9.13), where this gas is produced by the "roasting" of sulfide ores, as shown in the equation on page 342.

We now know that there is a wide variety of patterns of plant damage by air pollutants. For example, all fluorides appear to act as cumulative poisons to plants, causing collapse of the leaf tissue. Photochemical (oxidant) smog bleaches and glazes spinach, lettuce, chard, alfalfa, tobacco, and other leafy plants. Ethylene, a hydrocarbon that occurs in automobile and diesel exhaust, makes carnation petals curl inward and ruins orchids by drying and discoloring their sepals.

FIGURE 9.12 The Copper Basin at Copperhill, Tennessee. A luxuriant forest once covered this area until fumes from smelters killed all of the vegetation. (U.S. Forest Service photo. From Odum: ***Fundamentals of Ecology.*** **3rd Ed. Philadelphia, W. B. Saunders Co., 1971.)**

Chemical manufacturing, petroleum refinery, and storage areas, northern New Jersey. (From New Jersey Dept. of Environmental Protection.)

The fact that the symptoms of plant damage are usually characteristic of the specific causative pollutant helps us to monitor the distribution of the pollutant. In other words, the damaged plant serves as evidence that the pollutant was there. This is helpful because the pollutant itself is usually non-persistent and, in the absence of a dense network of sampling stations, there is often no other reliable way to get such information.

9.11 INJURY TO ANIMALS

Countless numbers of North American livestock have been poisoned by fluorides and by arsenic. The fluoride effect, which has been the more important,

FIGURE 9.13 A cow afflicted with fluorosis.

arises from the fallout of various fluorine compounds on forage. The ingestion of these pollutants by cattle causes an abnormal calcification of bones and teeth called **fluorosis**, resulting in loss of weight and lameness (see Fig. 9.13). Arsenic poisoning, which is less common, has been transmitted by contaminated gases near smelters.

9.12 DETERIORATION OF MATERIALS

Acidic pollutants are responsible for many dammaging effects, such as the corrosion of metals and the weakening or disintegration of textiles, paper, and marble. Hydrogen sulfide, H_2S, tarnishes silver and blackens leaded house paints. Ozone, as previously mentioned, produces cracks in rubber.

Particulate pollutants driven at high speeds by the wind cause destructive erosion of building surfaces. And the deposition of dirt on an office building, as on a piece of apparel, leads to the expense of cleaning and to the wear that results from the cleaning ac-

FIGURE 9.14 Old post office building being cleaned in St. Louis, Missouri, 1963. (Photo by H. Neff Jenkins. From Stern: *Air Pollution.* 2nd Ed. New York, Academic Press, 1968.)

tion (see Fig. 9.14). The total annual cost in the United States of these effects is very difficult to assess but has been estimated at several billion dollars.

9.13 AESTHETIC EFFECTS

Among the various effects of air pollution, climatic changes are slow, the deterioration of health is insidious, and the damage to plants, animals, or materials seems remote to those not directly involved. But smoke is visible to all, and the stinks of sulfides are pervasive. As a result, most complaints by individuals to air pollution agencies refer to something that can be seen or smelled.

A view of distant mountains through clear air is aesthetically satisfying, and an interfering acrid haze is therefore a detriment. Moreover, such aesthetic insults can be quantified by observing how they influence people's economic behavior. The depression of property values in polluted areas and the purchases of domestic air-cleaning devices, for example, have been used as indices of annoyance.

Unpleasant aesthetic effects cannot be neatly separated from the other disruptions caused by air pollution. The acrid haze referred to above is sensed not only as an annoyance, but also as a harbinger of more direct harm, somewhat as the smell of leaking gas forbodes an explosion. Thus the evident pollution engenders anxiety, and anxiety may depress our appetites, or rob us of sleep, and these effects, in turn, can be directly harmful.

9.14 AIR POLLUTION INDICES

(Problems 21, 22, 23)

The wind dies down and the air is still, but traffic continues to move, homes are heated, and industry operates. The air becomes hazy, murky, then uncomfortable. How dangerous is it? Shall we close down the factories, stop the traffic? Clearly, to answer such questions we must measure something. But what? SO_2? Particulates? Ozone? Oxides of nitrogen? All of them? And if we measure more than one, how shall we weigh the different measurements?

One reasonable approach is to adhere to a standard for *each* pollutant, and then consider the atmosphere to be hazardous when it exceeds any *one* of the standards. The degree of hazard depends on the degree to which the concentrations of pollutants exceed the tolerable levels. Alternatively, we may assign relative weights to the pollutants on the basis of their relative potentials to produce harmful effects. When these separately weighted numbers are added together, we get a total which serves as an overall **air pollution index**. A sequence of alerts is often keyed to the magnitude of the index.

Of course, various kinds of judgments are involved in the formulation of the index. We have seen that air pollution affects people, animals, plants, materials, and climate. How shall we rank these? And some consideration is necessarily, if not explicitly, given to the cost of control. Dare we construct an index so severe that it will demand curtailment of traffic a few days each month? Then how will people get to work? Finally, we may question whether it is reasonable to *add* the various pollution numbers. Mere addition fails to recognize adverse synergistic effects, such as the baleful combination of particulate matter and SO_2 described on page 364.

Well, we have to start somewhere, and as we mentioned above, the reasonable point is a set of standards. These refer to "tolerable" concentrations of pollutants in the outside air, and they are called **ambient air quality standards**. They differ from what we allow to come out of a chimney; the latter levels are called **emission standards**. Remember, we don't expect to find people at chimney tops; what we want to do is to regulate the emissions so that, by the time they reach the ground, they will conform to the ambient air quality standards.

Table 9.2 displays the ambient air quality standards (for six pollutants) as set by the United States Environmental Protection Agency, effective April 30, 1971. Note that the possibilities of synergistic effects are not recognized. Table 9.3 shows the standards for the State of California.

The (San Francisco) Bay Area Air Pollution Control district has constructed an index based on the California standards. This index adds the values for four of the pollutants after assigning the relative weights to them as shown in margin at left.

Oxidant	200
NO_2	100
CO	1
Coefficient of Haze (a measure of particulate matter)	10

The daily air pollution index of the New York City Department of Air Resources is based on the assumption that the air quality is as bad as its *worst* pollutant. The index is expressed by any of four adjectives: "good," "acceptable," "unsatisfactory," and "unhealthy." Five pollutants are measured: SO_2, NO_2,

TABLE 9.2 FEDERAL AIR QUALITY STANDARDS (AS OF APRIL 30, 1971)

	CONCENTRATION
SULFUR DIOXIDE	
Arithmetic Mean (annual)	0.03 ppm
24-hour concentration not to be exceeded more than once per year	0.14 ppm
SUSPENDED PARTICULATES	
Geometric Mean (annual)	$75 \mu g/m^3$
24-hour concentration not to be exceeded more than once per year	$260 \mu g/m^3$
CARBON MONOXIDE	
8-hour concentration not to be exceeded more than once per year	9 ppm
1-hour concentration not to be exceeded more than once per year	35 ppm
PHOTOCHEMICAL OXIDANTS	
1-hour concentration not to be exceeded more than once per year	0.08 ppm
HYDROCARBONS	
3-hour concentration not to be exceeded more than once per year	0.24 ppm
NITROGEN DIOXIDE	
Arithmetic Mean (annual)	0.05 ppm

CO, oxidants, and "smokeshade" (a manifestation of particulate pollution), and a series of ratings is given for each, as shown in Table 9.4. Carbon monoxide values are used only for the "unhealthy" index, and oxidant values only for "good," "acceptable," and "unhealthy." The daily index is then the worst single rating.

Chicago has two systems of alerts, one for combinations of SO_2 and particulate matter, the other for carbon monoxide. The first system recognizes the synergistic effect of its two components, the second recognizes the fact that CO effects are largely independent of the presence or absence of other pollutants.

More complex systems that attempt to take more

TABLE 9.3 CALIFORNIA AMBIENT AIR QUALITY STANDARDS

POLLUTANT	STANDARD
Oxidant	0.1 ppm for 1 hr.
Carbon monoxide	40 ppm for 1 hr., 10 ppm for 12 hrs.
Sulfur dioxide	0.5 ppm for 1 hr., 0.04 ppm for 24 hrs.
Visibility-reducing particles	Visibility reduced to 10 miles when relative humidity is less than 70%.
Suspended particulate matter	$100 \mu g/m^3$ for 24 hrs.
Lead (particulate)	$1.5 \mu g/m^3$ for 30 days.
Hydrogen sulfide	0.03 ppm for 1 hr.
Nitrogen dioxide	0.25 ppm for 1 hr.

TABLE 9.4 NEW YORK CITY AIR POLLUTION INDEX*

AIR POLLUTANT	*GOOD*	*ACCEPTABLE*	*UNSATISFACTORY*	*UNHEALTHY*
SO_2 (24-hr average)	0–0.03 ppm	0.04–0.05 ppm	0.07–0.10 ppm	0.10 ppm
NO_2 (24-hr average)	0–0.05 ppm	0.06–0.09 ppm	0.09–0.12 ppm	0.12 ppm
CO (max. 8-hr average)				9 ppm
Oxidants (max. 1-hr average)	0–0.03 ppm	0.04–0.08 ppm		0.08 ppm
"Smokeshade"	0–0.5 COH**	0.6–1.0 COH	1.1–1.6 COH	1.6 COH

*Note: *Good* means that the levels on this day did not exceed the Federal standards for the annual mean. *Acceptable* means that, if these were the *maximum* levels for the year, the Federal standards for the annual mean would probably be met, although they are exceeded on this particular day. *Unsatisfactory* implies the likelihood that the Federal standards would not be met if there were many days as bad as this one. *Unhealthy* means that there would be *chronic* health effects if all the days were as bad as this one.

**COH, or "coefficient of haze," is a measure of the reduction of visibility caused by particulate matter.

factors into account have been proposed. But the variability of human values mocks the precision of measurements. The rich woman sits in her penthouse and smokes a cigarette. The carbon monoxide concentration generated at the street level cannot compete with that in her smoke. What she wants is a clear view of the twinkling city lights below. But the sick old man who lives on a meager pension down there among the noise and traffic and has difficulty catching his breath doesn't care about the view. As far as he is concerned, traffic and industry could stop, because a slight rise in the SO_2 concentration is a "red alert" for him. Thus, no index can serve equally well for all.

9.15 THE CONTROL OF AIR POLLUTION–INTRODUCTION

(Problem 24)

It is easy to think of air pollution control as something one does to the smoke stack, or to the tail pipe, to stop the emissions. This is, of course, a valid concept, but it is not the only one. Instead, you could put the control device on your own face; such equipment is called a gas mask. Could you put a gas mask on your house? In a way, yes. Houses are leaky, and the air filters in and out in patterns that depend on the wind and on the construction of the walls, windows, and doors. But you could blow enough air into your house through some air-cleaning device at a sufficiently high rate that a slight pressure built up inside, and all leaks were directed outward.

We could also make cultural choices of many

kinds that serve to reduce air pollution. Many of these were discussed in the sections of Chapter 5 that deal with energy sources; for example, solar energy is non-polluting and, of course, we could simply use *less* energy.

We could also combine various social and technical choices; for example, we could eliminate sulfur from fuel before we burn the fuel; there would then be no need to eliminate it at the stack. We could modify manufacturing processes with a view to reducing air pollution, as by using chlorine-free catalysts, or by pretreating dusty materials so that they agglomerate into larger, more controllable particles, or by eliminating fluorides in the production of aluminum. Finally, manufacturing plants could be relocated away from areas that are densely populated or subject to atmospheric inversions.

We will first consider the technical methods of air-pollution control. Later, in Section 9.17, we will choose a single case (the automobile) to help us appreciate some of the complexities of social-technical interactions.

9.16 CONTROL OF AIR POLLUTION EMISSIONS

(Problems 26, 27)

There are two general classes of methods for controlling air pollution at the source: The pollutants are separated from the harmless gases and disposed of in some way other than by discharge into the atmosphere; or the pollutants are somehow converted to innocuous products that may then be released to the atmosphere.

Control of Pollutants by Separation

Before we consider specific methods, we should realize that the separation of pollutants from a gas stream cannot be a final step in pollution abatement—the collected material does not disappear, and therefore it must be handled in some way. If the disposal of this residue is not considered, the solution of an air pollution problem may create a solid waste or a water pollution problem. However, such a conversion does at least ease the situation, because it is much more convenient to handle a small volume of solid matter than a large volume of air. Typically, the volume of polluted matter in a contaminated air stream is in the range of 1/1000 to 1/10,000 of the total volume. Remember (page 336) that most of the volume occupied by a gas is empty space.

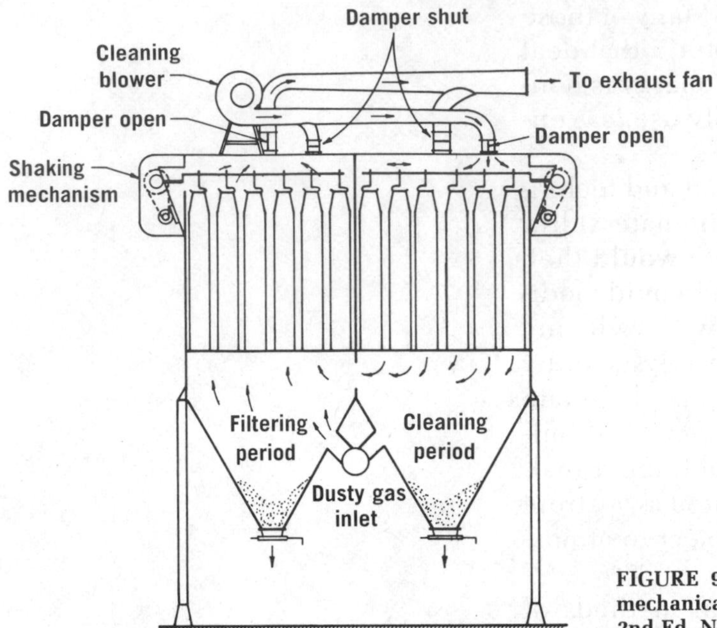

FIGURE 9.15 Typical bag filter employing reverse flow and mechanical shaking for cleaning. (From Stern: *Air Pollution.* **2nd Ed. New York, Academic Press, 1968.)**

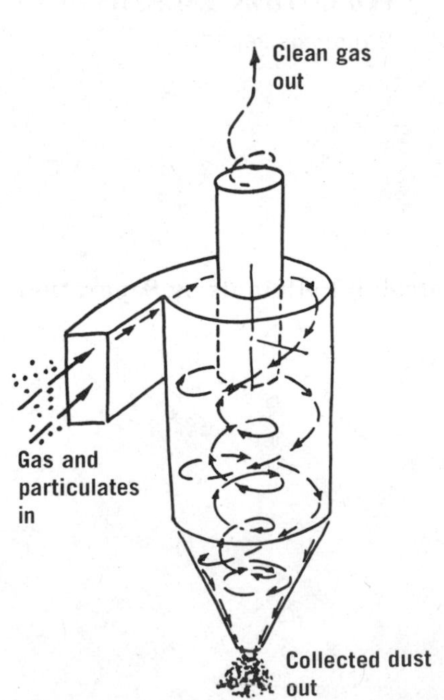

FIGURE 9.16 Basic cyclone collector. (From Walker: *Operating Principles of Air Pollution Control Equipment.* **Bound Brook, N.J., Research-Cottrell, Inc., 1968.)**

Particulate matter may be retained on porous media (filters) which allow the gas to flow through. Such separations are possible because particles are much larger than gas molecules. For handling large gas streams, the filters are often in the form of cylindrical bags, somewhat like giant stockings, from which the collected particulate matter is periodically shaken out (Fig. 9.15).

There are various mechanical collection devices that depend on the fact that particles are *heavier* than gas molecules. As a result, particles will settle faster and can be collected in a chamber that allows enough time for them to settle out. However, as evidenced by the data on settling rates shown in Figure 9.1, such methods are practicable only for very large particles. More important than their settling rate is the fact that heavier particles have more *inertia.* As a result, if a gas stream that contains particulate pollutants is whirled around in a vortex, the particles may be spun out to locations from which they may be conveniently removed. A device of this sort, called a **cyclone**, is shown in Figure 9.16.

Particles may also be electrically charged, and a collecting surface that bears an electric charge of the opposite sign therefore attracts them. Devices of this

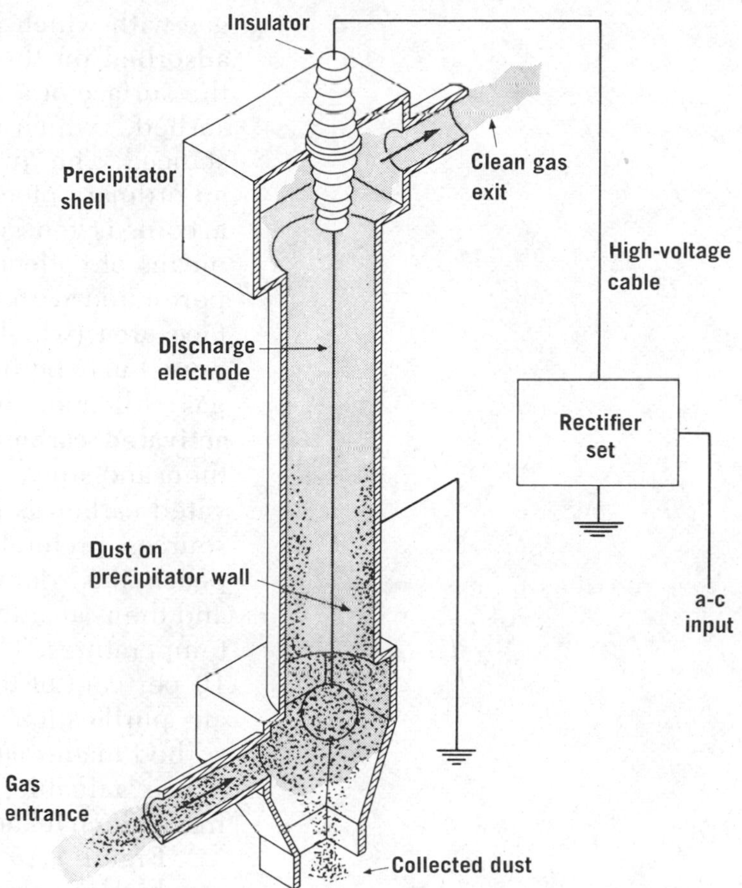

FIGURE 9.17 Basic elements of an electrostatic precipitator. (From White: "Industrial Electrostatic Precipitation." Reading, Mass., Addison-Wesley Publishing Company.)

sort, called **electrostatic precipitators** (Fig. 9.17), are used on a very large scale, notably for reducing smoke from power plants that burn fossil fuels.

Pollutant gases cannot feasibly be collected by mechanical means, because their molecules are not sufficiently larger or heavier than those of air. However, some pollutant gases may be more soluble in a particular liquid (usually water) than air is; they may therefore be collected by a process that brings them into intimate contact with the liquid. Devices that effect such separation are called **scrubbers** (Fig. 9.18). The methods of making contact between gas and liquid include spraying the liquid into the gas and bubbling the gas through the liquid. Ammonia, NH_3, is an example of a gas that is soluble in water and can be scrubbed out of an air stream.

Gas molecules adhere to solid surfaces. Even an apparently clean surface, such as that of a bright piece of silver, is covered with a layer of molecules of any

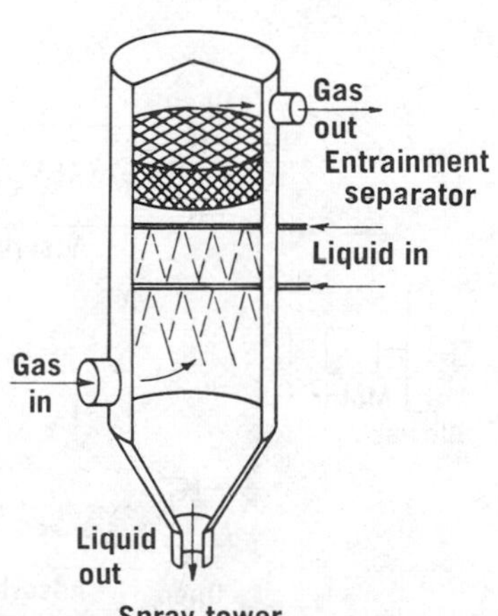

FIGURE 9.18 Schematic drawing of a spray collector, or scrubber. (From Stern: *Air Pollution*. 2nd Ed. New York, Academic Press, 1968.)

gas with which it is in contact. The gas is said to be **adsorbed** on the solid. "Adsorbed" means "held on the surface of a substance," and is different from "absorbed," which means "held in the interior of a substance." The quantity of gas that can be adsorbed on an ordinary piece of non-porous solid matter, such as a coin, is too small to be of any consequence as a means of collecting pollutants. However, if a solid is perforated with a network of fine pores, its total surface area (which includes the inner surfaces of the pores) may be increased so much that its capacity for gas collection becomes significant. Such a solid is **activated carbon**, which can have several hundred thousand square feet of surface area per pound. Activated carbon is made from natural carbon-containing sources, preferably hard ones, such as coconut shells, peach pits, dense woods, or coal, by charring them and then causing them to react with steam at very high temperatures. The resulting material can retain about 10 per cent of its weight of adsorbed matter in many air purification applications. Furthermore, the adsorbed matter can be recovered from the carbon and, if it is valuable, recycled into the process or product from which it had escaped.

Figure 9.19 shows a two-stage adsorption system in which the contaminated air stream is flowing down through adsorber 1, from which the purified air is exhausted to the atmosphere. Meanwhile, adsorber 2,

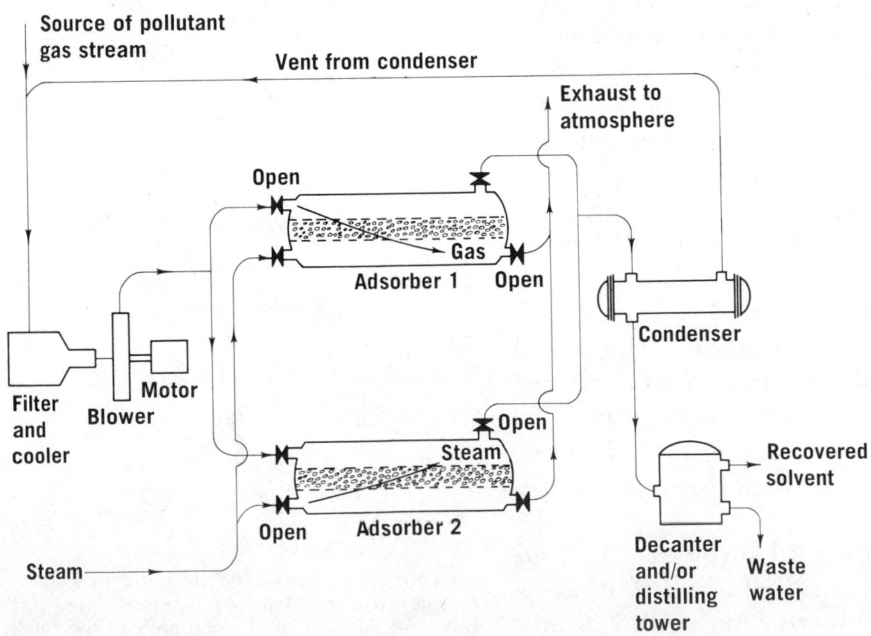

FIGURE 9.19 Two-stage system for adsorbing and collecting air-polluting gases.

which had been previously saturated, is being subjected to an upward flow of clean steam to regenerate it. The steam is then condensed and the adsorbed material is recovered. The adsorption bed consists of carbon granules about 1/8 inch in diameter, and it may seem strange that gases, which pass through the finest filter, can be effectively purified by a bed of loose granular material. But remember that on page 336 we mentioned that gas molecules move very rapidly between collisions. This rapid darting motion, when superimposed on the much slower speed on the entire current of gas, enables the molecules to reach some surface of the granular adsorbent particles in a very short time.

Control of Pollutants by Conversion

By far the most important conversion of pollutants is by oxidation in air. Oxidation is applied most often to pollutant organic gases and vapors, rarely to particulate matter. When organic substances containing only carbon, hydrogen, and oxygen are completely oxidized, the sole products are carbon dioxide and water, both innocuous. However, the process is often very expensive because considerable energy must be used to keep the entire gas stream hot enough (about 700°C) for complete oxidation to occur. If the pollutant is sufficiently concentrated, its own fuel value may contribute a large part of this energy. Also, the required combustion temperature may be reduced by using a catalyst.

There are a number of possible chemical conversions of pollutants other than combustion in air. These include the chemical neutralization of an acid or a base and the oxidation of pollutants by agents other than air, such as chlorine or ozone.

9.17 CONTROL OF AIR POLLUTION IN ENCLOSED SPACES

(Problems 20, 25)

Most people in developed areas of the world spend most of their lives indoors. Many, in fact, spend very little time outdoors at all—from home to car perhaps, and from car to factory or office. Air pollution emissions are usually higher in winter than in summer, because it is in winter that fuels are burned for warmth; it is also in winter that the least time is spent outdoors. We should be concerned, then, with the air quality of the indoor spaces that we occupy, and with methods of indoor air purification.

The relationship between indoor and outdoor air pollution is neither mysterious nor surprising—the more polluted the outside air, the more polluted it will be inside. Of course, the two atmospheres are not exactly the same; some outdoor pollutants are decreased on their way in, and some new pollutants are generated by indoor activities, such as cooking, space heating, aerosol spraying, smoking, and the growth of bacteria. Thus, indoor bacterial concentrations are governed by the presence and activities of people indoors more than by outdoor pollution. On the other hand, the total concentration of indoor particulate matter, especially in buildings with open windows, is determined in large measure by the outdoor concentration. Indoor concentrations of SO_2 are about the same as those outdoors when the outdoor concentrations are low (between 0.05 and 0.1 ppm), but drop to about one-third of the outdoor concentrations as the outdoor levels approach about 0.5 ppm. SO_2 is a reactive gas and tends to be absorbed by the inside and outside walls of buildings; such phenomena account in large measure for the observed differences.

The control of indoor air pollution, then, presents two aspects: the purification of the air that enters the indoor space, and the control of pollutants that are generated by indoor activities. The two situations are represented in Figure 9.20. There is a fundamental difference between the two modes. In the first case, namely the outside air intake (Fig. 9.20*A*), an air

FIGURE 9.20 Control of air pollution in enclosed spaces. *A*, **Purification of outdoor air intake;** *B*, **purification of recirculated indoor air.**

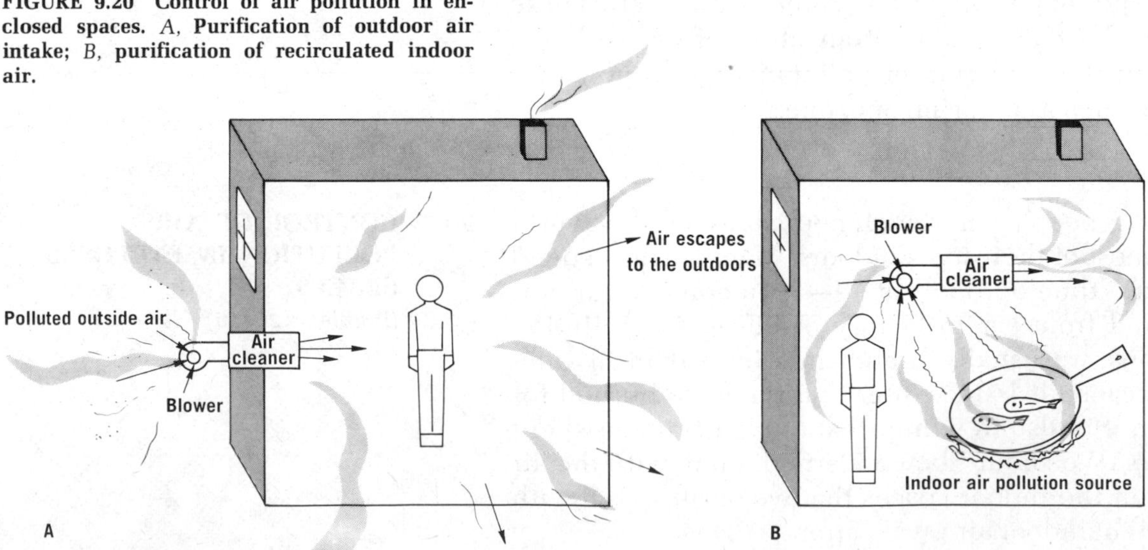

cleaner that is 100 per cent effective will allow *no* outdoor pollutants to enter the space. But in the second case (Fig. 9.20*B*), no matter how efficient the unit, pollutants will exist in the space from the time they are generated until they enter the air cleaner.

The choice of control methods is more limited for such applications than it is for outdoor emission sources of the type described in the preceding section. We do not set scrubbers, incinerators, or cyclones on window sills or in living rooms. For the removal of particulate matter, the best general type of equipment is a filter; electrostatic devices can also be used but they suffer from the disadvantage that they produce some ozone. For removal of gaseous contaminants, the medium of choice is a granular bed of an adsorbent. Activated carbon is most commonly used, and it works well if it is replaced before it reaches saturation. Another medium is a basic form of aluminum oxide, which is highly effective in removing SO_2 and other acidic gases.

9.18 THE AUTOMOBILE

(Problems 28, 29, 30 33)

Remember the mythical magic lamps and their genies? One had only to rub the lamp and speak one's wish, and it was done. The modern "fully equipped" automobile scarcely seems to ask for more. A touch of the finger, hardly even a push, brings warming, or cooling, or music; it opens or closes windows, locks or unlocks the car doors, or even the garage doors. A touch of the toe brings a surge of acceleration, or a swift stop. Perhaps many people find these comforts and conveniences too pleasurable to share, but whatever the reason, most moving automobiles are occupied by only one person, though they may have room for five or six. Of course, there is a price to pay and its coin is energy. But not all forms of energy are equally suited to the purpose. We have learned, for example, that nuclear reactors can never be tiny, and that solar energy can never be continuous. An automobile needs, instead, an energy supply and a working engine that can move around with it, and that will function whenever they are needed. And when the supply of energy runs out, there must be a network of outlets to buy more.

With these constraints, several conceivable energy sources are still available, such as coal, wood, alcohol, compressed hydrogen, electric batteries, or gasoline, and various possible types of engines for each

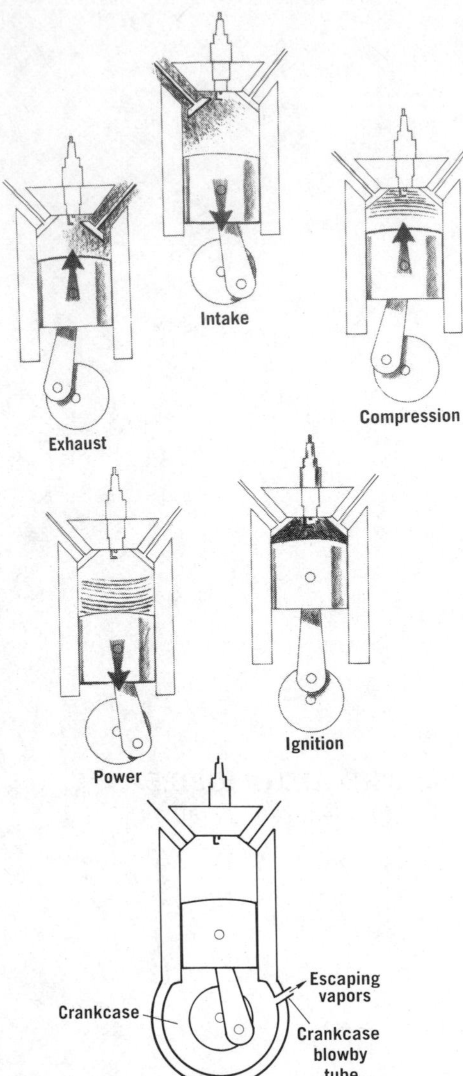

FIGURE 9.21 Schematic diagram of gasoline internal combustion engine, four-stroke cycle.

source. We all know that the overwhelming choice has been the gasoline-powered internal combustion engine, and that one of the consequences of its use has been the discharge of pollutants to the atmosphere. Let us consider, then, what gasoline is and why it works so well, what are the air pollution problems associated with it, and what are some possible solutions.

The source of gasoline is petroleum, which is a natural mixture of many hydrocarbons. At the prevailing temperatures of the Earth's surface, some of these hydrocarbons are true gases, some are highly volatile liquids that boil away as one looks at them, some are liquids that evaporate more slowly, some are thick sluggish liquids that seem hardly to evaporate at all, and some are tarry solids. To fuel a carry-along engine, by far the most convenient kind of material is a free-flowing liquid, because it is so easy to transfer through tubes and to control by valves. The objective of a compact engine is realized by the method of internal combustion, in which the working parts (piston and cylinder) and the fuel-burning chamber are one and the same (see Fig. 9.21). But this system demands a particular pattern of burning of a *gaseous* mixture of fuel and air. (We will say more about this pattern later.) Therefore, our liquid fuel must also be able to be readily vaporized—but not *too* readily, because then it would be both inconvenient and dangerous to handle.

All of this means that we cannot pump oil from a well directly into our fuel tanks; we must be far more selective in our choice of components. It turns out that a workable combination of properties is realized in mixtures consisting largely of hydrocarbons whose molecules contain eight carbon atoms each. These compounds are clear colorless liquids whose boiling points are not far from that of water (see page 344). The 18 compounds whose formulas are C_8H_{18} are called **octanes**; they have almost no odor, and, as noted earlier, their vapors are not toxic.

Gasoline actually includes a C_4-to-C_{10} range of hydrocarbon molecules to yield smooth operation over various driving conditions.

When a mixture of air and gasoline vapor burns in the cylinder, the volume of gas expands because (a) the gas gets hotter, and (b) more molecules of gas are produced. It is this *expansion* that pushes the piston and thereby does the work that drives the car. Now, this expansion is quite rapid; a piston's up-and-down motion reaches speeds of about 40 meters/second (90

$$\underbrace{2C_8H_{18} + 25O_2}_{\text{27 molecules}} \rightarrow \underbrace{16CO_2 + 18H_2O}_{\text{34 molecules}}$$

miles per hour). But this is much, much slower than a true explosion. The shock wave in a piece of exploding TNT, for example, travels at a rate of about 5000 meters per second (11,000 mph). We could not run a car by exploding bits of TNT in the cylinder; we would only destroy it.

The pioneers of gasoline chemistry found that not all hydrocarbons were alike. Some behaved quite properly; the geometry of their burning was like that of a cigarette—a steadily advancing flame front that gave the piston an efficient push. Others were impossible; their combustion was closer to an explosion that expended its energy wastefully by knocking on the engine. The best was a hydrocarbon called "iso-octane," which was honored with a score of 100. One of the worst was normal heptane, which was assigned a rating of 0. Any other hydrocarbon, or for that matter any gasoline, was rated as though it were a simple mixture of "iso-octane" and normal heptane. Thus, a gasoline that behaved like a mixture of 85 per cent "iso-octane" and 15 per cent normal heptane was said to have an octane number of 85.

```
       H
      HCH
   H   |   H   H   H
  HC———C———C———C———CH
   H   |   H   |   H
      HCH     HCH
       H       H
```

"Iso-octane"

```
   H   H   H   H   H   H   H
  HC———C———C———C———C———C———CH
   H   H   H   H   H   H   H
```

Heptane

But it was not easy to make gasolines with high octane numbers. A possible approach to the problem was to seek some other substance, not necessarily a hydrocarbon, that might be added to gasoline to improve it. It was found, for example, that iodine worked fairly well, but it was too corrosive. Several other materials also showed promise but were too expensive or too toxic, or too malodorous for practical use. Finally, shortly after the end of World War I, Thomas Midgley and his colleagues, working at the laboratories of the General Motors Company in Detroit, reasoned from their previous observations that some form of lead might be helpful. Lead does not dissolve in gasoline, but tetraethyl lead does, and when it was tried, it was found to be an *extremely* effective antiknock agent.

Let us look now at the chemistry of the oxidation. Recall the equation for the burning of methane:

$$CH_4 + 2O_2 \rightarrow CO_2 + 2H_2O$$

The balanced equation that represents the burning of octane uses larger numbers, but shows the same products:

$$2C_8H_{18} + 25O_2 \rightarrow 16CO_2 + 18H_2O$$

But engines don't read textbooks, and they don't balance equations so well. During the fraction of a second that is required for the piston to complete its stroke, some of the hydrocarbon completely escapes the advancing flame front and does not burn at all, and

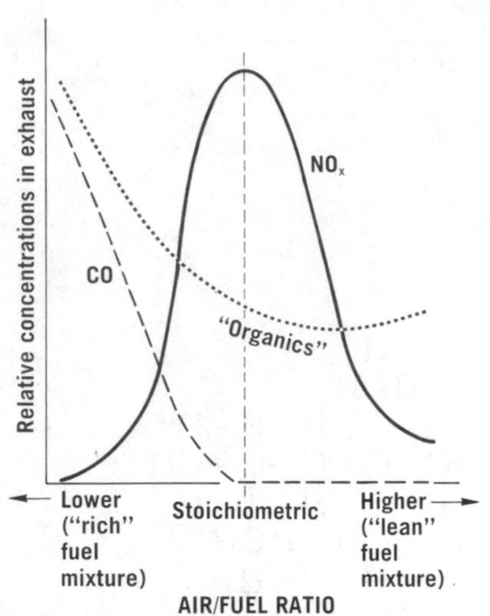

FIGURE 9.22 Emissions from automobile exhaust.

some is incompletely burned, like a piece of wood that is charred but not ashed.

The products of incomplete combustion are "oxygenates" (page 344), aromatic hydrocarbons, soot, and carbon monoxide (CO). Finally, some of the nitrogen in the air is oxidized, largely to NO and NO_2. The chemical reaction pathways are greatly influenced by the ratio of air to fuel that is introduced into the cylinder. These relationships are shown in Figure 9.22. The "stoichiometric" ratio is that mixture of air and gasoline which would just correspond to the relative amounts specified by the balanced equation. In other words, this is the ratio that would produce only CO_2 and H_2O, with nothing left over, if the reaction followed the equation. If we increase the air/fuel ratio, which means that the mixture becomes "lean" in fuel, more oxidation takes place, up to a point. It is like blowing air into a fire—the fire burns more brightly and therefore less smokily. Thus, the emissions of organic matter decrease. But the emission of NO_x, which is also produced by oxidation, increases. However, excess air has a cooling effect (like blowing out a fire) that slows down the oxidation; hence organic emissions start to rise and NO_x to fall. Carbon monoxide represents a special case; it can be produced both by oxidation of hydrocarbon and by reduction of CO_2. Neither of these reactions occurs at the low temperature of a very lean mixture; hence CO concentration remains low when the air/fuel ratio is high. Suppose we go the other way—to a low air/fuel ratio that is "rich" in fuel. In that direction both organic vapors and CO increase, and NO_x falls off. Other operating factors, such as ignition timing, sparking pattern, distribution of fuel among the cylinders, coolant temperature, geometry of the combustion flame and of the combustion chamber, and mode of engine operation—cruising, idling, acceleration, or deceleration—are also important determinants of the composition of the exhaust.

This statement is the truth, but not the whole truth. Much of the problem here lies in the fact that the distribution of fuel to the cylinders of a multi-cylinder engine is far from uniform. The result of this variability is that from time to time a cylinder gets a charge that is too lean to burn. This unburned fuel then gets exhausted. One of the efforts to control auto exhaust emissions is the development of techniques to make the fuel distribution more uniform.

$$C + CO_2 \rightarrow 2CO$$

We shall now see that the problem of air pollution from automobiles involves a complex set of interactions among technical, social, and economic factors (see Fig. 9.23).

Gasoline and Automobiles

Gasoline is volatile; therefore, whenever any of it comes in contact with air, some of it will evaporate.

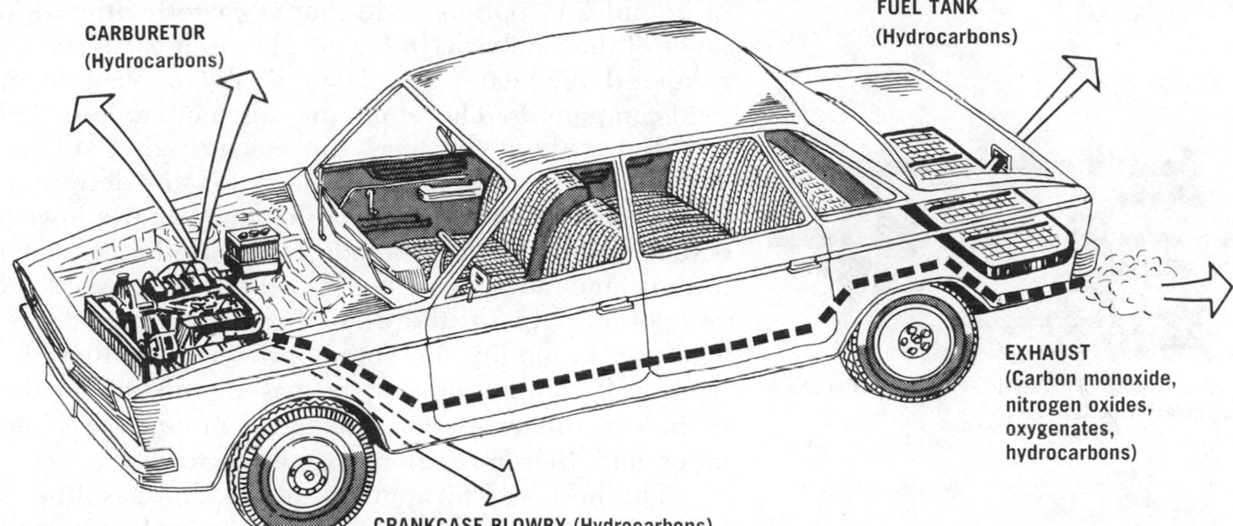

FIGURE 9.23 Potential sources of air pollutants from an automobile.

These occasions include any open transfer of gasoline from one container to another, and any storage of gasoline in a container that is vented to the outside atmosphere. The obvious remedy is to eliminate or minimize such exposures, usually by providing a closed path to recycle the vapors. The combustion exhaust, however, is more difficult to deal with. Modifications of factors such as the air/fuel ratio can provide a partial solution. Another method is the use of an auxiliary thermal reaction chamber in which the exhaust gases are mixed with additional air and held long enough at a high temperature to facilitate more complete oxidation (see Fig. 9.24). However, these measures cannot satisfy the stringent conditions that Federal standards demand for automobiles by 1976. These conditions are equivalent to about a 95 per cent reduction of emissions as compared with those of automobiles before any controls were mandated. The method that is considered to be the most promising for achieving these objectives is the use of a noble metal catalytic converter.

A catalyst is a substance that speeds up a reaction without itself being consumed. Back in the days of gas lighting it was known that metals such as platinum and palladium (said to be "noble" because they are so inert) catalyzed the ignition of a mixture of gas and air. In spite of all the research in the intervening years, no better catalysts have been found. For use with automobiles, the platinum is embodied in a "catalytic muffler" that is used in place of the ordinary exhaust muffler. The exhaust gas passes through a bed

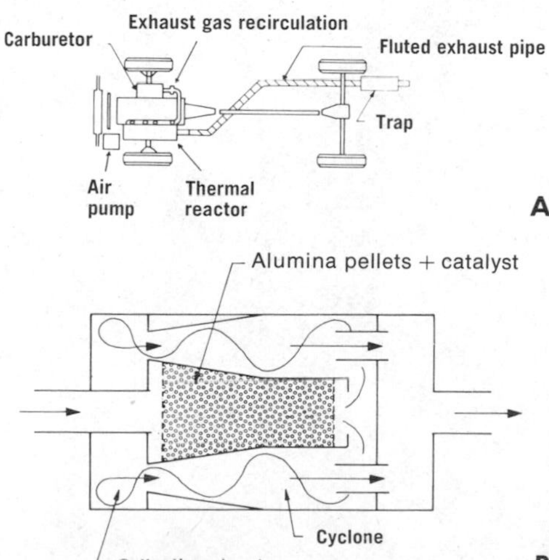

FIGURE 9.24 A. Thermal reactor system. B. Catalytic muffler system.

of granules of porous solid that is coated with a thin layer of the catalyst. However, platinum catalysts are rendered ineffective ("poisoned") by deposition of lead compounds. Therefore, they demand non-leaded gas. But without the lead, the octane rating suffers. There are two ways to compensate for this deterioration: (a) manufacture automobiles that can use lower-octane gasoline, and (b) modify the gasoline so that a high octane rating is achieved without the use of lead compounds. Again, there are further complications. The way to modify the automobiles is to reduce the degree of compression of the gas-air mixture in the cylinders; this reduction makes the engines less efficient, and, therefore, more expensive to run.

Cutaway view of catalytic exhaust muffler showing catalyst pellets.

The best way to upgrade non-leaded gasoline is to increase its proportion of aromatic components, but this change means that the exhaust emission will contain a higher proportion of carcinogenic hydrocarbons. This change has other effects as well. A "subregular" (91-octane) non-leaded gasoline is really a premium grade hydrocarbon blend minus its anti-knock additive. Its manufacture requires a higher degree of refining than a "regular" or "intermediate" grade of leaded gas. Refining utilizes only a portion of the crude petroleum. Therefore, the more a gasoline is refined, the more petroleum reserves must be depleted. Furthermore, for reasons we don't yet understand very well, the deposits left in the combustion chamber by unleaded gas increase the octane requirements of their engines more than the old-style leaded deposits did. As a result, the new automobiles that were designed to run on 91-octane fuel soon demand (by "knocking," of course) about 95-octane fuel. Some drivers pacify their engines with an occasional tankful of leaded gas.

Health and Money

Spokesmen for the automobile and gasoline industries believe that the proposed federal standards are unnecessarily stringent. The Chrysler Corporation, for example, estimates that it would cost 66 billion dollars *more* to meet them than to meet the somewhat less stringent standards requested by the State of California. The estimated costs per car are shown in Figure 9.25. The opposing sets of viewpoints may be summarized as follows:

1. Representatives of the automobile and gaso-

line industries maintain that the adverse health effects associated by the Environmental Protection Agency (EPA) with the proposed emission standards are not substantiated by the data. If there are no effects at the industry's suggested levels, then more restrictive standards offer no benefits. Furthermore, these industrial representatives contend, nature creates more pollutants than man—15 times as much NO_x, 10 times as much CO, and six times as much hydrocarbons; and nature provides its own mechanisms for cleaning the air.

2. Representatives of governmental agencies and of consumer groups take the position that protection of the public health must take precedence over economic considerations. They argue that we cannot accept the thesis that adverse health effects for each air pollutant for defined concentrations and exposure times must be proved beyond the shadow of a doubt to everyone's satisfaction before protective action can be taken. Moreover, global figures ignore local concentrations. Research is underway to supply missing data, and when more information is available, the standards may be revised up or down. Meanwhile, the argument continues, adverse effects must be prevented; they must not be permitted to occur.

The automobile industry's contention that nature

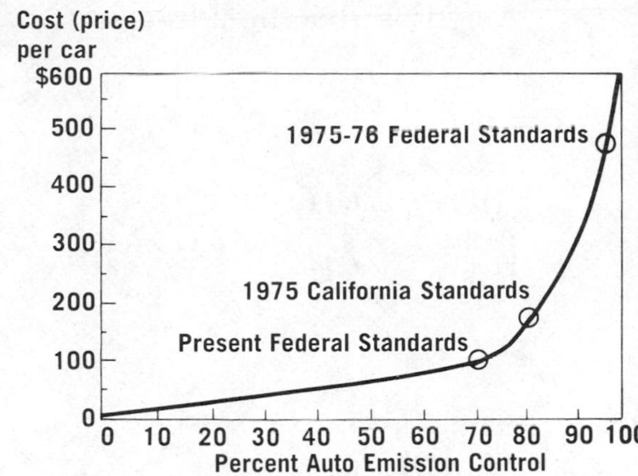

FIGURE 9.25 Estimated cost of automobile emission control. (Courtesy of Chrysler Corporation Research and Editorial Services, Detroit, Michigan.)

YOU DRIVE IT BACKWARD SO IT WILL TURN HYDROCARBONS INTO PURE GASOLINE.

(From *Keystone Motorist*, **June, 1972.)**

A hydrocarbon from trees:

```
      CH3
      |
      C
    //  \
  HC     CH2
  |       |
 H2C     CH
    \   //
      C
      |
      CH
    /    \
  CH3    CH3
```

terpinene

out-pollutes man, and the implied conclusion, "So why worry?", is refuted by the following arguments: (a) The natural sources differ in kind or in effect from those produced by the automobile. The natural oxidation of atmospheric N_2, for example, leads to nitrogen fixation by plant matter, not to air pollution. Natural hydrocarbons, such as those from anaerobic decomposition (methane) or from trees (terpenes) do not have the same potential for producing smog as the hydrocarbons in auto exhaust. (b) Damage is caused by locally high concentrations of pollutants; worldwide averages are usually irrelevant.

Drivers and Maintenance

Cars get old and people get careless. Many drivers do not have their engines tuned up, or even change their spark plugs, as long as their cars will run. None of the proposed anti-pollution devices will work if the cars are not properly maintained. If the emission control system fails, the driver will not notice it; in fact, the performance of his car might even improve slightly. Therefore, if maintenance is left to the vigilance of the owners of automobiles, the quality of the ambient air will be affected more by the number of cars with faulty emission controls than by the legislated emission standards. Compliance requires inspection, and even inspection procedures do not guarantee proper maintenance. The State of New Jersey, for example, requires an annual safety inspection, but about 40 per cent of the cars fail it once each year and six per cent fail it twice.

Other Engines and Other Energy Sources

How about gas turbine engines, or the rotary-piston Wankel engine, or external combustion engines that use steam or air? Why not use diesels, or electric cars powered by batteries or by fuel cells? All of these alternatives have been suggested, and some indeed promise to be considerably less polluting than today's familiar automobile.

Why, then, are non- or low-polluting cars so conspicuously absent from our highways? The answer lies in a complex of technical, social, and economic issues. Some insight into the complexity of the problem can be gained by considering two examples

FIGURE 9.26 Electric car. (United Press International photo.)

among the alternatives cited above: the steam car and the electric car.

The steam car runs on fuel that burns steadily *outside* the engine. There is no explosion, and, therefore, no need to control the geometry of the advancing flame by lead additives. Any of a variety of liquid fuels can be burned efficiently. These circumstances make the steam car potentially an extremely low polluter. However, because of economic factors associated with the automobile industry of the 1920's, the development of external combustion was not pursued. In fact, the automobile industry has been derelict in promoting any significant innovations in engine design to reduce air pollution.

The battery-driven electric car (Fig. 9.26) is an extreme and possibly even unrealistic example of an alternate design. We will discuss it here to illustrate the extent to which any change from the use of the automobile in its present form could have complex economic and social consequences. Certainly the electric car would be essentially non-polluting as it is driven along the road. However, the switchover would create various problems, such as the following:

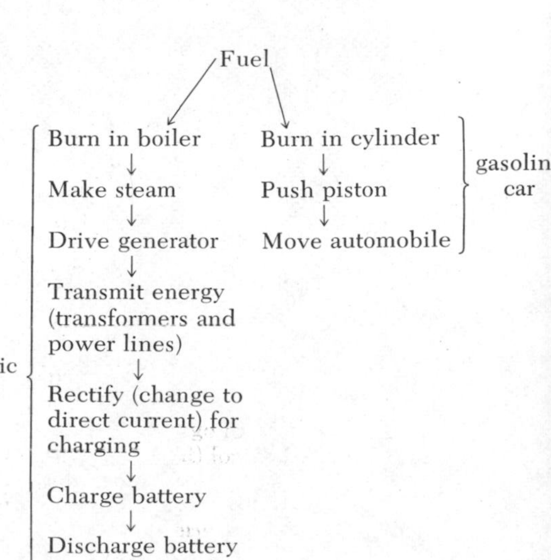

NEED FOR MORE CENTRAL POWER PLANTS. The energy for the electric cars isn't free. It would necessarily result in increased output by central power plants, which would be the source of the power required for recharging batteries. Furthermore, there are many more steps in the power plant sequence, each of them suffering some loss of useful energy. Generating capacity would have to be increased by about 50 per cent at a cost of over 20 billion dollars. Power plants are experiencing difficulty now in keeping up with the demand, especially during the summer (air conditioning) season. The required increase in capacity would take some time to establish, so the changeover would have to be gradual. Furthermore, power plants create their own pollution. This fact would be strong persuasion in favor of nuclear power plants, whose problems are discussed in Chapter 8. In general, however, it is expected to be easier to control emissions from a small number of large power plants than from millions of automobiles. Therefore the overall result should be a decrease in air pollution.

ECONOMIC DEPRESSION IN THE OIL INDUSTRY. It is difficult to predict accurately how much loss would be suffered by the oil industry; this would depend in part on how many of the new central power plants used oil. A fair guess is that total sales would be cut in half. The result would necessarily be pockets of unemployment and depression, and the shift to new industries would involve dislocations.

CHANGES IN AUTOMOBILES AND THE AUTOMOBILE INDUSTRY. It is not feasible to manufacture electric cars that are powerful enough to resemble the heavy, luxury-type gasoline-driven automobile. It is therefore likely they would resemble the new "mini" subcompacts. There would be fewer "extras" sold because there would be less room for accessories in the smaller cars. This loss of profit for the manufacturer, added to the fact that battery components are expensive, might make the small electric cars cost about as much as or more than the larger gasoline-driven ones. The entire "parts" industry would have to change. Instead of furnishing spark plugs, mufflers, carburetors, and so forth, it would have to supply motor commutators, electrical controls, and the like. During the changeover, which would take years, both kinds of parts would be needed. In fact, the changeover might require the development of a hybrid vehicle—one that used both a battery-driven motor and a gasoline engine. The gasoline engine in such a vehicle would be small and would operate continuously, while the battery would provide peaks of power for acceleration and would absorb braking power from the motor. It is estimated that such a system would reduce noxious emissions by at least 50 per cent compared with an all-gasoline vehicle of the same power. The use of intermediate designs of this type would greatly prolong the ultimate transition to an all-electric car.

CHANGES IN SERVICE STATIONS. Mechanical repairs would partially give way to electrical repairs. Gasoline sales would have to be replaced by battery sale, rental, exchange, and recharging services.

CHANGES IN HIGHWAY DESIGN. The current high-speed interstate highway system is designed for the rapidly accelerating gasoline-driven automobiles. The requirement for fast pickup is especially critical when an automobile enters the stream of swiftly moving highway traffic. The existing acceleration lanes would be insufficient for safe entry of the slower electric cars. This difference might prove to be especially difficult during the transition period when both types of vehicles were on the road.

TAXATION. The Federal government and the states gain revenue from gasoline taxes. The money from these taxes is generally earmarked for the building and maintenance of roads; thus, the people who pay for the highway system are those who use it. With the use of electric cars and the consequent decrease in gasoline purchases, this income would have to be replaced, most probably by an increase in tax on electricity. This increase, however, would be imposed on everyone, and the non-driver would be paying money which would primarily benefit the operators of automobiles. Moreover, the resulting increased cost of electricity would influence people to avoid its use where possible. This would result in the use of fewer electric stoves and electric hot water heaters, for example. Instead, gas and oil would be used, leading to the same kind of pollution that the electric car was designed to eliminate.

PROBLEMS

1. **Air pollutants.** What is air pollution? List 10 gaseous and 10 particulate air pollutants.

2. **Concentrations.** A concentration of 1 ppb of ethylene, (C_2H_4) gas in the air cannot be smelled and has no demonstrable effect on people, animals, or materials. However, it does produce dried sepal injury in growing orchids – to such a degree that they become unfit for sale. Is 1 ppb of C_2H_4 a high concentration, or a low one? Would your answer depend on whether or not you were an orchid grower?

3. **Air pollutants.** Distinguish between primary and secondary air pollutants. Give three examples of secondary air pollutants.

4. **Gases.** If gas molecules move so rapidly, why don't they settle quickly to the ground?

5. **Definitions.** Define aerosol; dust; mist.

6. **Vocabulary.** Differentiate between "oxygenate" and "oxidant." What is a hydrocarbon?

7. **Vocabulary.** Differentiate among the terms organic, viable, and nonviable.

8. **Carbon monoxide.** ^{14}C is a radioactive form of carbon, with a half-life of 5700 years, that is continuously produced in the earth's atmosphere. As a result of this production, atmospheric CO_2 has a slight but constant level of radioactivity. The ^{14}C content of coal and petroleum, on the other hand, is negligible.

   (a) Using these facts, outline a method by which you could determine what proportion of the earth's CO is natural and what is man-made. State the assumptions implicit in your method.

   (b) If it were established that the ^{14}C content of atmospheric CO is steadily decreasing, what would this finding imply as to the source of the CO?

9. **Inversion.** Under some conditions, two inversion layers may exist at the same time in the same vertical atmospheric structure. Draw a diagram of temperature vs. height that shows one inversion layer between the ground and 600 feet, and another aloft, between 3000 and 3500 feet, while the lapse rates at other altitudes are superadiabatic.

10. **Chimneys.** "Since tall chimneys do not collect or destroy anything, all they do is protect the nearby areas at the expense of more distant places, which will eventually get all the pollutants anyway." Argue for or against this statement.

11. **Meteorology.** Define lapse rate. Describe the meteorological conditions most conducive to the rapid dispersal of pollutants. Describe those that are least conducive.

12. **Climate and air pollution.** Imagine that a given factory, with a fixed rate of emission of gaseous air pollutants, could be located either in the tropics or in the Arctic. Assuming that the population densities in the two areas were equal, which location would you choose? Defend your answer.

13. **Climate.** What is the "greenhouse effect"? Why is the term considered to be a misnomer when applied to the atmosphere?

14. **Albedo.** As the winter ends, the snow generally starts to melt around trees, twigs, and rocks. The line of melting radiates outward from these objects. The snow in open areas melts last. Explain.

15. **Albedo.** Refer to Figure 9.9. (a) What percent of the incident solar energy is received by the Earth? (b) Does the Earth's surface receive any additional energy? If so, from what source? (c) Is the amount of energy emitted by the Earth greater, less, or the same as that which it receives from incident solar radiation? Explain.

16. **Climate.** Imagine that you must determine whether some particular climatic effect, such as increased fog or rainfall in a given area, is caused by human activity. Which of the following experimental method(s) would you rely on? Defend your choices. (a) Compare current data with that of previous years, when population and industrial activity were less. (b) Compare the effects during weekdays, when industrial activity is higher, with those on weekends, when it is low. (c) Compare effects during different seasons of the year. (d) Compare effects just before and after the switch to or from Daylight Saving Time, to see whether there is a sharp one-hour shift in the data. (e) Compare effects in different areas where populations and industrial activities differ.

17. **Air pollution effects.** Of the six categories of air pollution effects cited in Section 9.7, how many do you think could be produced by sulfur dioxide, SO_2, or by secondary air pollutants derived from SO_2? Justify your answer for each category.

18. **Health.** In Chapter 7, we saw that the United States Government has banned the domestic use of DDT. Since cigarettes have been implicated in a rather frightening extent of chronic disease causation, discuss the proposition that they too should be banned.

19. **Cigarettes and air pollution.** Which of the following groups of subjects would you study to learn the separate effects of cigarettes and air pollution on human health? Explain. (a) Urban smokers vs. urban non-smokers; (b) urban smokers vs. rural smokers; (c) urban smokers vs. rural non-smokers; (d) urban non-

smokers vs. rural smokers; (e) urban non-smokers vs. rural non-smokers; (f) rural smokers vs. rural non-smokers.

20. **Indoor/outdoor air pollution.** Draw an indoor/outdoor graph of SO_2 pollution in which the vertical axis is the indoor/outdoor ratio, in percent, and the horizontal axis is outdoor SO_2 concentration in ppm. Let the curve be a smooth approximation of the statement that appears on page 382. If the outdoor concentration approaches zero, but some sulfur-containing fuels are burned indoors, do you think the indoor/outdoor ratio might exceed 100 per cent?

21. **Air pollution index.** Fulton County (Atlanta), Georgia, has instituted an index which reports the levels of three pollutants as fractions of their standards and then adds these numbers to obtain the index. (a) Construct an index of this sort from three of the pollutants listed in Table 9.2. Defend your choice of pollutants. (b) Formulate an alert system that is based on your index.

22. **Air pollution index.** The New York and San Francisco air pollution indices are based on different pollutants. What are these differences? Do you think it would be reasonable for the two cities to exchange indices? Defend your answer.

23. **Air pollution index.** Supply an air pollution index rating for New York City for each of the following hypothetical conditions: (a) SO_2, 0.042 ppm, 24-hr average; NO_2, 0.04 ppm, 24-hr average; CO, 2 ppm, maximum 8-hr average; oxidant, none; smokeshade, 1.0 COH. (b) SO_2, 0.031 ppm, 24-hr; NO_2, 0.13 ppm, 24-hr; CO, 5 ppm, 8-hr; oxidant, 0.07 ppm, 1-hr, smokeshade, 1.1 COH. (c) SO_2 reached a maximum of 0.70 ppm between 7 P.M. and 8 P.M., but the 24-hour average was 0.03 ppm; all other pollutants were in the "good" range.

24. **Control strategy.** Charts such as that shown in Figure 9.3 are occasionally cited to support the contention that elimination of man-made sources is not urgent because they are only a minor portion of the total. What is your opinion of such an assertion? Defend your answer.

25. **Control methods.** Suppose that you keep some animals in a cage in your room and you are disturbed by their odor. Comment on each of the following possible remedies, or some combination of them, for controlling the odor:

    (a) Spray a disinfectant into the air to kill germs.

    (b) Install a device that recirculates the room air through a bed of activated carbon.

    (c) Clean the cage every day.

    (d) Install an exhaust fan in the window to blow the bad air out.

(e) Install a window air conditioning unit that recirculates and cools the room air.

(f) Install an ozone-producing device.

(g) Spray a pleasant scent into the room to make it smell better.

(h) Light a gas burner in the room to incinerate the odors.

(i) Keep an open tub of water in the room so that the odors will dissolve in the water.

26. **Control methods.** Distinguish between separation methods and conversion methods for source control of air pollution. What is the general principle of each type of method?

27. **Control equipment.** Explain the air pollution control action of a cyclone; a settling chamber; a scrubber; activated carbon; an electrostatic precipitator; an incinerator.

28. **Automobile.** Outline the factors that led to the widespread use of the gasoline-powered internal combustion engine.

29. **Automobile emission controls.** Present arguments in your own words in favor of either the industry or government viewpoints on auto emission controls.

30. **Social and economic factors.** Discuss some of the social and economic problems that would accompany a transition from an automobile-based transportation system to a cheap and efficient system of public transportation.

The following questions require arithmetic computation or reasoning.

31. **Particles.** Is the speed of settling of particles in air directly proportional to their diameters? (If the diameter is multiplied by 10, is the settling speed 10 times faster?) Justify your answer with data from Figure 9.1. Is a settling chamber a good general method of air pollution control? Explain.

32. **Concentrations.** One liter of carbon monoxide, CO, is mixed with 99,999 liters of air, without changing the temperature or pressure. What is the concentration of CO in the mixture in ppm by volume? in ppb by volume? in molecules of CO per million molecules of air?

33. **Gasoline.** (a) Define octane number. (b) Brand X gasoline causes a certain engine to knock faintly. Ten parts of normal heptane must be added to 110 parts of "isooctane" to produce the same engine knock under identical operating conditions. What is the octane number of Brand X?

Answers

32. 10 ppm; 10,000 ppb; 10 molecules CO per million.

33. 92 octane.

BIBLIOGRAPHY

The basic text on air pollution is a three-volume work:

Arthur C. Stern: *Air Pollution.* New York, Academic Press, 1968. Volume I, 694 pp.; Volume II, 684 pp.; Volume III, 866 pp.

Two good one-volume texts are:

Samuel J. Williamson: *Fundamentals of Air Pollution.* Reading, Mass., Addison-Wesley Publishing Co., 1973. 473 pp.

Samuel S. Butcher and Robert J. Charlson: *An Introduction to Air Chemistry.* New York, Academic Press, 1972. 241 pp.

Among the air pollution technical publications of the U.S. Environmental Protection Agency are two important series of documents on specific air pollutants. One series deals with "air quality criteria," which are established from our knowledge of the effects of air pollutants. The other series outlines "control techniques." The titles of the first series are:

Air Quality Criteria for Particulate Matter
Air Quality Criteria for Sulfur Oxides
Air Quality Criteria for Carbon Monoxide
Air Quality Criteria for Photochemical Oxidants
Air Quality Criteria for Hydrocarbons
Air Quality Criteria for Nitrogen Oxides

Two of the control documents are:

Control Techniques for Particulate Air Pollutants
Control Techniques for Sulfur Oxide Air Pollutants

For a very brief introductory text, refer to:

National Tuberculosis and Respiratory Disease Association: *Air Pollution Primer.* New York, 1969. 104 pp.

Various popular books that warn of the dangers of air pollution have appeared in recent years. Some representative ones follow:

L. J. Battan: *The Unclean Sky.* New York, Doubleday and Co., 1966. 141 pp.

D. E. Carr: *The Breath of Life.* New York, W. W. Norton & Co., 1965. 175 pp.

Howard R. Lewis: *With Every Breath You Take.* New York, Crown Publishers, 1965. 322 pp.

10

WATER POLLUTION

10.1 THE NATURE OF WATER POLLUTION

(Problems 1, 2, 5)

The terms water pollution and air pollution both imply the presence of undesirable foreign matter in an otherwise "pure" or "natural" substance. However, the concept of pure water is quite different from that of pure air. As we learned in the preceding chapter, air is a mixture of several components, and "pure air" is therefore considered to be a particular mixture that represents a sort of ideal terrestrial atmosphere. Water, however, is a single compound, not a mixture. Therefore, the chemist thinks of "pure water" as a substance consisting of molecules of only one type—those represented by the formula H_2O. However, most drinking water contains small quantities of dissolved mineral salts; these substances often contribute to its taste. Thus one speaks of "pure spring water" in the sense of a natural mixture of water and small amounts of harmless, and perhaps tasty, mineral matter.

We are ignoring differences in mass numbers that are due to the presence of different isotopes.

The pollution of water, then, is the addition of undesirable foreign matter which deteriorates the *quality* of the water. Water quality may be defined as its fitness for the beneficial uses which it has provided in the past—for drinking by man and animals, for the support of a wholesome marine life, for irrigation of the land, and for recreation. Pollutant foreign matter may be either non-living, such as compounds of lead or mercury, or living, such as microorganisms. Again, there is an important difference between air and water. The pollution of air can usually be considered to be the discharge of foreign matter into empty space—

We are not considering *secondary* air pollutants, which are formed by subsequent chemical reactions such as those of organic matter with oxygen and nitrogen under the influence of sunlight.

the space not occupied by the air molecules. The air molecules neither invite nor resist foreign matter; they simply leave ample room for it.

Consider a very different situation. It is not easy to pollute an iron bar, except on its surface. Immerse it in water or oil, expose it to bacteria or viruses, and its internal composition is unchanged. The reason for this resistance is that the atoms of iron are strongly and closely bonded to each other, and are very difficult to displace. This strong tendency for the atoms to retain their respective positions and not to be dislocated is what makes iron a solid. (For that matter, it is not easy to pollute ice either; polluted ice is produced by the freezing of polluted water, not by the pollution of pure ice.)

Liquids are intermediate between gases and solids in their readiness to accept contamination. The attractive forces between molecules in the liquid state are strong enough that a sample of liquid (for example,

"We should be thankful. What if oil and water *did* mix!" (American Scientist, November-December, p. 745, 1972.)

a raindrop) holds itself together. However, the attractive forces are not so strong as they are in solids; they are not strong enough to prevent the molecules from sliding past one another. Such molecular relocations manifest themselves in the familiar phenomenon of liquid flow. Now, when a molecule of a liquid relocates, it leaves behind a vacant site, or a "hole." This vacancy can be occupied by another molecule of the same type, or by a molecule of a foreign substance.

What determines whether a given liquid will accept or reject a given type of foreign matter? We drop a lump of sugar into a glass of water and note that it dissolves. But a piece of lead does not. Since sugar is a solid, the sugar molecules must be attracted to each other—otherwise they would fly apart and sugar would be a gas. But the sugar molecules are also strongly attracted by water molecules, and can be pulled away from other sugar molecules to occupy sites surrounded by molecules of water. Thus, sugar dissolves in water. Atoms of lead are attracted to each other much more than they are to water molecules; thus, lead is insoluble. Therefore, the ease with which a liquid can become contaminated by dissolved foreign matter depends on the chemical relationships between the molecules of the liquid and foreign molecules.

A contaminant may be harbored by a liquid without being dissolved in it. If we grind our piece of lead to a fine powder and stir it into the water, the suspended lead is a pollutant. However, the ease with which foreign matter can be suspended in a liquid also depends to some extent on the mutual attraction between the foreign particles and the liquid molecules.

Water is not a typical liquid. Corn oil is more like cottonseed oil, kerosene is more like gasoline, and grain alcohol is more like wood alcohol than *anything* is like water. One of the consequences of the unique physical and chemical properties of water is that it invites or accepts pollution readily, sometimes through mechanisms that are quite unexpected. Of course, water is the universal liquid medium for living matter; it is therefore uniquely prone to pollution by living organisms, including those that carry disease to man. Contamination pathways that involve suspension, solution, and biochemical change are not necessarily separate and distinct from each other, and many of these complex processes can occur *only* in water.

Therefore, to understand water pollution, we must first consider the nature of water itself.

10.2 WATER
(Problems 8, 38)

Wipe a drinking glass with a dry towel until it is sparklingly clean. Hold the glass upside-down just over a candle flame for five to ten seconds. The inside surface of the glass becomes clouded and wet. You have produced water from the hydrogen in the candle and the oxygen in the air. The oxygen in the air is oxygen gas, and the hydrogen in the candle is chemically bonded to carbon. The chemical composition of water can be established by weighing the ingredients that combine to produce it, although candle wax plus air is not the best choice for such an experiment. We find that water contains eight parts of oxygen to one part of hydrogen by weight. For example, nine pounds of water contains eight pounds of oxygen and one pound of hydrogen. The atomic weight of oxygen is 16, that of hydrogen is 1. Thus, an oxygen atom is 16 times as heavy as a hydrogen atom. The composition of water, and the atomic weights of oxygen and hydrogen, are reconciled by the formula, H_2O shown at left. The sum of the atomic weights, $1 + 1 + 16 = 18$, is the molecular weight of water. The oxygen is bonded to the two hydrogens, and the molecule has a bent shape with an angle between the bonds of 105°.

Strictly speaking, this ratio refers to the mass numbers of the more abundant isotopes. It is a good approximation, however, of the relative average atomic weights.

$$\text{Weight ratio} = \frac{O}{H_2} = \frac{16}{1+1} = \frac{16}{2} = \frac{8}{1}$$

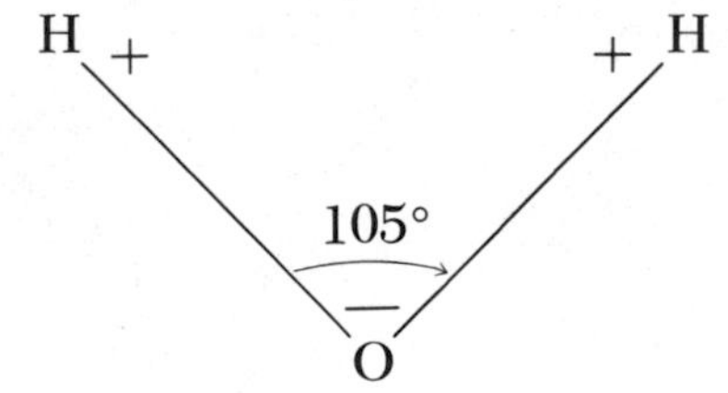

The negative charges (electrons) are crowded somewhat closer to the oxygen atom than the positive charges (protons) are. The effect of this is a separation of charges with the negatively charged part of the molecule nearer the oxygen atom and positively charged parts nearer the hydrogens. These electrical charges attract their opposites in other water molecules, with the result that liquid water consists of aggregates of H_2O molecules bonded to each other as indicated to the left.

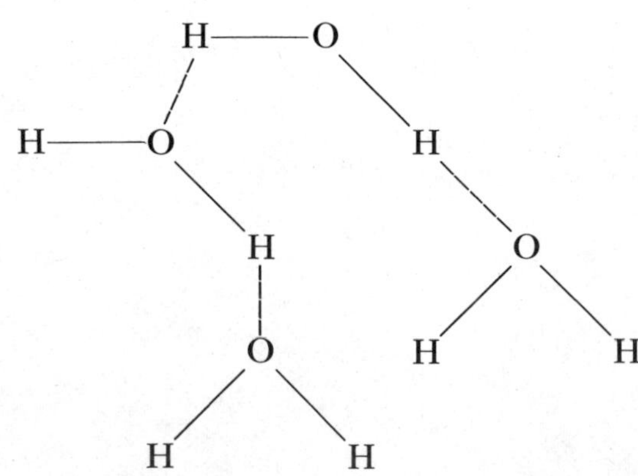

This strong aggregation accounts for the fact that water remains a liquid up to 100°C at normal pressure, in sharp contrast to the behavior of other substances of similar or even higher molecular weight. (For example, ether, molecular weight 74, boils at only 35°C.) The electrical forces that bind water molecules to each other can also serve to bind water molecules to those of foreign substances. Therefore, water is an unusually good solvent, especially for substances which have separated centers of positive and negative electrical charge. Such substances are, typically, inorganic compounds, such as the compounds of the metallic elements. Water is a poor solvent for substances whose molecules do not have separated centers of positive and negative charge; examples are the hydrocarbon substances derived from petroleum, such as gasoline, oil, and grease.

It is energetically costly to heat water, or to boil it, or to melt ice. This statement may not ring true because these operations are so familiar to us. But consider the following experiment:

Place equal weights of water and of olive oil at the same temperature (say, 10°C) in each of two pots. Put energy into both liquids at the same rate. This can be done most easily by passing electric current through two identical heaters (Figure 10.1). Continue the heating until the olive oil temperature has risen by 50 degrees (10°C → 60°C). Now look at the water temperature. It will be only 33°C, which means that its temperature has risen only 23 degrees. (If we compare water with practically *any* other liquid, it is always the water that warms more slowly.) Does this mean that the water has received less energy than the olive oil? Of course not! Equal amounts of energy went into the two liquids, and energy cannot be lost. What the experiment does tell us is that water stores heat energy

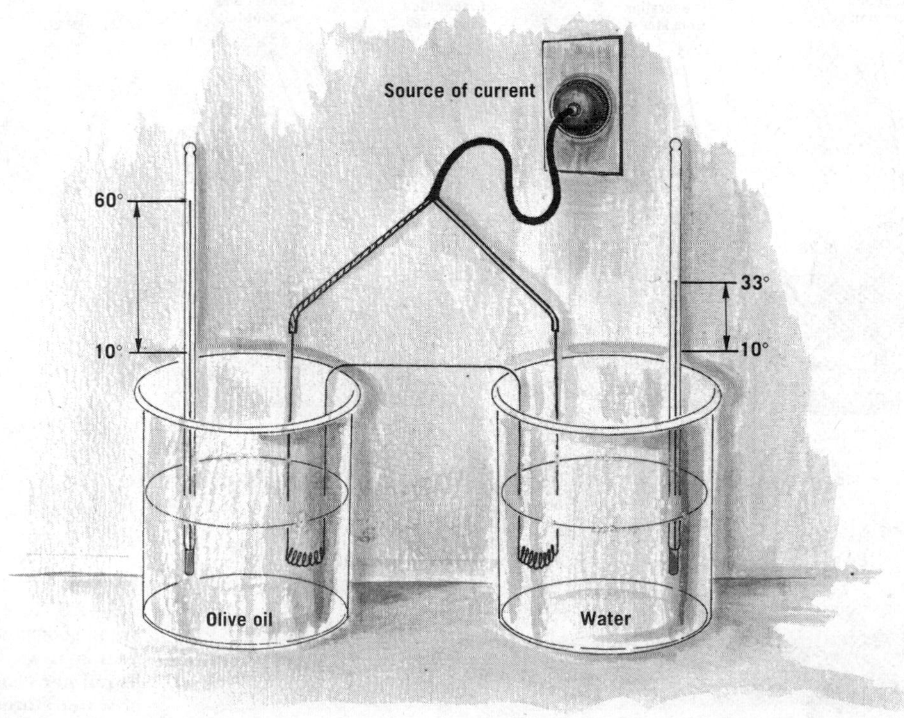

FIGURE 10.1 Heating water and olive oil.

more effectively. It is somewhat like lifting a 50-pound weight as compared to a 100-pound weight. If you put the same amount of work into each, you will lift the 100-pound weight only half as high. Thus, we say that water is a better heat sink; it is harder to heat and is therefore a better cooling agent. Conversely, hot water cools more slowly than other liquids. Similar experiments show that more energy is needed to melt ice or to boil water than to melt or boil almost any other substance. These properties of water are important in consideration of thermal pollution (Chapter 5).

10.3 THE HYDROLOGICAL CYCLE (WATER CYCLE)

(Problem 7)

Most of the water that moves through the biosphere does so in response to physical forces—the movement of air and ocean currents, the flow of rivers, the fall of rain, the creep of glaciers, evaporation from surfaces, and transpiration through the porous outer barriers of plants and animals. Human technology has contributed other drives, such as the thrust of pumps

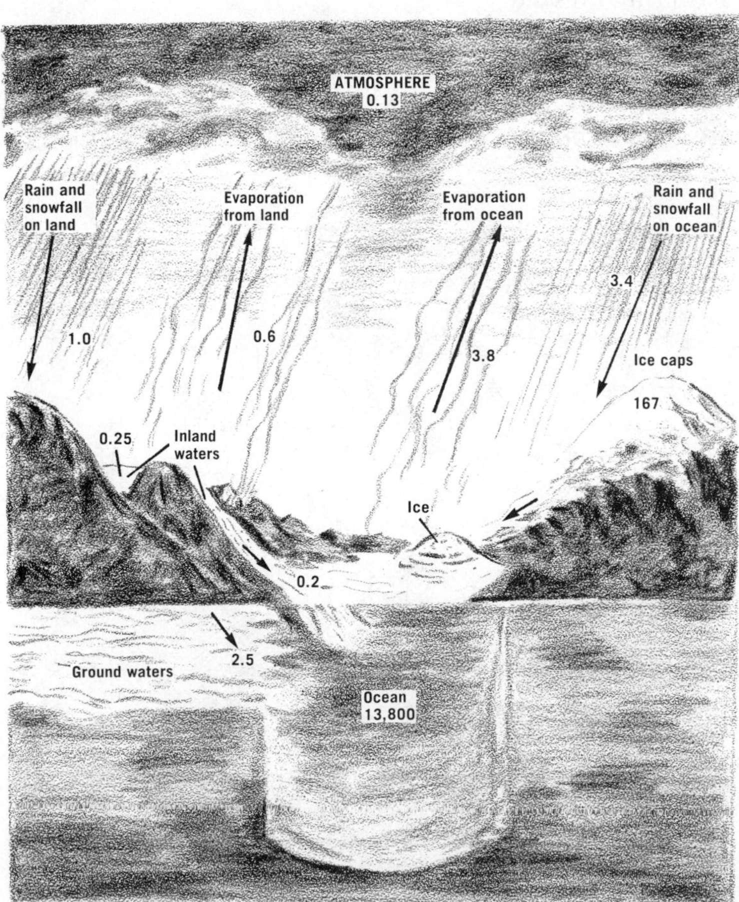

FIGURE 10.2 Hydrological cycle. Numbers are in geograms (10^{20} grams) of water. Numbers near arrows refer to amount of water transferred per year. Other numbers refer to amount of water stored.

and the flush of toilets. Some water is cycled by chemical changes, notably by photosynthesis, which rearranges the atoms of water molecules so that they become incorporated into the structure of plant matter, and by oxidation, which produces water again and releases it back to the hydrosphere.

Now, look at Figure 10.2. Note that water leaves the atmosphere as rain and snow, and returns by evaporation. It also moves by flow of liquid water and ice on the earth's surface and through the ground. Note also that the amount of water in the atmosphere is small compared to that in other locations, such as the oceans and the ice caps, and more important, small compared to the quantity passing through the atmosphere each year. This fact implies that, if man's activities might ever trigger serious disruptions of the hydrological cycle, the most likely means would be by manipulation of the water content of the atmosphere.

The concentration of dust particles or water-soluble gases in the upper atmosphere is comparatively small. Therefore, the purest natural water on Earth is in the droplets (or ice crystals) in high clouds, and the expression "pure as the driven snow" is an apt one. As rain or snow falls to earth, it intercepts various atmospheric contaminants. By the time it reaches the ground, therefore, it contains foreign matter from natural sources such as dust particles, carbon dioxide, microorganisms, and pollen. It may also contain traces of industrial pollutants such as sulfur dioxide, sulfuric acid, or pesticides. As the water runs over or percolates through the land, its load of impurities continuously increases. These substances are thus fed into the ocean, where they accumulate. The ocean is thus the Earth's ultimate sink for water-borne impurities. The concentration of dissolved matter in the ocean rises very slowly, however, because its volume is so large.

10.4 TYPES OF IMPURITIES IN WATER

(*Problems 3, 37*)

It is useful to classify foreign substances in water according to the size of their particles, because this size often determines the effectiveness of various methods of purification. Figure 10.3 shows a spectrum of particles arbitrarily divided into three classes: suspended, colloidal, and dissolved. Let us consider each in turn as we refer to the figure.

Suspended particles, which have diameters above

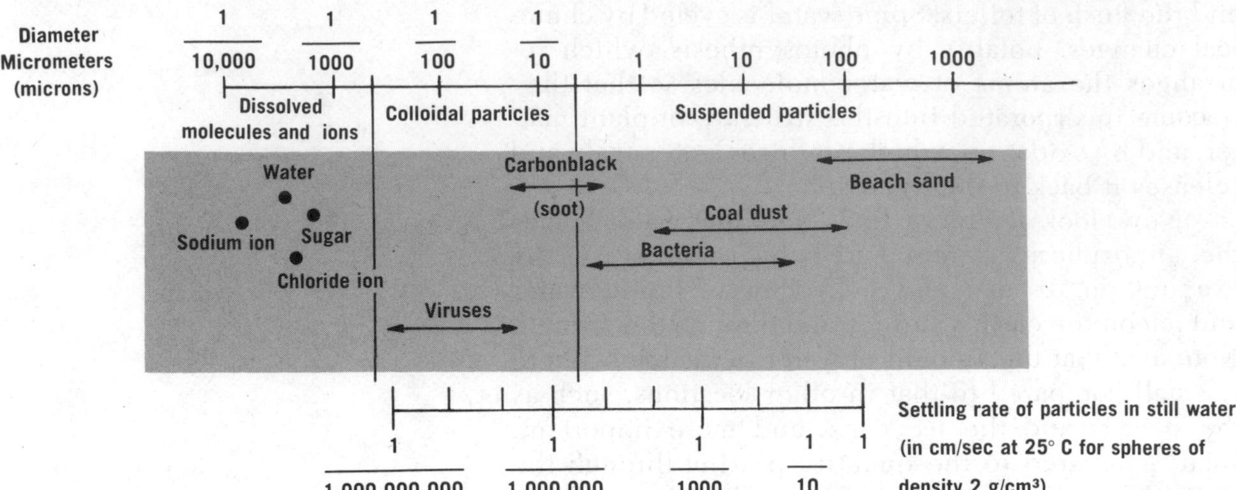

FIGURE 10.3 Small particles in water.

about one micrometer, are the largest. They are large enough to settle out at reasonable rates and to be retained by many common filters. They are also large enough to absorb light and thus make the water that they contaminate look cloudy or murky.

Colloidal particles are so small that their settling rate is insignificant, and they pass through the holes of most filter media; therefore, they cannot be removed from water by settling or ordinary filtration. Water that contains colloidal particles appears cloudy when observed at right angles to a light beam. (The same phenomenon occurs in air—colloidal dust particles can best be seen when observed at right angles to a sharply focused light beam in an otherwise dark room.) The colors of natural waters, such as the blues, greens, and reds of lakes or seas, are contributed largely by colloidal particles.

Dissolved matter does not settle out, is not retained on filters, and does not make water cloudy, even when viewed at right angles to a beam of light. The particles of which such matter consists are no larger than about 1/1000 of a micrometer in diameter. If they are electrically neutral, they are called molecules. If they bear an electric charge, they are called ions. Cane sugar (sucrose), grain alcohol (ethanol), and "permanent" antifreeze (ethylene glycol) are substances that dissolve in water as electrically neutral molecules. Table salt (sodium chloride) on the other hand, dissolves as positive sodium and negative chloride ions.

Of course, foreign substances may be classified by properties other than particle size, for example, they may be living or nonliving, organic or mineral, radioactive or non-radioactive, toxic or harmless, natural or added by man. It will be of most interest to us to focus our attention on those impurities that adversely affect the quality of water, to study their harmful effects and to consider the problems involved in removing them. Since natural water is never chemically pure, let us look first at impurities that occur in natural water, whether or not they are considered to be pollutants.

10.5 THE COMPONENTS OF NATURAL WATERS

(Problems 9, 10, 39)

Natural waters range in quality from tastily potable to poisonous; in saltiness, from fresh rainwater (non-salty) to brackish (partly salty, as where river water starts to mix with sea water), to ocean water, to the heavy concentrations of a landlocked evaporation sink such as the Dead Sea or the Great Salt Lake.

Table 10.1 outlines the sources and chemical compositions of impurities in natural waters. Let us examine some of them in more detail.

Acidity

Hydrogen ions (shown as H^+ in the table) render water acidic. The original meaning of acid is "sour," referring to the taste of substances such as vinegar, lemon juice, unripe apples, and old milk. It has long been observed that all sour or acidic substances have some properties in common, notably their ability to corrode (rust, or oxidize) metals. When the attack on a metal by an acidic solution is vigorous, hydrogen gas (H_2) is evolved in the form of visible bubbles. Acidic solutions also conduct electricity, with evolution of hydrogen gas at the negative electrode (cathode). These circumstances imply that acid solutions are characterized by the presence of positive ions bearing hydrogen. We know that a hydrogen ion, or proton, designated H^+, cannot exist as an independent entity in water because it is strongly attracted (in fact, chemically bonded) to the oxygen atom of the water molecule. The resulting hydrated proton is formulated as $H(H_2O)^+$, or H_3O^+. The simpler designation H^+ may therefore be regarded as an abbreviation.

Some slight transfer of protons occurs even in pure water.

$$H_2O + H_2O \leftrightarrows H_3O^+ + OH^-$$

TABLE 10.1 IMPURITIES IN NATURAL WATERS

SOURCE	*PARTICLE SIZE CLASSIFICATION*				
	Suspended	*Colloidal*	*Dissolved*		
			Molecules	*Positive ions*	*Negative ions*
Atmosphere	←Dusts→		Carbon dioxide, CO_2 Sulfur dioxide, SO_2 Oxygen, O_2 Nitrogen, N_2	Hydrogen, H^+	Bicarbonate, HCO_3^- Sulfate, SO_4^{2-}
Mineral soil and rock	←Sand→ ←Clays→ ←Mineral soil particles→		Carbon dioxide, CO_2	Sodium, Na^+ Potassium, K^+ Calcium, Ca^{2+} Magnesium, Mg^{2+} Iron, Fe^{2+} Manganese, Mn^{2+}	Chloride, Cl^- Fluoride, F^- Sulfate, SO_4^{2-} Carbonate, CO_3^{2-} Bicarbonate, HCO_3^- Nitrate, NO_3^- Various phosphates
Living organisms and their decomposition products	Algae Diatoms Bacteria ←Organic soil (topsoil)→ Fish and other organisms	Viruses Organic coloring matter	Carbon dioxide, CO_2 Oxygen, O_2 Nitrogen, N_2 Hydrogen sulfide, H_2S Methane, CH_4 Various organic wastes, some of which produce odor and color	Hydrogen, H^+ Sodium, Na^+ Ammonium, NH_4^+	Chloride, Cl^- Bicarbonate, HCO_3^- Nitrate, NO_3^-

Hydroxyl ions, OH^-, can neutralize H_3O^+ ions by reacting with them to produce water, as indicated by the arrow pointing left. The concentration of H_3O^+ and of OH^- in pure water at 25°C is 1.0×10^{-7} moles/liter. This solution is said to be *neutral*, because the concentrations of the two ions are equal. When the hydrogen ion concentration is greater than 1.0×10^{-7} moles/liter at 25°C, the solution is *acidic*. Hydrogen ion concentrations are usually expressed logarithmically as pH values, where pH = $-\log_{10}$ (hydrogen ion concentration).

If you are not familiar with moles, then it will be sufficient to consider that 1 mole = 1 gram of hydrogen ions. Therefore, a concentration of 1 mole/liter = 1 gram of hydrogen ions/liter.

Recall that the logarithm of a number "to the base 10" is simply the *number of times 10 is multiplied by itself* to give the number. As we can quickly see, when the number is a multiple of 10, its log is simply the number of zeros it contains. Thus,

10 has 1 zero, therefore log 10 = 1
100 has 2 zeros, therefore log 100 = 2
100,000 has 5 zeros, therefore log 100,000 = 5

When the number is 1 divided by a multiple of 10, its log is *minus* the number of zeros in the denominator. Therefore, when the number expresses hydrogen ion concentration, the pH is *plus* the number of zeros in the denominator:

Concentration of H^+ (moles/liter)	Calculation of pH	
1/10	log 1/10 = log 10^{-1} = −1. pH = 1	more acidic ↑
1/10,000,000	log 1/10,000,000 = log 10^{-7} = −7. pH = 7	neutral
1/1,000,000,000	log 1/1,000,000,000 = log 10^{-9} = −9. pH = 9	↓ more basic

Any pH less than 7 connotes acidity. The lower the pH of a body of water, the more prone it is to be corrosive and, thereby, to become polluted with metallic compounds. Acidic waters coursing through lead pipes will therefore become more toxic than pure water. And grapefruit juice standing in an iron cup develops a terrible taste.

Metal Ions

The ions of metallic elements are electrically positive. Examples are Na^+ (sodium ion), Ca^{++} (calcium ion), and Al^{+++} (aluminum ion). Like the proton, these ions are attracted to the negative ends of the water molecules, that is, to the oxygen atoms. A Na^+ ion, therefore, does not exist in water as an independent entity, like a dust particle in air, but is loosely bonded to a cloud of water molecules. A more highly charged ion, such as Al^{+++}, is more strongly bonded to water molecules. In fact, the linkage to oxygen is so strong that the bond between oxygen and hydrogen is weakened, and hydrogen ions are released to the solution:

$$Al(H_2O)_6^{+++} \rightarrow [Al(H_2O)_5OH]^{++} + H^+$$

or

$$\left[-Al-O{<}^{H}_{H} \right]^{+++} \rightarrow \left[-Al-O-H \right]^{++} + H^+$$

Therefore, solutions of highly charged metal ions tend to be acidic.

$$\begin{matrix} O \\ \| \\ S \\ \| \\ O \end{matrix} + H-O-H \rightarrow \left[\begin{matrix} O \\ \| \\ S-O-H \\ \| \\ O \end{matrix} \right]^- + H^+$$

bisulfite ion

$$CO_2 + H_2O \rightarrow HCO_3^- + H^+$$

bicarbonate ion

Acidic Gases

Atmospheric sulfur dioxide, SO_2, dissolves in water, rendering the water acidic. The SO_2 molecule is more negative nearer the oxygen atoms and more positive nearer the sulfur atoms. The positive end of the SO_2 molecule becomes bonded to the water molecule with the release of a hydrogen ion, rendering the water acidic.

Carbon dioxide reacts with water in an analogous manner, as shown at left.
However, the reaction of water with CO_2 proceeds to a much lesser extent than with SO_2, and hence the resulting solution is less strongly acidic. Therefore, when carbon dioxide gas is bubbled into water, almost all that dissolves remains in the water in the form of free CO_2 molecules. The availability of CO_2 in water is important because aquatic plants utilize this dissolved gas in photosynthesis, and, as we shall see, the total mass of water-borne plant matter is a prime determinant of water quality.

Oxygen

All animals and most plants require oxygen for metabolism of food. Aquatic animals utilize the oxygen dissolved in the waters they occupy. The solubility of oxygen in water is low: one liter of air at 25°C contains 0.27 gram of oxygen; one liter of water in contact with air at 25°C contains 0.0084 gram of oxygen. Therefore, one liter of aerated water contains only about one-thirtieth as much available oxygen as one liter of air. Furthermore, when oxygen has been removed from water, it is not rapidly replaced unless there is turbulent mixing with air, as in the "white water" of shallow rapids. The ecological consequences of the loss of dissolved oxygen in water will be discussed in Section 10.6.

Nitrogen

Table 10.1 shows that nitrogen can be present in natural waters as the gaseous element N_2, or as nitrate ion, NO_3^-, or as ammonium ion, NH_4^+. Nitrogen gas is inert and only slightly soluble in water (0.018 g N_2 per liter of water at 25°C), while nitrates and ammonium salts are highly soluble. Therefore, any conversion of soluble nitrogen compounds to gaseous N_2

is, in effect, a means of removing nitrogen from water (**denitrification**). Nitrogen compounds in water significantly affect aquatic life, and therefore are important determinants of water quality, as we shall see later in this chapter. The conversions among the various nitrogen compounds fall within a group of chemical reactions called **oxidation** and **reduction.** For our purposes here, it will be best to consider oxidation as a reaction in which oxygen is *added* to a substance, and reduction as a reaction in which the amount of oxygen is *reduced.* These definitions are old ones but they are correct and adequate in the context in which we are using them. The oxidation-reduction relationships among nitrogen compounds in water are therefore closely tied to the availability of oxygen in water, hence to the life forms that find the aquatic environment hospitable, and hence to the quality of the water as judged from a human point of view.

Reduction is also defined as the addition of hydrogen, because the hydrogen removes oxygen by converting it to water, and thus is equivalent to the *reduction* of the amount of oxygen in the substance under consideration. The broader concept of oxidation as a loss and reduction as a gain of electrons was first suggested by the German chemist Wilhelm Ostwald in 1903.

10.6 NUTRIENTS, MICROORGANISMS, AND OXYGEN

(Problems 11, 12, 13, 16, 17, 40, 41)

You don't have to read a book to be convinced that clear running water in which trout and salmon thrive is better for people than stagnant cloudy water that is slimy with plant growth and hospitable to sludge worms. Of the two ecosystems, the flowing trout stream is the more delicately balanced; its homeostatic mechanisms are the less effective, and, therefore, it is the more susceptible to change. As a result, a relatively small shift in environmental conditions may readily dislocate the ecosystem of such a stream in a way that adversely affects its water quality. We must understand the nature of the aquatic energy systems to be able to protect our waters.

Now, look at Figure 10.4. Note the three classes of substances shown in this highly simplified representation of the aquatic energy flow: nutrients, small organisms, and oxygen.

What about the fish and the whales? Figure 10.4 shows them to be relatively unimportant because their role in the cycling of oxygen and CO_2 is insignificant compared with that of the much larger total biomass of small organisms. Furthermore, much of the mass of small organisms is near their surfaces, and, therefore, they have closer access to dissolved oxygen and can utilize it more rapidly. Thus, a pound of one-celled primary consumers uses more oxygen in an hour than a one-pound trout, and there are more pounds of one-celled organisms in a pond than of

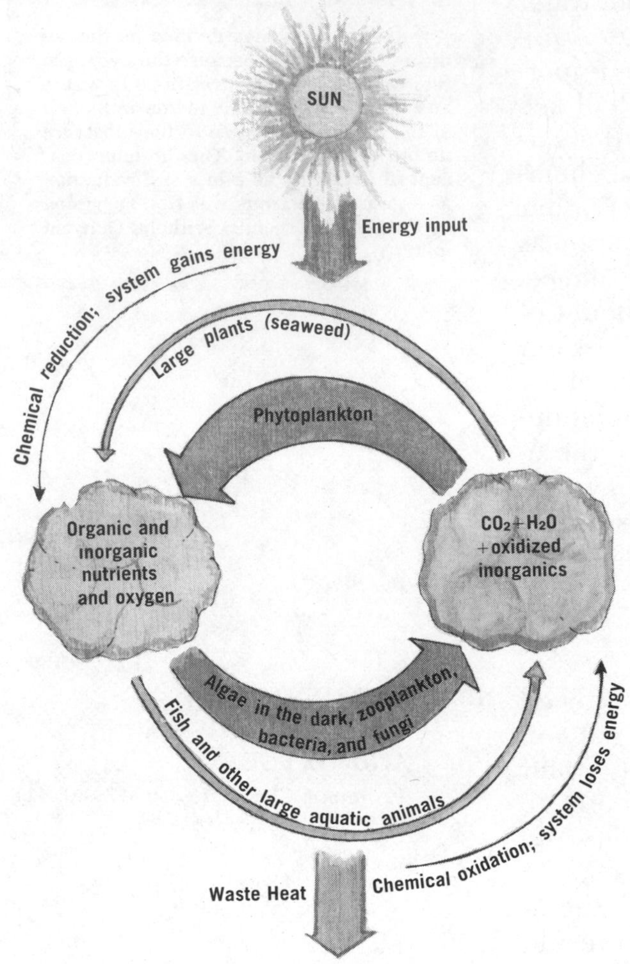

FIGURE 10.4 Aquatic energy flow in the presence of oxygen.

large organisms. It is for these reasons that gas exchange among large aquatic animals is a relatively insignificant portion of the total.

Additionally, microorganisms can survive a much greater variety of environmental conditions than can more complex organisms, and, therefore, the cycle shown in Figure 10.4 is almost always operative. The adaptability of microorganisms is largely due to the fact that a great many different species are present in a sample of pond life. Therefore, collectively they provide a great variety of chemical reaction pathways—some can survive in the presence of substances, such as heavy metals, that would be fatally toxic to larger organisms. Finally, if all goes wrong, microorganisms can suspend their life processes by becoming spores and thus survive by awaiting a return to a more hospitable environment.

All organisms contain carbon; therefore, they require a source of carbon for growth and tissue repair. Of course, they also need energy. Most bacteria get their carbon from organic compounds, and are therefore classified as heterotrophs (see page 4). When dissolved oxygen is available, the heterotrophic life processes convert the energy-rich organic matter to carbon dioxide and water, which are energy-poor substances. This transformation can be compared to the burning of a fuel such as wood, or coal, or candle wax, or methane gas, with the resulting release of energy to the environment. Such a conversion is illustrated by the chemical equation for the combustion of methane, CH_4, shown at right.

$$CH_4 + 2O_2 \rightarrow CO_2 + 2H_2O$$

Is this comparison meaningful? Bacterial decomposition does not occur at flame temperatures, and the specific sequences of alterations of chemical bonds are therefore quite different from those that take place in combustion. The overall release of energy, however, depends only on the starting materials and the final products, and not on the intermediate steps. Therefore, it is valid to consider overall simplified chemical equations in the study of the relationship between energy production and water pollution.

This is one of the statements of the First Law of Thermodynamics. See Chapter 5.

Bacterial decomposition in the presence of air is called **aerobiosis**, and it is this process that yields the most energy from a given weight of nutrients. For example, the complete aerobiosis of glucose (a sugar, $C_6H_{12}O_6$) may be represented by the equation:

$$C_6H_{12}O_6 + 6O_2 \rightarrow 6CO_2 + 6H_2O + 4100 \text{ calories per gram of glucose}$$

Proteins contain sulfur and nitrogen as well as carbon, hydrogen, and oxygen. The aerobic decomposition of a protein is represented by the following unbalanced equation, in which the very complex protein molecule is represented only by a general formula, $C_XH_yO_ZN_qS$:

$$C_XH_yO_ZN_qS + O_2 \rightarrow CO_2 + H_2O + \underset{\text{ammonium ion}}{NH_4^+} + \underset{\text{sulfate ion}}{SO_4^=} + 4100 \text{ calories per gram of protein}$$

The reactions represented above are typical of the first stage of bacterial action in the deoxygenation of polluted waters. When organic nutrient is exhausted, the environment becomes favorable for other bacteria that can facilitate the oxidation of ammonium salts to yield additional energy:

$$\underset{\text{ammonium ion}}{NH_4^+} + 2O_2 \rightarrow \underset{\text{hydrogen ion}}{2H^+} + H_2O + \underset{\text{nitrate ion}}{NO_3^-} + 4350 \text{ calories per gram of ammonium ion}$$

This process is called **nitrification**.

All of these reactions require oxygen, and the organisms that carry them out are said to be **aerobic**. They also require nutrients, either organic or inorganic. These reactions, therefore, must stop when *any one* of the essentials is depleted. Whichever one is depleted faster becomes the limiting factor. Recall (page 410) that oxygen is replaced in air much more rapidly than it is in water—therefore, insects, birds, and mammals may compete for grain, but oxygen is available to all. The oxygen dissolved in waters, however, can be depleted faster than it is replaced from the atmosphere, and, therefore, oxygen becomes the limiting factor when organic nutrients are plentiful. Small organisms are more efficient energy converters than fish are, in part because of their ability to reproduce extremely rapidly in response to a sudden influx of nutrients. When the run-off of sewage or agricultural fertilizers makes waters so rich in nutrients that oxygen becomes the limiting factor, then the denitrification bacteria, the sludge worms, and some protozoa become more plentiful, while the fish decline in population. The quality of the water must be regarded as deteriorated, and the added nutrients are therefore pollutants.

Suppose, instead, that there is plenty of oxygen and it is the nutrients that become the limiting factor. Again, the primary consumers get the bulk of the nutrient and accumulate the greatest biomass. But the consumers of higher orders now get their share of the oxygen, and they can eat some of the worms and the bugs. And the trout jumping in the sparkling waters do look good to us.

Let us now return to the nutrient-rich, or polluted, conditions, and consider what happens to aquatic life when the oxygen content of waters is depleted. Does it all go dead? No, bacterial action does not stop when the molecular oxygen is gone. Instead, a new series of decompositions called **anaerobiosis** occurs. The anaerobic decomposition of sugars and other carbohydrates is called **fermentation**, and that of proteins is called **putrefaction.**

These processes are represented by the following simplified equations (the second one is unbalanced).

$$\underset{\text{glucose, a sugar}}{C_6H_{12}O_6} \rightarrow \underset{\text{ethyl alcohol}}{2C_2H_6O} + 2CO_2 + 100 \text{ calories per gram of sugar}$$

$$\underset{\text{protein}}{C_xH_yO_zN_qS} + H_2O \rightarrow NH_4^+ + CO_2 + \underset{\text{methane}}{CH_4} + \underset{\text{hydrogen sulfide}}{H_2S} + 370 \text{ calories per gram of protein}$$

Some anaerobic bacteria convert carbohydrate matter to methane:

$$C_6H_{12}O_6 \rightarrow 3CH_4 + 3CO_2 + 220 \text{ calories per gram of sugar}$$

Note that all of these reactions yield much less energy (fewer calories) than oxidation, but are still energetically profitable. Methane is very insoluble in water, and practically all of it is evolved as a gas. Hydrogen sulfide is a highly odorous gas, resembling rotten eggs. Putrefaction, therefore, makes water bubble with foul smells and makes it unlivable for fish or other oxygen-breathing animals. It may be regarded as the worst condition of bacterial pollution. The production of methane is not in itself an undesirable process (except when it accumulates in confined spaces, like sewers, to such a concentration that it becomes explosive). Its evolution means that the aquatic system has rid itself of a quantity of organic matter that would otherwise have demanded oxygen for its conversion to CO_2 and H_2O. And methane is not a toxic substance in its natural concentrations in air.

Biochemical Oxygen Demand

We have seen that nutrient matter pollutes water because it serves as food for microorganisms. Microorganisms, including any pathogens that may be among them, multiply; the oxygen is exhausted and thus becomes unavailable for forms of life (such as fish) that man prefers; and finally, the stinks of putrefaction set in. What is the measure of such pollution? One might think that the analysis of a sample of water to determine the total amount of organic matter it contained would provide such an index. But not all organic matter is equally digestible by bacteria. In fact, some organic matter that is manufactured by industrial processes and is foreign to natural food chains may not be able to function as a nutrient at all. Such matter is said to be **non-biodegradable.** Some material, such as petroleum oil, may decompose only very slowly, so that it cannot be considered equivalent to, say, sugar as a nutrient. An appropriate measure of pollution of water by organic nutrients therefore must somehow recognize the *rate* at which the nutrient matter uses up oxygen, as well as the total quantity that can be consumed. Of course, the rate of biochemical oxidation depends on the temperature of the environment and on the particular kinds of microorganisms and nutrients present. If these factors are constant, then the rate of oxidation can be expressed in terms of the half-life of the nutrient. This concept is exactly the same as that applied to radioactive decay (see Chapter 8). The half-life is the time required for half of the nutrient to decompose, and the continuously decreasing rate can be represented in the form of a decay curve ("decay" here has both its biological and mathematical meanings) like that of Figure 10.5.

The chemical equation for the reaction can be considered to correspond to the "chemical oxidation" arrow of Figure 10.4.

$$\text{Nutrients} + \text{Dissolved oxygen} \xrightarrow{\text{microorganisms}} CO_2 + H_2O + \text{Oxidized inorganics}$$

Now, imagine that, in the presence of oxygen at 20°C, the half-life of the nutrient matter is one day. (This is in fact the situation shown in Figure 10.5 and it is typical of experimentally observed values.) Then half of the nutrient will remain in the oxygenated water after one day; half of that, or one-fourth of the original

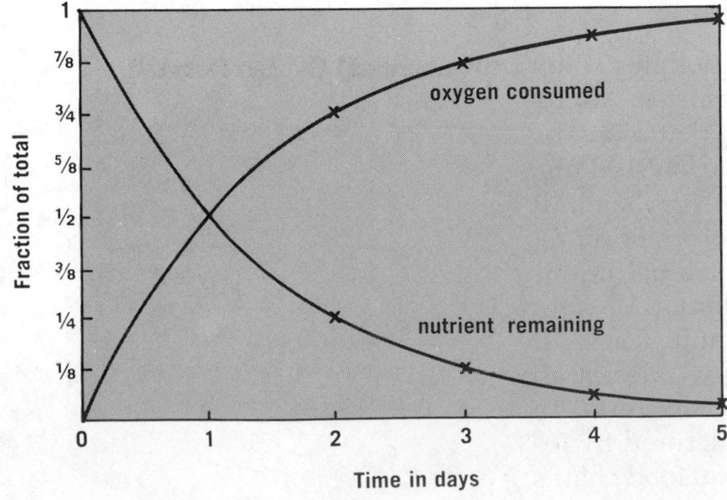

FIGURE 10.5 The biochemical oxygen demand (BOD) curve. The "nutrient" refers only to the maximum quantity that could be oxidized if all the oxygen were utilized. Any quantity in excess of this amount would necessarily remain unchanged.

amount, after two days; one-eighth after three days; and so on. This means that one-half of all the oxygen that is going to be used up will be used up in one day, another one-fourth (or a total of three-fourths) after two days, another one-eighth after three days, and so on.

The ideal shapes of the curves are shown in Figure 10.5. Note that at the end of five days almost all the nutrient is gone and almost all the oxygen that is going to be used has already been used. This length of time is considered to be a good compromise between completion of the oxidation and not having to wait forever. Therefore, a standard test is carried out by saturating a sample of the polluted water with oxygen at 20°C and determining how much oxygen has been used up after five days. The amount of oxygen thus consumed per liter of contaminated water is called the **biochemical oxygen demand** or BOD.

10.7 THE POLLUTION OF INLAND WATERS BY NUTRIENTS

(Problems 6, 14, 15, 18 19)

Streams and Rivers

As we have learned from the preceding discussion, unpolluted water is characteristically rich in dissolved oxygen and low in oxygen-demanding nutrients (BOD). Typical desirable levels are 8 mg of dissolved oxygen per liter and no more than about 2 mg per liter of BOD. When sewage is discharged into an unpolluted stream, the organic nutrient in the sewage creates an instantaneous increase in BOD; that is to say, the stream is polluted the moment the sewage enters it (see Fig. 10.6). But the demand on the oxygen is not instantaneous, as we have seen from the shape of the decay curve. Therefore, the dissolved oxygen falls off gradually, not instantaneously. Organic sewage that consists of human and animal wastes does not, of itself, kill fish; in fact it nourishes them. What does kill fish is lack of dissolved oxygen (in concentrations less than about 4 mg per liter, depending on the species); therefore, when the dissolved oxygen concentration dips below this level (about 15 miles from Pollutionville, Fig. 10.6), the fish begin to die. As the river continues to flow, it recovers oxygen from the atmosphere and from photosynthesis by its vegetation, and thus it repurifies itself. Of course, if additional sewage is discharged before recovery is complete, as by closely spaced cities, the pollution becomes continuous. A river in such a condition, which unfortunately can be found near densely populated areas all over the world, supports no fish, is high in bacterial content, usually including pathogenic organ-

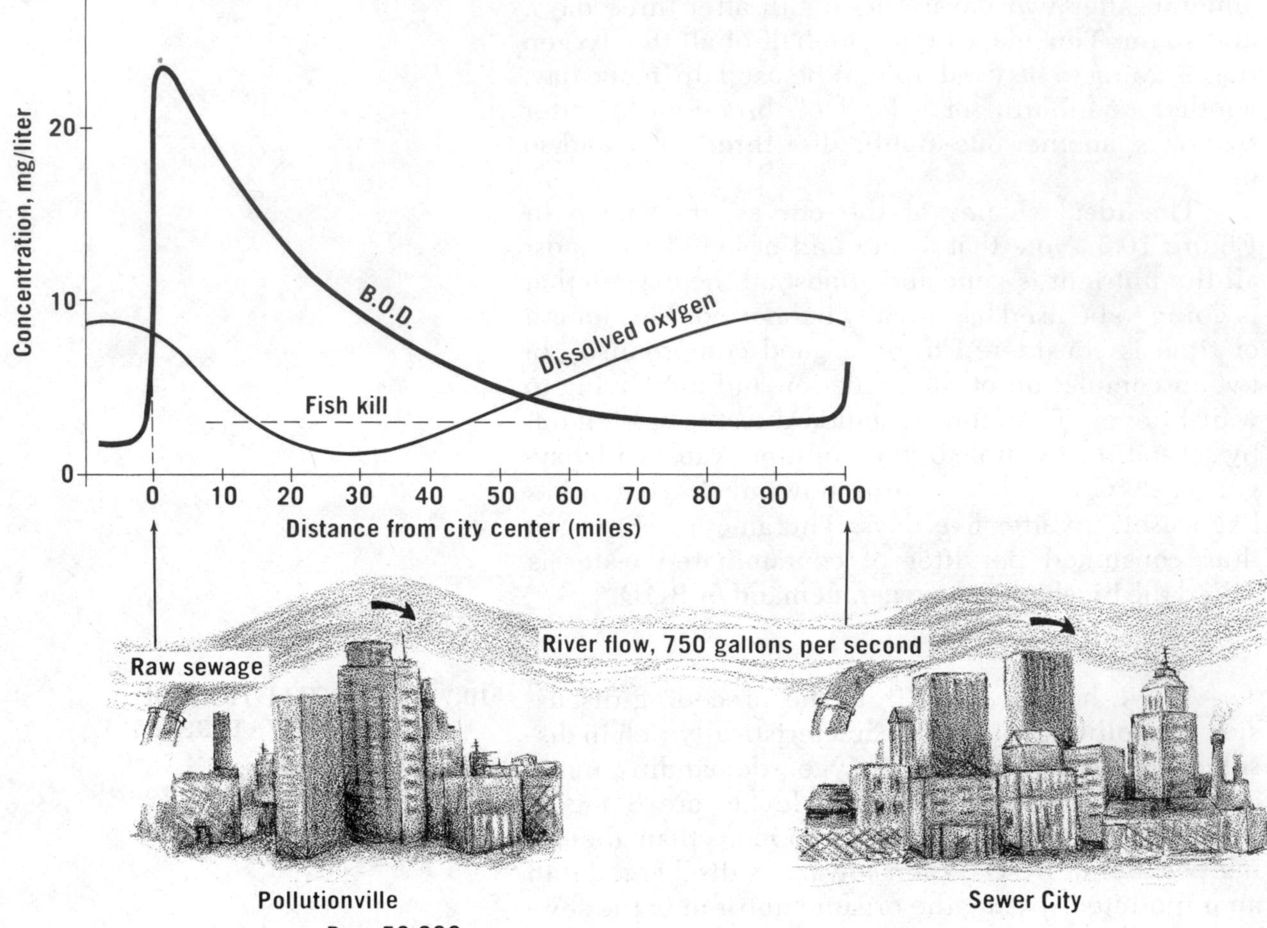

FIGURE 10.6 River pollution from hypothetical cities.

isms, appears muddily blue-green from its choking algae, and, in extreme cases, stinks from putrefaction and fermentation.

Lakes

Look at Figure 10.7 to see the typical summer and winter conditions in a small lake in a temperate climate, such as New England. In the summer, the upper waters, called the **epilimnion** (the "surface lake") are warmed by the sun. These warmer waters, being lighter than the colder ones below, remain on top and maintain their own circulation and oxygen-rich conditions. The lower lake waters (the **hypolimnion**) are cold and relatively airless. Between the two lies a transition layer, the **thermocline**, in which both temperature and oxygen content fall off rapidly with depth. As winter comes on, the surface layers cool and become denser. When they become as dense as the lower layers, the entire lake water circulates as a unit

> **Limnology is the study of the physical phenomena of lakes. The prefix *epi-* (Greek) means in addition to, or resting upon. *Hypo-* (also Greek) means under, or beneath.**

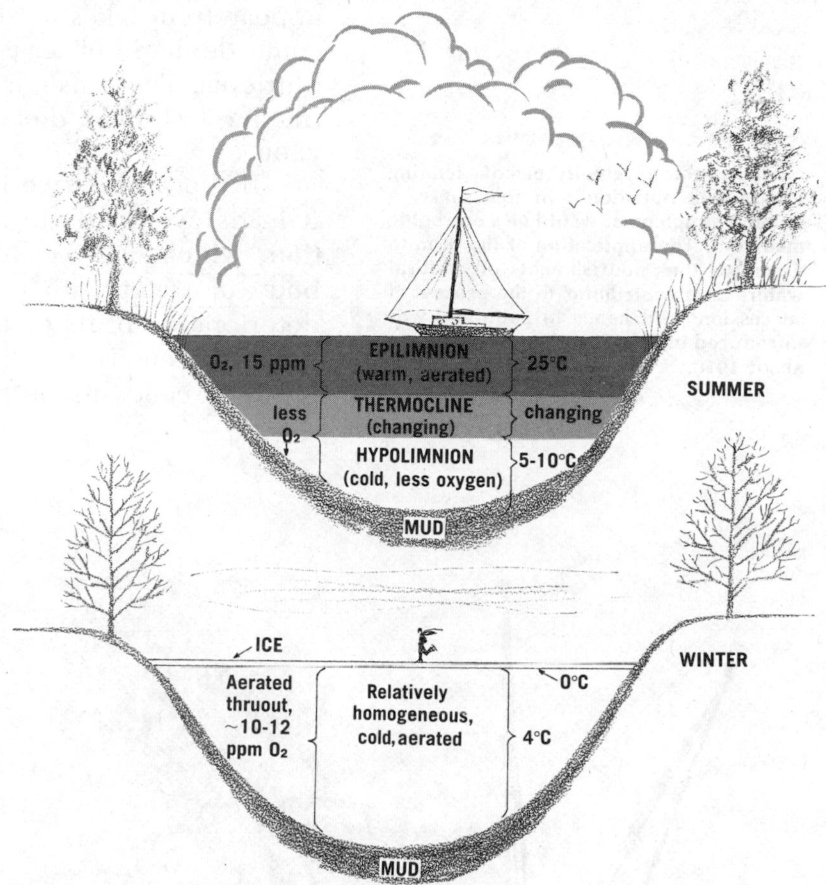

FIGURE 10.7 Thermal stratification in a small, temperate lake.

and becomes oxygenated. This enrichment is, in fact, enhanced by the greater solubility of oxygen in colder waters. Furthermore, the reduced metabolic rates of all organisms at lower temperatures result in a lesser demand for oxygen. When the lake freezes over, then, the waters below support the aquatic life through the winter.

With the spring warmth, the ice melts and the surface water becomes heavier, and again the lake "turns over," replenishing its oxygen supply.

Recall that water reaches its maximum density at 4°C, so any approach to this temperature, from above or below, is accompanied by an increasing density.

Now, what are the effects of oxygen-demanding pollutants on these processes? During the summer, increased supplies of organic matter serve as nutrients in the oxygenated upper waters; the oxygen is replaced as needed by physical contact with the air and from photosynthesis by algae and other water plants. But some organic debris rains down to the lower depths which are reached neither by air nor by sunlight. Therefore, in an organically rich or "polluted" lake, the bottom suffers first. Fish that live best at low temperatures are therefore the first to disappear from lakes as the cold depths they seek become depleted of oxygen by the increased inflow of nutrients. These fish are frequently the ones most attractive to human diets, such as trout, bass, and sturgeon.

Eutrophic originally meant "tending to promote nutrition"; in this sense, a vitamin supplement would be a eutrophic medicine. The application of the term to describe the nourishment of natural waters as a contributor to the process of succession (and hence to pollution) was introduced into the literature of limnology about 1910.

If the process we have just described continues, it leads eventually to a condition called **eutrophication**, which may be defined as the enrichment of a body of water with nutrients, with the consequent deterioration of its quality for human purposes. This process occurs naturally in any lake whose nutrient inflow exceeds its outflow. Such **natural eutrophica-**

"Never mind the weather report. What's the eutrophication report?" *(American Scientist,* **September-October, p. 599, 1971.)**

tion, which is closely associated with natural succession, is a slow process from a human point of view, frequently taking place over periods of thousands of years. In contrast, the discharge of untreated sewage and agricultural or industrial wastes into a lake hastens the process greatly, often shrinking millennia into decades. This accelerated process is called **cultural eutrophication**, in recognition of its civilized origins. Lakes in which the nutrient level is particularly high, which are characterized by abundant littoral (shore-dwelling) vegetation, frequent summer stagnation with algal blooms, and absence of cold-water fish species, are said to be **eutrophic.** Such a situation has been considered to be irreversible, and a lake that has reached it has been characterized as "dead." The justification for such a pessimistic outlook lies in the nutrients that accumulate in the muds of eutrophic lakes. These reservoirs of organic matter can supply many years of algal blooms and high oxygen demands. However, the situation is not always hopeless; more recent experience (Lake Washington, in Seattle, in the 1960's) has shown that cultural eutrophication can be reversed if the inflow of nutrients is greatly reduced.

Algae and Detergents—and How to Wash Your Clothes

We have spoken about algae and "algal blooms," and you have heard about soaps and detergents, and are aware that these agents of personal cleanliness are somehow also involved in aquatic pollution. Here we shall discuss these matters in some further detail.

Algae are aquatic plants; they are sometimes visible as a blue-green slime on the surface of still water. As plants, algae derive energy from photosynthesis. Therefore, they consume carbon dioxide, CO_2, in the presence of sunlight and release oxygen. Like other plants, algae also need various inorganic nutrients, such as compounds of nitrogen, potassium, phosphorus, sulfur, and iron. In natural systems, growth of algae is usually limited by the small quantities of inorganic nutrients dissolved in surface waters. If the nutrient supply becomes sufficiently lavish, algae grow rapidly and can cover the surface of the water as thick, slimy mats. As some of the algae die, either through exhaustion of some essential nutrients or for other reasons, they become, in turn, food for bacteria. As we have seen, bacterial decomposition consumes oxygen, with consequent polluting effects. Such a sequence of

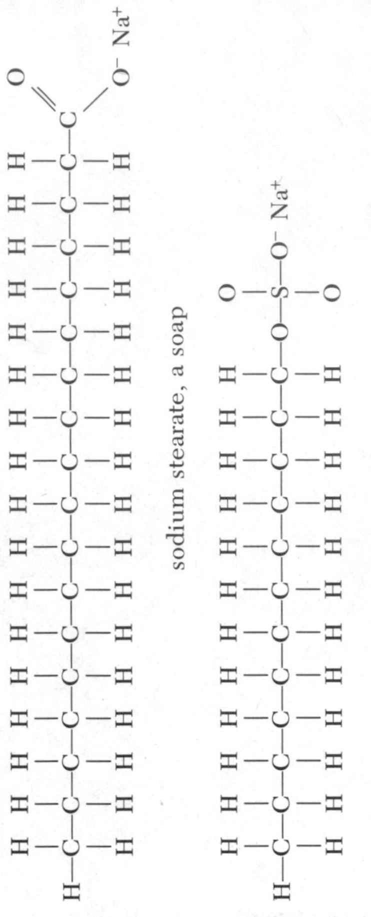

processes results in a condition in which the water beneath the slimy surface is deficient in oxygen and therefore unable to support forms of life that are useful to man. Trout and bass give way to less desirable scavenger varieties such as catfish, and to leeches and worms. As we have seen, this is the process of eutrophication.

The use of modern detergents has contributed to the overfeeding of algae. To appreciate this somewhat complex situation, it will be helpful to understand something about the nature and mode of action of detergents. We mentioned earlier in this chapter that the ease with which a foreign substance can dissolve in a liquid depends on how strongly the molecules of the two different substances attract each other, relative to the mutual attractions of like molecules. Sugar dissolves in water, and if your hands are sticky with honey or lollipops they can be washed clean by rinsing them in pure water. Vegetable oil and animal fat, however, are insoluble in water, and the pure water that rinsed away the honey will leave any grease on your hands behind. It was known in ancient Rome that heating a mixture of animal fat and wood ashes produced a substance that could dissolve both in water and in grease and that could somehow bring these two otherwise incompatible substances together. Therefore, if you rinse your greasy hand with

"Notice how bright and white Brand X gets your clothing because of the harmful chemicals and enzymes it contains. Pure-O, on the other hand, containing no harmful ingredients, leaves your clothes lacklustre gray but protects your environment." **(Drawing by Dana Fradon: © 1971 The New Yorker Magazine, Inc.)**

a mixture of water and this new substance, which we now call **soap,** the grease can be washed away. Soap functions in this manner because it is made up of long molecules that have, on one end, separated regions of positive and negative electrical charge that are strongly attracted to water molecules, and, on the other end, a hydrocarbon character that is attracted to grease molecules. This action of the soap molecule is called **detergency**.

The use of soap does not present a serious threat to the quality of inland waters unless large quantities are discharged directly into rivers and lakes. Soap is a nutrient for bacteria, not for plants, and is normally degraded quite satisfactorily by the bacterial action in sewage treatment plants. However, soap has not been a satisfactory detergent in all respects. The mineral matter in ground water contains metal ions that make soap insoluble, and thereby rob it of its detergency. This insoluble soap manifests itself as a "ring around the bathtub" or a "tattletale gray" on an otherwise white textile fabric. Water that contains such mineral matter is called "hard water." Water from which it is absent, such as rain, is said to be "soft." The years since World War II have witnessed increasing development and use of synthetic detergents that are effective in hard water and that have various other properties that offer advantages over soap.

Some of the synthetic detergents manufactured during the 1960's were not biodegradable, that is, they did not serve as nutrients for bacteria. As a result, they persisted in all their sudsiness through the sewage works, along the rivers, and occasionally right into the drinking water. This foamy problem was solved by the development of the straight-carbon-chain synthetic detergents used in recent years. These modern formulations also contain various other components—some to improve the detergent action, some to obstruct the redeposition of dirt, and some to provide bleaching and brightening actions. These additives typically include phosphates, which, as we learned in Chapter 6, are also used as agricultural fertilizers; they are plant nutrients. More important, phosphates are frequently in short supply in natural waters, and this deficiency serves as a limit to algal growth. The crucial barrier to aquatic plant blooms may thus be removed by supplying phosphates or other needed nutrients, with results that far exceed

Bacteria can degrade unbranched hydrocarbon chains, but they find branched chains much less digestible.

Phosphates are ions that contain the elements phosphorus and oxygen, arranged in linear or cyclic sequences such as:

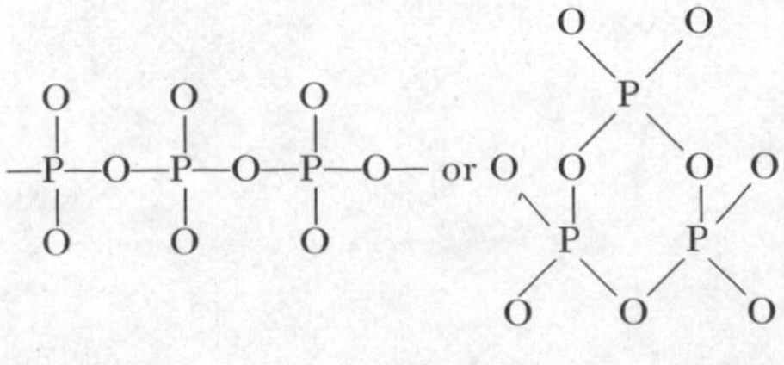

$$-\underset{\displaystyle O}{\overset{\displaystyle O}{\underset{|}{\overset{|}{P}}}}-O-\underset{\displaystyle O}{\overset{\displaystyle O}{\underset{|}{\overset{|}{P}}}}-O-\underset{\displaystyle O}{\overset{\displaystyle O}{\underset{|}{\overset{|}{P}}}}-O- \quad \text{or} \quad \text{(cyclic: three } PO_2 \text{ units joined in a ring through O atoms)}$$

unsophisticated expectations. In many areas of the world, especially in the great rivers and lakes of the tropical and subtropical regions, aquatic weeds have multiplied explosively (Fig. 10.8). They have interfered with fishing, navigation, irrigation, and the production of hydroelectric power. They have brought disease and starvation to communities that depended on these bodies of water. Water hyacinth in the Congo, Nile, and Mississippi rivers and in other waters in India, West Pakistan, Southeast Asia and the Philippines, the water fern in southern Africa, and water lettuce in Ghana are a few examples of

A

B

FIGURE 10.8 The choking of waters by weeds. *A.* **The dam on the white Nile at Jebel Aulia near Khartoum, Sudan. The area was clean when photographed in October 1958.** *B.* **The same area in October 1965, showing the accumulation of water hyacinth above the dam. (From Holm: Aquatic Weeds.** *Science 166*:**699–709, Nov. 7, 1969. Copyright 1969 by the American Association for the Advancement of Science.)**

such catastrophic infestations. Man has always loved the water's edge. To destroy the quality of these limited areas of the Earth is to detract from his humanity as well as from the resources that sustain him.

How, then, should you wash your clothes? That depends on how you balance your concern for the environment with your requirements for personal hygiene and for effective cleaning. Environmental impact would be avoided if you used no detergent at all, but neither personal hygiene nor esthetics would then be adequately served. However, the use of a combination of soap and washing soda (hydrated sodium carbonate, $Na_2CO_3(H_2O)_{10}$, also called sal soda) is minimally disrupting. Low-phosphate or non-phosphate synthetic detergents are also acceptable alternatives. Another very effective action is simply to use less detergent. Manufacturers often recommend quantities that are appropriate (or even excessive) for *extreme* conditions of water hardness. Frequently, one-half of the recommended amount gives very good cleaning, and as little as one-eighth gives fairly good cleaning. As far as personal health is concerned, it is futile to try to achieve a higher degree of sterilization in home laundry operations than is obtainable by soap alone, and the use of bleaches or other antiseptic agents should therefore be regarded as an effort in esthetics, not in hygiene. The final point is your personal criterion of "effective" cleaning. The various additives in modern detergents do contribute whitening and brightening effects, and you must therefore decide how important it is to have "dazzling" underwear.

10.8 INDUSTRIAL WASTES IN WATER

Industrial activity, especially pulp and paper production, food processing, and chemical manufacturing, generates a wide variety of waste products that may be discharged into flowing waters. Some of these wastes are known to be poisonous to man; the effects of others are obscure. Some have been known since antiquity; many are quite recent, and new types of wastes continue to appear as new technology develops. Many industrial wastes are organic compounds that can be degraded by bacteria, but only very slowly, so that they may carry unpleasant odors and tastes along a watercourse for considerable distances. (Even domestic sewage contains significant quantities of non-biodegradable substances of unknown origin.) To complicate matters still further, some of these wastes react with the chlorine that is used as a disinfectant for drinking water. The result of such reaction is the production of chlorinated organic compounds that smell and tase much worse than the original waste product.

$$Cd + 2HCl \rightarrow Cd^{++} + 2Cl^{-} + H_2$$

cadmium

Acids corrode (rust, oxidize) metals, and, as we have seen, soluble compounds of metals can pollute water. Therefore, water that is acidic can become contaminated with metallic pollutants more easily than pure water can.

One of the oldest known waterborne industrial poisons is lead. Throughout history, the most prevalent source has been the lead piping formerly used in water distribution networks. More recently, the use of lead arsenate spray as an insecticide has contaminated surface and ground waters with both lead and arsenic. Lead is a cumulative poison, and even small concentrations, if continuously present in drinking water, may lead to serious illness or death. Arsenic, which sometimes occurs in natural waters that flow through arsenic-bearing minerals, is also a cumulative poison. The "safe" limits in drinking water for both lead and arsenic are recommended to be no higher than about 0.01 ppm.

The compounds of various other metals, such as copper, cadmium, chromium, and silver, have sometimes been implicated as industrial water pollutants. In very recent years, most attention has been given to the problem of mercury poisoning, and we shall devote the remainder of this section to that topic as an example of metallic pollution. Mercury has always been regarded with fascination and alarm. It is the only metal that is liquid at ordinary temperatures (hence its other name, quicksilver), and it is fun to play with. (But don't do it. Its vapor is poisonous, and at high temperatures it can vaporize rapidly enough to be deadly.) Some of its compounds, whose toxicity has been well known since the Middle Ages, have been used as agents of murder and suicide. Until very recently, however, mercury was not considered a dangerous water pollutant, for the following reasons: Although mercury is widely distributed over the Earth, it generally occurs only in trace concentrations. Natural waters typically contain only a few parts per billion of mercury. Metallic mercury, itself, although poisonous in vapor form, is not particularly hazardous when taken by mouth as a liquid. The use of mercury as a component of dental fillings has been shown to be harmless; the mercury in the teeth does not migrate to other parts of the body. Many mercury compounds are very highly insoluble; for example, it is calculated that it would require about 25 gallons of water to dissolve one molecule of mercuric sulfide, HgS!

These considerations imply that mercury in water is not a potential pollutant and perhaps account for the previous lack of concern over the fact that half of the total amount of mercury mined annually is released into the environment. (About 10,000 tons are mined, of which 5000 tons are somehow "lost.") These discharges occur as waste effluents from manufacturing plants or else as the incorporation of traces of mercury into products in which it does not belong. An example is the electrochemical conversion of brine, NaCl dissolved in water, into chlorine and sodium hydroxide, as represented by the equation at right.

$$2NaCl + 2H_2O \rightarrow Cl_2 + 2NaOH + H_2$$

chlorine (Cl₂); sodium hydroxide (caustic soda) (2NaOH); hydrogen, which is released to the atmosphere (H_2)

"The devil with the food chain. I *like* mercury." (*American Scientist*, **September-October, p. 549, 1971.**)

Mercury does not appear in the equation, but it flows along the bottom of the reaction cell as an electrical boundary (electrode) at which the sodium hydroxide and hydrogen are produced. When the salt solution (brine) becomes too weak, it is discarded. This waste contains mercury, which then follows whatever watercourse is available to it. The sodium hydroxide product also is contaminated with mercury, and carries it into many products for which sodium hydroxide is a raw material. Finally, the hydrogen discharged to the atmosphere also carries some mercury vapor with it.

The complacent notion that such discharges are tolerable has been destroyed by various instances of acute mercury poisoning. Most notable was that which occurred in the 1950's in a coastal area of Japan known as Minamata Bay, where fishermen, their families, and their household cats all became stricken with a mysterious disease that weakened their muscles, impaired their vision, led to mental retardation, and sometimes resulted in paralysis and death. What the people and their cats had in common was a diet of fish, and what the fish had in their bodies was a high concentration of mercury that came from the bay waters. Minamata Bay received the mercury-containing effluent from a local plastics factory. Moreover, the mercury in the fish was present in organically-bound forms that are especially hazardous to man. These compounds are all related to methyl mercury, $H_3C \cdot Hg \cdot CH_3$. Such mercury compounds are sometimes used as pesticides and fungicides, and the discharge of these residues into waters is therefore a serious hazard. Following this episode, research led to the disturbing finding that metallic mercury and inorganic mercury compounds can be methylated (converted to methyl mercury) by anaerobic bacteria in the mud of lake bottoms, as well as by fish and mammals. The mercurial wastes that have accumulated in muddy lake bottoms therefore cannot be regarded as inert sludges; they are potential sources for biochemical conversion into forms of mercury that can enter and pass through the food chain in increasing concentrations and thus become poisons for man.

It would be easy to draw the conclusion from these disturbing circumstances that it would be good to "abolish" mercury, if that were somehow possible. That conclusion would be incorrect. Mercury has been present in the environment, including the food chains, throughout the history of life on Earth, and man has necessarily developed a tolerance for the concentrations to which he has been exposed during his evolution. "Tolerance" is not a passive detachment; it is a biochemical adjustment, and such an adjustment usually leads to dependence. It is therefore likely that man requires small amounts of mercury, as he does traces of other metallic elements that would be poisonous in higher concentrations. To say that mercury is the danger is therefore only a limited truth; the more general conclusion is that the danger lies in heedless disruption of a delicately balanced ecosystem.

10.9 THE POLLUTION OF THE OCEANS

(*Problems 23, 24, 25, 26*)

In Chapter 1 we included the oceans among the stable ecosystems of the Earth. The great mass of water implies stability, by virtue of its capacity to dilute foreign matter down to inconsequential concentrations. The total oceanic area is about 140 million square miles. and the average depth of the major oceans (Atlantic, Pacific, and Indian, total area, 130 million square miles) is about 13,500 feet. Of course, living organisms are not distributed uniformly in all this water. The organically most productive areas are the shallow waters near the shorelines, which make up about one-tenth of the total oceanic surface and much less of the total volume. Furthermore, we have learned that the magnitudes of ecological disruptions are not at all proportional to the masses or concentrations of polluting substances, but rather depend on the extent to which homeostatic mechanisms are upset. Let us, therefore, look at the various classes of pollutants now being discharged into the "world ocean," and try to assess their possible environmental impacts.

Oil

When we say oil we mean petroleum oil; either the crude oil as it comes from the ground or one of its products derived from industrial or natural processing. For example, the Kuwait crude oil carried in 1967 by the ill-fated tanker *Torrey Canyon* is a dark brown liquid that smells like diesel fuel and is about as thick as engine oil. Its specific gravity is 0.87; that of pure water is 1, and of sea water is 1.03; therefore, the oil floats on water. The crude oil is "crude"

$$\text{Specific gravity} = \frac{\text{density of the substance}}{\text{density of pure water}}$$

At 4°C, the density of water is 1 g/ml, and therefore the specific gravity of any substance is numerically equal to its density.

indeed, in the sense that it consists of many thousands of components of widely differing molecular weights. Most of the oil is hydrocarbon, but there is an appreciable proportion of sulfur, and there are trace concentrations of metals such as vanadium and nickel. If the oil is heated to 100°C, about 12 per cent of its volume boils off; if it is heated to 200°C, an additional 13 per cent boils off. The total (25 per cent) may be considered to be the volatile fraction that will evaporate from the floating oil surface within a few days. (Of course, the evaporated matter will not stay in the air indefinitely, but will eventually be returned to earth by mechanisms we have discussed in Chapter 9.) The remaining oil is slowly metabolized by bacteria, and some of it slowly evaporates. As we noted earlier when we discussed detergents, bacteria prefer straight chains of carbon atoms to branched chains, and the linear molecules are therefore degraded first. After about three months, practically all the material that can evaporate has evaporated and all that can be eaten has been eaten. The persistent remainder is an asphaltic residue that represents about 15 per cent of the original oil. These leftovers occur as small tarry lumps all over the Earth's seas. We may count ourselves fortunate for having them rather than an ocean-wide monomolecular layer of oil.

The wrecks of tankers are not the only sources of oil pollution, even though they are among the most dramatic. The same intractable tarry residues that eventually accumulate from crude oil in the open ocean also build up in the storage tanks of the tankers, and in the fuel tanks of oil-burning ships. Furthermore, an appreciable quantity of ordinary crude oil adheres mechanically to the walls of tanks even after they are drained. These oily ballasts must be dealt with somehow. In spite of mounting legal pressures, the tanks are often cleaned at sea and the waste dumped into the water, usually when the dark of the night matches the color of the oil.

Offshore drilling operations, now being conducted on continental shelves in many parts of the world, are subject to accidents that result in the direct release of oil from well to sea. A notable incident of this type occurred off the coast of Santa Barbara, California, in 1969; it is estimated that between 20,000 and 50,000 gallons of oil per day were released for 11 days.

"Look at the bright side—employment for thousands trying to clean it up." (© Sidney Harris as it appeared in *Bull. Atom. Scientists,* **Dec., 1971.)**

The total quantity of oil that finds its way into the

sea each year is very large. It has been estimated by various investigators that about one million tons of oil are spilled into the ocean each year from shipping and oil drilling operations alone. But there are also myriads of "mini-spills" – sludges from automobile crankcases that are dumped into sewers, routine oil-handling losses at seaports, leaks from pipes, and the like. Some oily aerosols also settle into the sea from the atmosphere. The grand total from all these sources is difficult to estimate, but it could well reach 10 million tons per year or more.

What are the effects of oil on the earth's ocean, and what protective measures are available? We may gain some insights into both questions by retracing the Torrey Canyon incident, to which we referred earlier in this section. At 9 A.M. on March 18, 1967, this tanker, bound for Milford Haven (England) from the Persian Gulf, ran aground on a reef 15 miles west of Land's End. She was carrying 117,000 tons of crude oil. The impact tore open six of her 18 tanks, and within two days, about 30,000 tons of escaped oil had created a slick about 20 miles long (see Color Plate). On March 28, 29, and 30, after all efforts to move her or to stop the flow of oil had failed, and after her back had been broken by the seas, and there was no hope of salvage, she was bombed, and her remaining oil was set afire, but most of the volatile components had already evaporated and the fires could not be maintained, even with the addition of gasoline. The ship finally sank completely under the water around the end of April.

The oil at sea moves in response to wind and currents. During the first few days a northerly wind kept the oil well offshore, but by Monday the wind had turned and it was apparent that both the English and French Channel coastlines were threatened. The main danger was considered to be the threat of damage to the amenities of shore and beach life, as evidenced by the ugliness and stickiness of the oil, the smell, and the depression of the value of resort properties.

But the actual damages went beyond esthetics. The major biological effect of the oil on the open sea was the destruction of sea birds. They could not survive the interference with their swimming, flight, and insulation that resulted from the matting of their feathers. There was no evidence of any other severe disruption of mid-ocean aquatic life. Along the shore, the biological effects were more extensive; this finding was consistent with various other studies that have shown that oil can cause massive kills of shellfish and other shore-dwelling marine animals. The main control method used by the British in the Torrey Canyon case was to disperse the oil with detergents. The object of the treatments was to break up the oil film into small droplets so that it could be more readily washed away from the shoreline and more easily attacked by the oil-consuming bacteria. The detergent itself does not destroy the oil. The treatment was, indeed, mechanically effective, but the detergent itself, including its aromatic hydrocarbon solvents (benzene and toluene), contributed significantly to the poisoning of marine life and therefore did more biological harm than good.

Situations of this sort present complexities at many levels. If your home were at the shore, and your sea wall became slippery with oil, it would seem reasonable for you to wash it off. The toxic solvents will soon evaporate, the bacteria will gradually degrade the oil, the shore will eventually become repopulated with marine life, and the inert globs of asphalt will go away somewhere. But nagging questions remain. Are the findings in temperate waters applicable to other environments, such as the tropics, where life is more abundant, or the arctic, where it is more fragile? Some of the intractable heavier hydrocarbons with cyclic structures are not merely inert, they are carcinogenic. Will they accumulate in the bodies of marine animals? Remember that when a human eats fish, he is a consumer of a higher order than when he eats beef, and so he is more subject to the magnification effect of the food chain. Communication among animals in the sea is carried out by means of chemical signals—fish can smell substances that tell of food, alarm, sex attraction, and the approach to a home territory. Will the foreign stinks of oil disrupt this delicate information network?

Other Chemical Wastes

How does one dispose of highly toxic chemical wastes, such as by-products from chemical manufacturing, chemical warfare agents, and pesticide residues? There is no easy answer. Biological treatment systems are inapplicable because the microorganisms are poisoned by the substances they are supposed to oxidize. Chemical conversions, including burning, are sometimes applicable, but the materials are fre-

A fish kill caused by water pollution.

quently expensive and dangerous to handle, and the conversion processes may produce noxious air pollutants. It is tempting, then, to seal the material in a drum and dump it in the sea. But drums rust, and outbound freighters do not always wait to unload until they reach the waters above the sea's abysmal depths. During the last few years, for example, hundreds of such drums were caught by Dutch fishermen operating in the continental shelf fisheries of the North Sea, or were washed ashore along the Dutch coast. Some of these materials were found to be toxic to fish even after being diluted to one part in 100,000,000. It is estimated that tens of thousands of such drums have been dropped into the sea.

Of course, all the river pollutants enter the same sink: the world ocean. The organic nutrients are recycled in the aqueous food web, but the chemical wastes from factories and the seepages from mines, including the mineral matter and the stubbornly resistant organic chemicals, are all carried by the streams and rivers of the world into the ocean.

And where do the air pollutants go – airborne lead and other metals from automobile exhaust, and mercury vapor from electrolysis operations, and the fine particles of agriculture spray dusts that ride the winds? Again, to the ocean—perhaps 200,000 tons of lead and 5000 of mercury per year, as well as many tons of other toxic materials. In ocean regions near large cities, such as New York, waste accumulations have rendered large areas unfit for any marine life at all, and these have come to be known as "dead seas." In other areas the riverborne organisms invade shellfish and make them carriers of diseases like infectious hepatitis. Radioactive wastes, discussed in Chapter 8, are another class of ocean pollutants. Still another threat comes from the possible future course of undersea industrial operations, such as the mining, concentration, and processing of ores from the ocean floor. All of the process wastes would go directly into the waters.

Is there an overall threat to life in the sea?

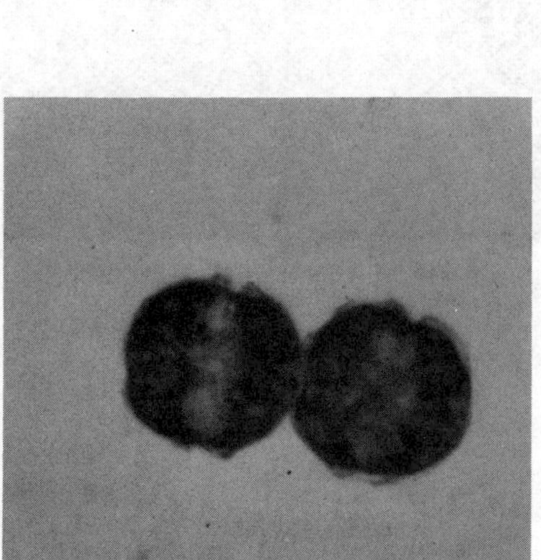

Red tide organism.

It is, in general, very difficult to predict the reactions of homeostatic mechanisms to environmental stresses. We are surprised at times by the extreme fragility and at other times by the apparent stability of ecosystems. Something, we are not sure just what, goes out of adjustment and a massive "red tide" (page 39) devastates a productive shoreline, or an unchecked bloom of starfish devours a coral reef. Or conversely, thousands of tons of crude oil are dumped in mid-ocean, and in a few months, go quietly away (or seem to, since we can hardly be sure of the possible long-term effects). The toxins that accumulate in the ocean have no other place to go; the earth's ocean is its ultimate sink. Complex biotic societies are typically adaptable to changing conditions, but no one can guarantee that the organisms of the ocean will continue to survive the present rate of influx of exotic chemical wastes. The earth's oxygen is continuously replenished by photosynthesis, and a large portion of that activity is carried out by the vegetation of the oceans. Some investigators have cautioned that the destruction of the phytoplankton or the impairment of its photosynthetic activity might seriously reduce the oxygen content of the atmosphere (not to speak of the capacity of the oceans to supply food for man). At this point, recall Lovelock's hypothesis (page 42) that the Earth's oxygen is maintained by Gaia, its overall biotic community, not by inorganic mechanisms. If this is so, a threat to life in the sea is a threat to life on Earth.

10.10 THE EFFECTS OF WATER POLLUTION ON HUMAN HEALTH

(*Problem 32*)

On a worldwide scale, the pollution of water supplies is probably responsible for more human illness than any other environmental influence. The diseases so transmitted are chiefly due to microorganisms and parasites. Two examples will illustrate the dimensions of the problem. Cholera, an illness caused by ingestion of the bacterium *Vibrio cholerae*, is an illness characterized by intense diarrhea which results rapidly in massive fluid depletion and death of a very large percentage of untreated patients. Though its distribution in the past was virtually worldwide, it has been largely restricted during the twentieth century to Asia, and particularly the area of the Ganges River in India. During the nine years from 1898 to 1907, about 370,000 people died from this disease, and thousands of Indians continue to die each year even up to the present. In 1947, a severe epidemic occurred in Egypt with about 21,000 cases, half of whom died.

Most Americans have never heard of schistosomiasis. This is actually a group of diseases caused by infection with one of three related types of worms. (Which worm you get depends on where in the world you live.) Current estimates are that over 100 million people are infected with schistosomiasis; these cases are distributed throughout the African continent, in parts of Asia, and in areas of Latin America. Estimating the amount of human suffering caused by schistosomiasis is much more difficult than for a disease like cholera, because unlike cholera it is a cause of much chronic, as well as acute, disease in endemic areas. For both these illnesses the main mode of transmission is from contamination of water supplies with the feces of infected individuals. Other bacterial illnesses, such as the salmonelloses (of which typhoid fever is a leading example), and viral infections like poliomyelitis and hepatitis may also be disseminated in this way. In the case of the bacterial and viral illnesses, the organisms themselves are shed in the stool, and must be ingested by others to cause disease.

In the case of schistosomal infections, however, the eggs of the organisms are shed. They then hatch into forms which must find a certain type of snail to complete their life cycle. Once safely in the snail, the worm develops into a free-living form which leaves the snail and may infect man if ingested in drinking water. Alternatively, it may penetrate human skin on contact and enter the bloodstream.

When we look at the United States, however, the picture is very different; in fact, nowhere is the contrast between developed and underdeveloped countries starker than in the comparison between the health effects of water pollution on the respective populations. During the decade 1961–1970, there were 130 reported outbreaks of disease attributable to contaminated water supplies in the United States; of these, all but a very few were probably due to the presence of microorganisms rather than chemicals. A total of 46,000 people became ill, but only 20 died. While the existence of such outbreaks in a technological society such as ours is deplorable, one can immediately see that water pollution is a very minor source of acute fatal illness in the United States. In this age of extreme mobility made possible by international travel, the possibility always exists that a disease such as cholera could spread to the United States and attain epidemic proportions here. That this could happen on a large scale, however, seems unlikely since about three-quarters of the American population derive their water from sources which are monitored by state and federal agencies.

The usual measure of microbiologic purity of a water supply is the so-called coliform count (coliforms are the class of bacteria present in the human intestine); therefore, the concentration of coliforms in a water supply is a measure of the amount of human fecal contamination, and not a direct measure of the number of disease-causing microorganisms. Water is generally considered safe if it contains fewer than 10 coliforms per liter. Though this method generally serves to safeguard the purity of water, its major pitfall is that some steps in water purification, notably chlorination, may destroy bacteria without killing viruses; hence viral disease may be transmitted by water that satisfies rigid bacteriologic standards.

As we have noted earlier, water supplies may become contaminated with a wide variety of chemical substances. It is surprising, therefore, to realize that although the potential for disease production from this source exists, actual accounts of major illness due to chemically contaminated water are few; the Minamata Bay disaster mentioned earlier is a devastating, but fortunately rare, example. But the simple fact that acute illness is uncommon does not rule out the possibility of chronic illness, about which there is very little definite information. Over the years the United

States Public Health Service has suggested standards for drinking water in the form of maximal allowable concentrations of various substances, including copper, lead, zinc, manganese, arsenic, chromium, silver, selenium, barium, cadmium, cyanides, nitrates, fluorides, phenols, alkyl benzene sulfonates (from detergents), and total organic substances, which include pesticides. All these substances are of some toxicological concern except for the alkyl benzene sulfonates, which are undesirable for both their bad taste and the excessive foaming that they induce. Moreover, several of these substances may result in chronic disease in man. The possible health effects of chronic ingestion of pesticides are discussed in Chapter 7.

The real authority of the Public Health Service relates only to communicable diseases and not those due to chemical contamination; this law is archaic and should be reformed. In addition, the P.H.S. exercises authority over only those water supplies that serve interstate carriers. It turns out that these supplies serve only about 80 million people, or somewhat less that 40 per cent of the population. However, most states have adopted the P.H.S. guidelines.

Recently, nitrates have come under close scrutiny. These substances are interesting in part because their concentration is actually *increased* by certain methods of water treatment. Trickling filter systems designed to oxidize organic molecules may sometimes also increase nitrate concentrations by oxidizing ammonium salts (see equation on page 414). Under certain circumstances, when relatively large amounts of nitrates are ingested, they may be reduced to form nitrites. The toxological significance of nitrites is twofold: (a) They can interfere with the ability of hemoglobin to bind oxygen. Severe cases of this condition have produced death in infants. (b) Nitrates may react with amines in the body to form nitrosamines, some of which have been shown to be carcinogenic in experimental animals. However, most of us probably ingest far more nitrates from foods than from drinking water; indeed, potassium nitrate (saltpeter) is actually added as a preservative to certain foods. Also, the extent of nitrosamine formation as a result of ingested nitrites is unknown.

The Department of Health, Education, and Welfare is continuing its efforts to identify possible contaminants of significance to human health and to set limits on their allowable concentrations in water supplies. The difficulty in doing so, however, is made clear when one realizes that about 12,000 toxic chemicals are used today by industry, and about 500 new chemicals are developed each year. This problem will be far more difficult to resolve than that of communicable disease prevention, for which the technology already exists and is highly effective.

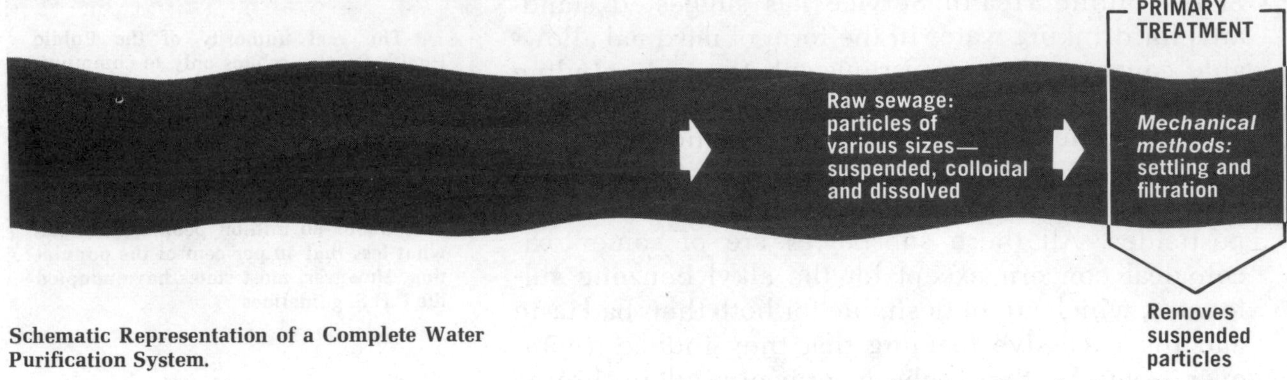

Schematic Representation of a Complete Water Purification System.

(Figure continued on opposite page.)

10.11 WATER PURIFICATION

(Problems 27, 28, 29, 30, 31, 33, 34, 35)

We must recognize that water molecules have no memory,* and therefore it is silly to talk about the number of times that the water you drink has been polluted and repurified, as if the molecules gradually wore out. All that is important is how pure it is when you drink it.

The purification of water has developed into an elaborate and sophisticated technology. However, the general approaches to purification should be comprehensible, and in some cases even obvious, from a general understanding of the nature of water pollution. If water contains impurities that will settle out, hold the water still long enough for settling to occur, or filter them if they can be retained on a filter. If the particles are too small for either process, make them larger by causing them to stick together, or coagulate, in some way, so that settling or filtration becomes feasible. If the water is so acidic that it is corrosive, neutralize the acid. Oxidize the organic waste. Get rid of the metallic poisons by some appropriate physical or chemical procedures. Kill the microorganisms. "Soften" the hard water so that non-polluting detergents can be used effectively.

Let us now look into these procedures in some detail. The first step in municipal water treatment is the collection system. Waterborne wastes from sources such as homes, hospitals, and schools contain food residues, human excrement, paper, soap, detergents, dirt, cloth, other miscellaneous debris, and, of course, microorganisms. This mixture is called sani-

Domestic wastes were once collected in pits called cesspools, from which they were periodically shoveled out and carted away. As cities grew denser, the task became more onerous, and toward the end of the last century it became customary to connect series of cesspools with conduits so that they could all be flushed out with water in a single operation. The next obvious step was to eliminate the cesspools, and use the piping system alone with flushing water continuously available. Thus were sewer systems born.

*Except in some cases. Water from freshly melted ice retains some of the molecular patterns of the solid state, and therefore may freeze more readily than water which has been liquid for a long time.

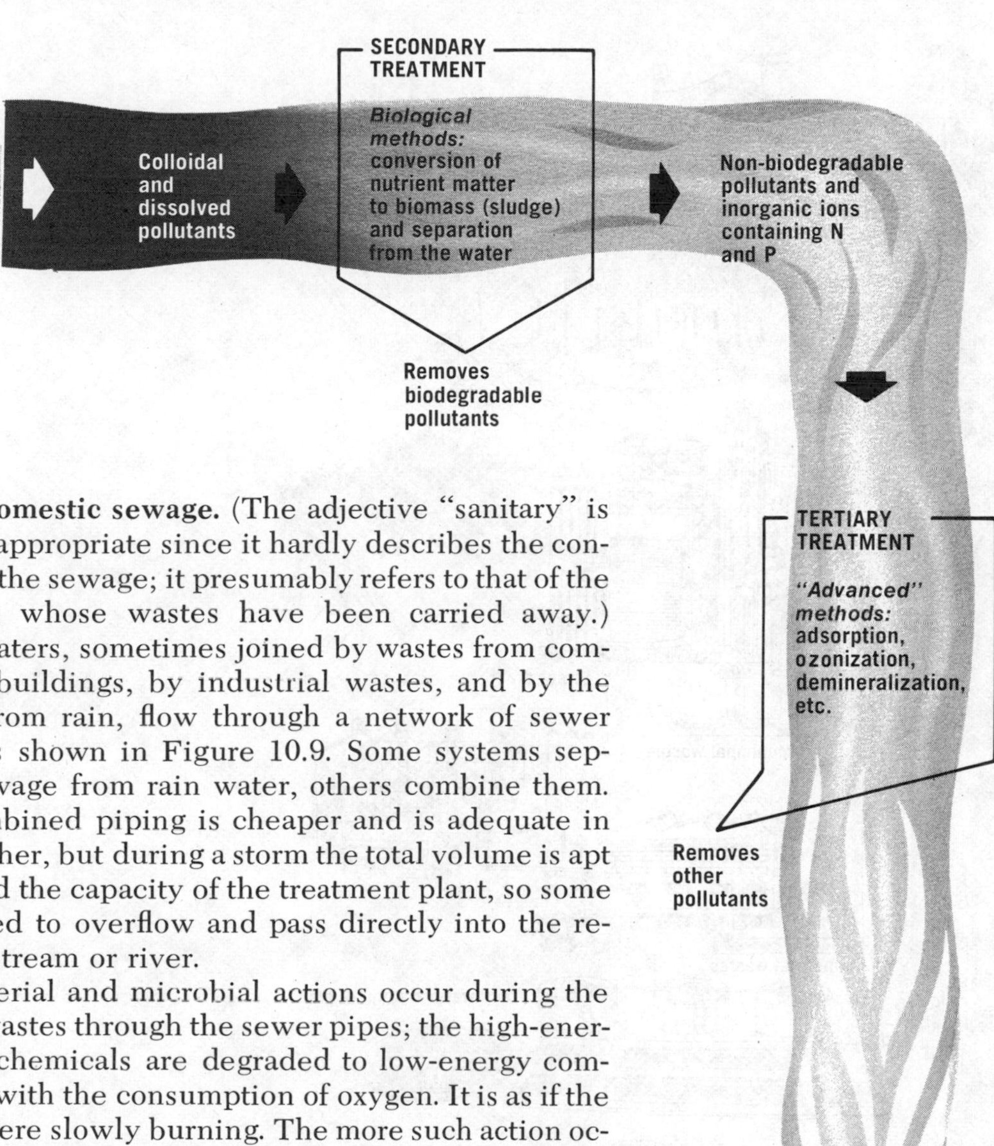

tary or **domestic sewage.** (The adjective "sanitary" is rather inappropriate since it hardly describes the condition of the sewage; it presumably refers to that of the premises whose wastes have been carried away.) These waters, sometimes joined by wastes from commercial buildings, by industrial wastes, and by the run-off from rain, flow through a network of sewer pipes, as shown in Figure 10.9. Some systems separate sewage from rain water, others combine them. The combined piping is cheaper and is adequate in dry weather, but during a storm the total volume is apt to exceed the capacity of the treatment plant, so some is allowed to overflow and pass directly into the receiving stream or river.

Bacterial and microbial actions occur during the flow of wastes through the sewer pipes; the high-energy food chemicals are degraded to low-energy compounds, with the consumption of oxygen. It is as if the waters were slowly burning. The more such action occurs before the sewage is discharged to open waters, the less occurs afterward; therefore, this process must be regarded as the beginning of purification.

Primary Treatment

When the sewage reaches the treatment plant (see Fig. 10.10), it first passes through a series of screens that remove large objects, such as rats or grapefruits, then through a grinding mechanism that reduces any remaining objects to a size small enough to be handled effectively during the remaining treatment period. The next stage is a series of settling chambers designed to remove first the heavy grit, such as sand that rain water brings in from road surfaces, and then, more slowly, any other suspended solids—including organic nutrients—that can settle

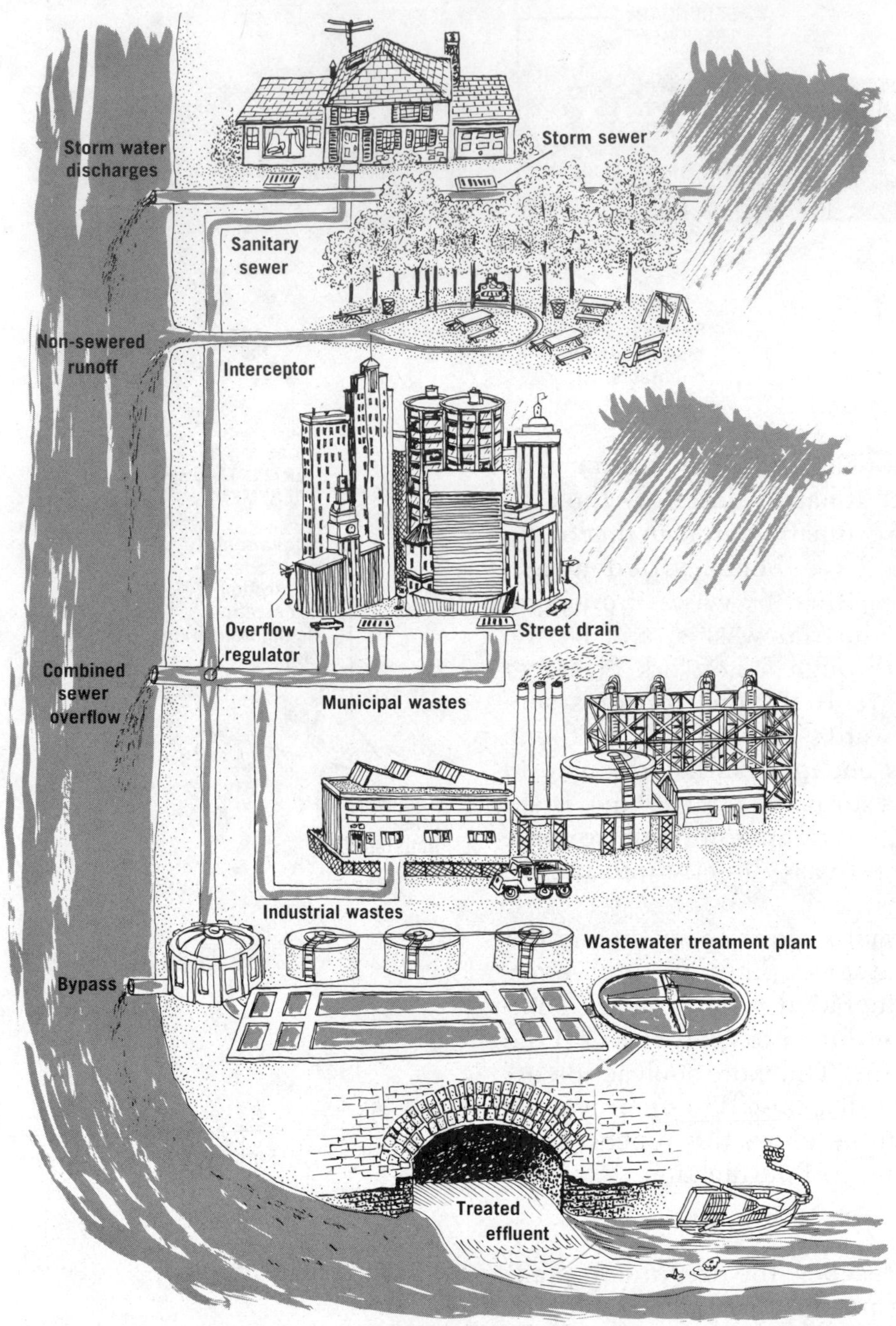

FIGURE 10.9 Sewer collection system.

out in an hour or so. (See Figure 10.3 for the relationship between settling velocity and particle size.) Up to this point the entire process, which is called primary treatment, has been relatively inexpensive but has not accomplished much. If the sewage is now discharged into a stream (as, unfortunately, is often the case), it does not look so bad because it bears no

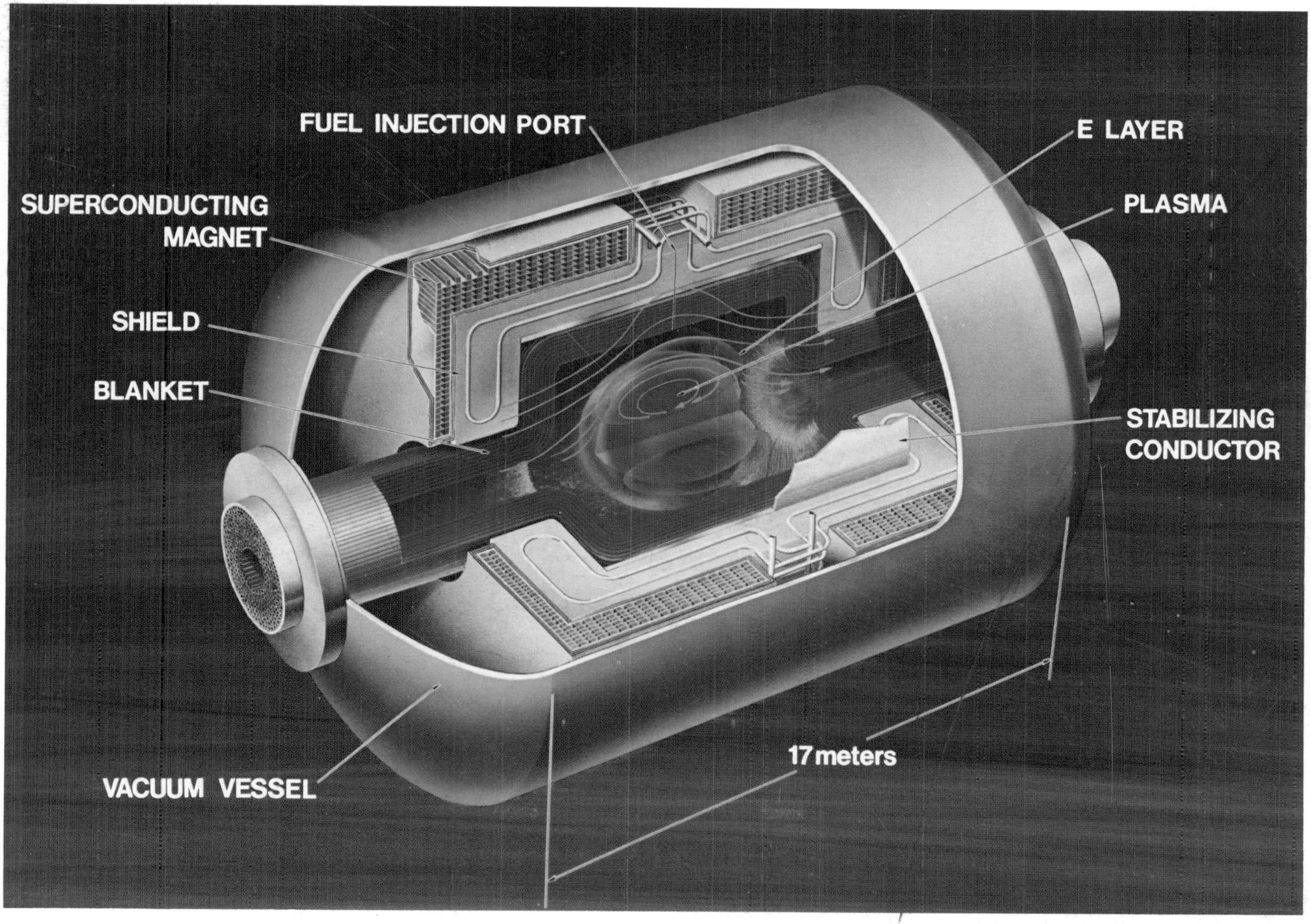

PLATE 9

Fusion reactor. (Courtesy of the University of California Lawrence Radiation Laboratory.)

PLATE 10

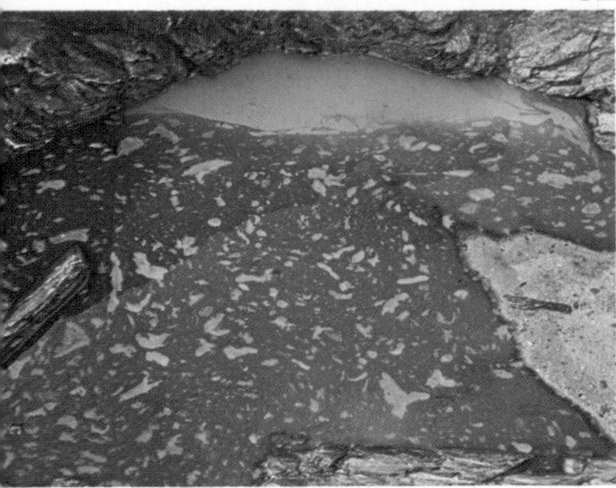

(A) Orange-brown partially emulsified oil floating on detergent solution in a high-water rock pool. (Photograph by Frank G. M. Spooner © Marine Biological Laboratory.)

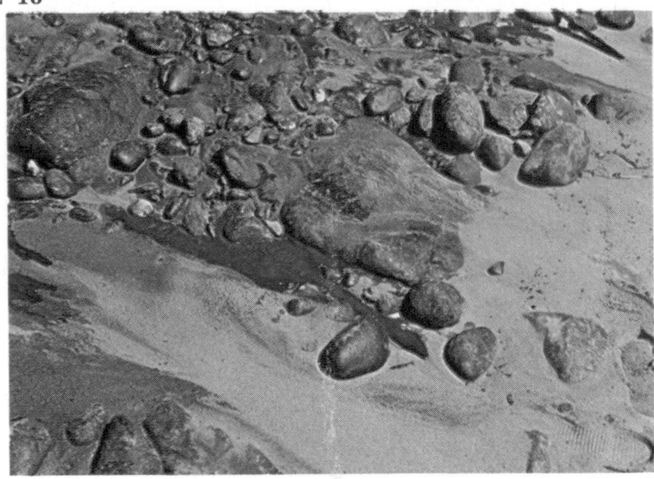

(B) Orange-brown emulsion after detergent treatment, oozing down from boulders near high-water mark. (© N.A. Holme.)

(C) A few days after area has been sprayed with detergent, dead and dying whelks (*Nucella lapillus*), top shells (*Gibbula umbilicalis*), and limpets (*Patella*) are found in a rock corner near a sewer outfall. (© M. Alison Wilson.)

(D) Damaged algae found in area that had been sprayed with detergent. Species of *Fucus* have been reddened, and subsequently little but the midribs of the fronds have survived. The coralline weed *Corallina officinalis* and the encrusting coralline *Lithothamnion sp.* have been killed and bleached. (© M. Alison Wilson.)

Effects of Oil Spill from the Torrey Canyon.

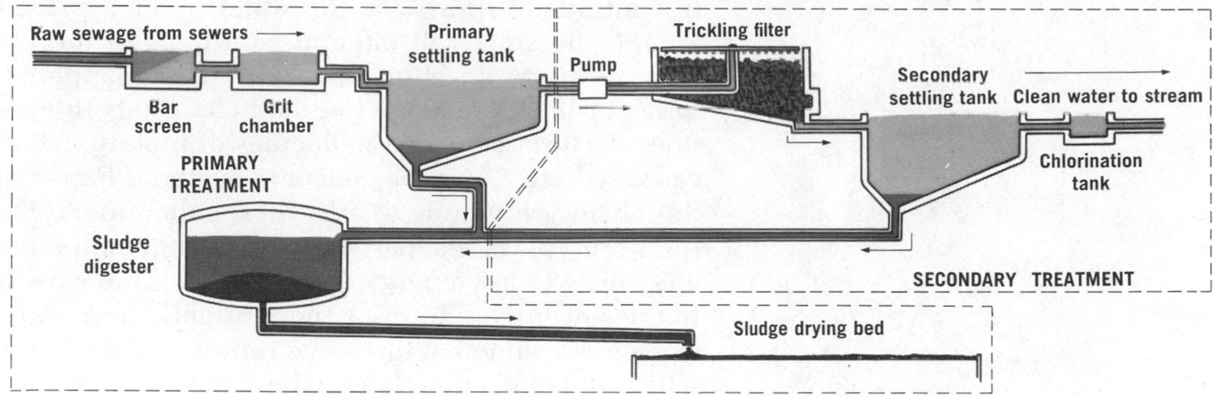

FIGURE 10.10 Sewage plant schematic, showing facilities for primary and secondary treatment. (From *The Living Waters.* **U.S. Public Health Service Publication No. 382.)**

visible solids, but it is still a potent pollutant carrying a heavy load of microorganisms, many of them pathogenic, and considerable quantities of organic nutrients that will demand more oxygen as their decomposition continues.

Secondary Treatment

The next series of steps is designed to reduce greatly the dissolved or finely suspended organic matter by some form of accelerated biological action. What is needed for such decomposition is oxygen and organisms and an environment in which both have ready access to the nutrients. One device for accomplishing this objective is the **trickling filter**, shown in Figure 10.11. In this device, long pipes rotate slowly over a bed of stones, distributing the polluted water

FIGURE 10.11 Picture of a trickling filter with a section removed so as to show construction details. (From Warren: *Biology and Water Pollution Control.* **Philadelphia, W. B. Saunders Co., 1971. Photo courtesy of Link-Belt/FMC.)**

in continuous sprays. As the water trickles over and around the stones, it offers its nutrients in the presence of air to an abundance of rather unappetizing forms of life. A fast-moving food chain sets in operation. Bacteria consume molecules of protein, fat, and carbohydrate. Protozoa consume bacteria. Farther up the chain are worms, snails, flies, and spiders. Each life form plays its part in converting high-energy chemicals to low-energy ones. It is as if the slow fire in the sewer were burning more brightly here. All the oxygen consumed at this stage represents oxygen that will not be needed later when the sewage is discharged to open water. Therefore, this process constitutes a very significant purification.

An alternative technique is the **activated sludge** process, shown schematically in Figure 10.12. Here the sewage, after primary treatment, is pumped into an aeration tank where it is mixed for several hours with air and with bacteria-laden sludge. The biological action is similar to that which occurs in the trickling filter. The sludge bacteria metabolize the organic nutrients; the protozoa, as secondary consumers, feed on the bacteria. The treated waters then flow to a sedimentation tank where the bacteria-laden solids settle out and are returned to the aerator. Some of the sludge must be removed to maintain steady-state conditions. The activated sludge process requires less land space than the trickling filters, and, since it exposes less area to the atmosphere, it does not stink so much. Furthermore, since the food chain is largely confined to microorganisms, there are not so many insects flying around. However, the activated sludge process is a bit trickier to operate and can be more easily overwhelmed and lose its effectiveness when faced with a sudden overload.

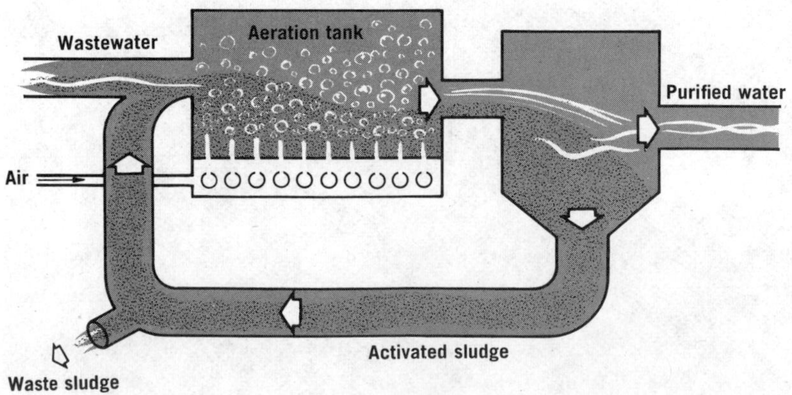

FIGURE 10.12 **Activated sludge process.**

The effluent from the biological action is still laden with bacteria, and so is unfit for discharge into open waters, let alone for drinking. Since the microorganisms have done their work, they may now be killed. The final step is therefore a disinfection process, usually chlorination. Chlorine gas, injected into the effluent 15 to 30 minutes before its final discharge, can kill more than 99 per cent of the harmful bacteria.

Let us now return to the sludge. Each step in the biological consumption of this waterborne waste, from sewage nutrients to bacteria to protozoa and continuing to consumers of higher orders (such as worms), represents a degradation of energy, a consumption of oxygen, and a reduction in the mass of pollutant matter. Also, and perhaps most important from a practical point of view, the process brings about an increase in the average size of the pollutant particles. Look at Figure 10.3 to see how dramatic this change can be. Sugar is dissolved in water in the form of molecules which never settle out. Partially degraded starch and protein occur as colloidal particles approximately in the same size range as viruses. Bacteria are much larger, growing up to about 10 micrometers. Protozoa are gigantic by comparison; some amoeba reach diameters of 500 micrometers and thus are comparable in size to fine grains of beach sand. Some agglomeration also occurs in the metabolic processes of the protozoa, so that their excreta are usually larger than the particles of food they ingest. Finally, when the microorganisms die, their bodies stick together to form aggregates that are large enough to settle out in a reasonably short time. This entire process of making big particles out of little ones is of prime importance in any system of waste water treatment. The mushy mixture of living and dead organisms and their waste products at the bottom of a treatment tank, constitutes the biologically active sludge. Typical sewage contains about 0.6 gram of solid matter per liter, or about 0.06 per cent by weight, while a liter of raw waste sludge contains about 40 to 80 grams of solid matter, corresponding to a concentration of four to eight per cent. Even after this magnification, however, the raw sludge is still a watery, slimy, malodorous mixture of cellular protoplasm and other offensive residues. The organic matter can undergo still further decomposition, but its high concentration engenders anaerobic conditions. Turn back to page 415 and look again at the equation which shows that anaerobiosis produces methane, CH_4, and carbon dioxide, CO_2, among other gases. Such conversions thus decrease the content of solid carbonaceous matter still further, although the process is necessarily accompanied by offensive nitrogenous and sulfidic odors. The final disposal of the sludge residue, whether by incineration, landfill, or other means, becomes a problem in the handling of solid wastes (see Chapter 11).

Tertiary or "Advanced" Treatments

As we have followed our waste waters through primary and secondary treatments, including chlorination, we have done a lot for them. However, these treatments are still inadequate to deal with many of the complexities of water pollution. First of all, many pollutants in sanitary sewage are not removed. Inorganic ions, such as nitrates and phosphates, remain in the treated waters; these materials, as we have seen, serve as plant nutrients and are therefore agents of eutrophication. If chlorination is incomplete, microorganisms will remain; in any case, chlorine will remain in some form or other, frequently as chlorinated organic matter which seriously impairs the taste of the water.

Waste treatment plant, Charleston, West Virginia. Grit basin: GB, Primary clarifier: PC, Sludge pumps: SP, Sludge thickener: ST, Vacuum filters: VF, Chlorinators: C, Aero accelerator: AA, Aeration basin: AB, Secondary clarifier: SC, Pump station: PS. (photo courtesy Union Carbide)

Additionally, many pollutants originating from specific sources such as factories, mines, and agricultural run-offs cannot be handled by municipal sewage treatment plants at all. In fact, most of these sources are not even subjected to any treatment. Some synthetic organic chemicals from industrial waste sources are foreign to natural food webs (that is, they are nonbiodegradable); they not only resist the bacteria of the purification system, but they may poison them, and thereby nullify the biological oxidation which the bacteria would otherwise provide. There are also inorganic pollutants, including acids and metallic salts, as well as suspended soil particles from chemical and mining operations and from natural sources. Some of these materials occur as very fine particles from sources such as roadways, construction sites, or irrigation run-offs. These sediments are troublesome before they settle, because they reduce the penetration of sunlight, and afterwards, because they fill reservoirs, harbors and stream channels with their silt.

The treatment methods that are available to cope with these troublesome wastes are necessarily specific to the type of pollutant to be removed, and they are generally expensive. We will mention a few of these techniques.

COAGULATION AND SEDIMENTATION. We mentioned earlier, in the discussion of biological treatment, that it is advantageous to change little particles into big ones which settle faster. So it is also with inorganic pollutants. There are various inorganic colloidal particles that are water-loving (hydrophilic) and therefore rather adhesive, and in their stickiness they sweep together many other colloidal particles that would otherwise fail to settle out in a reasonable time. This process is called **flocculation.** Lime, alum, and some salts of iron are among these so-called flocculating agents.

ADSORPTION. Go back to page 380 in the air pollution chapter and note the section on adsorption by activated carbon. Carbon is used also for water purification, particularly to remove chemicals that produce offensive tastes and odors. These include the biologically resistant chlorinated hydrocarbons, for which carbon is highly effective.

OTHER OXIDIZING AGENTS. Potassium permanganate, $KMnO_4$, and ozone, O_3, have been used to ox-

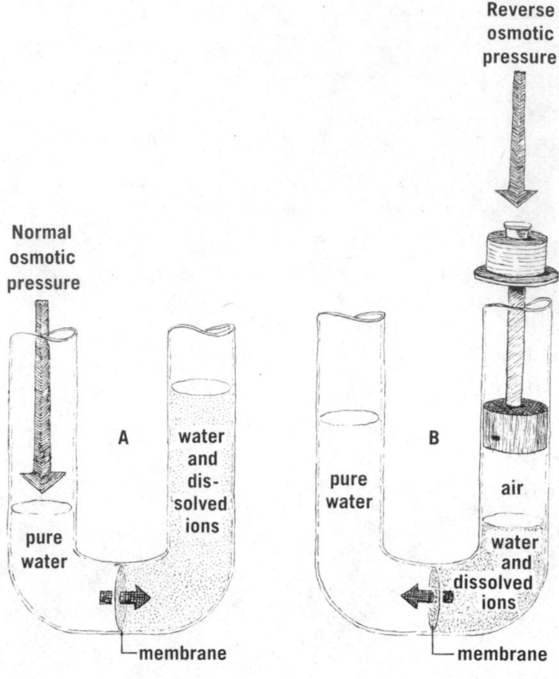

FIGURE 10.13 Reverse osmosis.

idize waterborne wastes that resist oxidation by air in the presence of microorganisms. Ozone has the important advantage that its only byproduct is oxygen.

$$2O_3 \rightarrow 3O_2$$

DEMINERALIZATION. Soluble metal ions, including those that may be toxic, are not removed by any of the methods cited above. Several processes are available, none of them very cheap. One of them is called **reverse osmosis**. Remember that osmosis is the process by which water passes through a membrane that is impermeable to dissolved ions. In the normal course of osmosis, as illustrated in Figure 10.13*A*, the system tends toward an equilibrium in which the concentrations on both sides of the membrane are equal. This means that the water flows from the pure side to the concentrated, "polluted" side. This is just what we don't want, for it increases the quantity of polluted water. However, if excess pressure is applied on the concentrated side (Fig. 10.13*B*), the process can be reversed, and the pure water is squeezed through the membrane and thus freed of its dissolved ionic or other soluble pollutants.

10.12 ECONOMICS, SOCIAL CHOICES, AND STRATEGY IN WATER POLLUTION CONTROL

The average total volume of water suppled to the United States per day by rain and snow is about 10^{12} (1 trillion) gallons. About 4×10^{11} (400 billion) gallons, more than a third of the total supply, is used daily by the manufacturing and power industries, by agriculture, and by cities and towns. What do these numbers mean? One can say that this volume of water would fill a swimming pool every day for every American; such a statement nags at us with the implication that anything we decide about water purification will involve a lot of money, but it doesn't help us make any decisions. Surely we must consider the fact that different modes of utilization affect water quality in vastly different ways, and therefore require different kinds of treatment for its restoration. Also important, but perhaps less obvious at first, is the fact that purification costs rise as the purity rises. Look at Figure 10.14 to see how great this effect is. Why should it be so? Think of a quantity of water that contains 100 pounds of pollutants, and a method of treatment that is 90 per cent effective each time it is applied. The first application of the treatment will therefore remove 90 pounds of pollutants, leaving 10 pounds in the water. The second application will remove 90 per cent of this remainder, or nine pounds, and so on. We can reasonably guess that the cost of each application will be about the same. The reason for this rough equivalence is that the same quantity of water must be handled each time, and when we deal with large quantities of water we need a lot of energy to lift it, or pump it, or force it through a filter, or to force something into it. It costs hardly any more to pump 1000 pounds of 97 per cent pure water than 1000 pounds of 99 per cent pure water, but in the former case we are handling 30 pounds of impurity and the latter only 10 pounds. Therefore, the handling cost *for each per cent impurity* is three times as great at the 99 per cent level.

(Problems 36, 42)

These relationships should not prevent us from asking for very pure water, but they do warn us that we had better direct our efforts toward devising rational strategies for water pollution control if we are not to be overwhelmed by the heavy financial burdens involved in cleaning up dirty water (see Fig. 10.15). Considerations of strategy demand choices at many different levels. Shall we wash our clothes less and preserve our lakes more? Or shall we wash our clothes dazzlingly, then apply expensive advanced treatment procedures at the sewage plant and preserve our lakes too? But then our taxes will go up, and we may not have enough money left to take our vacations at the clean lakes. Questions of this sort follow

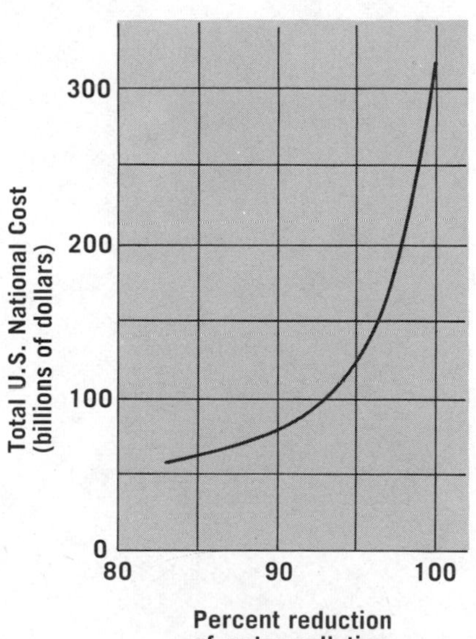

FIGURE 10.14 Cost of reducing water pollution. (Source: estimates from the Environmental Protection Agency, 1972.)

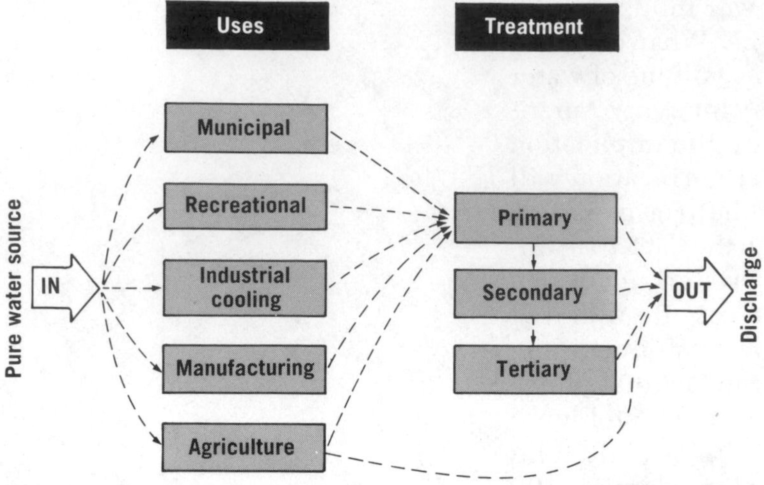

FIGURE 10.15 Strategy for water pollution control.

each other endlessly. Therefore, instead of pursuing a great number of them, let us consider an overall view of the strategy of water utilization (Fig. 10.15).

There are a great many ways to recycle water. If toilet drains were isolated from other domestic wastes, such as those from sinks and drinking fountains, these less-polluted waters could receive only primary treatment and be used for flushing toilets. Alternatively, sewer effluents after only primary treatment could be used for irrigation, if that didn't make the fields too odorous, or for industrial cooling, for as long as cooling water doesn't clog or corrode pipes, it doesn't need to be particularly pure.

Perhaps, after secondary treatment, the waters could be used to convert abandoned gravel pits into lakes suitable for swimming and boating. On the other hand, it may be necessary that some contaminants from chemical manufacturing be removed completely before they contaminate a river.

Another approach to water purification strategy is to classify the receiving watercourse (the river, lake, or estuary into which the treated effluents are released) according to its function, such as navigation, recreational use, or drinking water supply. This function will then dictate the degree of pollution that the receiving watercourse will be allowed to accept. The objectives of our strategies should be to give us the purest waters where we need them most, and to recycle our waters to save our money where we can.

PROBLEMS

1. **Vocabulary.** Define water pollution; water quality.
2. **Contamination.** Discuss the relative ease of contamination of gases, liquids, and solids.
3. **Vocabulary.** Define molecule; ion; colloidal particle; suspended particle.
4. **Water pollution.** A healthy person lives in harmony with bacteria in his digestive system. Why, then, should water that contains digestive bacteria be considered to be polluted?
5. **Water quality.** What are the criteria for water that is considered fit for drinking? Is such water always safe? Is water that does not meet these criteria always harmful? Explain.
6. **Water pollution.** Explain how a non-toxic organic substance, such as chicken soup, can be a water pollutant.
7. **Hydrological cycle.** Why is it conceivable that despite the great mass of water on earth, man's efforts to control climate by such means as cloud seeding might affect the hydrological cycle?
8. **Water.** On page 403 the heating of water and olive oil was compared to the lifting of a 50-pound weight and a 100-pound weight. In this example, is the water analogous to the 50- or to the 100-pound weight? Explain.
9. **Acidity.** Which do you think would be the more acidic in water, a one per cent solution of salt (NaCl) or of ferric (iron) chloride ($FeCl_3$)?
10. **Nitrogen.** Classify each of the following reactions among nitrogen compounds as denitrification, oxidation, or reduction:

    (a) $2NO + O_2 \rightarrow 2NO_2$

    (b) $\underset{\text{hydrazine}}{N_2H_4} + H_2 \rightarrow \underset{\text{ammonia}}{2NH_3}$

    (c) $NH_4NO_2 \rightarrow N_2 + 2H_2O$
11. **Vocabulary.** Define the terms aerobiosis; anaerobiosis; nitrification; fermentation; putrefaction.
12. **Oxygen.** What harmful effects on water quality result from the depletion of molecular oxygen?
13. **B.O.D.** Define biodegradability; biochemical oxygen demand. In what way is the latter a measure of water pollution?
14. **Eutrophication.** What is eutrophication? Explain how it occurs and why it is hastened by the addition of inorganic matter such as phosphates.
15. **Nutrients.** It has been suggested that the world food shortage could be alleviated if we cultivated sewage, and processed the final product in the form of "algaeburgers."

(a) Could such production be carried out on a 24-hour basis? Only during the daytime? Only at night? Explain.

(b) If the sewage were used as the food in a "fish farm," would the product be able to feed more people, or fewer people? Explain.

16. **B.O.D.** The B.O.D. curve of Figure 10.6 shows that the rise and fall occur sharply but not instantaneously. How would the curve look if both the rise and fall did start instantaneously? Which of the following is the more reasonable explanation for the non-instantaneous character of the changes: (a) some smaller discharges, such as those from individual homes or small farms, occur both before and after the main sewer effluent; (b) the sewage does not react instantaneously with oxygen. Defend your answer.

17. **Oxidation.** In the absence of molecular oxygen, O_2, some bacteria, called **facultative bacteria**, can use the oxygen content of various ions, such as sulfate and nitrate, for oxidation of organic matter. A simplified chemical equation for this process is:

$$\text{Nutrient} + NO_3^- \rightarrow \text{Oxidized nutrient} + NO_2^-$$

Would you consider such a change more closely analogous to aerobiosis or to fermentation? Defend your answer.

18. **Lakes.** (a) Construct a graph in which the vertical axis is depth and the horizontal axis is temperature. Draw one curve that represents a lake in summer, another that represents winter. (b) Construct similar graphs of depth vs. oxygen content. How would the shapes of these curves change with advancing eutrophication?

19. **Home laundry.** Write up a set of specific laundry instructions that embodies your personal decisions about pollution, hygiene, and washing effectiveness.

20. **Industrial wastes.** List seven metals whose compounds may have been implicated as water pollutants.

21. **Mercury.** The poisonous properties of mercury have been known since the late years of the Roman Empire, when mercury miners in Spain suffered chronic or acute symptoms, such as tremors and loss of hair and teeth, from exposure to mercury vapor. In more recent years, hatters exhibited the same symptoms (which they called "the hatter's shakes") from the mercury compounds used to convert fur to felt. Do you think these circumstances should necessarily have led to the conclusion that mercury would be a water pollutant if the metal were discharged into streams or lakes? What facts actually led to the opposite conclusion? Is

mercury dangerous as a water pollutant in the same manner that it is in mines or hat factories? Explain.

22. **Acidic pollutants.** The contents of our stomachs are acidic, and we drink acidic fruit juices without doing ourselves any harm. Why, then, are acids considered to be pollutants in drinking water?

23. **Oil spills.** Discuss some of the ecological consequences of the death of large numbers of sea birds.

24. **Oil spills.** The technique used successfully by the French in the Torrey Canyon disaster was to dump chalk dust on the slicks to absorb the oil. The chalk-laden oil is heavier than water, and sinks. Can you suggest advantages and disadvantages of this procedure compared with the detergent method?

25. **Ocean pollution.** What are the various categories of ocean pollutants? List them in what you consider to be the increasing order of their threat to life in the sea.

26. **Ocean pollution.** In its articles on "Sewerage," the Eleventh Edition of the Encyclopaedia Brittanica, published in 1910, states, "Nearly every town upon the coast turns its sewage into the sea. That the sea has a purifying effect is obvious. . . .It has been urged by competent authorities that this system is not wasteful, since the organic matter forms the food of lower organisms, which in turn are devoured by fish. Thus the sea is richer, if the land is the poorer, by the adoption of this cleanly method of disposal." Was this statement wrong when it was made? Defend your answer. Comment on its appropriateness today.

27. **Water purification.** An alternate method of waste water treatment is the **stabilization** or **oxidation pond**, which is a large shallow basin in which the combined action of sunlight, algae, bacteria, and oxygen purifies the water. It may be said that the stabilization pond trades time, space, esthetics, and flexibility for savings in capital and operating costs. Explain this statement.

28. **Water purification.** It was stated in Section 10.11 that biological treatment of waste water reduces the mass of pollutant. Where does the lost matter go?

29. **Water purification.** Distinguish among primary, secondary, and tertiary types of waste water treatment.

30. **Water purification.** The following advice has been offered to tourists who wish to avoid ill effects from drinking water in areas where sanitation is uncertain or where intestinal disorders are common: Do not drink cold tap water. *Never* drink water from pitchers, carafes, or bottles which may have been reused, even if washed. On the other hand, tap water that is hot enough to burn your hand may be directly used to rinse and fill a cup and, when it has cooled, is safe to

drink. Suggest a rational justification for each of these statements.

31. **Sewage.** In a combined piping system, some untreated sewage is dumped into the receiving watercourse during rainstorms. Is this procedure more acceptable than it would be in dry weather? Defend your answer.

32. **Health effects.** When people infected with schistosome worms excrete the eggs via the stools, the eggs hatch into forms which must find a certain type of snail to complete its life cycle. Once safely in the snail, the worm develops into a different form which leaves the snail and is again infective for man. In thinking about ways to decrease the incidence of schistosome infection, scientists have considered two major types of measures: (1) developing compounds which kill the snails, thus preventing the worm from completing its life cycle, and (2) increasing measures of sanitation to prevent human feces from getting to the water supply. Discuss what you think would be the pros and cons of these methods specifically with reference to relative costs, chemical pollution of the environment, and effect on other forms of life besides the snails.

33. **Sewage treatment.** What is flocculation, and how does it help to purify water?

34. **Sewage treatment.** List and explain five methods of "advanced" water treatment.

35. **Aquatic energy cycle.** If Figure 10.4, page 412, referred to the energy cycle in a trickling filter, should any additional organisms have been included? If so, what kinds?

36. **Strategy.** Represent each of the strategies suggested in the last sentences of this chapter by a flow chart similar to Figure 10.15, showing only the appropriate water uses and proper flow directions.

The following questions require arithmetic reasoning or computation.

37. **Water pollutants.** Is the speed of settling of particles in water directly proportional to their diameters? If the diameter is multiplied by 10, is the settling speed 10 times faster? (Justify your answer with data from Figure 10.3.) Is a settling pond a good general method of water pollution control? Explain.

38. **Water.** If the atomic weight of oxygen were 8 instead of 16, what would be the chemical formula for water?

39. **pH.** Could a solution be so strongly acidic that its pH is 0? A negative number? What is the pH of a solution whose H^+ concentration is one mole per liter? Ten moles per liter?

40. **Energy.** (a) Show that the relative energies obtained from complete oxidation and from fermentation of sugar as shown in the relevant equations of Section 10.6 are consistent with the values given in Figure 1.1 on page 4. (*Hint*: The nutrient in an apple is mostly starch, which has about the same energy content per gram as sugar. Evaluate the ratio of oxidation energy/fermentation energy from the two sources and determine whether or not they are roughly equal.) (b) What do you think of the prospects of generating electricity by the anaerobic oxidation of sewage sludge?

41. **B.O.D.** If the half-life of nutrient matter at 20°C is one day, what percentage of it will be consumed after five days? If you were in a hurry, would it be reasonable to approximate the B.O.D. by measuring the oxygen demand created in one day, then doubling your answer?

42. **Economics.** It was suggested in the last section that the fact that water purification costs more as the purity rises is accounted for in large measure because the cost of handling a given quantity of water hardly depends on what you do with it. Test this model against the data derived from Figure 10.14 by completing the following table. In which of the columns of the table are we making the central assumption of the model? Plot the data of column 5 in pencil directly over Figure 10.14 on page 447 and comment on how well or poorly the model agrees with the actual data.

(1) Per cent reduction of pollution	(2) Per cent pollution remaining (100 – col. 1)	(3) Total cost in billions of dollars (from graph)	(4) Cost per per cent pollution (60/col. 2)	(5) Predicted total cost (col. 4 × 16)
84	16	60	60/16	60
88				
92				
95				
98				

BIBLIOGRAPHY

There is a considerable volume of literature on water, on the analysis of its impurities, and on various aspects of its purification, including sewage treatment. A basic three-volume text on the properties of water itself is:

Felix Franks, ed.: *Water—A Comprehensive Treatise.* New York, Plenum Publishing Co., 1973. Vol. I, 596 pp.; Vol. II, 640 pp.; Vol. III, 430 pp.

An older but still valuable one-volume text is:

Ernest Dorsey: *Properties of Ordinary Water-Substance.* New York, Litton Educational Publisher, Van Nostrand—Reinhold Books, 1940. 704 pp.

The following texts are good sources of information on water pollution and its control:

Charles E. Warren: *Biology and Water Pollution Control.* Philadelphia, W. B. Saunders Co., 1971. 434 pp.

Metcalf & Eddy, Inc.: *Wastewater Engineering.* New York, McGraw-Hill, 1972. 782 pp.

John W. Clark, Warren Viessman, Jr., and Mark J. Hammer: *Water Supply and Pollution Control.* Scranton, Pa., International Textbook Co., 1971. 661 pp.

T. R. Camp: *Water and its Impurities.* New York, Litton Educational Publisher, Van Nostrand – Reinhold Books, 1963.

G. V. James and F.T.K. Pentelow: *Water Treatment.* 3rd Ed. London, Technical Press, 1963.

Note also the following books on special topics in water pollution:

National Academy of Sciences: *Eutrophication: Causes, Consequences, Correctives.* Washington, D.C., National Academy of Sciences Press, 1969. 661 pp.

———————————: *Beneficial modifications of the Marine Environment.* Washington, D.C., 1972. 166 pp.

J. E. Smith, ed.: *"Torrey Canyon" Pollution and Marine Life.* London, Cambridge University Press, 1968.

A number of recent chemistry texts emphasize various environmental topics, including water pollution. One such book is:

Stanley E. Manahan: *Environmental Chemistry.* Boston, Willard Grant Press, 1972. 393 pp.

The following are cited as representative of popular books that deal with the crisis of water pollution:

D. E. Carr: *Death of the Sweet Waters.* New York, W. W. Norton & Co., 1966. 257 pp.

F. E. Moss: *The Water Crisis.* New York, Encyclopaedia Britannica, Praeger Publisher, 1967. 305 pp.

G. A. Nikolaieff, ed.: *The Water Crisis.* New York, H. W. Wilson Co., 1969. 192 pp.

11

SOLID WASTES

11.1 SOLID WASTE CYCLES

Today's landscape is not littered with huge mounds of dinosaur bones or ancient fern spores, for debris from living things has been traditionally reused, and the chemicals of one organism's wastes have been incorporated into another organism's tissue. Occasionally, chemical elements are locked for long periods of time inside glaciers or geological deposits, but changing weather patterns, continental drift, upheavals of the earth's crust, and the actions of various organisms cause some of the long-inaccessible deposits to be returned into the rapidly recycled reserves of ecosystems.

Combustion of petroleum and coal in industrial society recycles ancient organic matter back into the active ecosystems of the planet. However, in many other respects recycling is quite inefficient in modern societies. An apple grown in an orchard in the State of Washington may be shipped to New York City to be eaten. The core is not left out to be consumed by scavengers; rather, it is stored in a garbage pail, trucked to a barge, and dumped in an area of the ocean too polluted to support the scavengers of the normal benthic community. Similarly, the feces of a person who has eaten the apple may ultimately be discarded into a polluted ocean, and the farmer in Washington must purchase manufactured fertilizers from an independent source.

To cite another example, coke, which is produced from coal, is used as a raw material for manufacturing

$$3C + CaO \rightarrow CaC_2 + CO$$

coke lime calcium carbide

$\downarrow H_2O$

$HC{\equiv}CH$
acetylene

$\downarrow$ catalyst

$HC{\equiv}C{-}\underset{H}{C}{=}CH_2$
vinyl acetylene

$\downarrow HCl$

$H_2C{=}\underset{Cl}{C}{-}\underset{H}{C}{=}CH_2$
chloroprene

$\downarrow$ catalyst

neoprene rubber

the gas acetylene, which in turn is used for making various plastics and synthetic rubber. The plastics and rubber eventually accumulate in some location such as a garbage dump; they do not return to the mine as coal. In fact, many new synthetic materials, particularly plastics and corrosion-resistant coatings for metals, were developed to be resistant to chemical changes so that they would not deteriorate during their useful lifetimes. Unfortunately, this resistance also persists after the products are discarded. The movement of matter through the industrial processes, unlike the movement through the life processes, therefore generates an ever-increasing quantity of waste, mostly in the form of solid material. This does not mean that *every* industrial product eventually becomes a dead-end waste. Some products are used as raw material for other manufacturing. Other industrial products—for example, soap—can be used as foods by living organisms. As previously stated, materials that can be consumed by living organisms are biodegradable. However, the fact that a waste product is biodegradable does not necessarily mean that it is benign to the ecosystem in which it is discarded. For example, although petroleum is degraded by bacteria, the process is very slow. Tarry residues dumped along a shoreline may disrupt a particular ecosystem long before bacteria consume the tar (see previous chapter).

In this chapter we shall discuss the sources of solid wastes, the extent to which they are recycled, and the problems and issues involved in their disposal. We exclude radioactive wastes, which were discussed in Chapter 8.

11.2 SOURCES AND QUANTITIES

Perhaps the most noteworthy characteristic of solid wastes is their variety. In our household garbage pails there are food scraps, old newspapers, discarded paper from miscellaneous sources, wood, lawn trimmings, glass, cans, furnace ashes, old appliances, tires, worn out furniture, broken toys, and a host of other items too numerous to mention. The total quantities of solid wastes are large and increasing. In the United States municipal solid wastes (which include household and commercial discards) averaged 2.75 pounds per person per day in 1920, 4.5 pounds in 1965, and about 5.5 pounds in 1970 (see Fig. 11.1). In other words, the waste disposal system of an average city must accommodate about 150 pounds of refuse per week for every family of four, and that is a lot of trash. In the year 1970 alone, about 200 million tons of municipal refuse accumulated in the United States. Americans produce more solid waste than any other people (see Fig. 11.2); in contrast to the 5.5 pounds per person per day in the United States, residents of Sydney, Australia, produce 1.75 pounds per person per day, and the average citizen in India produces about half a pound per day.

Affluence is responsible for most of this trash. Packaging materials amount to about one-fifth of the municipal refuse. Thus, the facts that over 50 per cent of the cost of soft drinks is in the bottles, and almost half the cost of many other items lies in their packages, do not prevent people from buying these products. But even without wasteful packaging, American citizens discard over twice as much as citizens of Sydney. For example, more food is discarded, cars are junked sooner, clothes are patched less often, and tires are recapped less frequently in North America than in Oceania.

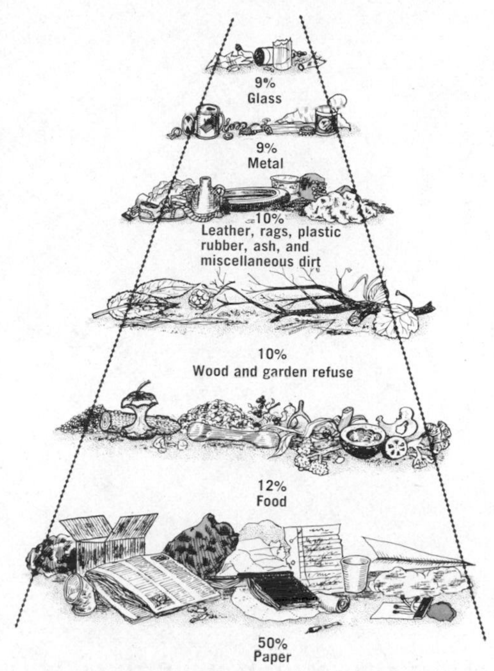

FIGURE 11.1 **Composition of municipal trash.**

Municipal sources contribute only a fraction of the types and amounts of solid wastes discarded in the United States. Agricultural activities, for example, produce each year over two billion tons of waste. About three-quarters of this is manure, and the balance is plant parts such as corn cobs, leaves and stems, forest slash from logging operations, culled fruits, slaughterhouse offal, pesticide residues, and containers. Mining operations are the second major contributor, with about 1.5 billion tons per year. Most of this material is rock, dirt, sand, and slag that remain behind when metals are extracted from the earth. More mineral wastes accumulate as concentrated ore deposits are depleted and as the mineral content of

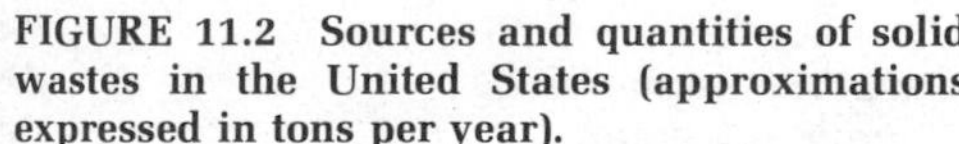

FIGURE 11.2 **Sources and quantities of solid wastes in the United States (approximations expressed in tons per year).**

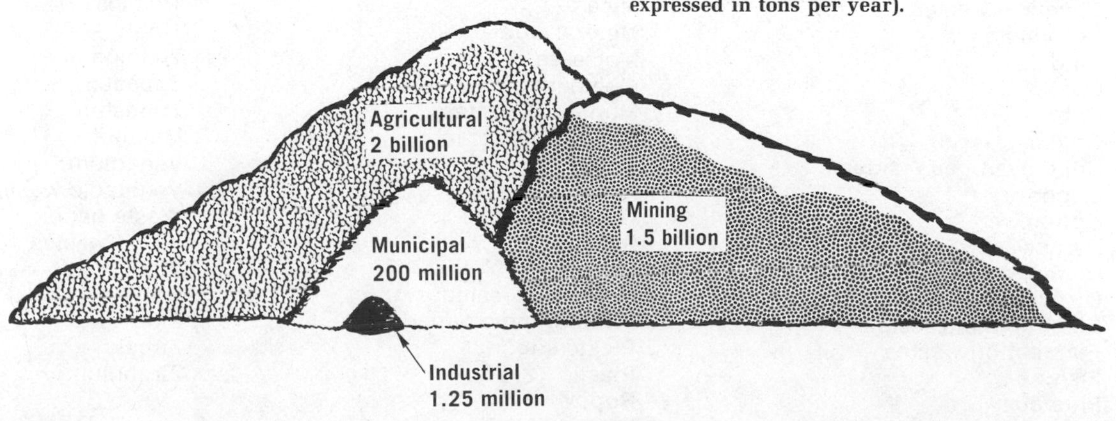

the raw ores decreases. Excess rock and dirt differ from any of the other wastes mentioned in this Chapter. The problems in coal mining associated with strip mining, acid mine drainage, and mine reclamation were discussed in Chapter 5. Waste problems of this sort are common to other types of mining as well.

The processing of raw materials such as metals, fossil fuels, or agricultural products to manufacture airplanes, shoes, beer cans, or balloons, always generates solid wastes. Look about the room you are in and note the number of manufactured objects in sight. Each different kind of object was produced by a series of industrial operations, and some solid wastes resulted from each stage of production. It should not be surprising, then, that industrial solid wastes are more varied in their categories than are municipal wastes. Table 11.1 is a classification of industrial

TABLE 11.1 MAJOR INDUSTRIAL SOLID WASTE CATEGORIES

Acetylene wastes	Fly ash	Precious metals
Agricultural wastes	Food processing wastes	Pulp and paper
Aluminum	Foundry wastes	Pyrite cinders and tailings
Animal-product residues	Fruit wastes	Refractory
Antimony	Furniture	Rice
Asbestos	Germanium	Rubber
Ash, cinders, and flue dust	Glass	Salt skimmings
Asphalt	Glass wool	Sand
Bauxite residue	Gypsum	Seafood
Beryllium	Hemp	Shingles
Bismuth	Hydrogen fluoride slag	Sisal
Brass	Inorganic residues	Slag
Brewing, distilling, and fermenting wastes	Iron	Sodium
Brick plant waste	Lead	Starch
Bronze	Leather fabricating and tannery wastes	Stone spalls (chips)
Cadmium	Leaves	Sugar beets
Calcium	Lime	Sugar cane fibers
Carbides	Magnesium	Sulfur
Carbonaceous shales	Manganese	Tantalum
Chemical wastes	Mica	Tetraethyl lead
Chromium	Mineral wool	Textiles
Cinders	Molasses	Tin
Coal	Molybdenum	Titanium
Cobalt	Municipal wastes	Tobacco
Coffee	Nonferrous scrap	Tungsten
Coke-oven gas residues	Nuts	Uranium
Copper	Nylon	Vanadium
Cotton	Organic wastes	Vegetable wastes
Dairy wastes	Paint	Waste paper
Diamond grinding-wheel dust	Paper	Wood wastes
Distilling wastes	Petroleum residues	Wool
Electroplating residues	Photographic paper	Yttrium
Fermenting wastes	Pickle liquor	Zinc
Fish	Plastic	Zircaloy
Flue dust	Poppy	Zirconium
Fluorine wastes	Pottery wastes	

sources compiled by the Bureau of Solid Wastes Management of the United States Department of Health, Education, and Welfare. Unfortunately, it is a rather uncritical compilation, with some repetitive, some obscure, and some trivial entries, but it does illustrate diversity. In 1970, industry in the United States produced 1.25 million tons of waste.

11.3 THE NATURE OF THE SOLID WASTE PROBLEM

Solid wastes present a many-faceted problem. The disposal of trash around the country creates litter. Its accumulation in cans on city streets attracts rats and flies, stimulates bacterial growth, and creates a collection problem. When large cities run out of space to dump the collected trash, a disposal problem is created. Finally, the accumulation of rusty old car bodies, cans, and other recyclable metal scrap hastens the depletion of non-renewable resources.

Litter is particularly vexing, because a small percentage of the population is responsible for a large nuisance. Moreover, there appears to be no solution. Advertising campaigns have been ineffective. Stiff fines, even when enacted into law, are difficult to enforce. One significant step is the development of plastic packaging materials that decompose in the sun. For example, the plastic plate of the type shown in Figure 11.3 disappears when left outdoors. (The atoms, of course, don't disappear; they redistribute themselves to produce gases and small solid particles.) The concept of the disappearing plate is not new; stale bread was used in medieval times. In fact, in our own era, the ice cream cone may perhaps be the perfect packaging material. But what should be the

The ideal biodegradable package.

Non-biodegradable containers.

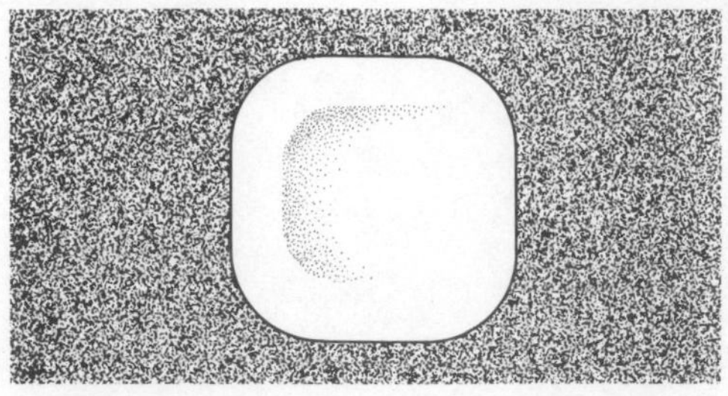

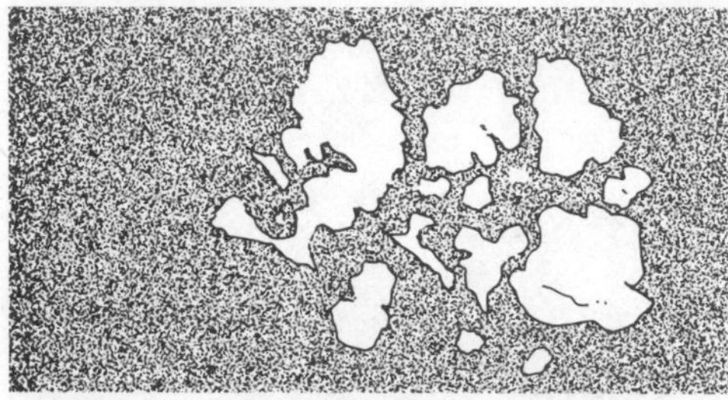

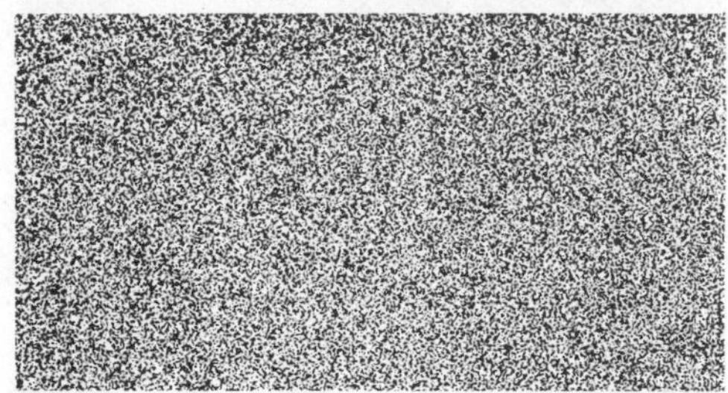

FIGURE 11.3 Disappearing picnic plate. (I. R. New Product Annual, 1972, p. 31.)

ideal life span for degradable plastics? If materials are constructed to decompose rapidly, then they may become activated by the fluorescent lights in supermarkets and disappear on the shelf. On the other hand, if objects decompose too slowly, then the false sense of security that people may feel in dealing with decomposable items may result in increased carelessness and more litter.

In most cases the solid waste problem begins with collection. Usually the contents of trash cans that are placed outside near a curb are loaded into trucks which compact the refuse to increase the hauling capacity of the vehicle. This system is expensive, noisy, and disruptive of traffic. For example, a high-rise apartment building built recently in New York City "solved" its trash collection problems by installing 400 individual garbage cans. The contents of each can are manually transferred into city sanitation trucks. If you think there must be a better system, you are right—there are several. Collection boxes with a three- to ten-cubic-yard capacity are used in some instances. When filled, these units can be loaded into trucks hydraulically in one quick, efficient operation. In suburban or rural areas where population densities do not warrant such large containers, automatic garbage trucks, such as the one shown in Figure 11.4, can reduce collection costs and provide faster service. But the ultimate in collection efficiency is found in some Swedish cities where the garbage truck has been replaced completely by pneumatic tubes. Apartment dwellers in areas served by the pipeline simply dis-

FIGURE 11.4 Automated collection vehicle (From Environmental Science & Technology 6:415, May, 1972.)

(From Danbury News-Times, March 19, 1973.)

pose of their trash in a receiving hopper, push a button, and everything is sucked into a central receiving station by a powerful vacuum system.

11.4 LAND AND OCEAN DISPOSAL

About 85 per cent of the municipal refuse discarded in the United States in 1970 was deposited on land or on the ocean floor. The most primitive waste repository is the **open dump.** Waste is collected and, to save space and transportation costs, is compacted. The compacted waste is hauled to the dumping site, usually in the morning, and spread on the ground, further compaction sometimes being effected by bulldozing. Organic matter rots or is consumed by insects, by rats, or, if permitted, by hogs. Various salvaging operations may go on during the day. Bottles, rags, knick-knacks, and especially metal scraps are collected by junk dealers or by individuals for their own use. In some communities, the accumulation is set

An open dump. (From Sanitary Landfill Facts. U.S. Department of Health, Education and Welfare, PHS 1970.)

afire in the evening (or it may ignite spontaneously) to reduce the total volume and to expose more metal scrap for possible salvage. Of course, the organic degradation, the burning, and the salvaging are recycling operations. However, there are serious detrimental features to the open dump. Its biological environment differs from those devised by man (agriculture and animal husbandry) and from those that have evolved in natural ecosystems, and is therefore not controlled by regulatory mechanisms common to either. The result is that the organisms that multiply at the dump are not likely to be the type that are benign to man. The dump is a potential source of diseases, especially those carried by flies and rats. The fires, too, are uncontrolled and therefore always smoky and polluting. Rainfall that circulates enters the dump and removes a quantity of dissolved and suspended matter, including pathogenic microorganisms, that are water pollutants. And, of course, the dumps are ugly.

Ocean dumping is practiced by many coastal cities and represents little advantage over disposal on land. Barges carrying the refuse travel some distance

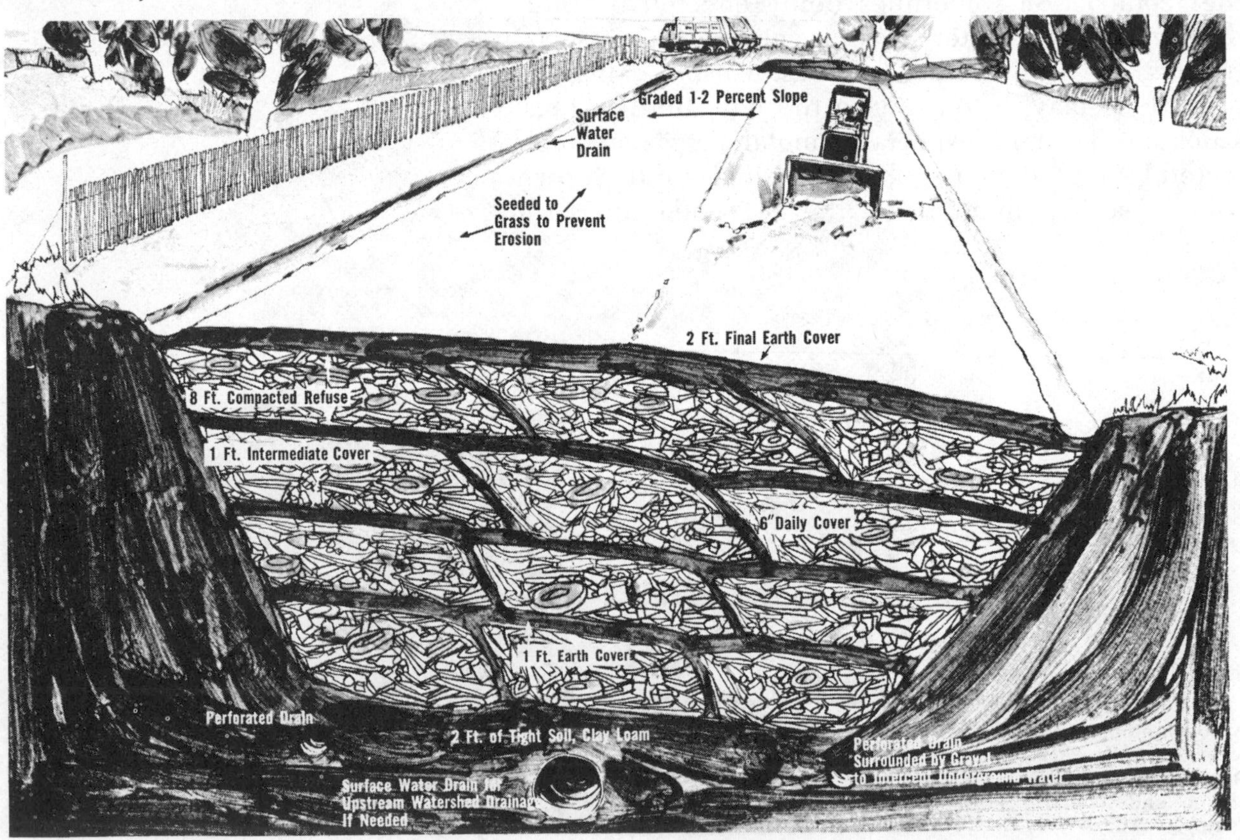

FIGURE 11.5 In ravines or valleys, the area method usually is best. Where the ravine is deep, refuse should be placed in lifts of six to 10 feet deep. Cover material may be obtained from the sides of the ravine. To minimize settlement problems, it's desirable to allow the first lift to settle for about a year. This is not always necessary, however, if the refuse has been adequately compacted. Succeeding lifts are constructed by trucking refuse over the first one to the head of the ravine. Surface and groundwater pollution can be avoided by intercepting and diverting water away from the fill area through diversion trenches or pipes, or by placing a layer of highly-permeable soil beneath the refuse to intercept water before it reaches the refuse. It's important to maintain the surface of completed lifts to prevent ponding and water seepage. (Courtesy of New York State Department of Environmental Conservation.)

from the harbor and discharge their loads into a natural trench or canyon on the ocean floor. In this way most of the trash is removed from sight, though not from the biosphere. Aquatic dumping areas are almost devoid of communities of benthic animals, and thus the normal food webs in the ocean are disrupted. Although plankton and fish may survive in dump areas, they are affected by the unusual environment. For example, flounder caught in the New York City dumping region have an off-taste. Analysis of the stomach contents of these fish reveals that old adhesive bandages and cigarette filters constitute part of the animals' diet, and such dietary aberrations might well be the cause of the foul taste.

The sanitary landfill is far less disruptive to the environment than is uncontrolled dumping onto open land or into the ocean. A properly engineered landfill should be located on a site where rainwater, leaching through the refuse, will not pollute the groundwater. If such drainage cannot occur naturally, the site must be modified by regrading or piping to redirect the flow of water. After waste is brought to a landfill, it is further compacted with bulldozers or other heavy machinery. Large objects, such as furniture, are sometimes shredded. Each day six to 12 inches of soil is pushed over the trash to exclude air, rodents, or vermin (see Fig. 11.5). In practice, however, the distinction between the sanitary landfill and the open dump is not always sharp. For example, a thin layer of earth may be an ineffective barrier against burrowing rats, flies developing from larvae, or gases evolving from decomposition.

Sanitary landfills permit inexpensive biodegradation without much pollution, disease, or unsightliness. With proper ingenuity, landfills may reclaim spoiled land or beneficially alter the topography of an area. (We already mentioned reclamation of strip mines in Chapter 5.) In some cases swamps and marshes have been filled and used as building sites for apartment complexes, parks, and athletic fields. In other cases, landfill mountains have been constructed and used as ski slopes, amphitheaters, and in one case as a "soap box derby" raceway.

On the other hand, several serious problems are associated with sanitary landfills. First, land conversion is not an unmitigated gain; as we have seen in Chapter 1, the loss of marshland itself leads to ecological disruptions. Second, many large city areas are

A botanical garden became the final layer of this completed landfill. (From Sanitary Landfill Facts. U.S. Department of Health, Education and Welfare, PHS.)

exhausting their available sites for landfill and will therefore soon be forced to transport their trash further into the countryside. In these instances, high transportation costs may well offset the low operating costs of a distant landfill. Third, and perhaps most serious, such disposal represents a drain of materials. Food wastes and sewage sludge which could be used to enrich surface land are buried deep underground. Paper and wood scraps which could be repulped are lost, and non-renewable supplies of metals are dissipated.

11.5 INCINERATION

A method increasingly used in many metropolitan areas is incineration. The process, as applied to waste disposal, is more complex than simply setting fire to a mass of garbage in an open dump. A modern incinerator unit is currently used in Montreal, Canada. A crane removes refuse from a 10,000 cubic yard storage pit and feeds it into the furnace at a constant rate. Burning occurs on a set of three inclined grates which are agitated to insure complete combustion and a constant movement of trash into and ash out of the chamber. The ashes are ultimately removed by a conveyor belt and cooled; the metals are salvaged, and the remainder is removed to a sanitary landfill. The furnace heats the boilers and the resultant steam is sold to industry. In the Montreal incinerator an auxiliary oil burner has been installed to ensure a constant supply of steam even if a crane were to break or the sanitation workers went on strike. Finally, the furnace gases are purified by an electrostatic precipitator.

In general, even if we credit it with the value of the steam it can produce, incineration is more expensive than the operation of a sanitary landfill. The reasons behind this expense lie in the difficulties inherent in handling so heterogeneous a fuel mixture as garbage. Incinerators operate most efficiently if constant furnace temperatures are maintained, but it is difficult to control the fire when the heat content of the fuel varies significantly from load to load. Also, municipal trash contains food scraps and other wet garbage, and special furnace design problems develop if this fraction is to be burned completely. Finally, the incineration of certain waste products produces acidic gases which corrode furnace walls and grates. Particularly notorious is polyvinyl chloride (PVC), a plastic used in the manufacture of consumer products such as rainwear, toys, containers, garden hoses, records, and credit cards.

$$\mathrm{H_2C{=}CHCl}\ \text{(vinyl chloride)} \longrightarrow \left(\!-\mathrm{CH_2}-\mathrm{CHCl}-\!\right)_n$$

vinyl chloride

polyvinyl chloride

The burning of PVC produces hydrogen chloride gas which, on contact with water, becomes a solution of strongly corrosive hydrochloric acid.

$$\left(\begin{array}{c} H\ \ H \\ -C-C- \\ H\ \ Cl \end{array}\right)_n + O_2 \rightarrow CO_2 + \underbrace{H_2O + HCl}_{\text{hydrochloric acid solution}} + \text{other gases}$$

polyvinyl chloride (PVC)

On the other hand, the future outlook of incineration might improve. In recent years, increasingly large quantities of dry paper and cardboard have appeared in refuse, and further trends in this direction would increase the quantities of salable steam from incinerators. (This fact should not be used as an argument in favor of increased paper consumption, for the combustion of wood products will deplete forests faster than it will alleviate the energy crisis.)

Despite these problems, however, incineration offers several notable advantages: (a) Steam from garbage may help alleviate the energy crisis (see Chapter 5). (b) Health problems associated with refuse accumulation are eliminated. (c) The volume of solid wastes is reduced by 80 to 90 per cent and thus much less land is needed for their eventual disposal. (d) A mixture of garbage and rubbish can be handled without prior separation. (e) The residue is inert and odorless and relatively easy to handle. (f) Reclamation of metals from the ash is relatively straightforward.

Nevertheless, incineration is disadvantageous for two major reasons: (a) By contrast to the well-designed plant in Montreal, probably 90 per cent of the incinerators currently in use pollute the air. Thus, while air pollution is not a necessary consequence of incineration, it is a fact of existence at the present time. (b) Although metals can be recovered from the furnace ash, many valuable materials such as paper, plastics, wood, textiles, and food wastes are burned rather than recycled.

11.6 RECYCLING—AN INTRODUCTION

Of course, there are many kinds of recycling paths. Consider, for example, the element phosphorus, which we discussed in Chapter 6. Much of the phosphate ore now being mined is used as a fertilizer and its phosphorus is incorporated into plant tissue, eaten by man, excreted into sewage systems, and concentrated again in sludge. In the United States most of the sewage sludge ends up in the ocean or in sanitary landfills; little is returned to the land as fertilizer. "So what," you may say, "can't we eventually mine the oceans or the old dumps?" The fallacy here, of course, is that we must think of the total environmental consequences of our waste-disposal practices. Ocean min-

"One evenin' he decided t'go out an' still hunt; try t'kill 'im a bear or something'. Sittin' at th' head a'th'swamp an' there's three bear come walkin' out. A small little bear in front, and they's a big he bear in th' center, an' they's a little cub behind this'um. An' he waited 'til this big bear got betwixt him an' a tree t'shoot it wi' his hog rifle so he could save his bullet—go cut it out of th' tree. And he shot this big bear.... An' he went home an' took a axe an' cut th' bullet out. He'd take it back an' remold it. Lead was hard t'get, so that's th' way they'd try t'save their bullets." (From Eliot Wigginton, ed.: *The Foxfire Book*. Garden City, N.Y., Anchor Books, Doubleday & Co., 1972. 380 pp.)

ing may disrupt aquatic food chains. In addition, it will be a very expensive process and thus may increase the cost of fertilizers prohibitively. The real crux of the problem, then, is to guarantee the greatest possible use of a substance in the most efficient manner. For this reason we begin this section with a discussion of planned obsolescence and of packaging practices. Less trash would be produced if items were built to last longer and if unnecessary packaging were eliminated.

The most effective way to reuse materials is by repairing and refurbishing items instead of discarding them. For example, if an automobile consumes too much oil and runs inefficiently, the engine can probably be rebuilt and the original performance restored. The body would remain untouched and other bulky components such as the engine block and the transmission could be remachined or readjusted rather than discarded. Similarly, when a refrigerator no longer cools efficiently, the fault generally lies with the compressor, not with the frame or the food compartment, and repair conserves metals and energy.

But not all items are reparable or reusable. A tire can be recapped only once with safety. Week-old newspapers and spoiled meat are useless to most people. When an item cannot be used in its present condition, it must be destroyed and treated somehow to extract its useful raw materials. For example, used tires can be shredded and converted to raw rubber, old newspapers can be repulped and converted to new paper, and spoiled meat can be rendered and converted to tallow and animal feed. The complex technology available for extraction of useful goods from refuse will be treated in Section 11.10.

At best, recycling processes conserve both energy and materials. However, not every item can be recycled efficiently. For example, imagine that a careless person throws a soft drink bottle onto a roadway in some isolated rural region, and that it falls under the wheels of a trailer truck and is crushed. Crushed glass is a recyclable item in principle, but the collection and concentration of all the crushed glass along rural roadways would consume so much energy as to be grossly inefficient. Therefore, recycling operations, despite their theoretical appeal, are not always less disruptive of the earth's ecosystems than non-recycling operations. Of course, the issues are usually much more complex than in the example just mentioned.

"Admit it. Now that they're starting to recycle this stuff, aren't you glad I didn't throw it out?" (From Saturday Review of Literature, July 3, 1971. By Joseph Farris.)

But in general, recycling operations, like any other industrial process, generate pollution and consume energy, and the environmental compromises must be examined in any particular instance.

We can easily envision an ideal sequence in which durable goods are used for a long time, repaired or patched to prolong their lives still further, and finally broken down into component parts for reuse as raw materials. Some materials would necessarily be drawn out of the cycle, but the size of the drain would be small. The fact is, however, that no such plan has operated effectively in modern society, for complex social and economic factors act as deterrents. Let us consider three examples: the recycling of glass containers, of automobiles, and of paper, in the United States.

11.7 GLASS CONTAINERS

A century ago, milk, beer, cooking oil, and other liquids were shipped to "general stores" where a re-

FIGURE 11.6 Bottling plant capable of filling 1000 bottles per minute.

tail customer's bottle or jar could be filled from a large drum or carboy. Many granular solid items such as coffee and nuts that are now available in glass containers were scooped from kegs into cloth sacks which could be used over and over again. In time, glass packaging became increasingly popular, especially for foods (see Fig. 11.6). Later the soft drink, dairy, and beer industries added a new approach to glass packaging—the deposit bottle. These generally well-constructed bottles cost more to manufacture than the two- or five-cent deposit values assigned to them. The bottling companies relied on the fact that the deposit provided sufficient incentive to the customer to ensure the bottle's return. This system worked so well at first that the average number of round trips that a deposit bottle made was about thirty. Those that were broken, lost, neglected, or used as flower vases and therefore never returned were, of course, replaced with new bottles. But gradually, consumers became so indifferent to the deposit that the average number of round trips declined to four, and the cost of washing bottles increased, so bottling companies initiated the shift to no-deposit, no-return containers. Raising the deposit would have been more environmentally sound.

In 1972, the energy used to manufacture nonreturnable bottles in the United States was 211 trillion Btu's *more* than that which would have been needed if returnables were used. That difference represents enough energy to supply almost ten million Americans with electric power for a year.

Although glass from a no-deposit, no-return bottle can be recycled, the reuse of crushed glass is inherently less efficient than refilling a deposit bottle. Of course, the shipping and washing of a deposit bottle pollute the environment, and both processes require energy, but these effects are far less than those involved in crushing, remelting, and remanufacturing a no-deposit bottle.

11.8 JUNK AUTOMOBILES

The scrap automobile presents a formidable problem. Traditionally, salvage dealers have stored inoperative cars in large lots (Fig. 11.7), and have sold individual parts. Salvage of this sort is an ecologically valuable operation, for functional automobile components are reused to replace worn-out components of operating vehicles. However, because junkyards are ugly, many localities are attempting to restrain their operation. In some towns, the owner of the yard must pay a tax on all vehicles in his lot; in others, high and expensive fences must be built to hide the auto hulks from public view.

After they are stripped, cars are usually compacted into bales and shipped, often by railroad, to the steel mill. Unfortunately, railroad freight rates for scrap usually exceed those for raw materials. This rate structure is not an anti-environmental conspiracy, but rather a reflection of the fact that the movement of scrap is less voluminous and more erratic than that of ore. If this situation changed, the rates could be equalized.

At the steel mill, some of the iron in the compacted cars is recovered. Recovery of a relatively pure metal, such as non-insulated copper wire, is technologically simple: You heat it to melting, and recast it into ingots. Recovery of metal is much more involved when a complex object like an automobile is to be

FIGURE 11.7 Auto wrecking yard enclosed by chain-link fence in Denver, Colorado. (Courtesy of Luria Brothers and Co., Inc., a division of the Ogden Corporation.)

TABLE 11.2 SOME FACTORS THAT CONTROL AUTOMOBILE RECYCLING*

1. Price paid by scrap processor to wreckers for stripped auto bodies.
2. Cost of transportation of a junk car to a wrecker or scrap processor.
3. Cost of transportation of scrap from processor to consumer.
4. Magnitude of scrap market within a competitive area.
5. Current level of scrap prices, particularly for No. 1 heavy melting steel, shredded scrap, electric furnace grades, cupola grades, No. 2 heavy melting steel, No. 2 bundles, bundled No. 2 steel, automotive slab, motor block, lead, and copper scrap.
6. Lack of consumer confidence in automotive scrap specifications and quality control.
7. Lack of small amounts of money at critical points in the disposal stream.
8. Price paid to owner for automobile at time it is no longer useful as a vehicle.
9. Cost of land for storage of junk cars.
10. Lack of financial cost to owner in leaving a junk car standing on his property, since an auto license is no longer required.
11. Copper content of automotive scrap.
12. Zoning ordinances and their enforcement.
13. Automobile inspection laws.
14. Automobile damage in minor accidents.
15. Price of used or rebuilt parts for maintaining a used car.
16. Cost of producing hot metal from iron ore.
17. Cost of pig iron fed to cupola furnaces in foundries.
18. Local governmental facilities for collecting and disposing of abandoned automobiles.
19. Cost of removing radiator, electric motors, wiring, and other copper-bearing parts, and preparing them for scale as scrap.
20. Cost of burning or otherwise stripping and disposing of upholstery, tires, and other non-metallic materials.
21. Cost of baling automotive scrap.
22. Cost of shredding automotive scrap.
23. Type and size of ironmaking or steelmaking process.

Table 11.2 continues on opposite page.

scrapped. Large non-ferrous components such as seats, tires, radiators, or batteries can be easily removed to leave behind a hulk that is purer in iron than the original whole vehicle. But it is expensive to remove small or firmly fixed objects such as copper wire, padded dash covers, and glass from a junk car. As a result, it had become customary to compact partially stripped vehicles into bales, and to ship them in this condition to steel mills. Because metallurgists customarily extract iron from ore that contains 70 per cent rock, sand, and other non-metallic matter, the technology is available to remove glass, plastics, asbestos, and related non-metallic impurities from melted auto bodies. On the other hand, many metals are chemically similar to each other and respond in the same way to processing procedures. As a result, the separation of small amounts of copper from large quantities of iron is not possible with conventional equipment and techniques. Moreover, copper is an undesirable component of most steels, for copper-iron alloys are not strong. In the past, autos were reused in steel manufacture despite the copper content. Recently, however, copper contamination has become more serious and the value of baled automobiles has decreased. To understand why, imagine that the steel used for manufacture of a car in 1940 contained no copper. When that car was scrapped and melted down, the steels contained, let us say, 0.1 per cent copper. Therefore, the body and engine block of the next automobiles were already contaminated, so when they were melted along with the electric wire used in them, the copper content of the second batch of scrap steel increased. Thus, over the course of time, steels have suffered increasing contamination. In addition, the many electrical devices in modern cars (radios, push-button windows, multiple tail lights, etc.) increase the total amount of copper per car. Therefore, the value of automobile scrap has decreased in recent years.

As a result of all these and many other factors affecting the economics of salvaging automobiles (see Table 11.2), the cost of moving an old car to a junkyard is frequently higher than it is worth. This situation has resulted in an increasing rate of abandonment of cars in the United States—1.8 million cars were abandoned in 1969.

But efforts are underway to increase salvage operation efficiency. One technique is to shred an automobile into myriads of small pieces in a giant grinder (Fig. 11.8), and then to separate the iron shreds from the strands of copper magnetically. Shredders are expensive, however, and must therefore be located in urban areas where a large supply of scrap is available to return the investment. Junk cars in rural areas cannot be shipped economically to the shredder plants. Alternatively, metallurgists have devised a process to extract copper from automobiles by immersing the whole vehicle in a molten salt bath. Contaminated waste salt from other industrial operations can be used, thus solving two solid waste problems simultaneously. A third solution relies on the fact that aluminum, a good conductor of electricity, separates easily from molten steel. Therefore, if aluminum wire replaced copper wire in cars, the problem would be alleviated. Of course, many other solutions are conceivable, such as levying a disposal tax on new cars, and then using that income to strip old cars.

TABLE 11.2 SOME FACTORS THAT CONTROL AUTOMOBILE RECYCLING* (*Continued*)

24. Ratio of blast furnace hot metal capacity to steelmaking capacity in an integrated operation.
25. Lack of commercial metallurgical method for removing copper and other unwanted metals from molten iron.

*Abridged from *Automobile Disposal—A National Problem,* U.S. Government Printing Office, 1967, p. 839.

FIGURE 11.8 Vehicle hulks being conveyed into a scrap shredder, and the resulting scrap. (Courtesy of Luria Brothers and Co., Inc., a division of the Ogden Corporation.)

11.9 OLD PAPER

Recycling of paper has also declined in recent years. Just after World War II, 35 per cent of the paper used by American industry was obtained from recycled sources, but that figure has dropped steadily over the years, and in 1970 only about 20 per cent of the total paper fiber produced originated from scrap. Recycled paper is often more expensive than virgin paper. Actually, several factors have operated to favor virgin over recycled paper: (a) the supply of used paper has been viewed as uncertain; (b) tax incentives, in the form of depletion allowances, favor timbering operations over recycling; and (c) public demand for shiny paper and more permanent inks have led to the production of paper products that are often difficult to de-ink and to repulp.

All these problems can be solved. Recycling operations can be given tax advantages. Ink need not be removed from all recycled paper. In fact, the residual ink from waste newspapers imparts a pleasing bluish-gray appearance to the recycled product. In addition, printers could revert to the old inks that can be applied to recycled paper.

11.10 RECYCLING – TECHNIQUES AND TECHNOLOGY

Much municipal, industrial, and agricultural trash can be neither reused nor repaired, and consequently must be reduced to raw materials suitable for remanufacture. Several techniques are available for this type of recycling.

Melting

Many materials such as metals, glass, and some plastics can be melted, purified, and recast or remolded. As with automobiles, however, problems arise if the substance is not pure.

Revulcanizing

Rubber is one plastic material that cannot simply be heated and remolded. Raw rubber is gooey and formless and must be reacted with sulfur to bind the individual rubber molecules together in a cohesive form. Used rubber goods can be shredded, broken down chemically, and then rebonded in a process known as revulcanization. Recycled rubber manufactured in this manner lacks the strength and resiliency of material made from virgin stock; therefore, for some applications it is useful only when mixed with more durable fibers.

Vulcanization is the process of heating raw rubber with sulfur or sulfur compounds to make it stronger and more durable.

Pulping and Converting to Paper

Any material containing natural cellulose fiber such as wood, cloth, paper, sugar cane stalks, and marsh reeds can be beaten, pulped, and made into useful fiber. The basic technology behind recycling of paper is as old as the technology of manufacturing paper. The initial step in reclamation of fiber is to mix three parts waste paper with 97 parts water in a hydropulping machine. Here the scrap is stirred and beaten vigorously with a device similar to an egg beater until a slurry forms. If paper from municipal sources is being pulped, de-inking chemicals are added to the pulping mixture. The de-inked pulp slurry is screened to remove large objects which might have contaminated the original stock and then rolled through wringers to remove inky water.

Small impurities are removed in a centrifugal separator and the fibers are then converted into paper by conventional procedures.

As the depletion of forest lands becomes severe, fibers from agricultural waste will undoubtedly become more attractive. When sugar is extracted from cane, the remaining fibrous stalks, or **bagasse**, are well suited to paper production. Bagasse currently contributes 50 million tons of solid wastes annually, and could easily be converted into a valuable resource.

Wastepaper cycle. (Courtesy of Container Corporation.)

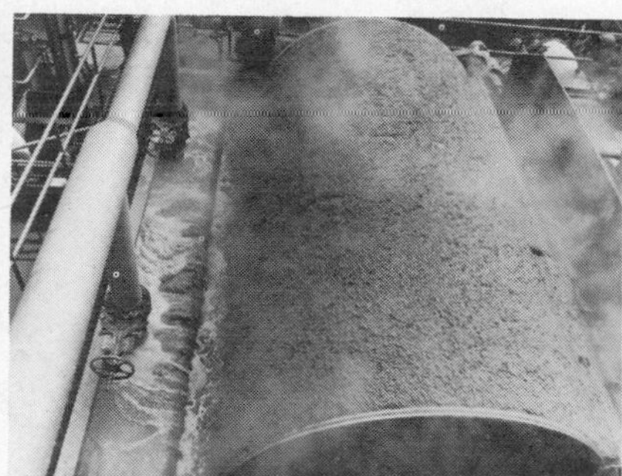

Formation of windrows in composting operations.

Composting

As we have noted in Chapter 6, the recycling of organic matter by decay organisms produces humus. The controlled, accelerated biodegradation of moist organic matter to a humus-like product that can be used as a fertilizer or a soil conditioner is known as **composting.** This process is a practical method of recycling organic wastes. Almost any plant or animal matter, such as food scraps, old newspaper, straw, sawdust, leaves, or grass clippings forms an excellent base for a composting operation. To increase the surface area available for decomposition, this stock should first be shredded or ground, and packed loosely. The organisms essential to composting need not only cellulose and starch, but also nitrogen, phosphorus, potassium, and trace elements. Thus straw, newspapers, and sawdust, which consist mainly of compounds of carbon, hydrogen, and oxygen, do not compost well by themselves but must be mixed with some source of additional chemicals. Manure is an excellent source of nutrients for composting, but chemical fertilizers or materials as diverse as peanut shells or dried blood, can substitute for manure.

Aerobic decomposition is generally quicker and more complete than anaerobic action, so a compost pile must be agitated or turned in some way to promote aeration. Additionally, water in appropriate amounts aids the growth of decay organisms.

The simplest such operations are known to backyard gardeners as **compost piles** and to industrial composters as **windrows.** In industrial operations refuse is shredded, and sometimes inorganic matter is removed. Sewage sludge is then added until a favorable chemical composition is obtained, and finally the solids are mixed and piled in long rows. As decay accelerates inside the windrows, heat is generated. In a properly regulated system, temperatures will reach 66°C (150°F), which is hot enough to kill pathogenic bacteria. These windrows are turned periodically by machine for a period of six to seven weeks. The final product is then dried and sold.

Although windrows are cheap and simple to operate, they generate odors and need a great deal of land to accommodate a large operation. Therefore, several different types of compost reactors have been built. One of these uses a large inclined drum which rotates slowly and pushes the refuse through the system at a controlled rate. If the temperature, airflow, agitation, water, and nutrient levels are carefully controlled, complete composting is possible in five to seven days.

Between 1950 and 1970, 17 out of 20 commercial composting operations that were established in the United States either went bankrupt or curtailed operations. The primary reason for the failures has been the lack of markets for the compost. In Chapter 6 we mentioned that farmers often find inorganic fertilizers cheaper to use than manure. Such circumstances have worked against the sale of compost. However, this strategy is not universal. In the Netherlands, for example, a great deal of organic refuse is composted because in that country farmers believe that soil conditioning with humus is valuable in the long run. Currently, at least one firm is trying to develop markets for humus in the wheat-growing areas of North America that are threatened with saline seep problems. (Refer back to page 233.)

"Compost" is an old form of the word *composed* or *composite*. Later, compost came to mean a composition or combination of anything, as in "to know what malice is . . . what villainy or treachery is, for Satan is but a compost of these" (Jackson, *Creed*, 1640). In cookery, the word became "compote," referring to a mixture of fruit preserved in syrup. In agriculture, compost means a mixture of various ingredients for fertilizing land, and this is the sense in which the word is used here.

A "windrow" is obviously derived from *wind* and *row*, and it means a row of agricultural matter, such as grass or hay, that is set out to be dried by exposure to the wind. In modern industrial agriculture, the *row* is more important than the *wind*, for it is the exposure to operations by machine, rather than by weather, that dictates the arrangement.

Rendering

Rendering is the cooking of animal wastes such as fat, bones, feathers, and blood to yield both a fatty product called tallow, which is the raw material for soap, and a non-fatty product that is high in protein and can be used as an ingredient of animal feed. The raw material for a rendering plant comprises wastes from a wide variety of sources—farms, slaughterhouses, retail butcher shops, fish processing plants, poultry plants, and canneries. If there were no rendering plants, these wastes would impose a heavy burden on sewage treatment plants, as well as add pollutants to streams and lakes and nourish disease organisms. At the rendering plant, the waste materials are sterilized and converted to useful products, such as tallow and chicken feed. But the cooking process generates odors. These odors can be controlled by methods such as incineration, but any control method is subject to occasional interruptions, such as those that might be caused by a power failure.

The community near a rendering plant may judge that the disadvantages outweigh the advantages, but the decision is based only on local considerations—the other communities whose animal wastes are being reprocessed do not participate in the decision.

Fermentation

In Chapter 6 we discussed how yeasts can be cultured on a mixture of sugars derived from agricultural wastes. In this way, such materials as straw, sawdust, orchard prunings, corncobs, and even hydrolyzed garbage can be converted to food. Collection of methane from anaerobic composting or yeast culturing is possible, although the process has never been tried on a large industrial scale.

Recently, news releases reported that an Englishman fueled his automobile with chicken manure. Of course, he didn't actually pump the manure directly into his carburetor; rather he fermented it first in the absence of air, and collected the sewery gas. Recall from Chapter 5 that methane gas, CH_4, is a clean and desirable fuel.

Destructive Distillation (Pyrolysis)

Destructive distillation, or pyrolysis, is the process by which a material is decomposed by heating it, in the absence of air, to about 1650°C (3000°F). If a substance like rubber or plastic that is composed of organic molecules is pyrolyzed, the molecules fragment, re-react, and re-fragment many times until an equilibrium is reached. This equilibrium mixture contains a wide variety of different valuable chemical compounds. The pyrolysis products of municipal refuse are shown in Figure 11.19. This technique appears particularly advantageous in that the pyrolysis

FIGURE 11.9 Destructive distillation. Organic fractions of municipal wastes can be pyrolyzed thermally to yield valuable byproducts. (From Environmental Science and Technology 7:597, July, 1971. Courtesy of American Chemical Society.)

equipment is essentially a closed system and therefore does not discharge pollutants into the atmosphere.

Industrial Salvage

Industrial salvage comprises a wide variety of operations in which industrial wastes are reused. Some common examples include manufacture of particle board and chipboard from sawmill waste, manufacture of building bricks from fly ash, mine tailings or glass, and manufacture of felt from fur scraps.

Table 11.3 summarizes various recycling routes of some common wastes. All of the processes illustrated in Table 11.3 are useful if more or less homogeneous waste is available, but there is no one-step operation which can process a mixture as varied as municipal waste. For this reason the most important recycling operations carried out in North America today involve the reuse of industrial scrap. The printing department of a large newspaper company generates many tons of waste or unsold paper every week. This paper is clean, concentrated, and thus ideal for recycling. Similarly, a metal fabricating shop generates large quantities of uncontaminated scrap which can be sold.

The problems in handling mixed municipal refuse have received considerable attention. Some people have proposed enactment of legislation requiring residents and businessmen to separate their trash into homogeneous categories, but such a law could hardly be enforced. Policemen would have to examine trash

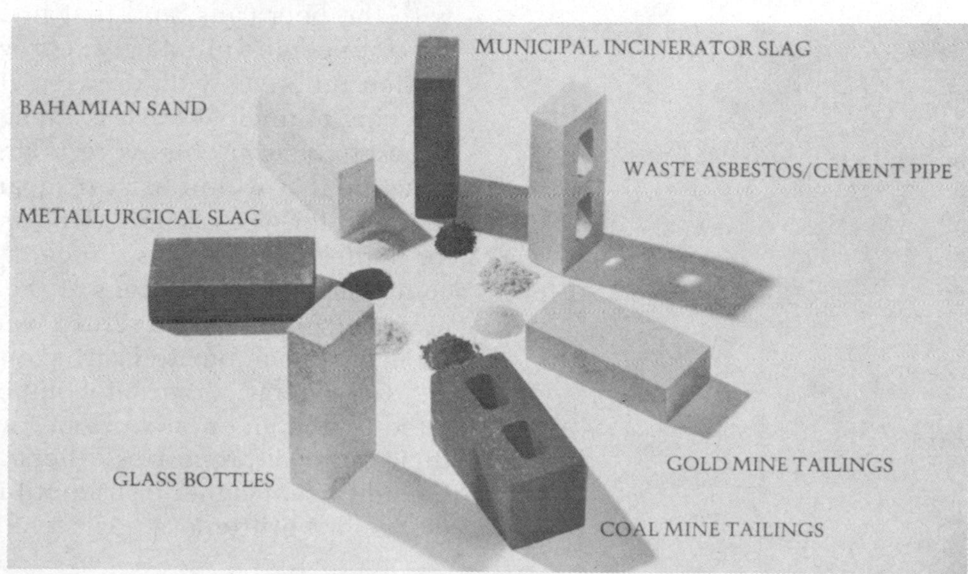

Building materials from solid wastes. (Courtesy of House and Home Magazine.)

TABLE 11.3 VARIOUS RECYCLING ROUTES OF SOME COMMON WASTES

WASTE	*RECYCLING POSSIBILITIES*
Paper	Write on the back of stationery; lend magazines and newspapers to friends, etc. Repulp to reclaim fiber. Compost. Pyrolyze. Incinerate for heat.
Glass	Purchase drinks in deposit bottles and return them; use other bottles as storage bins in the home. Crush and remelt for glass manufacture. Crush and use as aggregate for building material or anti-skid additive for road surfaces.
Tires	Recap usable casings. Use for swings, crash guards, boat bumpers, etc. Grind and revulcanize. Pyrolyze. Grind and use as additive in road construction.
Manure	Compost or spread directly on fields. Ferment to yield methane, use residue as compost. Convert to oil by chemical treatment. Treat chemically and reuse as animal feed.
Food scraps	Save for meals of left-overs. Sterilize and use as hog food. Compost. Use as culture for yeast for food production. Pyrolyze.
Slaughter house and butcher shop wastes	Sterilize and use as animal feed. Render. Compost. Pyrolyze.

cans and try to ascertain who put the chicken bone in with the beer cans, and that just wouldn't work. But even if people voluntarily segregated their trash, collection problems would arise. The present system of trucking municipal trash is inefficient and costly, and it poses traffic problems in a large city. Imagine the situation that would exist if separate collections were required for wet kitchen garbage, for papers, for cans, for bottles, and for rags. Since it is unthinkable to accomplish such separations at the curb, any successful recycling of municipal refuse would be dependent on some kind of automated garbage-separating plant.

In general, different components can be separated from a given mixture because one or another of their physical properties differs. Such properties may include size, magnetic susceptibility, inertia, resiliency, color, brittleness, or specific gravity. Thus, a screen can remove old appliances and tires from municipal trash, and a magnet can remove pieces of iron from non-ferrous wastes. Several devices are available to remove heavy materials from light ones. In an air classifier (Fig. 11.10*A*) mixed refuse is passed over a screen and compressed air is forced up through the system. The force of the air jet is adjusted so that light particles such as lettuce leaves and newspapers will travel up a tube to a conveyor belt while heavy particles such as glass bottles and cans will remain behind. A ballistic separator consists of a rotor which throws objects into a chamber with a constant force. Separation occurs because dense objects meet less air resistance and travel farther than light, fluffy ones (Fig. 11.10*B*). (Imagine throwing feathers and pennies.) Another type of separator operates by dropping mixed

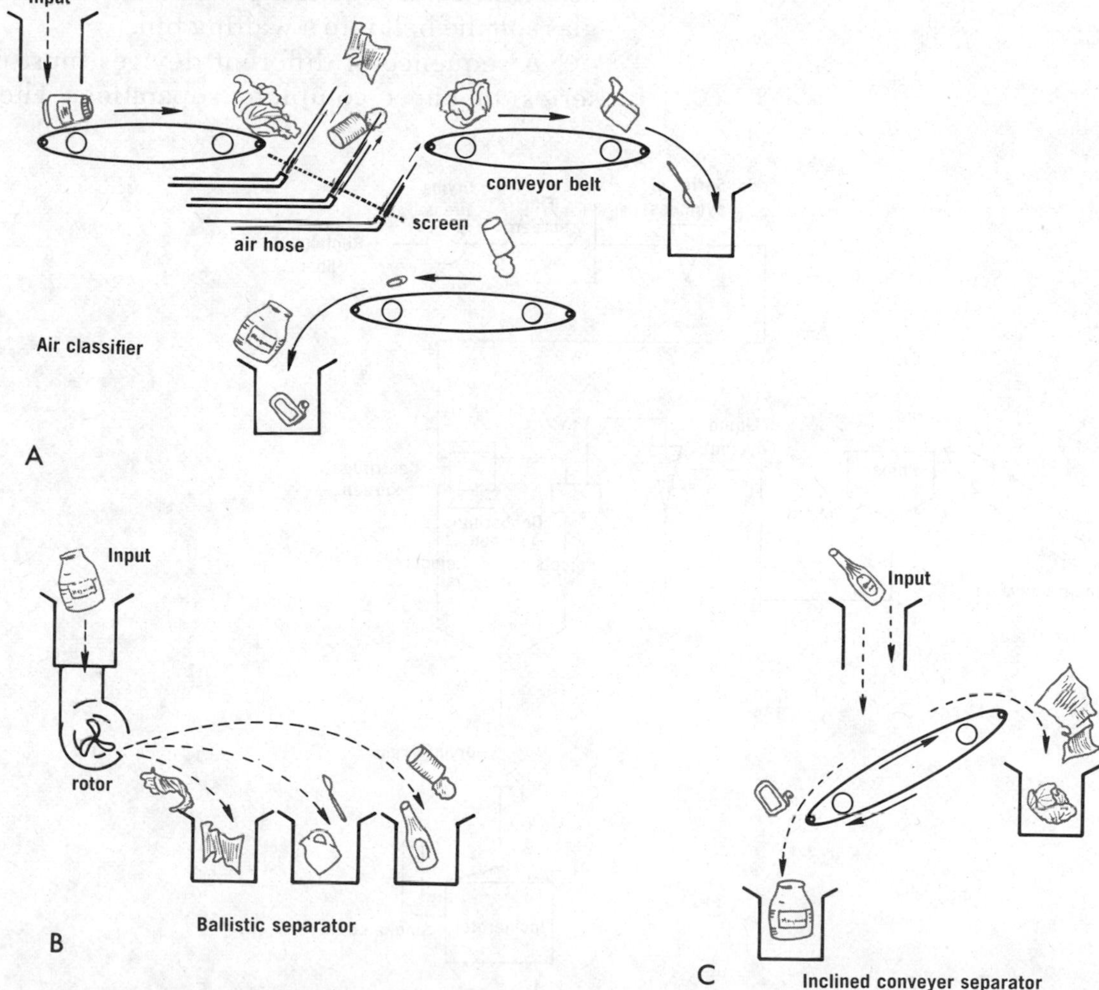

FIGURE 11.10 Types of waste-material separators. *A*. air classifier. *B*. ballistic separator. *C*. inclined conveyor separator.

refuse on an inclined conveyor belt (Fig. 11.10*C*). Heavy and elastic objects bounce and roll downhill against the movement of the belt, while light and sticky ones are carried up. (Imagine dropping bowling balls and banana peels on an "up" escalator.) The cyclone was discussed in Chapter 9 as a method for removing heavy dust particles from an air stream by inertial differences. The same principle can be used to separate glass or metal from a water slurry of finely ground refuse. Another device isolates materials by their differential flotation in water. A mixed glass separator developed by the Bureau of Mines is particularly ingenious. The refuse is spread out along a narrow conveyor belt and passes by a series of automatic optical sensors. Each sensor is adjusted to respond to a certain color. When the electric eye sensitive to green, for example, detects a piece of light green glass, it activates a pulse of compressed air which forces the glass off the belt into a waiting bin.

A sequence of different devices must operate in series to effect complete separation. The town of

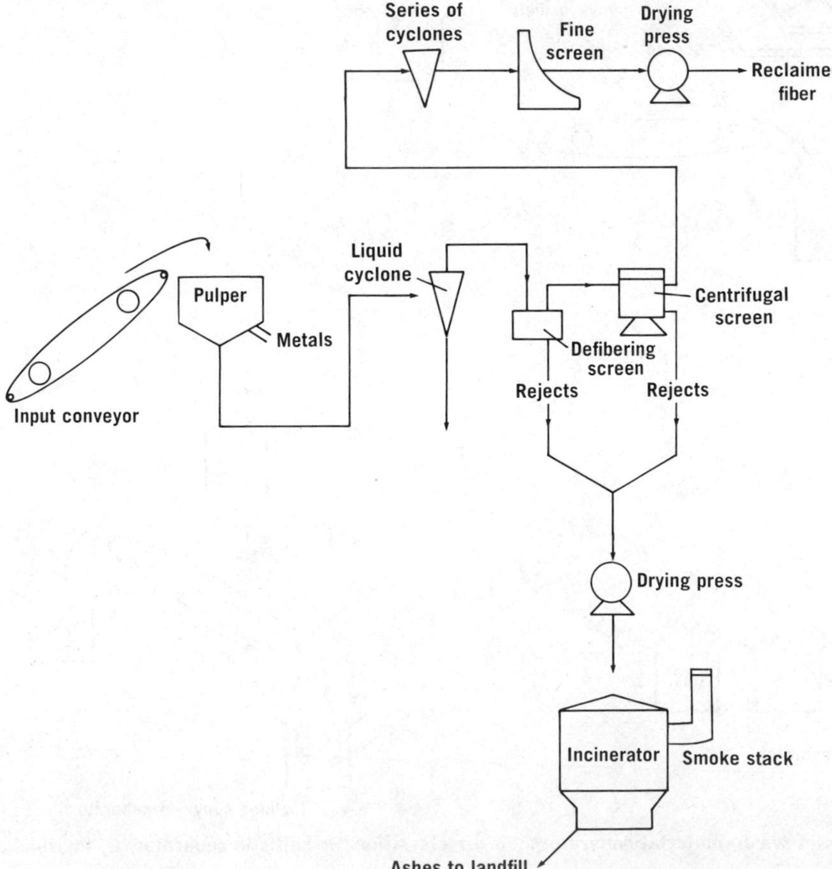

FIGURE 11.11 Schematic illustration of the Franklin, Ohio, recycling system.

Franklin, Ohio, has built a recycling system that is shown schematically in Figure 11.11. Large items like old swing sets and engine blocks are removed manually and the remaining refuse is fed into a specially designed hydropulper. The fiber component of the refuse is pulped, brittle material such as glass and bones is pulverized, but hard solid objects are not affected. Water is added during the pulping operation and the fiber slurry is drained off through a screen with a 3/4 inch mesh size. Anything that has resisted pulping and remains larger than 3/4 inch in diameter, such as large metal objects or rocks, is removed and washed. The iron is separated magnetically, and the non-ferrous metals are separated manually for sale as scrap.

Meanwhile, the fiber slurry is pumped to a liquid cyclone which separates the small, dense inorganic matter that passed through the screening. Glass, bones, small pieces of metal, especially aluminum from cans that break up in the pulper, and miscellaneous dirt, pebbles, and particles of sand are removed in this operation.

The slurry now consists mainly of paper, food wastes, plastics, and textiles. The Franklin plant is designed to extract useful paper fiber from this mixture. This goal is realized in two more pulping and screening operations. The last screening removes any particles larger than 1/16 inch. The slurry is cleaned again in cyclonic washers and finally long, useful fibers are separated from poor-quality, short ones to be dried and sold. Organic rejects from the various screens and cyclones are considered to be non-recyclable and are dried and incinerated. The heat generated by the incineration step can be converted to salable steam, although the Franklin plant is not now equipped for this operation.

While the major operation of this plant has been designed to reclaim paper fiber, other plants emphasize humus production or reclamation through pyrolysis. Usually these other centers start operations by grinding or shredding the trash and then separating it by some combination of the techniques already outlined.

11.11 THE FUTURE OF RECYCLING

In general, technology is available for the recycling of solid wastes, but the social climate is not favorable for immediate large-scale implementation of

TABLE 11.4 SOME FACTORS THAT CURRENTLY DISCOURAGE RECYCLING IN THE UNITED STATES

1. Public apathy toward bringing waste to recycling centers.
2. Scarcity of recycling centers.
3. Poor recyclability of many manufactured goods.
4. Tax incentives that favor mining and logging over recycling.
5. Freight rates that favor mining and logging over recycling.
6. Lack of markets for recycled goods owing to
   (a) farmer apathy towards compost, and
   (b) reluctance of many industries to try recycled goods.
7. Poor funding of research on recycling products from trees.
8. Lack of consumer pressure for recycled goods.
9. Zoning laws that discriminate against recycling.

the technology (see Table 11.4). Future improvements may increase the efficiency of operations so that municipal wastes can be handled profitably, at which time private industry might develop an interest in the process. We are still far from this stage. Even after government subsidy and sale of scrap, the Franklin facility is more expensive to operate than a sanitary landfill, but the engineers who designed the plant claim that a similar but much larger plant designed for a major city would reduce operating costs and become economically attractive.

A private corporation entering the solid waste business would charge individuals and municipalities a dump fee, presumably equal to the cost of operating a landfill, and then sell the recovered materials. Many large cities have nearly exhausted existing landfill sites. The purchase of new land and the expense of hauling a longer distance will increase the cost of this method of disposal. This increasing cost of landfill makes reclamation operations relatively more economically advantageous.

The price of raw materials will increase as natural resources are depleted. As logs and ore become more expensive, for example, salvage operators will be able to demand higher prices for their goods and salvage will become more lucrative. At such time industry will respond to a new economic situation. For example, a recent study in New York City concluded that paper recycling in the metropolitan area is decreasing because several repulping mills have gone out of business in recent years. When paper fiber becomes scarce enough, however, new repulping mills will open. In general, European reclamation practice and technology are more advanced than North American simply because fuels, fiber, and metals are scarcer there, thus reclamation affords a more compelling economic incentive.

Industry responds to a variety of pressures, not only economic but also social, political, and legal, all of which may be very effective in combination. For example, although the Franklin reclamation plant is slightly less economical than a landfill, the citizens of that city decided that the marginal increase in cost (about 25 cents per family per month) was more than offset by the benefits of the plant. Also, this plant was financed partly with federal grants to demonstrate the feasibility of recycling, so the choice to enter an "unprofitable" business was also a national legislative decision. Other examples of this sort are abundant.

The General Services Administration of the United States Government, the City of New York, Bank of America, Coca-Cola, and Canada Dry have all ordered large quantities of recycled paper for stationery or annual reports, although at present there is no economic advantage to the use of recycled paper. Finally, it was mentioned earlier that PVC containers are potentially troublesome because hydrochloric acid is generated when they are incinerated. Public pressure has impelled some food and cosmetic packagers to suspend the use of PVC containers in favor of bottles made of less-damaging plastics.

The treatment of waste matter was once considered to be one of the "offensive trades," which were characterized by offensive odors. These trades were generally associated with decomposition of animal matter, and therefore centered around dead animals and the products obtained from them (such as glue, fertilizer, and leather) and around the disposal of human wastes and human dead. The Massachusetts Supreme Court cited soap works, glue works, slaughterhouses, rendering plants, and tanneries among these industries. We may now categorize all these as recycling operations.

Another way to accelerate recycling rates is to pass laws which encourage reuse. Consider bottles, for example. The Environmental Protection Administration of the City of New York proposed a one- to three-cent tax on no-deposit bottles, but the bill was defeated. The States of Vermont and Oregon have imposed a mandatory tax on all non-returnable bottles.

As we shall see in Chapter 13, the current economic balance sheet is incomplete in assessing environmental effects of products. Hence, a five-cent deposit on bottles is too low. Considering the total environmental cost of a discarded bottle, including the energy and raw materials of manufacture as well as the litter and waste disposal problem, a bottle's true cost is closer to 25 cents. Perhaps a law requiring a 25-cent deposit on glass bottles would be an effective inducement for returning them.

The world cannot afford to wait to recycle a material until a resource is on the brink of exhaustion. Recall from Section 11.6 that complete recycling will never be realized, for some matter will be dissipated beyond the reaches of any reasonable recovery system. Therefore, some supply of raw materials will always be needed, and our natural resources must be conserved now to insure adequate ore, fertilizer, and fiber for future generations.

PROBLEMS

1. **Disposal of manufactured products.** Which of the following manufactured products will undergo biological recycling, industrial recycling, or accumulation as a solid waste? (a) a paste made from casein (a milk protein); (b) a polyethylene squeeze bottle used as a container for mustard; (c) a copper drainpipe from a wrecked house that is sold to a dealer in scrap metals; (d) a woolen sweater; (e) the gold filling in an extracted tooth; (f) the steel in an automobile body that is returned as scrap to the mill; (g) a "no-deposit" soda

bottle; (h) the coffee grounds you used to make this morning's coffee.

2. **Mine wastes.** Explain how mine wastes differ from municipal, industrial or agricultural wastes.

3. **Types of wastes.** Give an example of a biodegradable, a combustible, a toxic, an odorous, and an inert solid waste.

4. **Garbage collection and disposal.** Consider the following three commercially available devices: (a) the home garbage disposal unit, (b) the home garbage compacter, and (c) plastic garbage pail liners. Discuss the relative merits of each with respect to the litter problem, the collection problem, and the eventual disposal problem.

5. **Land disposal.** Explain the difference between an open dump and a sanitary landfill.

6. **Land disposal.** Compare economically and ecologically the use of low and high land as waste disposal sites.

7. **Sanitary landfills.** Glass forms a stable base for landfill areas because it doesn't decompose, settle, or release pollutants into groundwater. Since currently most trash accumulates in landfills do you feel that the beneficial aspects of glass as a landfill would provide a viable argument in favor of non-deposit bottles?

8. **Incineration.** Discuss the advantages and disadvantages of incineration as a method of waste disposal.

9. **Incineration.** Plastics, made from coal and oil, have a high heat content. If garbage incinerators were commonly used in the United States, and the resulting heat of combustion were generally used industrially, would you feel that plastic packaging would be an advantageous way to use fossil fuels twice? Explain.

10. **Recycling.** As mentioned in the text, broken or obsolete items can be repaired, broken down for the extraction of materials, or discarded. Which route is most conservative of raw materials and energy for each of the following items? (a) a 1948-model passenger car that doesn't run; (b) a 1968-model passenger car that doesn't run; (c) an ocean liner grounded on a sandbar and broken in two; (d) an ocean liner sunk in the central ocean; (e) last year's telephone directory; (f) an automobile battery that won't produce current because the owner of the car left his lights on all day; (g) an empty fountain pen ink cartridge?

11. **Planned obsolescence.** An old timer complains that years ago a man could store canned milk in the creek for three years before the can would rust through, but now a can will only last one year in the creek. Would you agree with the old man that cans should be made to be more durable? What about automobiles? Explain any differences.

12. **Recycling.** During World War II, when rubber was scarce in the United States, someone suggested that

industry should develop a process to scrape rubber off curves on roadways because that is where most tire wear occurs and then reform the reclaimed material into new tires. What do you think of this suggestion? Explain.

13. **Glass bottles.** Discuss some of the factors which led to the use of no-deposit, no-return bottles.

14. **Automobile salvage.** Outline some economic and technical problems of the automobile salvage business.

15. **Technology of recycling.** Describe the following processes: Revulcanization, repulping, composting, and rendering. State what materials can be treated by each of these processes, explain how they are treated, and describe what products are produced.

16. **Industrial salvage.** Why is industrial salvage more economically attractive than salvage from municipal refuse?

17. **Recycling municipal trash.** One of us (Jon) lives in a sparsely populated canyon in the Northern Rocky Mountains. His household trash is disposed of in the following manner: Papers are used to start the morning fire in the pot belly stove; food wastes are either fed to livestock or composted. Ashes are incorporated into the compost mixture; metal cans are cleaned, cut open, and used to line storage bins to make them rodent-resistant; glass bottles are saved to store food; miscellaneous refuse is hauled to a sanitary landfill. Comment on this system. Can you think of situations where this system would be undesirable? Do you think that it is likely that many people will adopt this system?

18. **Industrial salvage.** Manufacture of felt from fur scraps is a marginally profitable enterprise. One factory operates as follows: Fur scraps are first decomposed by heating them in a vat with dilute sulfuric acid. The useful fibers are extracted, and the remaining liquid, which consists of water, sulfuric acid, and decomposed animal skin is dumped untreated into a river. The cost of purifying the effluent would drive this particular business bankrupt. Discuss the overall environmental impact of this fur-scrap recycling center.

19. **Recycling and the Second Law.** Referring to the discussion on page 465 of the difficulty in recovering phosphate fertilizer from ocean sediments, Professor Peter Frank of the Department of Biology of the University of Oregon, who reviewed this book, wrote, "The Second Law of Thermodynamics comes in with a vengeance. . ." Explain what he meant.

20. **Recycling municipal trash.** The organic wastes that cannot be converted to paper fiber are incinerated at the Franklin, Ohio, reclamation center. Can you think of an alternate use for this scrap? Compare the desirability of your disposal method with incineration from an environmental standpoint.

21. **Recycling municipal trash.** Design a municipal refuse reclamation center which uses different equipment than the Franklin, Ohio, plant. Draw a flow sheet of your center and explain each operation.

22. **Recycling.** Environmental organizations have been active in establishing collection centers for old newspapers, cans, and so on. Can you think of other activities which these groups might engage in which would produce increased recycling?

23. **Recycling.** List the most efficient recycling technique and the resultant products for each of the following: (a) steer manure; (b) old clothes; (c) scrap lumber; (d) aluminum foil used to wrap your lunch; (e) a broken piece of pottery; (f) old bottle caps; (g) stale beer; (h) old eggshells; (i) a burnt-out power saw; (j) worn-out furniture; (k) old garden tools; (l) tin cans; (m) disposable diapers. How does your choice of technique depend on your location?

24. **Recycling.** Sand and bauxite, which are the raw materials for glass and aluminum, respectively, are plentiful in the Earth's crust. If we are in no danger of depleting these resources in the near future, why should we concern ourselves with recycling glass bottles and aluminum cans?

BIBLIOGRAPHY

A compilation of valuable articles on solid wastes is:

Solid Wastes. (Environmental Science and Technology Reprint Book.) Washington, D.C., American Chemical Society, 1971.

Most of the popular books on the environment that are cited in the bibliography of Chapter 1 include some discussion of solid wastes. For more technical information, refer to the following:

Andrew W. Breidenbach: *Composting of Municipal Solid Wastes in the United States.* Environmental Protection Agency Publication, Washington, D.C., U.S. Government Printing Office, Stock number 5502–0033, 1971.

Richard B. Engdahl: *Solid Waste Processing.* Public Health Service Publication No. 1856. Washington, D.C., U.S. Government Printing Office, 1969.

R. A. Fox: *Incineration of Solid Wastes.* New York, Robert A. Fox, 1967.

N. Y. Kirov: *Solid Waste Treatment and Disposal.* Ann Arbor, Michigan, Ann Arbor Science Publishers, 1972.

C. H. Lipsett: *Industrial Wastes and Salvage.* New York, Atlas Publishing Co., 1967.

Fred C. Price, Steven Ross, and Robert L. Davidson, eds.: *McGraw-Hill's 1972 Report on Business and the Environment.* New York, McGraw-Hill Publications, 1972.

R. D. Ross, ed.: *Industrial Waste Disposal.* New York, Van Nostrand Reinhold Co., 1968.

United States Bureau of Mines: *Automobile Disposal—A National Problem.* Washington, D.C., U.S. Government Printing Office, 1967.

12 NOISE

12.1 SOUND
(*Problem 1*)

To be most sensitively aware of sound, it is best to experience silence. Think of the quietest occasions of your life. Perhaps standing in the woods after a snow on a windless winter day was one such instance. Certainly you would not think of an occasion on which you were moving, whether walking, running, or riding, because motion itself produces sound. In fact, one relationship between motion and sound is suggested by the use of the word *still*, which means both motionless and quiet. Motion is related to energy —any moving body has an energy of motion that depends on its speed; the faster it moves, the more kinetic energy it has. Since sound is related to motion, and motion to energy, it is reasonable to think that sound is related to energy; and indeed, so it is.

The relationship is: Kinetic energy = $\frac{1}{2}$ × mass × $(\text{velocity})^2$

How is sound transmitted from a source such as a bell to a receiver such as your ear? We note that when the bell is struck, it vibrates. When the vibration stops, so does the sound. If the bell is struck in a vacuum, it will also vibrate, but no sound can be detected. From this experiment it can be deduced that the motion of the bell somehow moves the air, which then moves some receiving device in the ear. This transfer of sound energy through air occurs in the form of a wave. We usually visualize wave motion as water waves, especially ocean waves striking a beach, or ripples in a pond or swimming pool. Water that is still ("quiet" water) has a smooth, level surface. Water waves are disturbances; they alter the normal level so as to make it higher in some places and lower in others. The highest places are called **crests**; the lowest, **troughs**.

Water waves.

The distance between successive disturbances of the same type, such as between neighboring crests, is called the **wavelength** (Fig. 12.1). The rate at which a disturbance moves is the **speed of the wave**. The number of disturbances that pass a given point per unit time is the **frequency**. The relationship among these three attributes is given by the equation at left.

speed = wavelength × frequency

All of these characteristics of the water wave also apply to the sound wave, except that the nature of the disturbance is different. Instead of manifesting itself as crests and troughs, as in disturbances of the water level, the sound wave is a succession of compressions and expansions that disturb the normal density of the medium (such as air) in which they are propagated. This type of wave is called an **elastic wave**, and can be illustrated by the action of a coiled spring, as shown in Figure 12.2. Imagine that a bump on a rotating wheel hits the end of the spring twice per second. The resulting compressions, and the expansions which follow them, travel along the spring at a speed that depends on the properties of the spring (not on the rate of rotation of the wheel). Let us say that this rate is one foot per second. The frequency must be two beats per second, because that is established by the speed of rotation of the wheel. Therefore the wavelength is computed as follows:

$$\text{wavelength} = \frac{\text{speed}}{\text{frequency}} = \frac{1 \text{ ft per sec}}{2 \text{ per sec}} = 1/2 \text{ ft}$$

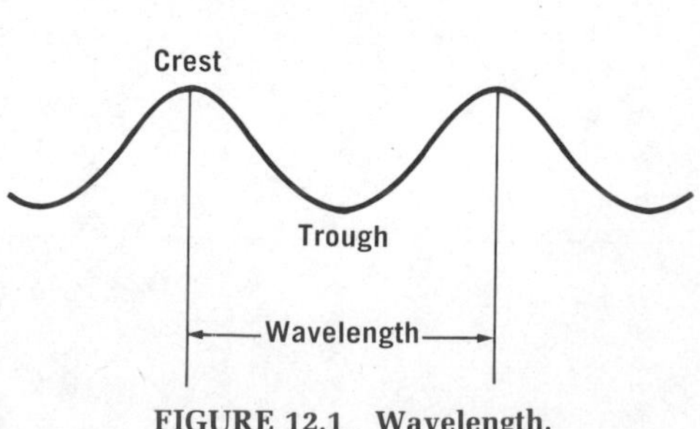

FIGURE 12.1 Wavelength.

Air is a springy substance; a squeezed balloon snaps back when released. Therefore, elastic waves can be propagated in air. This propagation is sound when the frequencies of the waves lie within the range that the ear can detect. Work must be done to beat out the successive compressions; thus, sound is a form of transmission of energy. The speed of sound in air under normal conditions on earth is about 1100 ft/sec. Any object, such as an airplane, that travels slower than sound is said to be **subsonic**; one that is faster is **supersonic.**

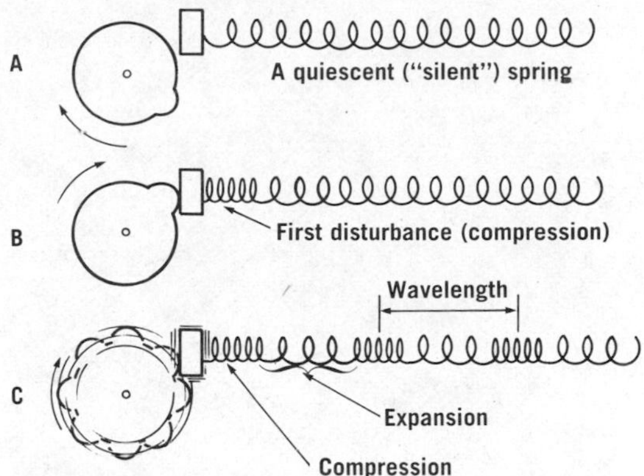

FIGURE 12.2 Elastic waves. *A.* **Wheel starts to rotate twice per second.** *B.* **First impact.** *C.* **Continuous production of waves. Frequency = 2 per second. Wave speed = 1 foot per second.**

The pitch of any given sound is determined by the frequency of the waves that produce it. The energy of a given sound, however, is not determined by the frequency, wavelength, or wave speed. Thus, if you slap the surface of a pond gently once per second with a spoon, you will make waves at a frequency of one beat per second. If you slap the water hard once per second with a paddle, you will still make waves at a frequency of one beat per second, and they will not travel any faster, but they will be bigger waves. The disturbances, or the heights of the crests and depths of the troughs, will be greater. The magnitude of the disturbance is called the **amplitude** of the wave (Fig. 12.3). The energy of a wave thus depends on the work that must be done to create the disturbance.

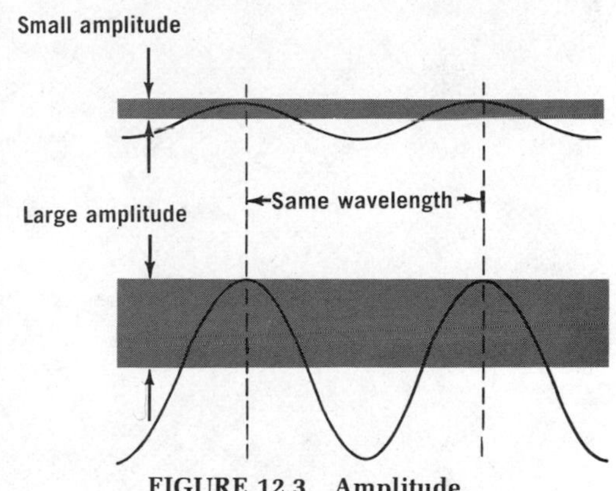

FIGURE 12.3 Amplitude.

To think of sound energy, imagine a four-engine plane with only one engine running. The frequency of the sound depends on the type of engine and its operation; the speed of the sound is a property of the air, not of the engine. Now we start the other three engines, so that they operate just like the first. The frequency of the sound has not changed, nor has the speed of the wave. But we are doing four times as much work. (Our fuel bill will be four times as high.) The sound energy being created is therefore four times as great. Sound energy is related to loudness, but the two are not the same. We shall discuss loudness in more detail in Section 12.5.

12.2 PURE AND IMPURE TONES

(*Problems 2, 10, 11*)

Think of a wheel that is mounted vertically, such as the wheel of an automobile or a bicycle. Let us attach a leaking pen to the rim of the wheel by means of a freely rotating pivot, so that the pen always hangs straight down. The pen will drip and make an ink spot on a piece of paper below it (Fig. 12.4*A*). Now we rotate the wheel; the pen rotates with it, but always re-

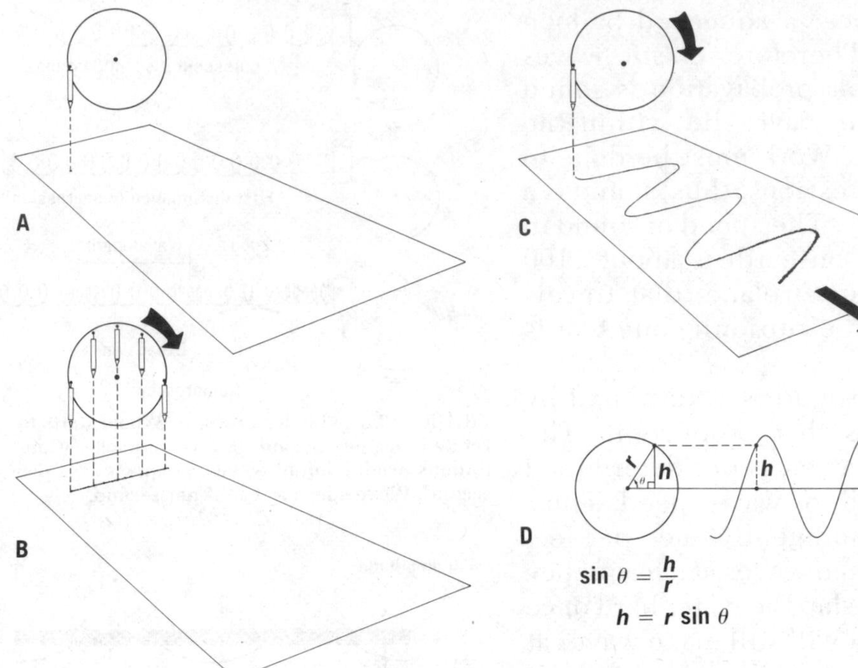

FIGURE 12.4 Generating a sine curve. *A.* **Wheel and paper stationary.** *B.* **Wheel rotating.** *C.* **Wheel rotating and paper moving.** *D.* **Sine curve.**

mains vertical. The dripping ink will therefore oscillate back and forth and will describe a straight line on the paper (Fig. 12.4*B*). Let us now slide the paper at constant speed at a right angle to the plane of the wheel. As the wheel continues to rotate, the pen continues to move back and forth, but it no longer leaves a straight line on the paper. Instead, we have a wavy line that looks very much like the silhouette of water waves, or like Figure 12.4*C*. This shape is said to be sinusoidal. What has all this to do with sound? It is this: If we replace our leaky pen with a rod that drives a piston that beats against a column of air, we will create a sound wave in which the compressions and rarefactions of the air will alternate sinusoidally (that is, like a sine curve). Such a sound is said to be a "pure tone" (see Fig. 12.5).

Recall that in a right triangle the sine of an angle is the ratio of the length of the opposite side to that of the hypotenuse. Therefore, the height, *h*, of any point *X* on the rim of the circle from the horizontal is proportional to the sine of the angle that it makes with the horizontal, as shown in Figure 12.4*D*.

Of course, not many natural sounds are pure tones. Any sound that can be represented by a wave form which repeats itself periodically, however, can be imagined to be made up of the sum of a series of

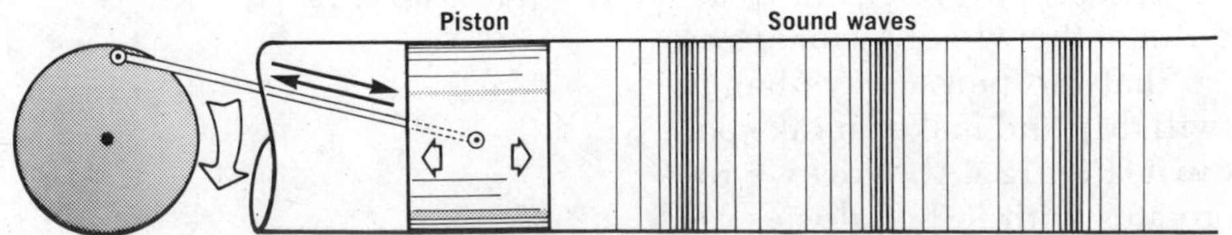

FIGURE 12.5 Generating a pure tone.

sine waves. The frequency of a tone, then, is the number of cycles that pass a given point per second; this unit used to be called, simply enough, "cycles per second," or cps. It is now called Hertz, Hz, after the German physicist Heinrich Rudolph Hertz. We also use the kilohertz, kHz. The relationships are:

The general statement of this theorem, first made by the French mathematician J. B. J. Fourier in 1822, is, "Every finite and continuous periodic motion can be analyzed as a simple series of sine waves of suitable phases and amplitudes."

$$\text{Hz} = \text{cps}$$

$$\text{kHz} = 1000\ \text{Hz} = 1000\ \text{cps}$$

What of sounds that are neither pure nor periodic, sounds that do not seem to contain any internal repetitions, such as the rustle of leaves in a tree, or the gurgling of water through a drain, or the hiss of steam from a pipe, or a rumble of thunder? Can we compare them with each other in terms of frequency? The answer is yes—a hiss of steam is obviously higher pitched than a rumble of thunder, because the pressure variations in the hiss, even though they are not rhythmic, occur much more rapidly. These differences are shown in Figure 12.6.

12.3 HEARING

By comparison with mechanical devices such as microphones, the ears of humans function with remarkable sensitivity and complexity. They respond to sounds over a ten-billion-fold range of intensities, they are sensitive to the individual pure tones into which complex sounds may be analyzed, they transmit to the brain complex information such as is contained in speech, they serve as a direction and range finder, and, almost incidentally, they help people balance on two feet.

A diagram of the human ear is shown in Figure 12.7. The **auricle** or **pinna** (called simply "the ear" in common parlance) serves the important function of funneling sound waves into the external ear canal. At the inside end of the canal is the tympanic membrane, or ear drum. Attached to the inside of the eardrum is a network of three tiny bones, the **malleus**, **incus**, and **stapes** (hammer, anvil, and stirrup), which are collectively called the **ossicles**. These bones, which lie in the middle ear, form a mechanical bridge between the eardrum and the inner ear. When a sound wave enters the ear, the eardrum vibrates; these vibrations cause the ossicles to vibrate synchronously and thus to transmit the energy of the sound wave to the window of the inner ear.

The inner ear is a complex structure that has two broad functions: (1) the conversion of sound-induced mechanical vibrations into nerve impulses, and (2) maintenance of a sense of balance and spatial orienta-

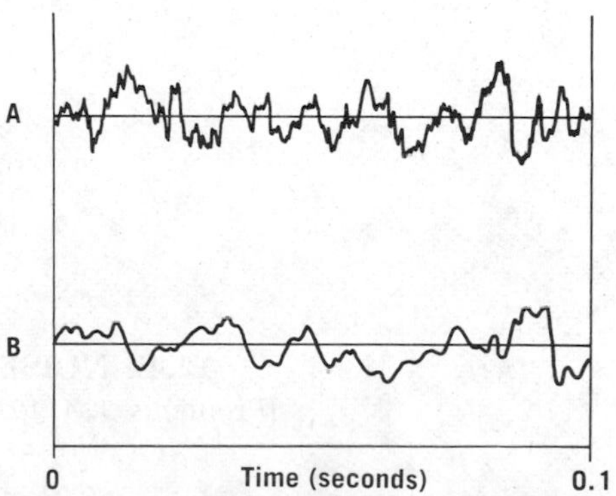

FIGURE 12.6 Wave forms of a hiss and a rumble: (A) hiss—higher frequency; (B) rumble—lower frequency.

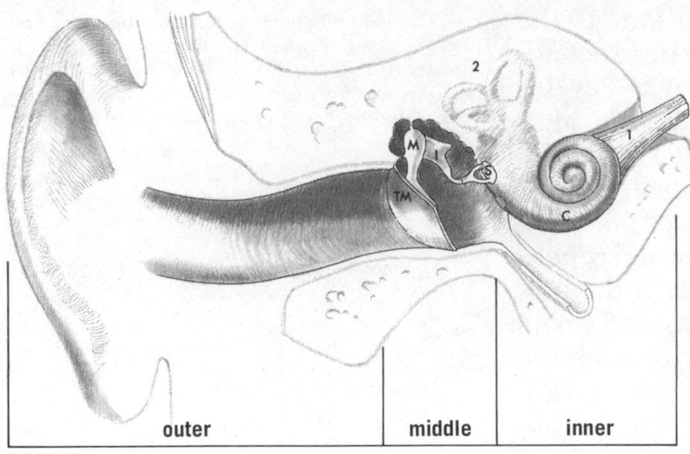

FIGURE 12.7 TM: tympanic membrane (ear drum); M: malleus (hammer); I: incus (anvil); S: stapes (stirrup); C: cochlea; 1. auditory nerve to brain. 2. semicircular canals.

tion. In this section we are concerned with only the first of these. The part of the inner ear that is responsible for the conversion of sound-induced mechanical vibrations into nerve impulses is called the **cochlea**. This structure, set deep in the temporal bone of the skull, has a shape similar to that of a tiny snail. When the energy of incident sound is transmitted to the inner ear as described above, the fluid inside the cochlea is also made to vibrate. These vibrations excite the basilar membrane which in turn excites the thousands of tiny hairs that form part of the **organ of Corti** (located in the cochlea), and, in ways that are not fully understood, the vibrations are converted into nerve impulses which travel via the auditory nerve into the appropriate part of the brain. These impulses are then interpreted as sound by the individual.

Now obviously "hearing" a sound is a complex process with many steps, and each step provides a place for something to go wrong. When any step in the sequence fails, hearing acuity may be diminished or even lost entirely. For example, an external canal full of wax, a perforated ear drum, a middle ear filled with pus such as happens with the common ear infections, ossicles whose joints are not functioning properly, or a diseased organ of Corti or auditory nerve—all these may produce varying degrees of hearing loss.

12.4 NOISE

(Problems 3, 4, 19)

The concept of noise is often related to the idea of randomness, but there are other connotations. The complete physical description of a given sound cannot establish whether you, as an individual, will like it or not. If you don't like a sound, that sound is noise. We now have two definitions of noise: (a) any erratic, intermittent, or statistically random oscillation, and (b) any unwanted sound.

The concept of "unwanted" sound seems simple enough, but is obviously difficult to quantitate for an entire community. A given sound may be music to one person but noise to another, pleasant when soft but noise when loud, acceptable for a short time but noise when prolonged, intriguing when rhythmic but noise when randomly repeated, or reasonable when you make it but noise when someone else makes it. Of all the attributes that distinguish between wanted and unwanted sound, the one that we generally consider the most significant is loudness. There is ample evidence that exposure to loud sounds is harmful in various ways, and, in any event, loudness tends to be annoying; therefore, the louder a sound is, the more likely it is to be considered noise.

There are yet other subtleties to complicate matters. People often associate noise with power, and in that context noise becomes desired. For example, some homeowners choose noisier lawnmowers and vacuum cleaners over quieter ones in the mistaken idea that the noisier ones work better or faster. And some drivers of "hot rods" or of trucks remove the exhaust mufflers from their vehicles to help "soup them up." In these instances it is true that such removal reduces the resistance imposed on the exhaust gases and thereby increases the efficiency of the engine, but the benefit is small (two to three per cent) and the cost is high (impaired hearing). Some of these attitudes date back to the 1940's, however, and have been undergoing a reversal in favor of quietness during the heightened environmental awareness of the present decade.

More difficult to modify, perhaps, is the association of noise with social recognition. A loud noise connotes authority, even though what it utters may be nonsense. Such is the case, all too often, with a loud motorcycle, or loud music, or roaring toys, especially toy guns and firecrackers.

Definitions and Units Related to Sound

Acoustic power, **or** *sound power,* **is the sound energy per unit time radiated by a source. It is measured in watts.** *Sound intensity* **is the sound power that passes through a unit area (watts/meter). A point source of sound that radiates, say, 10 watts becomes the center of an ever-expanding field of sound. The intensity of the sound diminishes as the radius of the field increases, though the power of the source remains unchanged. The "intensity of a barely audible sound" is usually taken to be one-trillionth of a watt per square meter, or 10^{-12} watt/m^2.** *Sound pressure* **is the force per unit area exerted by sound of a given intensity. Sound measuring instruments ("decibel meters") actually respond to sound pressure, but they convert this response to intensity by making use of the fact that sound intensity is proportional to the square of the pressure.** *Loudness* **is the subjective reaction of the listener to the intensity of the sound as it is received, not to the power of the source.**

12.5 THE DECIBEL SCALE

(*Problems 5, 13, 14, 15, 16, 17, 18*)

Loudness, we have said, is related to energy. Consider a machine that converts energy into sound—for example, a siren. Imagine that you have a 50-watt electric siren. This will consume as much energy in a given time as any other 50-watt device, such as a 50-watt light bulb. Your electric bill, which requests

Jack hammer noise.

payment for the electrical energy you consume, will be the same for a 50-watt bulb glowing all month as for a 50-watt siren wailing all month. But the bulb is more efficient at producing light, and the siren more efficient at producing sound. Of the two, the siren does its job better, because its actual sound power is about 10 watts, the other 40 being wasted as heat. In contrast, the actual light power of the 50-watt bulb is only about 1 watt. This difference explains, in part, why light bulbs feel hotter than sirens. We conclude from these facts that it would be correct to rate loudness in watts. For example, it has been determined that the sound power of a pneumatic drill is about one watt, that of a symphony orchestra, playing loudly, is about 10 watts, and that of a large rocket engine is about 10 million watts. What is remarkable about the ear is its ability to respond to a wide range of intensities. A very soft whisper of a human voice has an intensity equivalent to only one-billionth of a watt. Therefore, if we were to express intensities in watts, we would suffer the inconvenience of using both very small and very large numbers. Furthermore, the scale would be awkward, for it would not start at zero, but at some small number that represents the softest audible sound.

Consider the sound power numbers we have just cited:

Source	*Sound power, watts*	
rocket	10 million	$= 10^7$
orchestra	10	$= 10^1$
drill	1	$= 10^0$
whisper	one-billionth	$= 10^{-9}$

Motorcycle noise.

Physicists have created a unit that defines a *tenfold* increase in intensity, and named it a Bel, after Alexander Graham Bell, Then, if an orchestra sound is 10 times as intense as a drill, it is one Bel more intense. The rocket is one million times as intense as the orchestra, or 10^6 times as intense, and is therefore 6 Bels more intense. Thus,

$$\text{Difference in intensity between two sounds, X and Y expressed in Bels} = \log_{10}\left(\frac{\text{sound intensity of X}}{\text{sound intensity of Y}}\right)$$

It happens that the Bel is a rather large unit, so that

it is convenient to divide it into tenths, or decibels, dB, as follows:

$$1 \text{ Bel} = 10 \text{ decibels}$$

Then,

$$\text{Intensity in decibels} = \text{intensity in Bels} \times \frac{10 \text{ decibels}}{\text{Bel}}$$

Therefore,

$$\begin{array}{l}\text{Difference in intensity}\\ \text{between two sounds,}\\ \text{X and Y, expressed in}\\ \text{decibels, dB}\end{array} = 10 \log_{10} \left(\frac{\text{sound intensity of X}}{\text{sound intensity of Y}}\right) \quad (1)$$

Finally, it is very convenient to start our scale somewhere that we can designate zero, and the most convenient point is at the softest sound level which is audible to the human ear. We will call this level zero decibels. We can now write our equation for the decibel scale as follows:

Grinding noise.

$$\text{Intensity in decibels of any given sound} = 10 \times \log_{10} \left(\frac{\text{intensity of the given sound}}{\text{intensity of a barely audible sound}}\right) \quad (2)$$

Example 12.1 The sound of a vacuum cleaner in a room has 10 million times the intensity of the faintest audible sound. What is the intensity of the sound in decibels?

Answer:

$$\left(\frac{\text{sound intensity of vacuum cleaner}}{\text{faintest audible sound}}\right) = 10{,}000{,}000$$

$$\log 10{,}000{,}000 = 7$$

$$\begin{aligned}\text{Sound Intensity} &= 10 \times \log 10{,}000{,}000\\ &= 10 \times 7\\ &= 70 \text{ decibels}\end{aligned}$$

Example 12.2 What is the intensity in decibels of the faintest audible sound?
Answer: This had better turn out to be zero, which is what we promised the log scale would provide:

$$\left(\frac{\text{intensity of the faintest audible sound}}{\text{intensity of the faintest audible sound}}\right) = 1, \text{ and } \log 1 = 0$$

Thus,

$$\text{Sound Intensity} = 10 \times \log 1 = 10 \times 0 = 0 \text{ decibels}$$

Traffic noise.

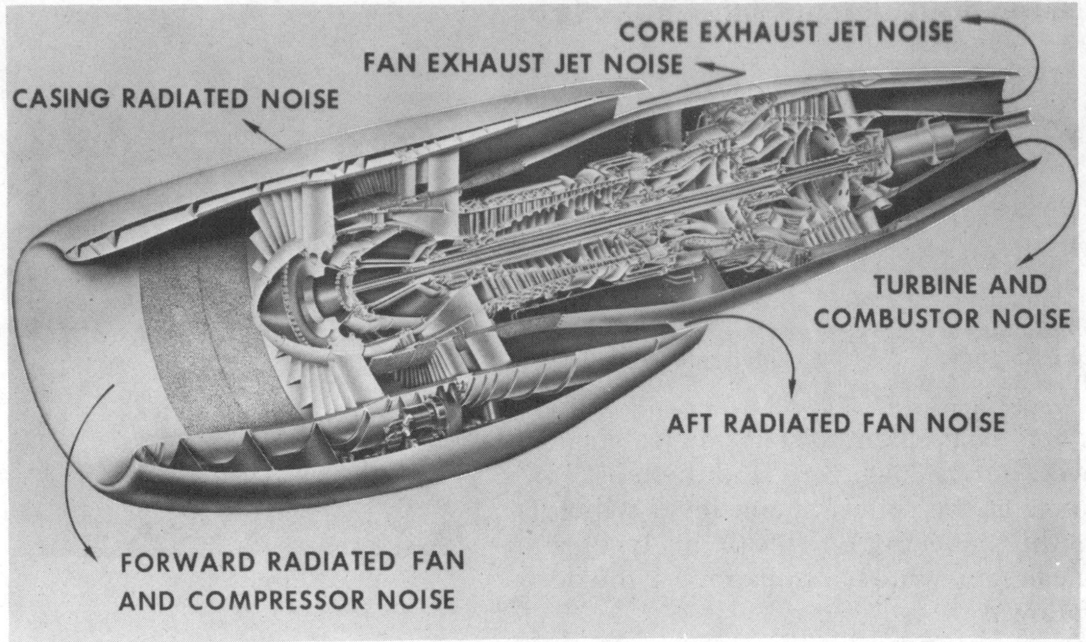

Jet engine.

Now let us pursue our arithmetic a bit further. If the sound of one jet engine is 100 dB, what is the sound of two jet engines? The answer is *not* 200 dB; we must not add these values because they are logarithms. Thus, $10^2 + 10^2$ is *not* 10^4. But, of course, one engine + one engine = two engines, and two engines generate twice as much power as one. It is therefore correct to add engines, or watts, or energy, or fuel bills—but not decibels. What we can do is to convert the decibels to watts, add the watts, then recalculate the decibel levels. To save you all this trouble, use the chart of Figure 12.8.

Example 12.3 A four engine jet plane stands on the runway. The sound of each engine is 100 dB. What is the sound level when one engine is running? Two? Three? All four?

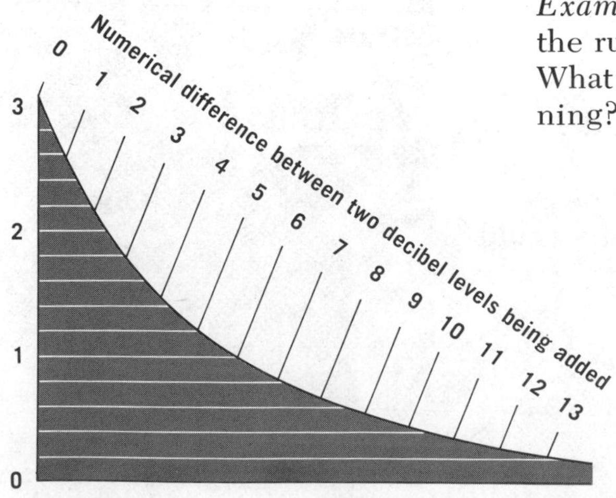

FIGURE 12.8 Nomograph for adding decibel levels.

Answer: With one engine, of course, it is 100 dB. For two engines, see Figure 12.8. Then, 100 dB − 100 dB = 0 dB.

We find 0 on the curved scale, which corresponds to 3 on the vertical scale. So we add 3 dB to the larger number (here they are both 100 dB) and 100 + 3 = 103 dB.

For three engines we are now adding another 100 dB to our previous 103 dB. Then, 103 − 100 = 3, which means that we add 1.8 dB to the larger number, and 103 + 1.8 = 104.8 dB.

Similarly, for four engines, we have 104.8 − 100 = 4.8, so we add 1.2, and 104.8 + 1.2 = 106 dB.

12.6 LOUDNESS

(Problem 12)

The sensitivity of the human ear depends on the *frequency* of the sound to which it is exposed. Very high or very low notes are not heard as well as those at intermediate frequencies. Figure 12.9 shows this relationship. Note that the ear is most sensitive to pure tones at about 4000 Hz. The important information that this curve provides about the ordinary decibel scale is that it does not match the human ear very well; it is fairly good in the 1000- to 4000-Hz range, but at much higher or lower frequencies the scale would predict a loudness that the ear does not hear. We can correct for this mismatch by subtracting the required number of decibels from the actual sound level at the various frequencies where such corrections are necessary. The most widely used scale of

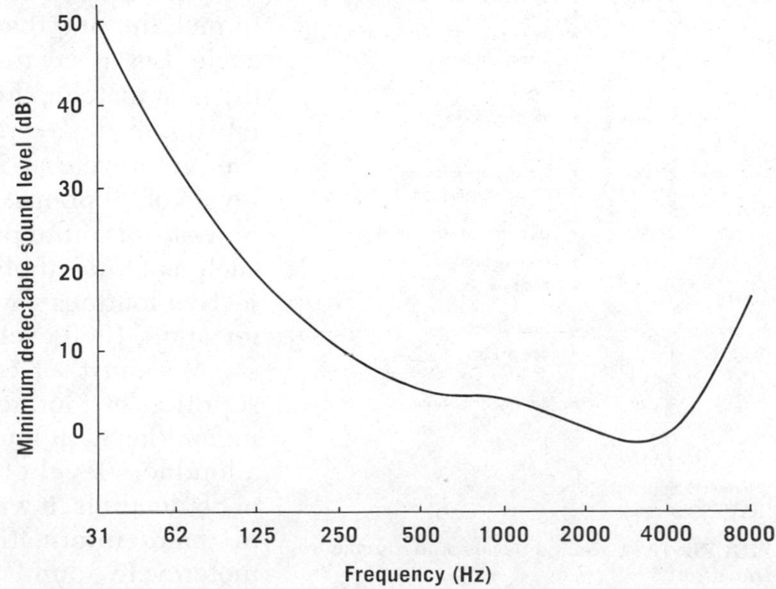

FIGURE 12.9 Minimum detectable sounds for human listeners.

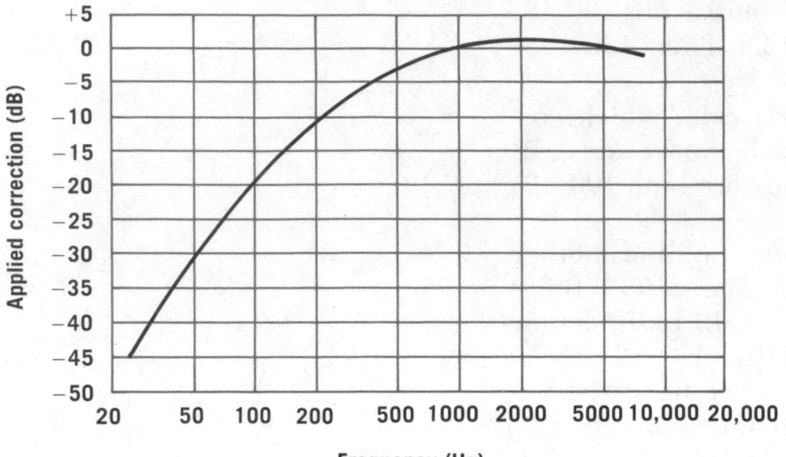

FIGURE 12.10 The dBA scale.

this sort, called the decibel-A, or dBA scale, is shown in Figure 12.10. Sound level meters, called decibel meters, can be constructed to filter out sound at different frequencies in accordance with the dBA scale.

Human beings, however, are not decibel meters; when they are exposed to a given sound and are asked, "How loud is it?" they introduce a certain subjectivity into their answers. To reconcile this difference between man and machine, a **subjective loudness scale** has been developed. This scale is based on a comparison of any given sound with that of a pure tone having a frequency of 1000 Hz. Let us say that we wish to rate the subjective loudness of some particular sound—for example, that of a motorcycle at a distance of 25 feet. A given person is offered a selection of pure 1000 Hz sounds of different loudnesses, and is asked to pick the one that sounds just as loud as the motorcycle. Let us say that he chooses a 90-decibel sound as the best match to his ears. If this choice represents the typical, or the average response of many people, then the motorcycle at 25 feet is said to have a "loudness level" of 90 **phons**, even though its own decibel rating in terms of sound power might be some other number, such as 88 decibels. The **phon** is thus a unit of subjective loudness, which is matched to a decibel scale for a pure 1000-cycle sound.

We must, of course, consider the subjective description of loudness. If you told someone who was unfamiliar with loudness scales that a motorcycle had a loudness level of 90 phons, or produced a loudness of 88 decibels, it would mean nothing to him. It would be more informative, if less precise, to say that a motorcycle sounds "very loud." Therefore it is con-

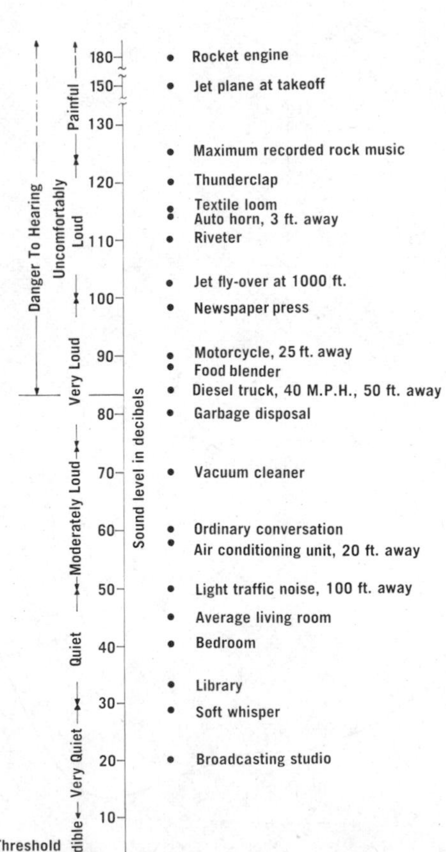

FIGURE 12.11 Sound levels and human response.

venient to append a scale of purely verbal description, such as "quiet," "very loud," etc., alongside a decibel scale of loudness. Figure 12.11 shows the decibel levels and verbal loudness descriptions of various sounds. The upper level is about 120 decibels or a little more; such sounds effectively saturate our sense of hearing, and greater sound powers are not perceived as being louder. Of course, they may be more harmful, but we are discussing only sensations. Note the extreme range of the scale: a 120-decibel sound is a trillion times more powerful than a zero-decibel sound.

12.7 THE EFFECTS OF NOISE

(Problem 7)

Noise can interfere with our communication, diminish our hearing, and affect our health and our behavior.

Interference With Communication

We have defined noise as unwanted sound. Let us think for a moment about the sounds that we do want: live or recorded speech or music, danger warnings such as the cry of a baby in a distant room or the rattle of a rattlesnake, pleasant natural sounds such as the chirp of a bird or the rustle of leaves in a gentle breeze. We want to hear these sounds at the right level—not too loud or too quiet—and without interference. Noise keeps us from what we want to hear, with the result that we do not hear it so well, or at all, or that the sound we want to receive must be unpleasantly loud for us to get the message.

It is true that modern technology enables us to compensate for its noisiness. We can turn up the volume of our radio or television sets if the traffic outside gets too noisy. Actors, musicians, and public speakers can likewise amplify their sounds. Even if such electronic aids fail to transmit the message clearly, the chances are that we may be able to read about it in the next day's newspaper. Such options were not always available. Before the advent of electricity there were no volume controls to turn up, and no high-speed presses to produce a flood of newspapers and other printed matter; in fact, there was very little artificial lighting at night to read by. Instead, people used their ears to gain information. The tolling of the church bells was not solely a musical recital; it provided information, announcing festivals, births, deaths, and the danger of attack. The town crier was the verbal newsman. Medieval manuscripts were written with a minimum of punctuation and separation between words; such visual aids were deemed unimportant because the text was intended to be read aloud so that others might hear, not to be merely scanned speedily with the eye while the scanner maintained both external and internal silence.

If you think that the quality of communication afforded by loud sounds superimposed over a noisy background is equivalent to that of soft sounds rising above silence, venture out some quiet winter day into a snowy woodland and hold a muted conversation there, or sing, or listen to the little natural sounds that noisier environments extinguish. Then take a ride on a subway with your girl or boy friend and yell sweet nothings at her or him.

Loss of Hearing

An occasional noise interferes with unwanted sounds, but we recover when quiet is restored. However, if the exposure to loud noise is protracted, some hearing loss becomes permanent. The general level of city noise, for example, is high enough to deafen us gradually as we grow older. In the absence of such noise, hearing ability need not deteriorate with advancing age. Thus, inhabitants of quiet societies, such as tribesmen in southeastern Sudan, hear as well in their seventies as New Yorkers do in their twenties.

It is important to understand that most instances of loss of hearing that result from environmental noise are not traumatic; in fact, the victim is often unaware of the effect. Let us picture a worker who completes his first day in a noisy factory. Of course, he recognizes the noisiness, and may even feel the effect as a "ringing in the ears." He will have suffered a temporary hearing loss that is localized in the frequency range around 4000 Hz, as shown in Figure 12.12A. This means that he will not hear moderately high frequencies so well, but low-frequency and very high-frequency sources will be unaffected. As he walks out of the factory, then, most sounds will seem softer. His car will seem to be better insulated, because he will not hear the rattles and squeaks so well. He will judge people's voices to be just as loud as usual, but they will seem to be speaking through a blanket. He will also feel rather tired.

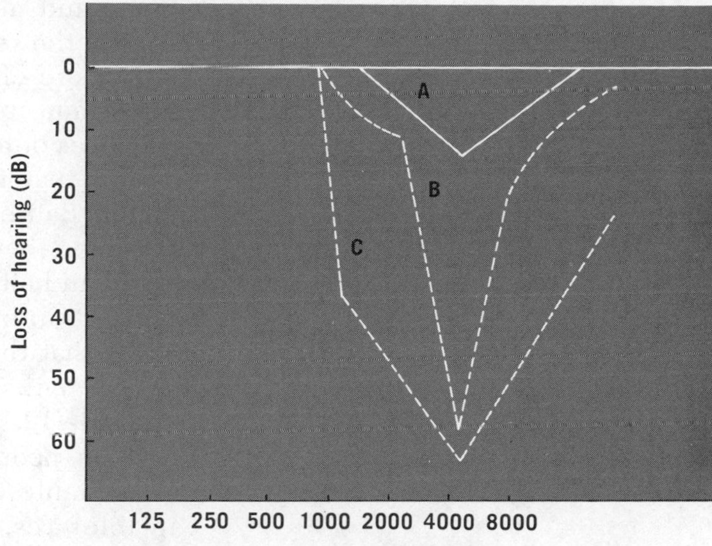

FIGURE 12.12 Patterns of hearing loss from exposure to industrial noise. *A.* **Temporary loss of hearing.** *B.* **After 20 years.** *C.* **After 35 years.**

By morning he will be rested, the ringing in his ears will have stopped, and his hearing will be partly but not completely restored. The factory will therefore not seem quite so noisy as it was on the first day. As the months go by he will become more and more accustomed to his condition, but his condition will be getting worse and worse. Can he recover if he is removed from his noisy environment? That will depend on how noisy it has been and how long he has been exposed. In many cases, his chances for almost complete recovery will be pretty good for about a year or so. However, if the exposure continues, hearing loss becomes irreversible, and eventually he will become deaf. Look at Figure 12.12*B* and *C* to see a typical downward progression caused by prolonged exposure to industrial noise.

In general, noise levels of about 80 decibels or higher can produce permanent hearing loss, although, of course, the effect is faster for louder noises, and it is somewhat dependent on the frequency. At a 2000-cycle frequency, for example, it is estimated that exposure to 95-decibel noise (about as loud as a power lawn mower) will depress one's hearing ability by about 15 decibels in 10 years. Occupational noise, such as that produced by bulldozers, jackhammers, diesel trucks, and aircraft, is deafening many millions of workers. The fact that women in technologically developed societies hear better than men is undoubtedly related to the fact that they are less exposed to occupational noise; in undeveloped, quiet societies, women and men hear equally well. This difference between the sexes in developed societies may diminish as more effort is exerted to make industry quiet, and as more people are exposed to noisy household appliances or to radio or television speakers turned up high so that they may be heard over the vacuum cleaner. Battle noises, such as those made by tanks, helicopters, jets, and artillery, are so deafening that more than half of the American soldiers who undergo combat training sometimes suffer enough hearing loss so that they could no longer meet the requirements for enlistment into combat units.

Table 12.1 shows some typical upper noise limits to which people may be exposed without suffering unacceptable damage. Is any damage to hearing acceptable? Yes, if we are to live in a developed society at the present state of technology; otherwise we had better choose to live elsewhere, and in that case, of course, we may be paying other kinds of prices. The "acceptable" damage level is therefore a matter of judgment; one reasonable criterion, which is very close to that on which Table 12.1 is based, is that the noise should not produce any permanent damage to hearing that exceeds the values shown in Figure 12.12A after a 10-year exposure. These limits refer to noises that cover a wide band of frequencies, usually spread over about one octave. Pure tones, however, can be more damaging, although we are not sure why. The effect has been attributed to the possibility that a given pure tone might just happen to resonate with some sections of the cochlea of the inner ear, or that its energy might be concentrated on a narrow section of the basilar membrane. A more recent study indicates that pure tones fail to produce a continuous reflex action of the inner ear muscles which protect the ear. In any case, for narrow-band noise that might contain pure tones, we subtract five decibels from the limits of Table 12.1. For noise which is known to be continuous over a range of several octaves, five decibels may be added.

An octave is a band of frequencies in which the highest frequency is twice the lowest.

What about extremely loud noises? Recent concern over exposure of people to rock music stems from the fact that such music is often indeed very loud. Sound levels of 125 decibels have been recorded in some discothèques. Such noise is at the edge of pain and is unquestionably deafening. Noise levels as high as 135 dB should never be experienced, even for a

TABLE 12.1 MAXIMUM ACCEPTABLE NOISE EXPOSURES

DURATION (per day)	*LIMIT (dBA)*
1.5 minutes or less	120
3 minutes	110
7 minutes	103
15 minutes	97
30 minutes	93
1 hour	90
2 hours	87
4 hours	85
8 hours	85

brief period, because the effects can be instantaneously damaging. Such an acoustic trauma might occur, for example, as the result of an explosion. If the noise level exceeds about 150 or 160 dB, the eardrum might be ruptured beyond repair, and the ossicles can be displaced or broken. In addition, the action of the inner ear may be so violent that the hair cells are destroyed.

Other Effects on Health and Behavior

As we have already discussed in many contexts, a living organism, such as a human being, is a very complicated system, and the effects of a stress or a disturbance follow intricate pathways that may be very difficult to elucidate. Having read this book thus far, you should be skeptical if you are told that a disturbance great enough to deafen you will have no other effects. Indeed, many investigators believe that loss of hearing is not the most serious consequence of excess noise. The first effects are anxiety and stress reactions or, in extreme cases, fright. These reactions accompany a change in the hormone content of the blood, which in turn produces body changes such as increased rate of heart beat, constriction of blood vessels, digestive spasms, and dilation of the pupils of the eyes. The long-term effects of such overstimulation are difficult to assess, but we do know that in animals it damages the heart, brain, and liver and produces emotional disturbances. The emotional effects on people are, of course, also difficult to measure. We do know that work efficiency goes down when noise goes up.

Hormones are biochemical regulatory substances that act in trace amounts. They are produced by various glands in the body.

12.8 NOISE CONTROL

Noise is transmitted from a source to a receiver. To control noise, therefore, we can reduce the source,

(Problem 8) interrupt the path of transmission, or protect the receiver.

Reducing the Source

The most obvious source reduction is simply the reduction of the sound power. Don't beat the drum so hard, or ring the bell so loud, or run so many trucks or motorcycles, or mow the lawn with a power mower so often. There are obvious limitations to this type of solution; for example, if we run fewer trucks, we will have less food and other essentials delivered to us.

Even if we do not reduce the sound power, we may be able to reduce the noise production by changing the source in some way. Our purpose in pushing a squeaky baby carriage is to move the baby, not to make noise. Therefore, we can oil the wheels to reduce the squeaking. Thus, machinery should be designed so that parts do not needlessly hit or rub against each other (see Fig. 12.13).

It might be possible to modify technological approaches so as to accomplish given objectives more quietly. Rotary saws instead of jackhammers could be used to break up street pavement. Ultrasonic pile drivers could replace the noisier steam-powered impact-type pile drivers.

FIGURE 12.13 Hydraulically operated shear has less impact noise than mechanically driven shear. (Courtesy of Pacific Industrial Manufacturing Company.)

We could also change our procedures. If a city sidewalk must be broken up by jackhammers, it would be better not to start early in the morning, when many people are asleep, but later in the day, when many have left for work. Aircraft takeoffs could be preferentially routed over less densely inhabited areas.

Control of noise is a complex and sophisticated technology, and it is most effective when it is applied to the original design of the potentially offending source. All too often a device or a machine or an entire industrial facility is designed with a view only to maximize its capacity to carry out its assigned function. If it turns out to be excessively noisy, an acoustical engineer may be called in to "soundproof" it. Under such circumstances, the engineer may be forced to accommodate to features of construction that should never have been accepted in the first place. Therefore, much of his effort may necessarily be applied, not to the source, but to the path between sound and receiver.

Interrupting the Path

We have learned that sound travels through air by compressions and expansions. It also travels through other elastic media, including solids such as wood. Such solids vibrate in response to sound and therefore do not effectively interrupt its transmission, as many residents of multiple dwellings will readily attest. However, we could use various materials that vibrate very inefficiently, such as wool, and absorb the sound energy, converting it to heat. (Very little heat is involved; the sound power of a symphony or-

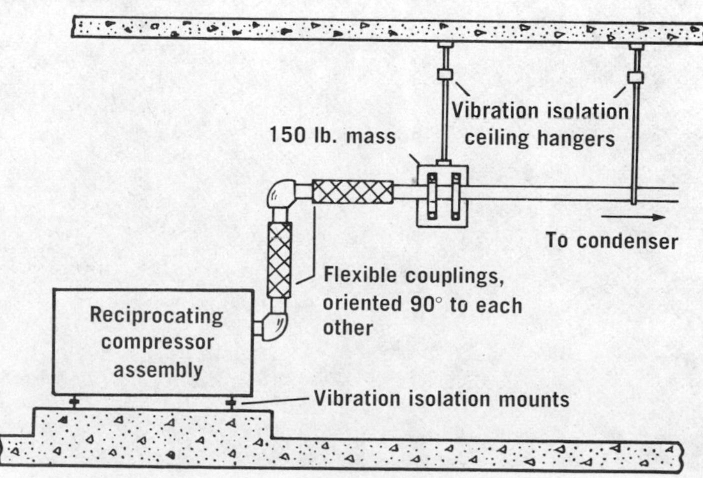

FIGURE 12.14 Mounting of compressor and piping to isolate vibration noise. (From Beranek, L. L. (ed.): Noise Reduction. Copyright © 1960 by McGraw-Hill Book Company. Used with permission of McGraw-Hill Book Company.)

FIGURE 12.15 Sound absorbers suspended close to noise sources. (Courtesy of Elof Hannson, Inc., & Sonosorber Corporation.)

FIGURE 12.16 Vibration isolation of printing presses required to reduce noise in office on floor below. (From Beranek, L. L. (ed.): Noise Reduction. Copyright © 1960 by McGraw-Hill Book Company. Used with permission of McGraw-Hill Book Company.)

chestra will warm up a room about as much as a 10-watt electric heater.) Sound-absorbing media have been developed extensively; they are called **acoustical materials**. We could also build interruption of the sound waves mechanically into more kinds of machinery; devices that function in this way are called **mufflers** (see Figs. 12.14, 12.15 and 12.16). Finally, we may be able to deflect the sound path away from the receiver, as by mechanically directing jet exhaust noise upward instead of down. Such deflection is, in effect, an interruption between source and receiver.

Protecting the Receiver

The final line of defense is strictly personal. We protect ourselves instinctively when we hold our hands over our ears. Alternatively, we can use ear plugs or muffs. (Stuffing in a bit of cotton does very little good.) A combination of ear plugs and muff can reduce noise by 40 or 50 decibels, which could make a jet plane sound no louder than a vacuum cleaner. Such protection could prevent the deafness caused by combat training, and should also be worn for recreational shooting.

We can also protect ourselves from a noise source by going away from it. In a factory, such reduction of exposure may take the form of rotating assignments so that different workers take their turns at the noisy jobs.

12.9 A PARTICULAR CASE: THE SUPERSONIC TRANSPORT (SST)

(Problems 9, 20, 21, 22, 23)

The SST is a passenger aircraft that travels faster than sound and at much higher altitudes than subsonic airplanes. Higher speed requires more power, and more power makes more noise. Near airports, the noise problem is associated with takeoff and the rapid climb shortly after, although the speeds at these times are subsonic. The engines on an SST must be small in diameter to provide optimal streamlining, and the noise from jet exhaust increases very rapidly (for a given engine thrust) as jet diameter is reduced and its speed increased. The statement has been made that the proposed American SST, on the runway, would sound like 50 ordinary jets taking off at the same time. This statement engendered considerable controversy because it was interpreted by some to mean that the SST would be 50 times as loud as an

In 1971, the U.S. Congress stopped federal funding for the SST.

ordinary jet. We can understand the situation better by analogy to Equation (1), page 495:

$$\text{Loudness (compared with one jet plane)} = 10 \times \log_{10}\left(\frac{\text{sound power of 50 jets}}{\text{sound power of 1 jet}}\right)$$

$$= 10 \times \log_{10} 50$$
$$= 10 \times 1.7 = 17 \text{ decibels}$$

The calculation tells us that 50 jet planes or their equivalent with reference to sound (one SST) are 17 decibels louder than one jet plane. However, the calculation does not tell you how much more annoyance this would cause you; nor does it tell what you would do about it. Experiments show that a sound 17 decibels louder than another is judged "three to four times as loud," but it sounds nevertheless just like 50 of the weaker sounds all sounding simultaneously. Experiments on airport noise show that a single aircraft 10 decibels louder than another produces about the same *annoyance* as 10 separate flights of the quieter craft spread throughout the day. This relationship is used by the government and by airport operators in planning land use around airports.

When the SST reaches supersonic speed in flight, another effect, the **sonic boom**, occurs. To visualize this effect, think of a speed boat moving rapidly in the water. Its speed is greater than that of the waves it creates, and it therefore leaves its waves behind it. Moreover, the wave energy is being continuously reinforced by the forward movement of the boat. The result is a high-energy wave, called a **wake**, that trails the boat in the shape of a V and that slaps hard against other vessels or against the shoreline. The sonic boom is a high-energy air wave of the same type. The tip of the wake moves forward with the airplane, while the sound itself moves out from the wake at the usual speed. The faster the airplane, the more slender is the wake.

To understand the geometry of the wake, study Figure 12.17. A stationary object (Fig. 12.17*A*) remains in the center of the circular waves it generates. The waves from a moving object will crowd each other in the direction of the object's motion (Fig. 12.17*B*). The object is, in effect, chasing its own waves. Sound travels in air at sea level at a speed of 334 meters/sec (1096 ft/sec, 747 miles/hr); any speed

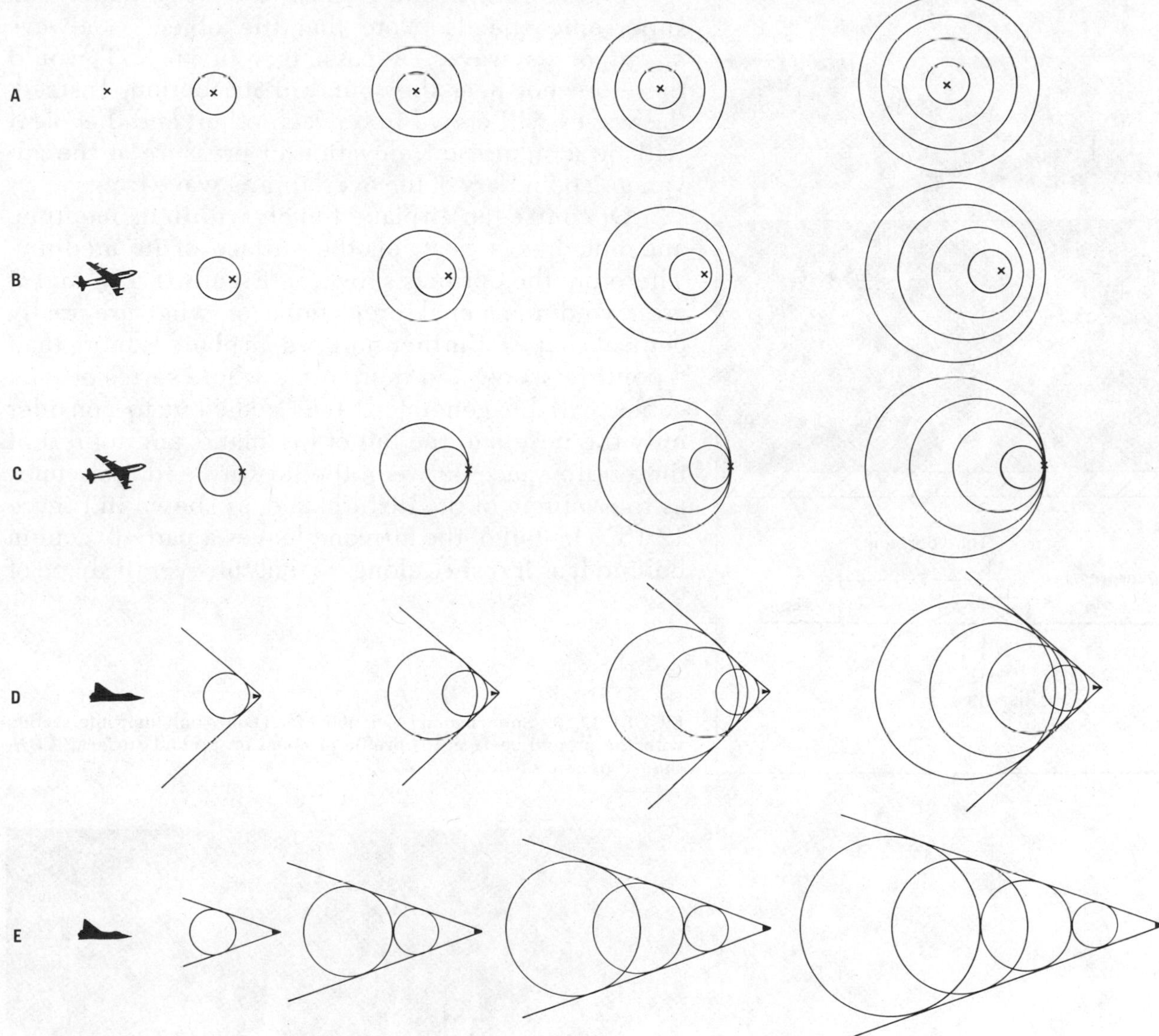

FIGURE 12.17 Wave patterns of subsonic and supersonic speeds: (*A*) stationary; (*B*) subsonic; (*C*) sonic, Mach 1; (*D*) supersonic, Mach 1.5; (*E*) supersonic, Mach 3.

less than this is said to be **subsonic**. When the speed of the object equals the speed of the wave, the object will not see any waves before it; it will just be keeping up with them (Fig. 12.17*C*). In air, such a speed is said to be **sonic**; and speeds greater than this are **supersonic**. Such speeds are usually measured in Mach numbers:* If an object is traveling at the speed of sound, then the numerator and the denominator of the equation are the same, and the Mach number equals 1. Mach 2 is twice the speed of sound, Mach 3 is three times the speed of sound, and so forth.

$$\text{Mach number} = \frac{\text{speed of object}}{\text{speed of sound}}$$

*After Ernst Mach, 1838–1916, a physicist who made important discoveries about sound.

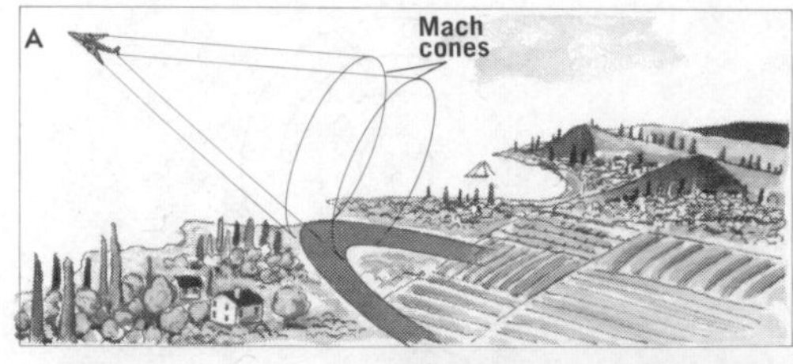

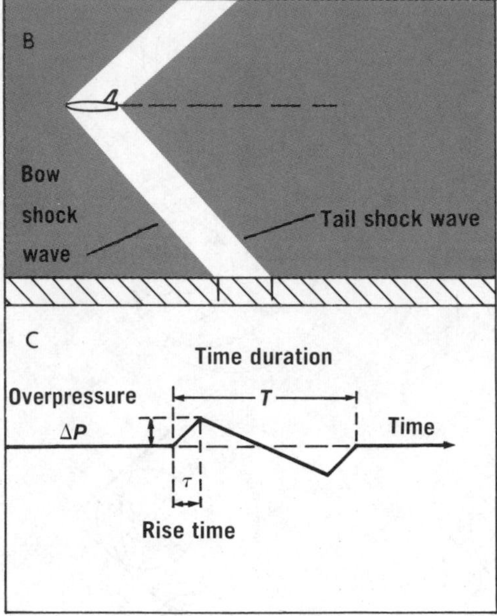

Figure 12.17*C* and *D* show the wave patterns at supersonic speeds. Note that the object is always *ahead* of its waves. A passenger in an SST would therefore not hear the sound of its motion. Instead, the waves will crowd in on each other, and the effect will be a significant elevation of pressure at the advancing boundary of the overlapping wave fronts.

Of course, an airplane travels within its medium, and not, like a boat, on the surface of its medium. Therefore the outlines shown in Figure 12.17*C* and *D* are two-dimensional projections of what are really conical shapes. Furthermore, an airplane is more than a point in space, and therefore a whole series of such cones will be generated. It is sufficient to consider only the nose and the tail of the plane, and represent the entire space between the forward and rear cones as the volume of the disturbance, as shown in Figure 12.18. The tail of the airplane leaves a partial vacuum behind it as it rushes along, so that the overall shape of

FIGURE 12.18 Sonic boom: (*A*) double Mach cones and their intersection with the ground surface; (*B*) profile of boom on ground surface; (*C*) N-shaped pressure pattern.

Insulated noise test chamber. (Courtesy of Lockheed California Company.)

the sonic boom is like the letter N: a rise above normal pressure, then a fall below, and finally a recovery, as shown in Figure 12.18*C*.

To be struck unexpectedly by a sonic boom can be quite unnerving. It sounds like a loud, close thunderclap, which can seem quite eerie when it comes from a cloudless sky. The duration of the boom is only about 1/10 to 1/2 second, and the two distinct pressure changes may be heard separately, or they may seem to merge into one if the time is very short. Depending on the power it generates, the sonic boom can rattle windows or shatter them or even destroy buildings. It is important to avoid the misconception that the sonic boom occurs only when the aircraft "breaks the sound barrier," that is, passes from subsonic to supersonic speed. On the contrary, the sonic boom is continuous and, like the wake of a speedboat, trails the aircraft all during the time that its speed is supersonic. Furthermore, the energy of the sonic boom increases as the supersonic speed of the aircraft increases.

12.10 EPILOGUE

(Problem 6)

Noise is not a substance; it does not accumulate in the environment. Noise is a form of energy, and when its echo dies, something will be the warmer for it. But, as we have learned, this effect is entirely trivial as compared with other sources of thermal pollution, and it may be ignored.

We have seen that noise affects the quality and sometimes the health of human life. The effects on other living species have been less severe, because animals do not work in factories, and in any event, man has more powerful means of displacing his nonhuman competitors. There are probably exceptions to this generalization, such as the scattering of sparrows by the gongs of Chinese villagers (page 282), or the disruption of animal life in winter by the roar of snowmobiles.

Noise, the "hustle and bustle" of industry and commerce, was once looked upon with favor as a sign of prosperity, like smoke from chimneys. Today, perhaps, it is most significant to think of the prevailing noise levels of our environment as an index of the price in human health, or in tranquility of spirit as well as of ear, that we have chosen to pay for the benefits of technology.

(From Saturday Review, May 13, 1972.)

PROBLEMS

1. **Vocabulary.** Define elastic wave; wavelength; frequency; amplitude.
2. **Pure tone.** What is a pure tone? Which of the following sounds would you consider to be the purest: the croak of a frog; the sound of an orchestra; the sound of a tuning fork?
3. **Noise.** What is noise? Do you think it would be feasible to develop an instrument that would indicate how noisy a given sound is? Defend your answer.
4. **Noise.** The first definition of "noise" on page 492 does not include the word "sound." Which of the following phenomena would you be willing to classify as noise? Defend your answers. (a) Mysterious blips on a radar screen that is scanning an area near an airport for approaching aircraft; (b) "snow" that interferes with your television picture but not with the sound; (c) smells from burning garbage that interfere with the chemical trail-making signals from the scent glands of the red deer.
5. **dBA Scale.** If you turned Figure 12.9 upside down and held it in front of a mirror, would it look something like Figure 12.10? Explain the relationship.
6. **"Acceptable" loss of hearing.** The statement was made on page 511 that some loss of hearing must be considered as an acceptable price to pay for living in a developed society. Do you agree? Defend your position.
7. **Loss of hearing.** Explain why curve *C* of Figure 12.12 is said to characterize deafness, even though it shows no loss of hearing below 500 Hz.
8. **Noise control.** A man carries a decibel meter with him for a day and records the following readings in his diary:

7:00 A.M.	Baby crying.	84 dB
7:30	Dishwasher in kitchen.	70
7:45	Garbage truck, 150 feet away.	90
8:00	Traffic noise while waiting for bus.	81
8:45	Arrived at entrance to office. Noise of jackhammer on sidewalk.	106
9:00–12:00 Noon	Average sound in office.	45
12:00–1:00 P.M.	Noise in restaurant–dishes, etc.	45
5:00–5:30	Rode home on subway (windows open).	90–111
6:00	Mowed lawn with power mower.	93

Offer suggestions for reducing the perceived loudness of each of these various noises.

9. **Definitions.** Define or explain: Mach number; sonic boom.

The following questions require arithmetic or algebraic reasoning or computation.

10. **Pure tone.** When we say that the shape of a "pure tone" is sinusoidal, which of the following statements do we imply? (a) The molecules move along in a wavy motion like water waves. (b) A graph of air density vs. distance will be a sine curve. (c) A graph of the degree of increase or decrease of air pressure above or below atmospheric pressure vs. distance will be a sine curve. (d) A graph of pressure vs. time at any one point will be a sine curve. (e) The molecules move back and forth like the inking of the paper in Figure 12.3*B*. (f) Since the air does not ripple, there is no real sine function, but rather we are using a figure of speech to compare a "pure" tone with the pure symmetry of a rotating wheel.

11. **Combinations of sine waves.** Given two sine curves of different frequencies: (a) Construct the combination of these curves by adding them, using the numerical values on the vertical scale. (b) If one of the curves were shifted slightly to the right or left, would the sum of the two curves still show a periodic variation?

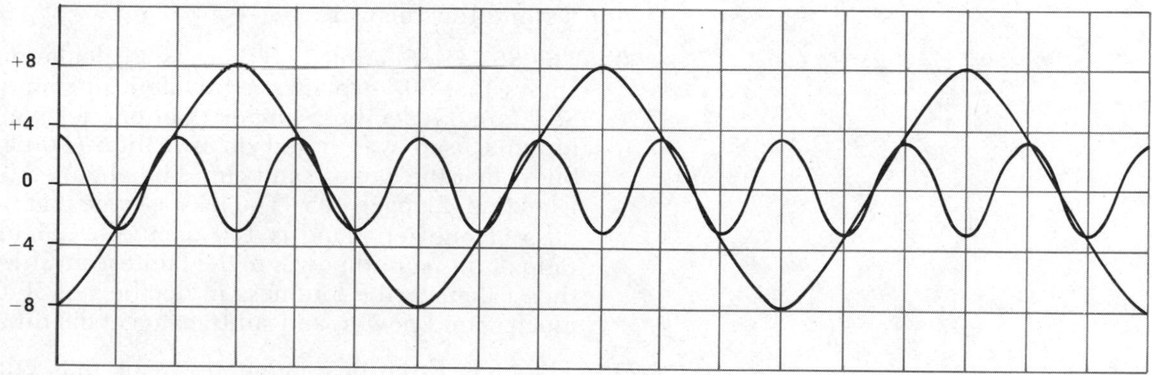

Combination of two sine waves.

12. **Loudness.** As stated in Section 12.5, the scale of subjective loudness is based on a pure tone of 1000 cycles per second. Calculate the wave length of this tone if the speed of the sound is 1100 feet per second.

13. **Decibels.** A person in a living room makes a singing sound that has 10 thousand times the power of the faintest audible sound. What is his sound power in decibels?

14. **Sound intensity.** The sound intensity of a motorcycle at a distance of 25 feet is 90 decibels. How many times more intense than the faintest audible sound is this motorcycle sound?

15. **Decibel scales.** The statement is made in Section 12.5 that a 120-decibel sound is a trillion times more powerful than a zero-decibel sound. Show that this statement is true.

16. **Decibel scales.** Can there be a negative decibel level? A dog can hear sounds inaudible to a human. Suppose a dog could just hear a 10^{-13} watt sound; what is the decibel level to which he would be sensitive?

17. **Decibel scales.** If you wanted to check the answers to Example 12.3, page 496, without using Figure 12.9, would it be correct to use Equation 1, page 495, and substitute numbers of engines for the parenthetical expression, such as $\log_{10}$ (3 engines/2 engines)? Try it and see.

18. **Sound pressure level.** As you have already learned, a sound wave produces regions of high pressure in the air through which it travels, and the sound intensity is proportional to the square of the pressure, p^2, or:

$$\text{Sound intensity} \propto p^2$$

Derive a formula for sound intensity in terms of sound-pressure level, SPL, expressed in decibels.

19. **Signal vs. noise.** Which of the following proportionalities do you think better expresses the discussion in Section 12.7 on "interference with communication"?
    (a) Communication $\propto$ (signal − noise)
    (b) Communication $\propto$ (signal/noise)
    Defend your answer.

20. **The SST.** In Section 12.8 there is a calculation which shows that 50 jet planes, equivalent in sound to one SST, are 17 decibels louder than one jet plane. The formula used was based on Equation (1), page 495. Show that the same result could be obtained by using Equation (2), page 495. (You may assume that the loudness of one jet plane is 100 decibels. Calculate the ratio of this sound power to the faintest audible sound; then calculate the loudness in decibels of 50 times as much sound power, and subtract to get the difference.)

21. **The SST.** From the statements made in Section 12.9, how many *separate* takeoffs, spread throughout the day, of an ordinary jet would be as annoying on the runway as one SST? How many separate takeoffs of an ordinary jet would be as annoying on the runway as 50 ordinary jets all taking off at once?

22. **Mach numbers.** A rocket is moving through the lower atmosphere, where the speed of sound is 1100 feet per second, at Mach 2.5. What is the speed of the rocket in feet per second?

23. **Mach numbers.** If you still remember your trigonometry, try this one: Derive a formula that expresses wake angle in terms of Mach numbers.

Answers:

12. 1.1 ft

13. 40 dB

14. 1,000,000,000

16. −10 dB

21. 17; 17

22. 2750 ft/sec

23. Wake angle = 2 arc sine (1/ Mach number)

BIBLIOGRAPHY

A delightfully written, non-mathematical, yet authoritative paperback book that covers the entire field very well is:

Rupert Taylor: *Noise.* Baltimore, Penguin Books, 1970. 268 pp.

A rather technical book on the effects of noise is:

Karl D. Kryter: *The Effects of Noise on Man.* New York, Academic Press, 1970. 632 pp.

For a basic text on noise control, refer to:

Leo L. Beranek: *Noise Reduction.* New York, McGraw-Hill, 1960. 752 pp.

Four recent popular books that take up the environmental aspects of noise are:

Theodore Berland: *The Fight For Quiet.* New York, Prentice-Hall, 1970. 370 pp.

Robert Alex Baron: *The Tyranny of Noise.* New York, St. Martin's Press, 1970. 294 pp.

Henry Still: *In Quest of Quiet.* Harrisburg, Pa., Stackpole Books, 1970. 220 pp.

Clifford R. Bragdon: *Noise Pollution: The Unquiet Crisis.* Philadelphia, University of Pennsylvania Press, 1971. 280 pp.

13

SOCIAL, LEGAL, AND ECONOMIC ASPECTS OF ENVIRONMENTAL DEGRADATION

13.1 INTRODUCTION

(Problems 1, 2, 3)

The pollution of Lake Erie in the United States (see Fig. 13.1), the transformation of the sparkling Rhine into a sewer flowing through central Europe, and the despoiling of the deepest lake in the world, Lake Baikal in Siberia, all attest to the polluting potential of industrial development under various economic, social, and political systems. Many segments of society contribute to the increasing environmental insult. Mining and timber operations ravage the landscape. Manufacturing produces dirty air, water, and earth. Municipalities dump raw or partially treated sewage into waterways, and use the air as a sink for wastes. National governments, too, have abetted environmental degradation. In the United States, for example, laws which permit and encourage the transfer of public property into the private domain and which aid industrialization have had, until recently, few provisions designed to protect the environment. The Homestead Acts of the nineteenth century, current laws governing mineral rights, such tax incen-

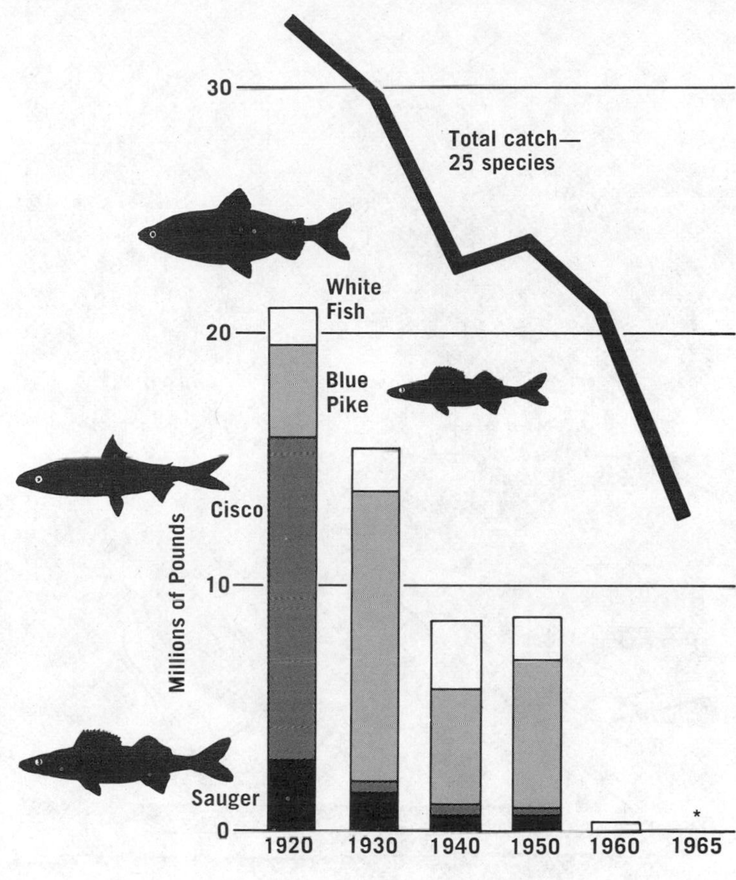

FIGURE 13.1 Commercial fish catch in Lake Erie. (Adapted from the New York Times, April 20, 1970.)

tives as oil, mineral, and timber depletion allowances, and contracts for bombers and space programs, as well as for housing projects and urban renewal, have all neglected consideration of environmental side effects. Automobiles, self-defrosting refrigerators, electric can openers, dripping water faucets, improperly insulated homes, overpackaging of goods, and the overheating of dwellings in winter and overcooling them in summer all add to the environmental burden.

The nations of the world today, faced with the massive task of cleaning up and of preventing further destruction, have begun to realize that technology alone has not, and probably cannot, salvage the environment. Instead, we must alter our values, in particular, our celebration of progress, if we seek for a relatively unspoiled environment. Our laws and economic structures must be changed so that productivity is no longer the ultimate measure of success. Legislators must determine what segments of society should bear the responsibility for the damage

"I ask you, what's wrong with the environment?" **(Drawing by Alan Dunn; © 1972 The New Yorker Magazine, Inc.)**

caused by environmental degradation. Preventive legislation must be written and methods for enforcement established. Although most of the discussion in this chapter is based on experience and law in the United States, the underlying principles are applicable to all industrialized nations.

Much of the damage caused by pollution is usually assessed in economic terms. For example, under current economic structures, the price of a product reflects mostly the cost of raw materials and labor, the amount of capital investment, and the demand. Any product entails not only costs internalized by the manufacturer, but also costs irrelevant to the accounting system. These costs, external to the manufacturer, are called **externalities**. They involve social, including environmental, costs. If the smoke from manufacturing (or from disposal) darkens nearby houses and dirties the clothes of local residents, the costs of more frequent repainting and laundering are part of the total environmental cost of the product. Hidden costs of products whose manufacture involves occupational health hazards include the cost of medical bills, loss of work because of illness, decreased feeling of well-being, and increasing deafness.

Whenever an effluent pollutes a river which must be purified by a town downstream, the cost of water is reflected elsewhere than in the price of the product. Loss of tourist trade because of pollution is another economic burden shouldered by parties other than the manufacturer or the purchaser. Somehow, costs of all these factors should be assessed, for all are part of the total cost to society for the manufacture, use, and eventual disposal of the product. The total cost of all externalities is formidable. The Environmental Protection Agency estimates that in the United States, the current cost of sickness and premature death related to air pollution is about six billion dollars per year. This figure includes only medical care and loss of work. Damage to crops and materials is estimated to cost five billion dollars each year. Depression of property values in polluted areas is estimated at another five billion dollars each year. Table 13.1 projects damage for 1977. Both the current figures and the projections are subject to large errors, for they represent only known dangers. Unknown effects, including possible long-term effects, are not included in these figures.

The "cost of death" cannot be assessed in terms of the worth of a human being, which is beyond our ability or right to judge. Instead, as a purely economic calculation, the "cost of death" is taken to be the loss of the value of income or other productivity that would have been expected from a normal life span.

Not all externalities are readily quantifiable, for many environmental effects involve the deterioration of the quality of human life. As such, they require subjective judgments and the willingness to assign numerical values to qualitative feelings. For example, recreational opportunities are changing as an increasing number of natural ecosystems across the globe are

TABLE 13.1 PROJECTED ANNUAL DAMAGE COSTS BY POLLUTANT FOR 1977 (IN MILLIONS OF 1970 DOLLARS)

CONDITION AFFECTED	PARTICULATES	SULFUR OXIDES	OZONE	NITROGEN OXIDES	CARBON MONOXIDE
Health	3,880	5,440	*	*	*
Residential property	3,330	4,660	*	*	*
Materials and vegetation	970	3,680	1,700	1,250	0

*Data not available. (Adapted from *The Economics of Clean Air,* Annual Report of the Administrator of the Environmental Protection Agency, March, 1972.)

An example of an economic externality. A home in Appalachia collapses as the overburden on which it stands subsides into an abandoned, unsupported deep mine shaft. The price of coal is not high enough to allow the coal industry to pay for the effort necessary to prevent such destruction. (Courtesy of Bureau of Mines, U.S. Department of the Interior.)

being transformed into farmland and cities, and as other ecosystems are being destroyed by pollution. Fifty years ago, for instance, on the south shore of Lake Erie, people fished, boated, swam, and explored the beach and marshes. Children chased turtles and lovers chased each other along the shore. Now swimming there is unsafe, the number of tasty game fish has declined dramatically, the wild secluded beaches have disappeared, and boats ply polluted waters. Alternative recreational opportunities are now available. Swimmers can find artificial swimming pools, children can play in approved parks, people who want to see turtles can drive to the Cleveland Zoo, and lovers can still find places to play. For many, these recreational changes involve poignant losses, but their dollar value is difficult to quantify.

One might argue that the industrialization of the past fifty years has brought unprecedented leisure, comfort, and health, affording the modern residents of the Lake Erie shore an easier and more relaxing life than their grandparents had. Moreover, the same industrialization that has polluted this area has increased wealth, thereby opening travel opportunities for residents. However, areas like Yellowstone Park, formerly relatively untouched, are now so popular that the natural fauna and flora are threatened by the large crowds of tourists.

On the other hand, many believe that the comforts of modern society are a poor substitute for the joys of nature, and that by losing certain pleasures we are losing an important part of our humanity. The anxieties, tensions, and social problems of the day, this argument runs, might be alleviated if man were able to enjoy natural things. Moreover, the poor do not have travel opportunities; they never get to see Yellowstone or Bermuda.

There is no way that we can quantitatively evaluate the relative validity of these two arguments. The answers become even more obscure when we try to evaluate costs in cases like the proposed Alaskan Pipeline, where few people will ever be directly affected (see Chapter 5).

"Which would you rather see built on this site? (A) An intercontinental jetport; (B) an atomic powerplant; (C) a mall-type shopping center; or (D) a 3,000-unit middle-income housing development."

(Reprinted from Audubon, the magazine of the National Audubon Society.)

13.2 COST OF POLLUTION CONTROL

(Problems 4, 6, 9)

The considerable expense of cleaning our environment will lead to many economic changes. In studying the economic impact of pollution control, **micro-** and **macroeconomic effects** have been distinguished. Microeconomic studies deal with the economic behavior of individual families, firms, or industries, particularly as they affect the working of the price system. Macroeconomic effects include effects upon the demand and supply of total production, and on general levels of prices, employment, income and economic growth. The Gross National Product, the balance of trade, and the Consumer Price Index are all examples of macroeconomic measures; the economic behavior of a family, the stock-market fluctuations of a single firm or group of firms are microeconomic effects. Imposition of pollution controls will certainly have both micro- and macroeconomic effects.

Macroeconomic Effects

Various studies of the probable effect of pollution control on American business have led to the following conclusions: (a) Only a portion of the additional cost of manufacturing need be ultimately reflected in consumer costs and (b) companies that have been on a strong economic footing will continue to prosper even under strict pollution guidelines, but (c) a large proportion of companies which are currently in economic jeopardy will fail. The consequences of economic failure will of course depend on the nature of the company. The failure of a small company in a large industrial area will cause relatively little unemployment; the failure of a company in a single-industry town will create severe local economic problems. (d) As a result of pollution controls and increased consumer prices, U.S. goods will be relatively more expensive than foreign imports. Unless foreign countries impose the same types of pollution control, protective tariffs may be fought for by industrial interests.

All these predictions are based on cost estimates that are very difficult to assess adequately. There are vast discrepancies in the estimates cited by those who seek to control pollution and those who are subject to control. For example, one study recently conducted for the Environmental Protection Agency estimated the cost of suppressing sulfur oxides from copper smelters at 87 million dollars. The copper industry claims the cost will be 345 million dollars if the control techniques are effective. If they are not, the industry contends, new smelters will be needed and the total cost will be over one billion dollars.

Some factors tend to make pollution abatement appear more costly than it is. Capital expenditures for pollution control can often be spread over 20 years, but, for income tax advantages, the depreciation can be compressed to as little as five years. When the control devices are long-lasting, industry benefits from such a tax structure.

In certain cases, pollution control methods may possibly be incorporated along with a more efficient manufacturing process. Many Federal and State agencies offer tax-exempt bond issues that indirectly afford industry lower finance rates than in other types of construction. All these economic advantages are paid for ultimately by the taxpayer or consumer, and therefore may be thought of as more externalities.

Investment and operating expenditures for pollution lead to various economic consequences (see Fig. 13.2). If we assume an otherwise constant level of

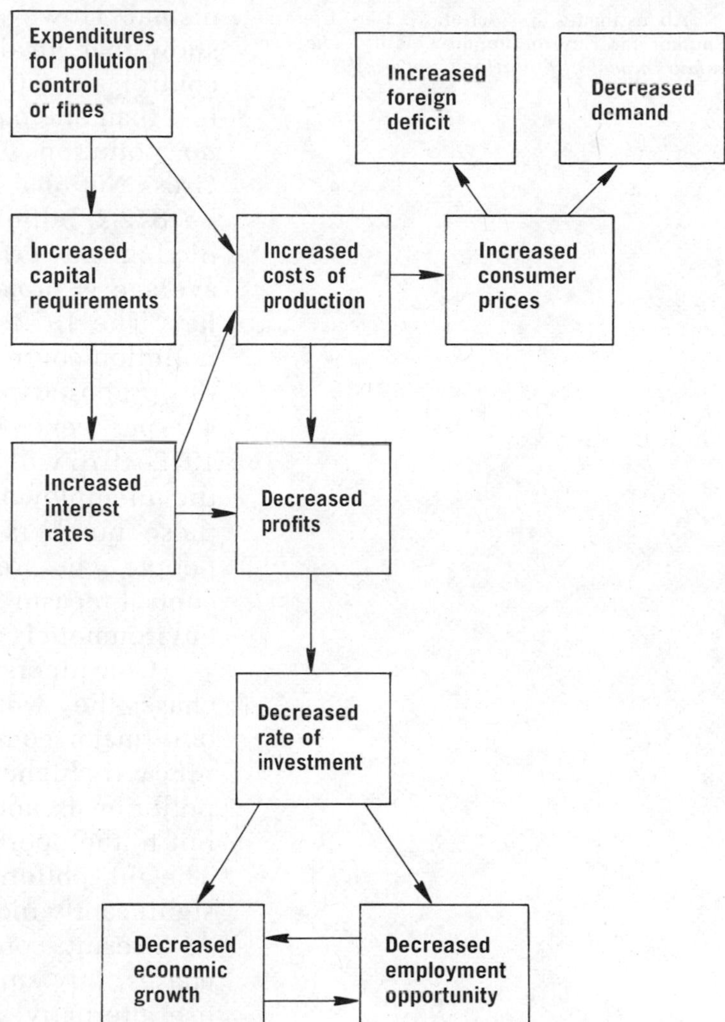

FIGURE 13.2 A simplified diagram of the relatively immediate effects of the cost of pollution abatement.

the economy, then the imposition of pollution abatement will lead to higher costs of production, which, in turn, engender higher consumer prices, and reduced demand for goods, both domestically and abroad. By contrast, foreign goods will be comparatively less expensive, and the balance of payments deficit will grow.

Higher investment cost creates a demand for more capital. Since the cost increase for pollution control will pervade the entire economy, the increase in capital needed will tend to drive the interest rates up. Both the higher cost of production and the increased interest rates will lead to lower profits, which, in turn, lead to lower rates of investment. Decreased investment, in turn, implies a lower rate of economic growth and increased unemployment. The picture sounds

All estimates in Section 13.2 are from the Council on Environmental Quality, *The Economic Impact of Pollution Control*, 1972.

dismal. However, estimates of the projected effects show that, in the aggregate, the proposed pollution control standards will increase manufacturing costs less than one per cent from 1972 to 1977. If there were no pollution control laws, the estimated average Gross National Product for 1972 through 1976 would be 872.2 billion dollars (in dollars valued at 1958 buying power). With pollution control costs, the average is projected to be 867.5 billion in 1958 dollars. The 1972–1976 average annual growth without pollution control is projected to be 5.3 per cent; with the proposed pollution controls, the average will be 4.9 per cent. The average unemployment rate in 1972–1976 will be 4.8 per cent; with pollution control, the unemployment rate will be 4.9 per cent. Although these numbers do not predict large macro-effects, people who are unemployed as a result of pollution control measures might feel that the quest for a clean environment is not egalitarian.

Consumers will probably make the same purchases they would have before pollution control. The only major commodity that will probably suffer significantly higher prices is the automobile, for here the pollution is not due primarily to the manufacturing, but to the operation of the car itself (see Table 13.2). The anti-pollution devices added to each new car will significantly increase the overall cost to the consumer. As a result, consumers can be expected to defer purchases, buy smaller cars, forego luxury options, and use alternative modes of transportation. These decisions necessitate investment in mass transit by Federal, State, and local governments.

Thus far in this section we have considered the economic effects of standards that are currently established, or under serious consideration, by the U.S. Federal and local governments. But these standards will not bring blue skies over Los Angeles, nor make Lake Erie once more fit for swimming. Indeed, some proposed and existing "anti-pollution" regulations are more permissive than current practice. For example, the laws regulating billboards on Connecticut highways permit one billboard every 500 feet, which would allow considerable construction of new billboards. If the United States commits itself to pure air and water, to quiet cities, airports, and airplanes, and to the preservation of wild areas, then manufacturing costs must necessarily increase much more than one

TABLE 13.2 ESTIMATED INVESTMENT COSTS FOR ALTERNATIVE MOBILE SOURCE EMISSION CONTROLS 1967–1977 (Autos and Light-Duty Trucks)

MODEL YEAR	*CONTROLS ADDED*	*COST PER NEW VEHICLE (DOLLARS)*	*CUMULATIVE COSTS (DOLLARS)*
1967	None	0.00	0.00
1968–69	Closed PCV system (above cost of open system). Carburetor changes. Ignition timing changes. Inlet air temperature control.	5.40	5.40
1970	Additional carburetor changes. Idle control solenoid. Ignition timing changes.	7.40	12.80
1971	Evaporative emission control. Improved idle control solenoid with overheat protection (above '68–'70 costs) including transmission spark control. Lower compression ratios. Additional carburetor changes.	19.70	32.50
1972	Valve and valve seat changes for unleaded gasoline.	2.00	34.50
1973–74	Exhaust gas recirculation for NO_x control, fixed orifice system. Speed-controlled spark advance.	48.00	82.50
1975	Catalytic oxidation of exhaust HC and CO (includes long-life exhaust system). Unitized ignition systems for 50,000-mile service-free performance. Air injection for catalytic unit needs.	164.00	246.50
1976–77	Dual catalyst units for HC, CO oxidation and NO_x reduction (additional cost above previous controls which are supplanted). Modified manifold reactors to reduce catalyst load.	105.00	351.50

From *The Economic Impact of Pollution Control,* Council on Environmental Quality, 1972.

per cent, and existing patterns of consumer behavior must change considerably.

Microeconomic Effects on Industries

Of course, the aggregate picture does not indicate changes in individual industries. The effects of pollution control legislation on a company specializing in scrubbers (see Chapter 9) will be advantageous, for demand will increase, and employment should rise. But for industries which must purchase pollution control devices, an unsatisfactory economic pattern similar to that depicted in Figure 13.2 may prevail. As an example, we consider pulp and paper mills in the United States.

During the period 1972–1976, pulp and paper mills will spend approximately 3.3 billion dollars to meet air and water pollution abatement requirements. As a result, cost of production will probably increase 3.5 to 10 per cent. If other countries impose similar anti-pollution requirements on their own paper industries, this cost increase will probably not affect sales of United States paper products on the international market. Seven hundred fifty-two pulp and paper mills are currently in operation in the United States. Of these, 329, accounting for 15 per cent of total production, are economically shaky. Because most of these are old, they need high pollution control expenditures, and, because of their low profitability, several may be unable to borrow the necessary capital. Even without pollution requirements, about 30 of these mills would probably close during 1972–1976. An additional 60 to 65 mills, affording 1600 jobs, may be forced to close because of the new abatement requirements. Although net growth in the industry will probably create more than 1600 new jobs in the more profitable plants, many of these jobs will be in different communities from those in which jobs will be lost. Several of the expected plant closures will be in rural areas where the impact of unemployment is particularly severe.

Thus, while the overall prospects in the paper

(N.J. Dept. of Environmental Protection.)

industry are optimistic, requirements for pollution abatement may lead to critical local recessions. Similar pictures are predicted for other industries in the United States. The economic burden of environmental protection will not distribute itself evenly; some people will be made jobless in the battle for a cleaner environment; some areas will suffer serious social readjustment.

Other micro-effects can be gauged only by studying some mechanisms for decision-making in large corporations. Many powerful industries in the U.S. are complaining that they cannot absorb the cost of pollution control. Yet, their approach is often grossly inefficient. For example, often in large industries composed of many plants each, the plant manager is rewarded with an annual bonus, the amount of which increases with the profitability of the plant during a specific year. From a corporate point of view, such a system of incentives encourages each plant to be highly profitable without requiring constant surveillance by the corporate management. By encouraging yearly computation of profits, however, efficient and frugal long-term approaches to pollution control are discouraged, for these may involve very high initial costs. If costs are high for a year or two, the plant's profitability, the manager's bonus, and even his job may all be jeopardized. Thus, the bonus system encourages the manager to opt for some piecemeal devices – an incinerator one year, a scrubber another, and some patchwork and chimney repairs a third. Moreover, the stockholders are more likely to be favorably impressed by high profits than by large expenses. Finally, the seemingly obvious solution – assigning to corporate management the task of pollution control – is fraught with difficulties, for the different operating procedures of each plant and the various legal standards of each location are best known by the individual plant managers and not by the corporate leaders.

Microeconomic Effects Upon a Hypothetical Family

Our hypothetical example concerns a family of two adults and two children. Both children are boys; the parents want a girl. The family is looking for a three-bedroom house. They plan to give each of the boys a room. If a new baby is born, it will have a private room and the older boys will share one room. The family finds the perfect all-electric house. The bill for heating to 72° in the winter and cooling to 70° in the summer averages 30 dollars a month. Before purchasing the house, however, the family learns that a new anti-pollution tax has been proposed. The law includes a steeply progressive tax on electricity; the cost per kilowatt hour for the first 500 kilowatt hours of each month is unaffected by the new tax. That amount should be ample for the family's lighting, cooking, and miscellaneous appliances. But, instead of 30 dollars per month, the new tax will increase the monthly heating and cooling bill to 150 dollars a month. (The proposers of the law know that the use of electricity is relatively insensitive to changes in prices, or **price inelastic**; they therefore propose a huge tax to discourage use.) What will our family do?

1. If 1800 dollars a year is immaterial to the parents' budget, they might buy the house and absorb the tax increase.
2. They could learn to live comfortably during the winter in a house heated to 68° at night. Also, they might decide to use fans instead of air conditioning to cool the house during the summer. (Their local utility company estimated their annual savings at 2500 kilowatt hours, or 400 dollars.)
3. They might obtain estimates of the price of improving the insulation; if too high, they might decide to find a house with better insulation.
4. If they want the house badly enough and the tax severely strains their budget, they might delay the third child.
5. They might search for a smaller house, an apartment, or a row house. All of these would be less expensive to heat or cool.
6. The wife might decide to work more hours.
7. They might search for a house which uses nonelectric forms of energy for temperature control.
8. They might decide to move to a town with lower taxes.
9. They might buy the house and install an auxiliary solar heater.

The sensation of cold is related to the rate at which the body loses heat, and this depends on bodily insulation as well as on the surrounding temperature. A bedroom temperature of 60°F is sufficient for comfortable sleep under good blankets. In fact, with highly efficient insulation such as that used in sleeping bags, all that would be necessary would be to keep the house warm enough so that water would not freeze in the pipes.

Most of these options lead to choices environmentally preferable to purchasing and using the house as is. The options are quite varied, and consequently we might expect that the response in a town which initiates such a program would be varied. People who were already settled in a house would be unlikely to move. They might complain, weatherproof their windows, buy warm sweaters, blankets, and fans, or launder less often, thus reducing their electricity usage. New houses would probably be constructed to minimize projected tax burdens; in particular, the popularity of electric heating would almost certainly decrease. Builders would tend to insulate their houses better, for a high electricity tax would put a premium on a well-insulated house. Architects would be encouraged to design houses which effectively utilize climatic factors to control temperature. The new law might encourage different types of housing. Instead of single-occupancy detached housing, clusters of row houses might be built. Such designs use less space, and less energy per family for heating and cooling.

The law would affect industry as well. In order to minimize air-conditioning costs, private firms might encourage informal clothing, including shorts for men. In hot seasons, working hours might be changed to take advantage of the cool times of the day, such as by adopting siestas.

Of course, if the town offered no advantages over the neighboring towns, and subjected its residents to a steep electricity tax not levied by nearby towns, many people and industries might move to less expensive municipalities or repeal the tax. Thus, to be effective, and not destroy the town, an electricity tax would have to be levied nationally or regionally, or the town would have to supply other inducements, for example, better schools, more parks, or safer streets.

13.3 POLLUTION CONTROL: ALTRUISM, LEGISLATION, AND ENFORCEMENT

(Problems 5, 14, 15, 16, 23)

Altruism

Altruism is devotion to the interests of others. We regard an altruistic approach to environmental control as the action of an individual person, firm, or government to help preserve the environment voluntarily. If taken by large numbers of individuals, such actions would benefit the environment. For example, if no woman bore more than two children, population size would decrease. From the point of view of limiting population growth, therefore, an altruistic act would be performed by a woman who decided to have only two children, though she wanted more. How have altruistic approaches succeeded, and what is their prognosis? Housewives recycle paper, cans, and bottles, carefully read detergent boxes and study books written to aid concerned individuals in keeping the environment healthy. In spite of all these efforts, the per cent of newspapers being recycled is still low, and consumption in general is very wasteful.

If we all acted in accordance with sound ecological principles, our environment would be much cleaner and more pleasant. But incentives operate against the altruistic approaches. Consider two housewives—one who acts with reasonable, but not fanatic, care toward the environment, and another who acts solely in her own interest. The first awakens, changes her baby's diaper and soaks it in soapy water. After breakfast, which includes oatmeal, bought in bulk, and fruits, eggs and milk, she washes her dishes by hand, and puts the baby's diapers in the washing machine. To dispose of the morning's wastes, she first removes the labels from any empty glass jars and flattens any used tin cans, and then she carries her garbage to her compost pile. Finally, she hangs her laundry on the line. Once a week, her morning is concluded with a visit to the neighborhood recycling center. On the other hand, her neighbor eats instant breakfast and makes frozen pop-up waffles for her husband and children. The toilet disposes of the baby's paper diapers, and the dishwasher cleans the few breakfast dishes. All newspapers, paper napkins, and paper towels are discarded along with the family's food garbage and placed outside for the garbage man. The second housewife finishes her tasks long before her concerned neighbor. Whose environment is cleaner? Both women share the same polluted air. Both sets of ears are shattered by the same jet overhead. Neither may swim in the dirty lake nearby. Moreover, the tax structure assesses them both at equal rates, and public money is used to clean the sewers stuffed with paper diapers. Indeed, all tangible rewards accrue to the environmentally careless housewife, for she gains extra time to use as she pleases. This example points out why it is unrealistic to rely on the good will of individuals to clean the environment. For altruism to work, the great majority of people must help. Perhaps public education to increase each person's notion of his own self-interest may encourage many individuals to act with care toward the environment.

Garrett Hardin's essay, "The Tragedy of the Commons," found in his book *Exploring New Ethics for Survival,* deals with the problems of relying on altruism for environmental protection.

Similarly, for a few concerned families to limit their number of children in order to curb the population "explosion" is futile. If population control is to be demographically successful, vast numbers of people must practice it.

There is yet another problem in leaving environmental protection to individual goodwill. Often the information necessary to make correct choices is not available. Consider, for example, the seemingly frivolous question of choosing the color of toilet paper to use in your bathroom. Much has been written about the polluting effect of the colored dyes, so environmentally concerned families have been using white toilet paper. They have not been told that this white paper has been bleached, and that the bleach itself causes serious ecological damages. Which is worse, dye or bleach? The ecologically sound product, unbleached and undyed white toilet paper, is simply not for sale.

The arguments against depending upon altruism to clean the environment may strike the reader as coldly hardheaded, and may appear to encourage profligate use of resources. That is not our intent. The point is simple: people who show little concern toward the environment are not punished, and people who act carefully are not rewarded. Since vast numbers of people do not act unselfishly in their daily lives, the environment cannot be cleaned by altruism alone.

Charity is analogous in many ways. Although donors of large sums of money receive little except tax deductions by way of compensation for their generosity, humanitarian and religious feelings lead many to give unselfishly. But many more do not. And the methods of disbursement of charity do not usually reach all pockets of need. One gives to a cause which moves one. A heart-rending story, a dread disease, an ethnic tie, or a sense of guilt may elicit kindly donations, but unglamorous poverty may be neglected. Charity is not enough to help all the needy. Instead, pockets of abject poverty disappear only in societies with effective methods for income redistribution.

Industrial firms cannot be expected to take altruistic action to prevent pollution any more than individuals can. As mentioned in Section 13.1 the costs related to pollution are not reckoned in our present system of social cost accounting. Hence, a company which decides to inaugurate pollution control devices puts itself at a competitive disadvantage. Often, the town in which a firm is located is unwilling to force emission standards upon a company for fear of putting the plant in economic jeopardy, and risking an increase in local unemployment.

Strip mining slag does not normally support vegetative growth (*left*); however, when irrigated with secondary-treated sewage effluent, the lush growth and resulting ground cover (*right*) can reclaim useless land. (From Environmental Science and Technology, October, 1972.)

Even when industry's intention is altruistic, pollution control often falls far short of optimum. Consider a large corporation with a dozen plants producing similar products. Such a corporation's approach to environmental clean-up is often to choose one plant as a prototype, and to develop air and water pollution control devices for that plant. When the systems are finally effective, the devices are installed in the remaining 11 plants. During the development stage, however, the 11 plants spew pollutants at an unabated rate. Moreover, the specific operating conditions within each plant often cause the pollution control devices to be less effective than they were in the prototype plant. In addition, conditions rarely remain constant even within a single plant, for the suppliers of raw materials may change, some processes may be altered, and the products manufactured may adjust to meet the needs of customers. All these changes will usually diminish the effect of pollution control devices. In practice, the incentives to alter such devices tend to be weaker than those which impel changes in the manufacturing process.

Municipalities themselves have been notorious polluters. Towns clean their own drinking water, but often do not treat their sewage. In 1971, for example, only 15 per cent of municipalities in Alabama disposed of their solid wastes in accordance with minimal state requirements. In the same year, the city of Holyoke, Massachusetts, pleaded guilty to a criminal pollution charge. During a labor dispute with waste department employees, the city was dumping raw sewage directly into the Connecticut River, and bypassing its own sewage treatment plant. In general, towns and cities find cleaning the environment very costly. It is notoriously difficult to float bonds or to impose new taxes by public referenda. Pollution control taxes have been particularly unpopular, for they are frequently viewed as a discouragement to growth. They raise costs and discourage growth, and yet dirty air may still be blown in from a neighboring town. (See p. 536 for recent trends.) With effective regional and national planning, many of these problems should assume less importance.

Legal Approaches

Damage suits, injunctions, and legislation provide three avenues of legal recourse. The Anglo-American legal system has developed judicial rules and procedures governing controversies in which one party, the plaintiff, alleges that he has been harmed by another party, the defendant. If the court judges that harm has indeed occurred, the defendant is often ordered to pay the plaintiff money to offset the damage. Sometimes the plaintiff wants not money, but an injunction, a legal order to the defendant requiring him to do something, or to stop doing something. Classical English examples were cases in which a plaintiff sought an injunction against a defendant to prevent the latter from hunting on the plaintiff's land.

When a plaintiff fulfills all necessary requirements to bring his case to court, he is said to have legal **standing**. Standing for damage suits is granted only if an individual's harm is distinguishable from that of the public in general, or, as the United States Supreme Court has recently stated, one is required to have a "personal stake in the outcome of the controversy." The question of standing has rendered the courts inaccessible to many environmental suits. For example, an individual in Chicago could not have sued Union Oil Company for the oil spill in Santa Barbara simply because of his own personal anguish over the killing of birds. However, if he had had a boat anchored in the harbor, he might have been able to sue for damage to his own property. Nor can an individual sue the Department of Interior for destroying a wilderness area by commercializing it, even though part of his taxes are being used, and he is thus being economically "hurt" by the development.

◆ Flast *v.* Cohen. 392 U.S. 83 (1968).

He lacks standing, for his harm cannot be distinguished from that of the public in general.

Sometimes a group is granted standing where an individual is not; some judges are particularly likely to grant standing in environmental cases. Recent statements by some Supreme Court judges suggest that perhaps valleys and mountains, like corporations, can be granted standing. Some cases are heard because they are considered "public interest" cases. For instance, a factory spewing noxious gases into the atmosphere might be sued by a town and legally declared a "public nuisance." Certain legal scholars declare that a pollution-free and healthy environment is one of the unenumerated rights of the Ninth Amendment, which says, "The enumeration in the Constitution, of certain rights, shall not be construed to deny or disparage others retained by the people."

An adversary approach to environmental protection is fraught with problems. The issue of standing is still in flux. Moreover, litigation is very expensive and slow. Another disadvantage is that lawsuits may be addressed to a very small problem. In addition, a given decision is not generally binding on future offenders. Each violation requires a new case. Environmentalist groups may be pitted against an array of giant corporations and government agencies, and an individual case may cost as much as half a million dollars. Often, however, the threat of a suit and adverse publicity will discourage the defendant.

Legislation Designed to Protect the Environment

Laws regulating pollution are very old. Indeed, the first smoke abatement law was passed in England in 1273. Today, in the United States, there are many laws regulating emissions. Perhaps the most dramatic law is the Refuse Act of 1899 which provides fines for polluters of navigable waterways and directs "in the discretion of the court one half of said fine to be paid to the person or persons giving information which shall lead to the conviction." In August, 1972, the Hudson River Fishermen Association was awarded 20,000 dollars for information leading to the conviction of Anaconda Wire and Cable Company, which was dumping toxic discharges into the Hudson River at Hastings, New York. Ten thousand dollars of the award went to a single individual, and many private

LOUIE

(From the New York Daily News, April 14, 1972.)

citizens have since then been enthusiastically searching for other polluters. Laws which regulate emissions, noise, effluents, heat, and other environmental nuisances do not in themselves prevent pollution. Twenty thousand dollars may be a windfall to a fishing club, but it is irrelevant to a huge corporation. Fines are often too small to affect corporate decisions. More important, to prove violation of the law requires a legal suit, which is, of course, expensive and time-consuming. Acceptable laws are extremely difficult to write. For example, one cannot ban destruction of every marsh. Also, weather conditions dictate that acceptable ambient air quality standards for Los Angeles should be different from those for New York. Moreover, acceptable levels vary seasonally, and even daily, within a city.

Emission control standards are very useful for environmental protection. However, sometimes the standards applied are vague. For example, if a factory is discharging sulfuric acid into a river, the relevant government regulatory agency may charge that the emissions shall be decreased to a degree consistent with the "available technology." Although one available step is a decrease in production, what is usually meant is to incorporate control devices. For a given factory, however, the available devices may not be useful. A regulatory agent may find a scrubber, say, in a catalog, and, after reading the description of the device, may demand that it be installed by the offending factory. If the catalog description is inaccurate or overenthusiastic, or if the processes of the factory are slightly different from those for which the scrubber was designed, installation may not effect much abatement of pollution. Moreover, the doctrine that pollution abatement should utilize the best "available technology" often engenders severe corporate anxieties. The "best technology" might be interpreted by some as that which yields the highest degree of abatement regardless of cost, and by others as that which yields the most abatement per dollar. One result of these differences in viewpoint is that industries are often reluctant to publicize all their know-how concerning methods for pollution control. Consequently, government agencies are not always aware of all the available technical alternatives and thus lack information that would make it possible to enforce stricter standards. This suppression of information operates internationally because of intricate international linkages among companies, especially among members of international conglomerates.

Legislative approaches do not all deal with emissions, for the role of the law is varied. For example, increased support of new pollution control technology would also tend to favor environmental repair.

Another legislative route is the establishment of municipal or government-owned recycling centers. The Franklin, Ohio, garbage recycling center, discussed in detail in Chapter 11, is one case. In that city, citizens legislated to assess themselves 25 cents per month to support a commercial garbage-sorting plant. Thus, for a small expense, a large number of people would be assured that their refuse was recycled without great inconvenience to individuals.

Recent electoral trends have tended to encourage some environmental legislation. In November 1972, New York State voters approved a bond issue of 1.15

Red Clay Creek, a small stream in Delaware, has fish living in it for the first time in 50 years. NVF Company, founded as a flour and saw mill in 1763, is located on the Creek; it manufactures 45 per cent of the world's vulcanized fiber. The zinc used in the manufacturing process has in the past been dumped into the creek. In the late 1960's, Delaware declared that the NVF fiber plant failed to meet state requirements for maximum acceptable zinc concentrations. Today, NVF recycles 50,000 pounds of zinc chloride per month at roughly the same price as fresh zinc chloride. (Reprinted with permission from Environmental Science and Technology, Oct., 1972. Copyright by the American Chemical Society.)

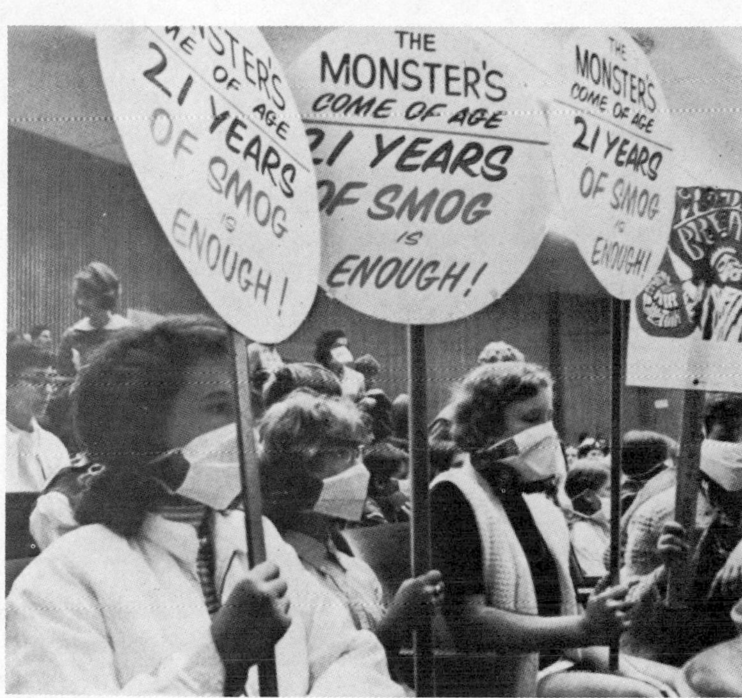

Female students on Los Angeles campus make an appeal for cleaner breathing room. (From All Clear, January, 1970.)

billion dollars for combating air and water pollution, controlling solid waste, and acquiring recreational land. Washington State voters approved 265 million dollars in environmental bonds, and Florida voters approved an issue for 240 million dollars for environmental purposes. Perhaps most important, the "environmental vote" is becoming a decisive political factor. In the 1972 election, the environment was considered by the New York Times to have had an impact in 50 congressional contests. Continuing public pressure will force changes in environmental policy.

13.4 ECONOMIC INCENTIVES FOR POLLUTION CONTROL

(Problems, 7, 8, 10, 11, 12, 13)

Legislation for control of pollution may take the form of economic incentives to persuade polluters to refrain from despoiling the environment. The theory is that by forcing pollution costs to be internalized by the polluter, such legislation would afford cities, towns, industries, and individuals with an economic incentive to protect the environment. Two major economic incentives have been proposed. First is a subsidy for cleaning-up; second is a tax on pollutants. Subsidies would operate as follows. An industrial firm declares that its normal operating procedures would yield a certain amount, say x units, of pollutants. However, by installing anti-pollution devices, the amount of pollution is only y units. Then the government rewards the firm $(x - y)S$ dollars, where S is an amount of subsidy per unit of pollution prevented. Such a system is open to abuse. In many cases, but not all, it may be relatively easy to measure y, the total amount of pollution, but very difficult to estimate x, the amount that would have occurred without controls. It is always to the firm's advantage to overestimate x and to the government's advantage to underestimate x. A corrupt government agent could conspire with the firm he is to regulate and deliberately overstate x. As a reward, he might be given special favors, or even bribes, by the firm he aided. Another criticism of the subsidy approach is that it uses tax money to fight pollution caused by industry.

Another approach is the so-called **pollution tax**, or **residual charge**. Each polluter is charged an amount proportional to the quantity of pollutant emitted. Such a scheme would encourage ecologically sound manufacturing processes, for it levels economic sanctions against processes which pollute. Surveys of industries have generally found that this approach is favored by most firms, for it would simultaneously internalize pollution costs for all companies. Of course, standards for particularly hazardous or toxic materials would still need to be established and enforced. The President's Council of Environmental Quality, many environmental groups, and several members of Congress have endorsed this approach. Its main disadvantage is that, like sales taxes and value-added taxes, a pollution tax is passed on to the consumer in a regressive fashion. Thus, under this system, the poor would ultimately pay a higher proportion of their income for pollution control than would the rich.

Steps taken to conserve energy are subject to regressive effects. For example, a 50c per gallon tax on gasoline would selectively oppress the low income worker driving to his job. On the other hand, gas rationing is designed to distribute the burden equitably among all drivers. However, the administration of a national rationing system is expensive, and prone to abuses.

Ideally, residual charges present the polluter with a choice—either dump waste and pay the fine, or deal with the waste in some other way—treat it, recycle it, store it, or minimize it in manufacturing. If the residual charge is too low, there will be little economic incentive to prevent waste, and pollutants will continue to be dumped untreated into environmental sinks.

Setting reasonable charges will be extremely difficult. A trial-and-error approach might be necessary. Suppose a city wanted to reduce by 80 per cent the rate of entrance of phosphorus into a river. A

This does not refer to the element alone; it includes any compound, such as phosphates, that contains phosphorus.

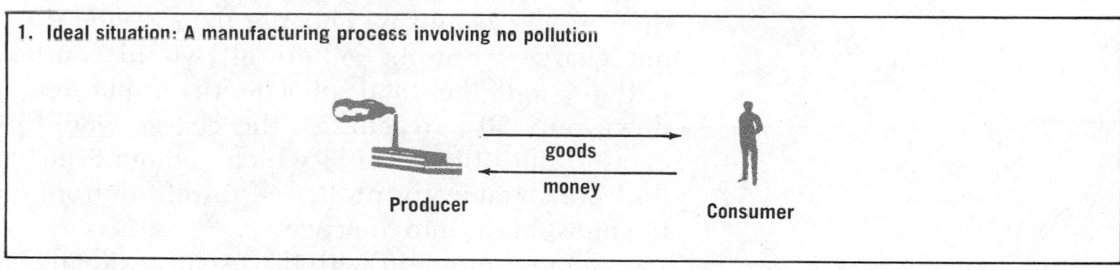

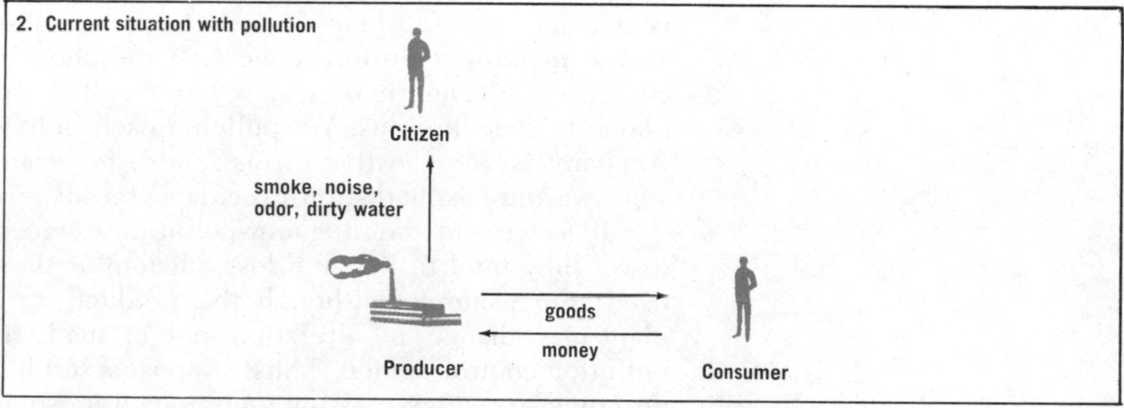

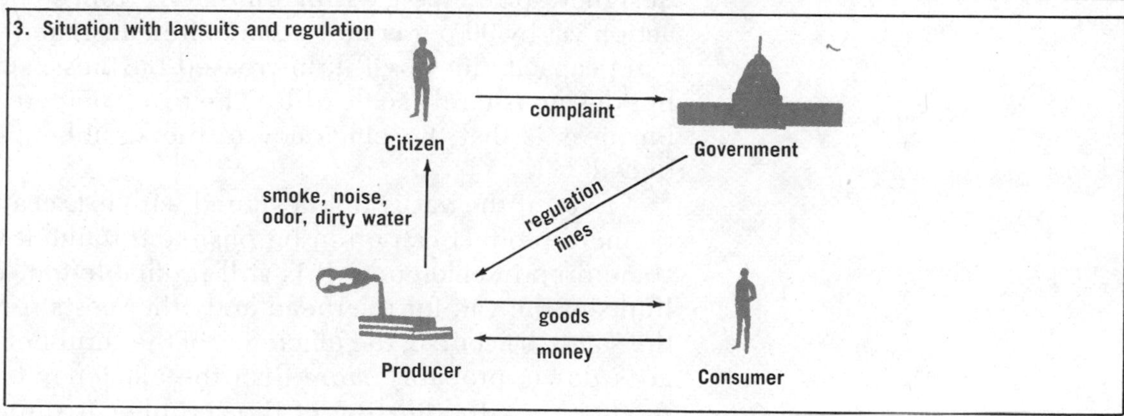

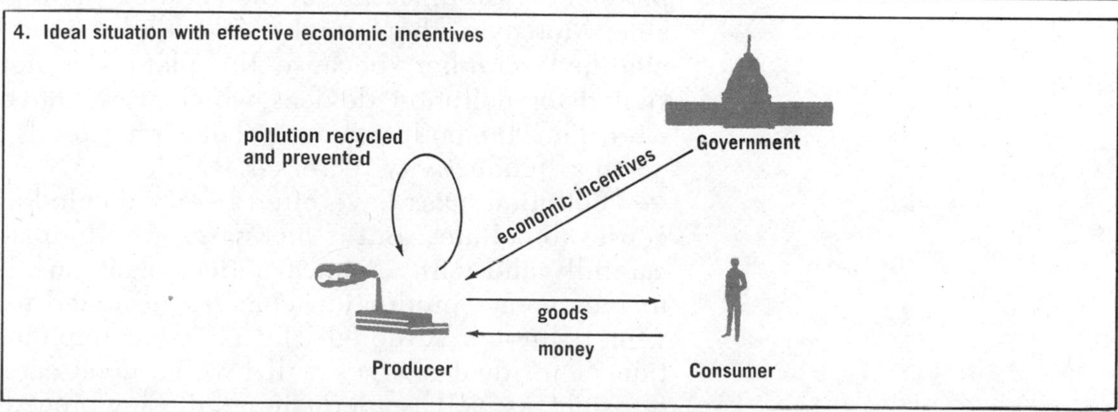

Simple economic models for pollution.

charge would be established. Some firms along the river would find that recycling or otherwise treating their wastes would be cheaper than paying the effluent charges. But the others all would continue to pollute, and the total phosphorus input might be down only 50 per cent. So the charge would be increased until the point at which enough firms would find it economically profitable to refrain from dumping phosphorus into the river.

Another problem with the use of residual charges is that methods must be devised to police effluents and to monitor pollution devices. If the phosphorus content of a river increases, it is difficult to assign blame to specific firms. Yet policing each individual company is very costly. Firms could, for example, store waste phosphates until the day after inspection.

It is wise to monitor anti-pollution devices because they tend to become less efficient as they are used. For example, although the productivity of a plant may change, no alteration may be made in the pollution control device. Thus, suppose a scrubber is designed to reduce existing emissions from some reaction vat by 90 per cent. Vats are often designed with extra capacity for possible increased business; scrubber design is rarely so flexible. The result of increased business is that the efficiency of the scrubber is reduced.

Even if the vat is not designed with extra capacity, more product can often be pushed through it with some drop in efficiency. It is still profitable to tax the limits of the vat, for overhead and other costs remain the same. Of course, the efficiency of the scrubber also goes down, probably *more* than the efficiency of the production. Modification of the scrubber is unprofitable. Moreover, there might not even be room for a modified scrubber, because the plant was not designed for pollution devices which were shoved in later. Thus the pollution control efficiency tends to go down as productivity increases.

Pollution taxes have often been ridiculed as "licenses to pollute," but if such taxes are administered carefully and with advance notice, plants are likely to incorporate production changes designed to prevent pollution. Although the effective implementation of residual charges will involve great care and foresight, as well as adequate monitoring procedures, the idea is supported by many as a general approach to environmental control.

A variation on the pollution tax is the establishment of methods for reducing consumption, and hence, indirectly, production. This approach is much less regressive than the pollution tax itself. Individuals could be given economic incentives, or punishments, to encourage reduction of pollution, waste, and other environmental stresses. Then pollution charges could be levied against individuals. The prices of many consumer goods do not adequately reflect their effect on nature. Three goods which are grossly underpriced if we consider environmental effects are water, electricity, and automobiles. Many of our environmental problems would be alleviated if people refrained from profligate use of water, excessive reliance on electricity, and unnecessary dependence on the automobile. Pollution taxes for these three commodities could greatly decrease their use. For water and electricity, a steeply increasing tax might be levied; the tax should allow a reasonable amount of usage at very low rates, but should tax heavy users severely. Rationing is another approach. If electric costs reflected externalities, individual homemakers might be quite reluctant to use electric heat and air-conditioning, and would rarely use an electric dryer for a very small load. A decrease in per capita demand of electricity would obviate the need for so many new power plants.

Some states, notably New Jersey, have already inaugurated the so-called "smog-tax" for cars. This tax is a residual charge. At its annual inspection, an automobile is evaluated in terms of the amount and quality of the emitted exhaust. Large cars are taxed at a higher rate than small cars; poorly tuned cars and those with faulty anti-pollution devices are not passed; the total tax increases proportionally with increasing annual mileage. Such a program would do much to encourage smaller cars with cleaner-burning engines. More important, if the taxes were steep enough, it might discourage multiple-car families and encourage car pools and alternate modes of transportation.

Of course, such a stringent tax program cannot be inaugurated suddenly. Only in very few places is public transportation an available alternative to the private automobile. Bicycling is dangerous in many parts of the country. The tax burden on the automobile must be increased gradually, and governments must simultaneously encourage the growth of public transportation, and establish bicycle paths and pedestrian malls. Some city planners have suggested establishing new towns in which the automobile is banned; if such towns are popular, existing towns and cities might be willing to experiment with areas free from motor vehicles.

13.5 RADICAL SOCIAL POLICIES LEADING TO POLLUTION CONTROL

The kinds of economic incentives and the legal paths discussed in the preceding sections would not lead to a sudden return of the environment to a former, unsullied state. Nor would they guarantee any improvement at all. Rather, they would probably lead to a lessening in the rate of deterioration of the environment in some areas, a gradual improvement in other regions, and an occasional dramatic improvement. If a government desires rapid improvement, it might opt for radical approaches, those that attack the root of the problem (see Fig. 13.3). Some radical approaches are authoritarian. The most effective way to limit babies is to prevent their birth, for example by compulsory sterilization after the second child, or compulsory vasectomy at age 35. Similarly, imposition of severe punishments for harming the environment would greatly assist the battle for a purer world. For example, Singapore's system of very heavy fines is generally credited with cleaning up the city. A 165-dollar fine, an amount equivalent to six months' pay, is levied against anyone caught throwing a cigarette on the ground. Similar harsh fines are imposed for other environmental infractions. The system of priorities in the U.S. simply does not place high values on the cleanliness of city streets.

At present, most Western nations encourage births through systems of tax deductions, welfare structures, and child-support payments. Family size might be more limited if tax rates increased with increasing family size, or if free education were avail-

FIGURE 13.3 Possible government policies relating to births.

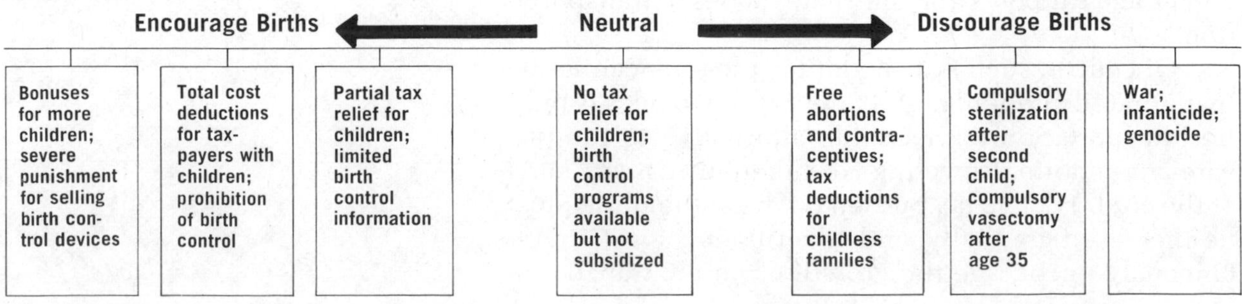

able only to the eldest two, or three, children in a family. Ethically, this kind of approach is suspect. Although punitive tax structures will undoubtedly discourage births, the most severe burden will be placed upon specific children. A fourth child never asked to be born. Why should it be deprived of education because of a parental decision or mistake? All these policies will reduce pollution, but would lead to a loss in individual liberty.

13.6 CASE HISTORY: AUTOMOBILES AND HIGHWAYS

(Problems 17, 18, 19, 22, 24)

Environmentalists often point to automotive transportation as the single most destructive influence on the environment. The automobile itself pollutes the air—smog in cities is largely due to automobile exhaust. Use of a car instead of another mode of transport is inefficient utilization of natural resources, for it represents extravagant use of petroleum products. Cars cause urban and rural noise; old cars are piled in junk yards, scarring the land. And cars bring highways which bring more cars. Highways themselves have a profound effect on housing patterns, for as remote areas become easily accessible to urban centers, suburbs scatter along the highways. Indeed, the recent declines in population of many American cities in the U.S. and the rises in suburban areas are often ascribed to automobiles and highways (see Table 13.3). And yet, even as the problems new highways pose become clearer and more acute, the United States continues to build highways, and spends very little on mass transportation (see Fig. 13.4).

The major source of money for interstate highway building is the Highway Trust Fund, which was established in 1956. The fund receives revenue from Federal taxes on the sale of trucks, tires, and other items related to highways. A four per cent levy on every gallon of gasoline sold in the United States is earmarked for the fund. In addition, the Treasury Department pays about five per cent interest on the unused balance. The money is to be spent solely for highway construction. Some Congressional action has been taken to attempt to divert a small amount of Trust money for mass transport.

Mass transportation systems that could replace a large number of cars would be very expensive to

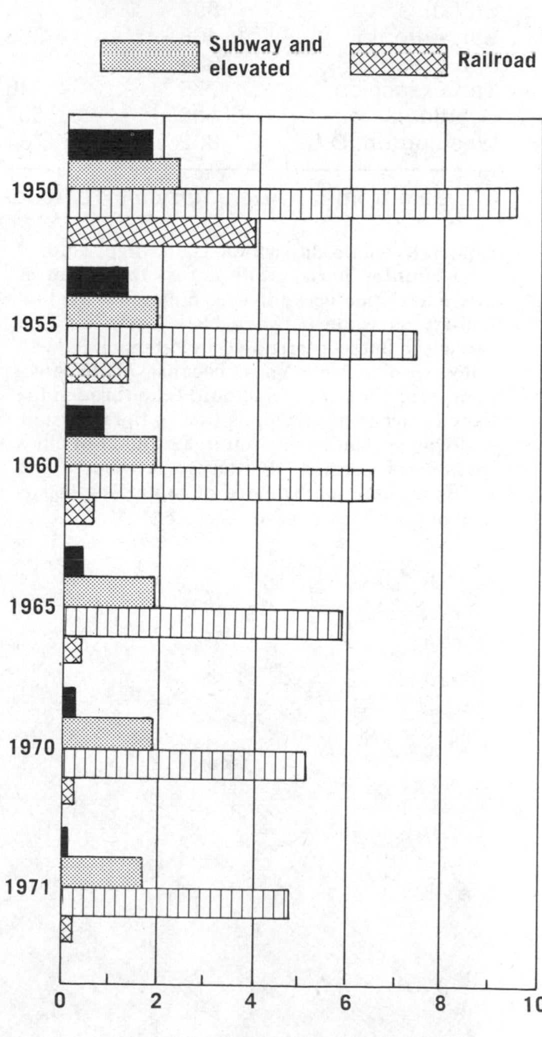

FIGURE 13.4 Passengers carried by public transportation: 1950 to 1971. (Adapted from Statistical Abstracts of the United States, 1972. Data from the American Transit Association, Washington, D.C.)

TABLE 13.3 CENSUS POPULATIONS IN 1950, 1960, AND 1970 OF ALL AMERICAN CITIES WITH POPULATIONS GREATER THAN 500,000 IN 1960

	POPULATION OF CENTRAL CITY (in thousands)			*POPULATION OF METROPOLITAN AREA (in thousands)*		
CITY	*1950*	*1960*	*1970*	*1950*	*1960*	*1970*
Baltimore	950	939	906	1405	1804	2045
Boston	801	697	641	2410	2595	2730
Buffalo	580	533	463	1089	1306	1334
Chicago	3621	3550	3367	5178	6220	6893
Cincinnati	504	503	453	904	1268	1373
Cleveland	915	876	751	1466	1909	2043
Dallas	434	680	844	743	1119	1539
Detroit	1850	1670	1511	3016	3762	4164
Houston	596	938	1233	807	1418	1958
Los Angeles	1970	2479	2816	4368	6039	6974
Milwaukee	637	741	717	957	1279	1393
New Orleans	570	628	593	685	907	1034
New York	7892	7782	7895	9556	10695	11448
Philadelphia	2071	2003	1949	3671	4343	4777
Pittsburgh	677	604	520	2213	2405	2384
St. Louis	857	750	622	1719	2105	2331
San Antonio	408	588	654	500	716	864
San Diego	334	573	697	557	1033	1318
San Francisco	775	740	716	2241	2649	3070
Seattle	468	557	531	845	1107	1404
Washington, D.C.	802	764	757	1464	2077	2836

Data from *U.N. Demographic Yearbook*, 1970, and *Statistical Abstract of the United States, 1972.*

build. Existing trains and subways are often in a state of terrible disrepair, and many areas are entirely lacking in effective public transport. Many of the former rights-of-way have been, and continue to be, sold. Few public systems could be expected to operate at a profit, and, under existing laws, losses would

Interstate Route 80, which is the light-colored road running horizontally across the center of this aerial photograph, was planned as a limited-access route between New York and San Francisco but is interrupted in Paterson, N.J., 10 miles west of New York, because of disagreement over the course it should take through the heavily urban area. Opposition to the interstate highway system has arisen in a number of cities because of conflict of interest between urban residents and convenience of motorists. (Courtesy of Soil Conservation Service.)

Some alternate modes of transportation. (Right: From Source Catalog—Communities/Housing, Chicago: The Swallow Press, Inc., 1972. Left: © David Plowden: Farewell to Steam. Brattleboro, Vermont, The Stephen Greene Press, 1966.)

have to be borne by local communities. On the other hand, highways are paid for by Federal, State, and city governments, and a highway is not expected to be profitable.

On June 15, 1973, as this book was going to press, the Environmental Protection Agency urged a sharp limit to the use of private automobiles in urban centers. Strong Federal and local support of these recommendations is needed for improvement of urban environments.

13.7 ECONOMIC PROBLEMS OF POLLUTION CONTROL IN DEVELOPING NATIONS

In the few developed nations that share most of the wealth and industry of the world, a decision to curtail economic expansion will not greatly affect the large majority of inhabitants. On the other hand, a developing nation, striving to industrialize and to obtain a larger share of international trade, is faced with an agonizing dilemma. Rapid industrialization and growth lead not only to rapid increases in personal wealth, but also to serious environmental degradation if no pollution controls are established. Controls are, unfortunately, expensive to install and operate. Mexico City in the early 1950's was clean and relative-

"I KNEW THERE'D BE A CATCH IN IT WHEN THEY SAID THE MEEK WOULD INHERIT THE EARTH"

(By John Fischetti, Courtesy of Publishers-Hall Syndicate.)

ly quiet, but poverty was everywhere evident. Today automobile traffic has contributed to making the city terribly noisy, and a blanket of air pollution from industry, automobiles, and dust hovers over the city. However, the desperately poor have become less obvious. Could the standard of living have risen so fast had care been taken to protect the environment of the city? Should a decision have been made to keep the air clean, and limit economic growth? Probably no active decisions or policies were set. The environment was simply not considered during the period of rapid industrialization. Those who live in smelly, noisy cities, or those who have seen rural areas deteriorate may feel the environment should be protected at all costs; those who are poor may want to share in the general wealth before cleaning the air.

13.8 CONCLUSION

(Problems 20, 21)

The physical character of our planet is changing as forests, jungles, and deserts become farms and cities, and as clear living spaces become polluted. Environmental degradation is rapid, even in the most remote parts of the world. Missoula, Montana, in "Big Sky Country," has a serious air pollution problem; on the Rat River, 200 miles north of the Arctic Circle in the Canadian wilderness, brush bridges and oil cans mar the muskeg, and Antarctica boasts its own unsightly garbage dumps. Just as the residents of the shore of Lake Erie adapt to a polluted world, we all are forced to make changes in our styles of life as the environment continues to deteriorate.

The degradation of the environment continues to accelerate in spite of the increasing current interest in pollution control. Pollution from manufacturing may decrease in some countries, although in some areas the trend is bleak. On the other hand, degradation arising from existing patterns of land use and housing, current modes of transportation, and affluent habits of living cannot reverse themselves quickly even if there were general agreement that radical changes are necessary. Changes in patterns of living for masses of people require technological and engineering skill, time, and money. For example, the Interstate Highway System is here. We cannot abandon our thousands of miles of concrete, both because abandonment would be wasteful and because we have no effective alternative ways of moving people and goods at the present time. Similarly, millions of people live in suburbs, which we now realize use space inefficiently, and require automobile transportation. We cannot move everyone back into cities and villages and start again. Nor can we easily build remedial measures like bicycle paths, sidewalks, and neighborhood stores, for all are very expensive. Urban centers, too, are trapped by former building patterns. A May, 1973 report to the city of New York proposes an aerial "people mover" to relieve congestion in the streets of Lower Manhattan. A three-mile loop would cost roughly 90 million in 1972 dollars and would take five and a half years to build. Even simple changes are costly and difficult. Encouraging walking up flights of stairs rather than riding elevators requires an architectural design that makes stairwells pleasant and accessible. (Go to the stairwell of a relatively new elevator building. Look at it, smell it, and try to decide what improvements would have to be made before you would choose it over the quiet, carpeted elevator.)

The consequences of widespread environmental problems continuing into the twenty-first century are unknown. Some pessimists predict dire catastrophe: crime, starvation, war, anarchy, and a drastic deterioration of man's life. Technological optimists predict large, contented populations housed in magnificent indoor cities with gymnasiums, movie theaters, swimming pools, and zoos. A middle view predicts more subtle and gradual changes. Life in a particular city may become so unpleasant that industry and residents will leave, and the area will suffer a local depression. The industrial flight will lead to cleaner air, but increased poverty. And the pollutants will move to the country. Or a nation will be unable to control its rapid population rise, and will experience rising death rates, malnutrition, and emigration of the healthy young. Eventually, these factors will decrease the growth rate. Such a view anticipates pockets of environmental destruction and wide areas of degradation, but not the sudden collapse of humanity or an idyllic technological world.

Environmental limits on the survival of man have been discussed by many studies. If air pollutants become too severe, some people will be poisoned as a result. Too large a population brings starvation to some. Unsanitary disposal of wastes will lead to deaths from waterborne disease. Yet, the limits within which man can live are constantly being redefined by a changing technology. Lethal levels of air pollution do not kill people wearing gas masks. Food manufactured from carbon dioxide, water, and the energy from fusion reactors could support many billions of people on earth. If medicine continues its rapid advancement, health may be maintained in an unsavory environment. Artificial food may have caloric and nutritive values, but unless food technology becomes much more sophisticated, it is unlikely to yield culinary delights. Today, hungry people are living in crowded, dirty, polluted conditions, but the limits on existence have not been reached, for such populations continue to increase. We have to ask not what are the limits to existence, but what qualities of human life are precious enough to be protected by legislative, legal, and personal actions.

PROBLEMS

1. **Vocabulary.** Define economic externality.
2. **Economic externalities.** On February 26, 1972, heavy rains destroyed a dam across Buffalo Creek in Logan County, West Virginia. The resulting flood killed 75 people and rendered 5000 homeless. The dam was built from unstable coal-mine refuse or "spoil." A United States Geological Survey had warned of the instability of many dams in Logan County, especially the one at Buffalo Creek. Soil stabilization and reclamation programs directed at mine spoil dams would have added to the cost of producing coal. After the flood, the

Bureau of Mines denied responsibility for corrective action, contending that although mine spoil is within the Bureau's jurisdiction, dams are not. (a) How would you relate the concept of economic externalities to this tragedy? (b) Suggest legislation designed to prevent future Appalachian dam disasters. (c) How will your legislation affect the price of coal and its competitive position in the fuel market?

3. **Economic externalities.** What information would you need to assess whether a cotton or a cotton-polyester shirt is more ecologically sound? In your answer, consider that pesticides are used to grow cotton, that the spinning of both natural and artificial fibers has environmental side effects, and that electricity is needed to iron the shirt. The usual blend in no-iron shirts is 35 per cent cotton, 65 per cent polyester. What do you think were the factors entering into this choice of blend? How would you establish the composition of a shirt that minimizes harm to the environment? How could your dress behavior affect this choice?

4. **Vocabulary.** Distinguish between micro- and macroeconomic effects.

5. **Pollution tax.** Why is it difficult for firms with small profits to borrow money for pollution control?

6. **Vocabulary.** Define price elasticity and inelasticity.

7. **Price elasticity.** The following consumer items involve environmental degradation of some kind: (a) electricity; (b) leopard coats; (c) luxury cars; (d) compact cars; (e) soda in non-returnable bottles; (f) paper napkins. How does the manufacture, use, or disposal of each harm the environment? Are the prices of each elastic or inelastic? What kinds of legislation would be needed to prevent the harm?

8. **Price elasticity.** On page 528, it was stated that only some of the effects of pollution control will be eventually passed to the consumer. For each of the six items in Problem 7, suggest a product that may be substituted. How is ease of substitution related to elasticity of demand for these items? (Can you generalize?) How is price elasticity related to the likelihood that pollution control costs can be passed on to the consumer?

9. **Usage of utilities.** Most gas companies charge less per cubic foot as the amount used increases. A typical rate schedule for a Northeastern city is:

Hundreds of cubic feet per month	*Cents per 100 cubic feet*
5.0 or less	minimum bill
Next 10	17.0
Next 15	15.0
Next 570	13.2
Next 5400	10.6
Next 94,000	9.5
Next 100,000	9.0
Over 200,000	8.5

How do you think this rate structure influences use of gas in homes? (Roughly 300 cubic feet of gas are sufficient for the monthly operation of a gas stove for a family of four.)

10. **Usage of utilities.** About 10 per cent of the gas consumed by a household kitchen stove with a pilot light is used to maintain the pilot light. Water that is simmering is just as hot as water that is boiling vigorously. An egg covered by half a pint of boiling water will harden just as rapidly as in a quart. Outline specific guidelines for a family of four that might cut its consumption of natural gas in half.

11. **Altruistic approaches.** Suggest one action you personally might take to help preserve the environment. How do you assess the importance of your own action? How would you estimate the number of people who would have to join you before the action would have a significant environmental impact?

12. **Damage suits.** How can legal suits against polluters be used? What are the drawbacks of depending on litigation as the primary social tool for environmental protection?

13. **Legislation.** Discuss the following alternative proposals for reducing water consumption. (a) Tax all water usage at a constant rate per gallon. (b) Tax water usage at a progressive rate, that is, allow the tax to rise with increasing use. (c) Shut off all water from 2 to 4 P.M. (d) Shut off all water from 2 to 4 A.M. (e) Ration water at some reasonable level. What is the purpose of the legislation? What are the side effects? Can you propose better legislation?

14. **Public policy.** If you were the mayor of a small town, and a prosperous factory, the largest single employer in the town, were illegally dumping untreated wastes into a stream, what action would you recommend? What if the factory were barely profitable?

15. **Vocabulary.** Define residual charge.

16. **Residual charge.** Do you think a pollution tax on automobile mileage should be proportional to the miles driven, or progressive? Should miles driven on each car be used, or miles per family? Can you think of a progressive tax structure that would encourage more than one car?

17. **Residual charge.** Explain why automobiles will probably become considerably more expensive as pollution controls are enforced. Why will paper prices not rise so steeply with increased pollution control?

18. **Highway construction.** Park land and slum neighborhoods are the two favorite land corridors used for highways. (a) Why are these areas desirable to highway builders? (b) What type of land do you think should be

used? Remember, the more expensive the land, the higher the costs of the highway, and the more tax money must be used.

19. **Automobiles.** Suppose legislation were introduced banning cars in the central shopping area of your town or city. Would you support the legislation? What advantages and disadvantages do you see for your city or town?

20. **Environmental policy.** Rendering plants convert animal wastes into useful products. However, the conversion is very smelly, unless costly odor-control systems are installed. Would you recommend that air pollution ordinances grant special exemptions to rendering plants? Why, or why not?

21. **Housing.** Compare the environmental impact of single-family homes with other types of housing.

22. **Highways.** The recent rapid growth of Vermont as a ski resort and a summer retreat has been called a byproduct of the Interstate Highway System. Discuss.

23. **Recycling.** On May 9, 1973, the *Wall Street Journal* reported that there is a market for only a portion of the materials returned to recycling centers. Most recycled products, with the notable exception of newsprint, cost more to manufacture than the virgin products. For instance, recycled glass costs as much as $23.77 a ton to produce, while the virgin product costs only $18.43. Suggest legislation that would encourage the use of all returned glass bottles. Do you think that people should continue to bring paper and glass to recycling centers even though much of the collected material is currently being discarded?

24. **Traffic.** Proposals to reduce urban air pollution invariably include measures to control automobile use. Do you feel that some of the problems of the city would be alleviated if highway tolls and gasoline taxes were increased and if center city parking lots were eliminated? Why or why not? What additional measures can you suggest to ease problems of urban transportation?

REFERENCES

A useful annotated bibliography covering general economic and social issues in environmental control is:

U.S. Department of Housing and Urban Development: *Environment and the Community: An Annotated Bibliography,* 1971. 66 pp.

In addition to the references there, two more recent well-written and provocative books discussing economy and environment are:

Gerald Garvey: *Energy, Ecology, Economy.* New York, W. W. Norton & Co., 1972. 235 pp.

Edwin G. Dolan: *TANSTAAFL: The Economic Strategy for Environmental Crisis.* New York, Holt, Rinehart, & Winston, 1971. 113 pp.

For economic projections, the following volume is useful:

Council on Environmental Quality, Department of Commerce, and Environmental Protection Agency: *The Economic Impact of Pollution Control,* 1972. 332 pp.

An engaging volume on social policies for the environment is:

Garrett Hardin: *Exploring New Ethics for Survival.* New York, Viking Press, 1972. 273 pp.

A guide to actions individuals may take to protect the environment is:

Paul Swatik: *A User's Guide to the Protection of the Environment.* New York, Ballantine Books, 1971. 312 pp.

A book which has had much impact in persuading others that nature cannot support mankind much longer unless patterns of growth and consumption change is:

Donella H. Meadows, Dennis L. Meadows, Jørgen Randers, and William W. Behrens, III: *The Limits to Growth.* New York, Universe Books, 1972. 205 pp.

A volume dealing with current social issues is:

Michael Micklin: *Population, Environment, and Social Organization: Current Issues in Human Ecology.* Hinsdale, Ill., Dryden Press, 1973.

For a survey of the ways in which State, Federal, and international bodies have responded to the growth of environmental problems, see:

Thomas W. Wilson, Jr.: *International Environmental Action—A Global Survey.* Cambridge, Mass., Dunellen, 1971. 382 pp.

APPENDIX

A.1 THE METRIC SYSTEM (INTERNATIONAL SYSTEM OF UNITS)

Systems of measurement are used for the acquisition of knowledge of quantities in terms of standard units. The metric system, now used internationally in science, was originally established by international treaty at the Metric Convention in Paris in 1875 and has since been extended and improved. In this book, we are concerned with four of the fundamental metric units. These are:

Quantity	*Unit*	*Abbreviation*
length	meter	m
mass	kilogram	kg
time	second	sec
temperature	degree	°C

Larger or smaller units in the metric system are expressed by the following prefixes:

Multiple or Fraction	*Prefix*	*Symbol*
1000	kilo	k
1/100	centi	c
1/1000	milli	m
1/1,000,000	micro	μ

Length

The **meter** was once defined in terms of the length of a standard bar; it is now defined in terms of wavelengths of light. A meter is about 1.1 yards.

One **centimeter**, cm, is 1/100 of a meter, or about 0.4 inches.

The unit commonly used to express sizes of dust particles is the **micrometer,** μm, also called the **micron.** There are one million micrometers in a meter, or about 25,000 micrometers per inch.

Mass

The **kilogram,** kg, is the mass of a piece of platinum-iridium metal called the Prototype Kilogram Number 1, kept at the International Bureau of Weights and Measures, in France. It is equal to about 2.2 pounds.

One gram, g, is 1/1000 kg.

Volume

Volume is not a fundamental quantity; it is derived from length.

One **cubic centimeter,** cm^3, is the volume of a cube whose edge is 1 cm.

One **liter** = 1000 cm^3.

Temperature

If two bodies, A and B, are in contact, and if there is a spontaneous transfer of heat from A to B, then A is said to be *hotter* or at a higher temperature than B. Thus, the greater the tendency for heat to flow away from a body, the higher its temperature is.

The Celsius (formerly called Centigrade) temperature scale is defined by several fixed points. The most commonly used of these are the freezing point of water, 0°C, and the boiling point of water, 100°C.

The Fahrenheit scale, commonly used in medicine and engineering in England and the United States, designates the freezing point of water as 32°F and the boiling point of water as 212°F.

Energy

The metric unit of energy is the **joule,** which is defined in terms of physical work. The energy of one joule can lift a weight of one pound to a height of about nine inches. An **erg** is a much smaller unit; there are 10 million ergs in a joule.

The unit commonly used to express heat energy, or the energies involved in chemical changes, is the **calorie,** cal. One calorie is about 4.2 joules. The energy of one calorie is sufficient to warm one gram of water 1°C.

The **kilocalorie,** kcal, is 1000 calories. This unit is also designated Calorie (capital C), especially when it is used to express food energies for nutrition.

A table of units of energy and power appears on page 166.

A.2 CHEMICAL SYMBOLS, FORMULAS, AND EQUATIONS

Atoms or elements are denoted by symbols of one or two letters, like H, U, W, Ba, and Zn.

Compounds or molecules are represented by formulas that consist of symbols and subscripts, sometimes with parentheses. The subscript denotes the number of atoms of the element represented by the symbol to which it is attached. Thus H_2SO_4 is a formula that represents a molecule of sulfuric acid, or the substance sulfuric acid. The molecule consists of two atoms of hydrogen, one atom of sulfur, and four atoms of oxygen. The substance consists of matter that is an aggregate of such molecules. The formula for oxygen gas is O_2; this tells us that the molecules consist of two atoms each.

Chemical transformations are represented by chemical equations, which tell us the molecules or substances that react and the ones that are produced, and the molecular ratios of these reactions. The equation for the burning of methane in oxygen to produce carbon dioxide and water is:

$$CH_4 + 2O_2 \rightarrow CO_2 + 2H_2O$$

Each coefficient applies to the entire formula that follows it. Thus $2H_2O$ means $2(H_2O)$. This gives the following molecular ratios: reacting materials, two molecules of oxygen to one of methane; products, two molecules of water to one of carbon dioxide. The above equation is balanced because the same number and kinds of atoms, one of carbon and four each of hydrogen and oxygen, appear on each side of the arrow.

The atoms in a molecule are held together by chemical bonds. Chemical bonds can be characterized by their length, the angles they make with other bonds, and their strength (that is, how much energy would be needed to break them apart).

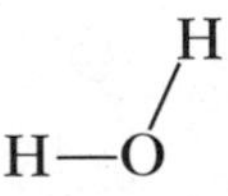

The formula for water shows that the molecule contains two H–O bonds. The length of each bond is about 1/10;000 of a micrometer, and the angle between them is 105°. It would require about 12 kcal to break all of the bonds in a gram of water. These bonds are strong, as chemical bonds go.

In general, substances whose molecules have strong chemical bonds are stable, because it is energetically unprofitable to break strong bonds apart and rearrange the atoms to form other, weaker bonds. Therefore, stable substances may be regarded as chemically self-satisfied; they have little energy to offer, and are said to be energy-poor. Thus, water, with its strong H–O bonds, is not a fuel or a food. The bonds between carbon and oxygen in carbon dioxide, CO_2, are also strong (about 1.5 times as strong as the H–O bonds of water), and CO_2 is therefore also an energy-poor substance.

In contrast, the C–H bonds in methane, CH_4, are weaker than the H–O bonds of water. It is energetically profitable to break these bonds and produce the more stable ones in H_2O and CO_2. Methane is therefore an energy-rich substance and can be burned to heat houses and drive engines.

A.3 ABBREVIATIONS

calorie, cal
Calorie, Cal (see also kilocalorie)
chronic obstructive pulmonary disease, COPD
cubic centimeter, cm^3
cycles per second, cps
decibel, dB
degree Celsius (Centigrade), °C
degree Fahrenheit, °F
deoxyribonucleic acid, DNA
foot, ft
gram, g
Hertz, Hz
hour, hr
inch, in
intrauterine device, IUD
kilocalorie, kcal (same as Calorie with a capital C)
kilogram, kg
phenylketonuria, PKU
pound, lb
ppb, parts per billion
ppm, parts per million
second, sec
total fertility ratio, TFR

LOGARITHMS TO THE BASE 10 (THREE PLACES)

N	*0*	*1*	*2*	*3*	*4*	*5*	*6*	*7*	*8*	*9*
1	000	041	079	114	146	176	204	230	255	279
2	301	322	342	362	380	398	415	431	447	462
3	477	491	505	519	532	544	556	568	580	591
4	602	613	623	634	644	653	663	672	681	690
5	699	708	716	724	732	740	748	756	763	771
6	778	785	792	799	806	813	820	826	833	839
7	845	851	857	863	869	875	881	887	892	898
8	903	909	914	919	924	929	935	941	945	949
9	954	959	964	969	973	978	982	987	991	996

TABLE OF RELATIVE ATOMIC WEIGHTS (1969)*

Element	*Symbol*	*Atomic Number*	*Atomic Weight*	*Element*	*Symbol*	*Atomic Number*	*Atomic Weight*
Actinium	Ac	89		Mercury	Hg	80	200.59
Aluminum	Al	13	26.9815	Molybdenum	Mo	42	95.94
Americium	Am	95		Neodymium	Nd	60	144.24
Antimony	Sb	51	121.75	Neon	Ne	10	20.179
Argon	Ar	18	39.948	Neptunium	Np	93	237.0482
Arsenic	As	33	74.9216	Nickel	Ni	28	58.71
Astatine	At	85		Niobium	Nb	41	92.9064
Barium	Ba	56	137.34	Nitrogen	N	7	14.0067
Berkelium	Bk	97		Nobelium	No	102	
Beryllium	Be	4	9.01218	Osmium	Os	76	190.2
Bismuth	Bi	83	208.9806	Oxygen	O	8	15.9994
Boron	B	5	10.81	Palladium	Pd	46	106.4
Bromine	Br	35	79.904	Phosphorus	P	15	30.9738
Cadmium	Cd	48	112.40	Platinum	Pt	78	195.09
Calcium	Ca	20	40.08	Plutonium	Pu	94	
Californium	Cf	98		Polonium	Po	84	
Carbon	C	6	12.011	Potassium	K	19	39.102
Cerium	Ce	58	140.12	Praseodymium	Pr	59	140.9077
Cesium	Cs	55	132.9055	Promethium	Pm	61	
Chlorine	Cl	17	35.453	Protoactinium	Pa	91	231.0359
Chromium	Cr	24	51.996	Radium	Ra	88	226.0254
Cobalt	Co	27	58.9332	Radon	Rn	86	
Copper	Cu	29	63.546	Rhenium	Re	75	186.2
Curium	Cm	96		Rhodium	Rh	45	102.9055
Dysprosium	Dy	66	162.50	Rubidium	Rb	37	85.4678
Einsteinium	Es	99		Ruthenium	Ru	44	101.07
Erbium	Er	68	167.26	Samarium	Sm	62	150.4
Europium	Eu	63	151.96	Scandium	Sc	21	44.9559
Fermium	Fm	100		Selenium	Se	34	78.96
Fluorine	F	9	18.9984	Silicon	Si	14	28.086
Francium	Fr	87		Silver	Ag	47	107.868
Gadolinium	Gd	64	157.25	Sodium	Na	11	22.9898
Gallium	Ga	31	69.72	Strontium	Sr	38	87.62
Germanium	Ge	32	72.59	Sulfur	S	16	32.06
Gold	Au	79	196.9665	Tantalum	Ta	73	180.9479
Hafnium	Hf	72	178.49	Technetium	Tc	43	98.9062
Helium	He	2	4.00260	Tellurium	Te	52	127.60
Holmium	Ho	67	164.9303	Terbium	Tb	65	158.9254
Hydrogen	H	1	1.0080	Thallium	Tl	81	204.37
Indium	In	49	114.82	Thorium	Th	90	232.0381
Iodine	I	53	126.9045	Thulium	Tm	69	168.9342
Iridium	Ir	77	192.22	Tin	Sn	50	118.69
Iron	Fe	26	55.847	Titanium	Ti	22	47.90
Krypton	Kr	36	83.80	Tungsten	W	74	183.85
Lanthanum	La	57	138.9055	Uranium	U	92	238.029
Lawrencium	Lr	103		Vanadium	V	23	50.9414
Lead	Pb	82	207.2	Xenon	Xe	54	131.30
Lithium	Li	3	6.941	Ytterbium	Yb	70	173.04
Lutetium	Lu	71	174.97	Yttrium	Y	39	88.9059
Magnesium	Mg	12	24.305	Zinc	Zn	30	65.37
Manganese	Mn	25	54.9380	Zirconium	Zr	40	91.22
Mendelevium	Md	101					

*Based on the assigned relative atomic mass of $^{12}C = 12$.

INDEX

Key to abbreviations:

ff = following pages
fig = figure
ftn = footnote
mrg = margin
prb = problem
tbl = table

ABO blood group, 66, 67 mrg
Abortion, induced, 149, 150
Achondroplastic dwarfism, 63
Acidity, 407 ff
Acoustical materials, 507
Activated carbon, 380–381
Adaptation, 55
  by learning, 71 ff
  genetic, 56, 75
Adiabatic lapse rate, 350
Aerobiosis, 413
Aerosol, 347
Agricultural wastes, 250–251
Agriculture
  disruptions of, 219–227
  energy flow in, 227–230, 229 fig
  fertilization in, 230–236, 231 fig
  irrigation in, 237–239, 238 fig
  mechanization in, 236–237
Air, 335
  composition of, 339, 339 tbl
  pure, 339 ff
Air pollutants, 335 ff, 339 ff
  gaseous, 340 ff
  particulate, 346 ff
  primary, 342
  secondary, 342
Air pollution
  acute effects, 364 ff
  aesthetic effects, 375
  from automobiles, 383 ff
  chronic effects, 367 ff
  costs of, 519, 519 tbl
  effects of, 354 ff
  effects on animals, 371–372
  effects on climate, 356 ff
  effects on human health, 361 ff
  effects on materials, 372, 372 fig
  effects on vegetation, 370–371
Air pollution control, 377 ff
  by adsorption, 379–380
Air pollution control (*Continued*)
  by cyclones, 378, 378 fig
  by electrostatic precipitators, 379, 379 fig
  by filtration, 378, 378 fig
  indoor, 381 ff
  by oxidation, 381
  by scrubbers, 379, 379 fig
Air pollution indices, 373 ff
Air quality standards, 374 ff, 375, 376 tbls
Alaska pipeline, 178–181
Albedo, 357
Algae, 10 ff, 421 ff
  as food, 248–249
Amensalism, 34
Amino acids, 62, 62 fig, 65 mrg
Anaerobic bacteria, 415
Anaerobiosis, 414–415
Aquaculture, 247–248
Arithmetic growth, 110, 110 tbl, 111 fig
Aswan High Dam, 225–226
Atmosphere, 335
Atmospheric inversion, 351 ff, 351 fig
Atom, 293–294
Atomic nucleus, 292–293
Atomic number, 293 mrg
Atomic structure, 292–294
Atomic weights, 557
Automobile bodies, 469–471, 469 fig, 470–471 tbl, 471 fig
Automobile exhaust, 386 fig, 387 fig
  cost of control, 389 fig, 525 tbl
Automobiles, 383 ff
  alternate engines for, 390–391
  and highways, 543 ff
  electric, 391 fig, 391 ff
Autotrophs, 4, 5 fig
Avogadro, A., 337 mrg
Balanced polymorphism, 71
Barnacles, 30, 67
Becquerel, H., 295
Benthic species, 21, 22 fig
Biochemical oxygen demand, 415 ff, 416 fig
Biodegradability, 423
Biomass, 8
Biomes, 24
Biosphere, 11, 20
Birth cohort, 120
Birth rate, 125 ff
  age and sex specific, 129 fig
  crude, 115, 117 fig, 126, 127 tbl
  intrinsic, 158 prb
Blood groups, 66, 67 mrg
Bloom (population), 27, 39
BOD, *see* biochemical oxygen demand.
Boll weevil, 261 fig
Borlaug, N., 274 mrg

Calhoun, J., 141
Calorie, 161 mrg
Carbon cycle, 13 ff, 14 fig
Carbon dioxide, 340, 410
Carbon monoxide, 339 tbl, 341, 386
Carcinogenic hydrocarbons, 348
Carson, Rachel, 272 mrg
Caustic, 177 mrg
Cesspools, 438 mrg
Chain reaction, 297–298, 299 fig
Chelates, 233
Chemical symbols, 554
Chimneys, 353–354, 353 fig
Cholera, 435
Chromosomes, 59 fig, 60
Chronic obstructive pulmonary disease, COPD, 368 ff

Cigarette smoking, 366–367
Climax, 43, 45, 45 mrg
Coal mining, environmental problems, 175–177
Coal production, 172 fig
Codling moth, 277 fig
Cohort, birth, 120
Colloidal particles in water, 406, 406 fig
Commensalism, 36
Community interactions, 36
Competition, 34
Composting, 474–475, 474 mrg
Consumers, 6 fig, 7 fig
  primary, 4
Contraception, 150
Control of population, 150
Cooling towers, 207–208
Corn earworm, 275 fig
Critical mass, 308
Crude birth rate, 115, 117 fig
Crude death rate, 115, 117 fig, 119
Crude reproductive rate, 116
Cycles, *see* specific subjects, such as carbon cycle, mineral cycle, etc.
Cyclic disruptions, 39
Cyclones, 378, 378 fig

Darwin, Charles, 68 mrg
dBA scale, 498, 498 fig
DDD, 269
DDT, 259, 265–266, 270 fig, 271 fig, 271–274
Death rates, 125 tbl
  age and sex specific, 119, 121 fig, 126 fig, 129 fig
  crude, 115, 117 fig, 119
  intrinsic, 158 prb
Decay process, 7
Decibel scale, 493 ff, 496 fig, 498 fig
Demineralization, 446
Demographic data, accuracy of, 135
Demographic transition, 144
Demographic trends, in the United States, 138 ff
  worldwide, 145, 146 fig
Demography, 112 ff
Denitrification, 411
Deoxyribonucleic acid, 60, 60 fig, 61, 61 fig, 323
Deserts, 26, 27 fig
Detergents, 421 ff, 422 mrg
Detritus, 7
Detritus feeders, 7
Deuterium, 317 tbl
Diabetes, 77, 77 mrg
Disorders, genetic, 63 ff
Disruptions, cyclic and sporadic, 39
Distribution of population, 115
DNA, *see* deoxyribonucleic acid.
Dominance, 58 mrg
Donora, Pa., 364–365
Doubling time, 128
Dumps, open, 461–462, 461 fig
Dust Bowl, 221

Ear, 491–492, 492 fig
Ecological niche, 28
Ecology, operational definitions, 48 mrg
  population, 112
Economic externalities, 518 ff
Ecosystem, 1
  and natural balance, 17 ff
  definition, 48 mrg
  energy relationships within, 4
  estuary, 23
  freshwater, 24
  ocean, 21 ff
  role of people in, 47 ff
  stable, 18 ff, 18 fig
  terrestrial, 8, 24
Ecosystem homeostasis, 17
Ecotones, 24, 47
Electrical appliances, 210 tbl
Electricity, consumption of, 195–197, 195 fig
  taxation of, 528
  transmission of, 197–198
Electrostatic precipitators, 379, 379 fig
Elements, 12, 12 mrg
Endangered species, case histories, 96 ff
Endrin, 267–268
Energy, 2 ff, 3 tbl, 4 fig, 554
  domestic and commercial consumption, 191–195
  flow of, 8 fig, 9, 10 fig
  from burning garbage, 186
  geothermal, 184
  hydroelectric, 183–184
  production of, in the U.S., 171 fig
  solar, 183
  tidal, 185
  units of, 166 tbl
  U.S. consumption of, 209 fig
  wind, 185–186
Energy balance of the Earth, 356 fig
Energy consumption, 161–164, 163 tbl
  for industry, 190–191, 191 mrg
  for transportation, 188–191, 189 mrg
Energy crisis, 187–188
Energy flow in aquatic systems, 412 fig
Energy relationships within an ecosystem, 4
Enzymes, 62
Epidemic, 367 mrg
Estuary, 23, 23 fig
Euphotic zone, 21
Eutrophication, 420–421, 420 mrg
Everglades, 46 mrg
Expectation of life, 123, 124, 124 tbl
Externalities, economic, 518 ff
Extrapolation, 109

Faculative bacteria, 450 prb
Fermentation, 415, 475–476
Fermi, E., 309 mrg
Fertilization for agriculture, 230–236, 231 fig
Filters, 378, 378 fig
Fire climax, 46
Fire-dependent systems, 27
First Law of Thermodynamics, 166–167, 413
Fish kill, 433 fig
Fission, nuclear, 298–308
Fitness, biological, 67, 70, 76
Flea, 257 fig
Fluorides, 371–372
Fluorosis, 371 fig, 372
Food, from the sea, 246–248
  supplies of, 244–252
  synthesis of, 251–252
Food chain, 8, 10 fig
Food cycle, 12 fig
Food pyramid, 9 fig
Food web, 8, 8 fig
  aquatic, 6 fig
  land based, 5 fig
Fossil fuel supply, 170–175
Frequency, 488
Freshwater systems, 24
Fuel supply, 170–175

Garbage, burning of, 186
Gasoline, 384 ff
Genes, 56–57
Genetic adaptation, 56
Genetic disorders, 63 ff
Genetic variation, 56
Genetics, Mendelian, 56, 58 mrg
  molecular basis of, 59 ff
Genotypes, 58
Geometric growth, 110, 110 tbl, 111 fig
Geothermal energy, 184
Glass containers, 467–468, 468 fig
Global interactions, 40
Grasslands, 26, 26 fig
Green Revolution, 239–244, 241 fig
Greenhouse effect, 358 ff, 358 mrg
Growth,
  arithmetic, 110, 110 tbl, 111 fig
  cyclical, 111, 112 fig
  geometric, 110, 110 tbl, 111 fig
  sigmoid, 111 fig

Habitat, destruction of, 86
Half-life, 295–296
Hearing, 491–492
Herbicides, 283–285
Heterotrophs, 4
Heterozygous, 58
Highway Trust Fund, 543–544
Home range, 28
Homeostasis, 17
Homozygous, 58
Horse, 83 fig
Humus, 19, 233–234, 234 fig

Hydrocarbons, 344, 344 mrg, 348, 348 mrg, 390
Hydroelectric energy, 183–184
Hydrogen, 410–411
  isotopes of, 317 tbl
Hydrogen fluoride, 346
Hydrogen sulfide, 342 fig, 343, 343 fig
Hydrological cycle, 404 ff, 404 fig
Hypolimnion, 418, 419 fig

Incineration of wastes, 464–465
Induced abortion, 149, 150
Infant mortality rate, 120, 120 tbl, 125
Inland waters, 417 ff
Insect pests, 257–259
Insecticides, 271, (*see also* pesticides)
  formulas of, 259 tbl, 260 mrg
Intelligence, 78
Interactions, community, 36
  global, 40
Interactions between species, 33 ff
Intrinsic rates, 158 prb
Inversion, atmospheric, 351 ff, 351 fig
Irrigation, 237–239, 238 fig

Japanese beetle, 264 fig
Jet aircraft, effects on climate, 360–361

Kelvin, Lord (William Thompson), 168 mrg

Lake Erie, commercial fish catch in, 517 fig
Lakes, pollution of, 418 ff, 419 fig
Land fill, sanitary, 462 fig, 463, 463 fig
Lapse rate, atmospheric, 350
Leaching, 233
Lead, 349, 426
Legal standing, 533–534
Lemmings, 32, 39 fig, 40
Lethal recessive traits, 70
Lichen, 41
Limnology, 418 mrg
Logarithms, table of, 556
London air pollution, 365–366
Loudness, 497–499
Louse, 258 fig
Lovelock, J. E., 42, 42 mrg
Lye, 177 mrg

Mach, E., 509 ftn
Mach number, 509 mrg
Macroeconomic effects, 521 ff
Magnetohydrodynamics (MHD), 204–205
Malthus, T. R., 109 mrg, 110
Marsh, 44
Mass, 294 mrg, 554
Mass number, 293 mrg
Mass transportation systems, 543–545
Meadow, 44
Melanism, 68 fig, 69, 69 mrg
Mencher, J. P., 157 prb
Mendel, J. G., 56, 57 mrg
Mendelian genetics, 56
Mercury pollution, 426–429
Meteorology of air pollution, 349 ff
Methane, 339 tbl, 413
Metric system, 553
Microclimate, 19
Microeconomic effects, 521 ff, 525 ff
Microorganisms in water, 411 ff
Migration, international, 116, 116 tbl, 154 prb
Mineral cycles, 15 fig, 16 ff
Morgan, T. H., 101 mrg
Mortality, 117
Mutation, 62 ff
Mutualism, 36

Natality, 125
Natural balance, 17, 17 mrg
Natural increase, rate of, 115, 116
Natural selection, 67 ff
Natural succession, 43 ff
Nematodes, 7, 7 fig, 8 fig
Neutralism, 34
Neutron, 293 tbl, 298–301, 317 mrg
Newton, I., 3 mrg
Niche, ecological, 28
  competition within, 32 tbl
  overlap and diversity in, 29 ff, 30 fig
Nitrification, 414
Nitrogen cycle, 15 ff, 15 fig
Nitrogen oxides, 339 tbl, 343–344, 386
Noise, 492–493
  effects on communication, 499–500
  effects on health, 503
  effects on hearing, 500–503, 501 fig, 502 tbl
Noise control, 503 ff
  by interrupting the path, 505–507
  by protecting the receiver, 507
  by reducing the source, 504–505
Noise pollution, definition of, 199 mrg
Nuclear fission reactors, 300–308, 302 fig
  breeders, 307–308, 308 fig
  containment structure, 306 fig
  environmental hazards from, 308–316
Nuclear fuels, 181–182
Nuclear fusion, 316–319
Nuclear fusion reactors, 318 fig
  environmental hazards of, 319–320
Nuclear reactions, 297–299
Nutrient cycles, 10 ff
Nutrients in water, 411 ff, 417 ff
Ocean systems, 21 ff
Oceans, pollution of, 429 ff
Octane number, 385
Octanes, 384
Oil production, 173 fig
  environmental problems from, 175–177
Oil shale, 174
Oil spills, 429 ff
Omnivores, 7
Organic, definition of, 232 mrg
Organic gardening, 232–233
Ostracodes, 7, 8 fig
Oxidants, 344 ff
Oxidation, 411
Oxygen, 410
  atmospheric concentration of, 41
  in water, 411 ff
Oxygen cycle, 13, 14 fig
Oxygenates, 344, 386
Ozone, 339 tbl, 344 ff, 346 mrg
  action on rubber, 345 fig

PAN, 346 mrg
Paper recycling, 472
Parasitism, 35
Particulate matter in air, 338 fig
Passenger pigeon, 94 fig
Period measure, 122
Pest control
  by cultivation, 281–282
  by insect hormones, 278, 279 fig
  by integrated methods, 282–283
  by natural enemies, 274–276
  by resistant strains of crops, 280–281
  by sterilization, 276–278, 277 tbl
  economic factors in, 285–286
Pesticides
  biological concentration of, 266–271, 267 fig, 270 fig, 271 fig
  chemical action of, 260–264
  dispersion of, 265–266
  effects on human health, 271–274
  organochloride, 265–271
  persistence of, 264–265
pH, 408–409
Phenotype, 58
Phenylketonuria, 64, 64 fig, 65, 77
Pheromones, 278
Phon, 498
Phosphates, 423, 423 mrg
Photosynthesis, 13, 13 mrg
Phytoplankton, 6 fig, 21 fig
PKU, *see* phenylketonuria.
Pleistocene Age, mammals of, 84 fig
Pollution, economic models of, 539 fig
Pollution control
  altruism in, 525 ff
  costs of, 521 ff, 523 fig
  economic incentives, 537 ff
  economic problems, 545 ff
  government policies in, 542 fig
  legal approaches, 533 ff
  radical social policies in, 542–543

Pollution tax, 538
Polymorphism, balanced, 71
Polyvinyl chloride, 464 mrg, 465 mrg
Population,
  size of, 107 fig
  stabilization of, 149
  stable, 158 prb
  U.S., 544 tbl
Population control, 150
Population density, 141 ff, 142 mrg, 143 tbl
Population distribution, 115, 130, 132 ff
Population dynamics, 112
Population ecology, 112
Population growth, 132 ff, 147
Population projections, 153
Population stabilization, 149, 158 prb
Power, 164–165
  units of, 166 tbl
Prairie, 26, 26 fig, 45
Predation, 34
Predators, 6 fig
  extermination of, 89
Price elasticity, 528
Primary consumers, 4
Producers, 6 fig
Protein, 413
Protocooperation, 36
Pulp and paper mills, 525 ff
Punnett cross, 58
Putrefaction, 415

Radiation, biological effects of, 321–329
Radiation sickness, 324–327
Radioactive decay, 296
Radioactive pollutants, problems and issues in, 329–331
Radioactive wastes, 320
  disposal of, 311–316, 311 fig, 312 fig, 313 fig, 314 fig, 315 fig
  routine emissions of, 310–311
Radioactivity, 294–297
  units related to, 325 tbl
Radioisotopes, 294–295
Radium, 295, 295 mrg
Rain forest, tropical, 26
Rate, arithmetic, 110
  geometric, 110
Rate of natural increase, 115
Recessive trait, 57
Recycling, economics of, 481–483
  of wastes, 465 ff
Recycling technology, 478 tbl
  composting, 474–475
  destructive distillation (pyrolysis), 476–477, 476 fig
  fermentation, 475–476
  industrial salvage, 477 ff, 477 fig
  melting, 472
  pulping, 473
  rendering, 475
  revulcanizing, 472
  waste separation, 479 fig, 480 fig
Red tide, 40, 434
Reduction, 411
Rendering, 475
Replacement level, 127
Reproductive advantage, 76
Reproductive rate, crude, 116
  intrinsic, 158 prb
Residual charge for pollution, 538
Respiration, 8, 8 fig, 10 fig, 13 mrg, 44 mrg
Respiratory tract, 362–363 fig
Reverse osmosis, 446, 446 fig
River pollution, 418 fig
RNA, 62
Rotenone, 258
Rutherford, E., 297

Salt marsh, 23, 23 fig
Sanitary land fill, 462 fig, 463, 463 fig
Saprophytes, 7
Savanna, 27
Schistosomiasis, 435
Screwworm fly, 276 fig
Scrubbers, 379, 379 fig
Seaborg, G. T., 331 mrg
Second Law of Thermodynamics, 167–170
Selection, natural, 67 ff
Sewage, 439 ff
Sewage treatment, 439 ff, 440 fig, 441 figs, 442 fig, 444 fig
Sex attractants, 278–280
Sickle cell anemia, 64, 66 fig, 71, 71 mrg
Sigmoid curve, 111
Sine curve, 490 fig
Smog, 345 fig, 346
Snail, 8 fig
Soil, forest, 18 fig
Soil mite, 7–8
Solar collectors, 183 fig, 193 fig
Solar energy, 183
  cost of, 194
Solid waste cycles, 455–456
Solid wastes, collection of, 459 ff
  incineration of, 464–465
  sources of, 456–459, 457 fig, 458 tbl
Sonic boom, 508, 510 fig
Sound, 487–489
Sound levels of various sources, 498 fig
Sound pressure level, 514 prb
Species
  critical extermination level of, 93 ff
  current trends in extinction of, 98 ff
  definition of, 67 mrg
  development of, 82
  displacement of, 100
  extinction by hunting, 88
  extinction of, 81 ff, 86 ff
  extinction rates, 86 tbl
  interactions between, 33 ff
  replacement of, 82
Spruce budworm, 263 fig
SST, *see* supersonic transport.
Standing, legal, 533–534
Steady state, 8
Sterilization, 150
Strip mines, 176–177
Succession, natural, 43 ff
Sulfur dioxide, 341 ff, 410
Sulfur trioxide, 341 ff
Supersonic transport (SST), 507 ff, 509 fig, 510 fig
Survival of the fittest, 76
Survival rate, 125 tbl
Survivorship function, 121, 122 fig, 123 fig
Synergism, 364, 364 mrg

Taiga, 24
Tar sands, 174
Temperature, 554
Terrestrial ecosystems, 24
Test cross, 58
Tetraethyl lead, 349
Thermal pollution, 196 fig, 198–203
  definition of, 199 mrg
  effect on climate, 203–204, 204 tbl
  solutions to the problem, 204–208
  sources of, 203 tbl
  utilization of waste heat, 206–207
Thermocline, 418, 419 fig
Thermodynamics, 165
  First Law of, 166–167
  Second Law of, 167–170
Thermonuclear reaction, 317
Thompson, William (Lord Kelvin), 168 mrg
Tidal energy, 185
Tones, impure, 489–491
  pure, 489–491, 490 fig
Torrey Canyon, 429 ff
Total fertility rate, 126, 127 tbl
  replacement level of, 127
Tritium, 317 tbl
Trophic levels, 4
Tropical rain forest, 26
Tundra, 24, 25 fig

United States, demographic trends in, 138 ff
Uranium, 298–301
Uranium reserves, 182 tbl
Urbanization, consequences of, 148

Vital event, 113
Vital rate, 113 mrg, 114
Voles, 99 mrg
Volume, 554

Water, 402 ff, 402 mrg
  impurities in, 405 ff, 406 fig, 407 ff, 408 tbl

Water (*Continued*)
industrial waste in, 425 ff, 432–434
microorganisms in, 411 ff
nutrients in, 411 ff, 417 ff
oxygen in, 411 ff
Water cycle, 404 ff, 404 fig
Water pollution, 399 ff
effects on human health, 435 ff
Water purification, 438 ff
activated sludge treatment, 442, 442 fig
Water purification (*Continued*)
adsorption, 445
advanced treatment, 444–446
chlorination, 443
coagulation, 445
economics of, 446 ff
oxidation ponds, 451 prb
primary treatment, 439–441
reverse osmosis, 446, 446 fig
secondary treatment, 441–444
strategy of, 448 fig
Water purification (*Continued*)
tertiary treatment, 444–446
trickling filter, 441
Water quality, 399
Wave length, 488, 488 fig
Waves, 489 fig
Work, 165–166

Zoo and pet trade, 91
Zooplankton, 6 fig, 21, 21 fig